Sustainable Development Goals Series

The **Sustainable Development Goals Series** is Springer Nature's inaugural cross-imprint book series that addresses and supports the United Nations' seventeen Sustainable Development Goals. The series fosters comprehensive research focused on these global targets and endeavours to address some of society's greatest grand challenges. The SDGs are inherently multidisciplinary, and they bring people working across different fields together and working towards a common goal. In this spirit, the Sustainable Development Goals series is the first at Springer Nature to publish books under both the Springer and Palgrave Macmillan imprints, bringing the strengths of our imprints together.

The Sustainable Development Goals Series is organized into eighteen subseries: one subseries based around each of the seventeen respective Sustainable Development Goals, and an eighteenth subseries, "Connecting the Goals", which serves as a home for volumes addressing multiple goals or studying the SDGs as a whole. Each subseries is guided by an expert Subseries Advisor with years or decades of experience studying and addressing core components of their respective Goal.

The SDG Series has a remit as broad as the SDGs themselves, and contributions are welcome from scientists, academics, policymakers, and researchers working in fields related to any of the seventeen goals. If you are interested in contributing a monograph or curated volume to the series, please contact the Publishers: Zachary Romano [Springer; zachary.romano@springer.com] and Rachael Ballard [Palgrave Macmillan; rachael.ballard@palgrave.com].

Fran Humphries
Editor

Decoding Marine Genetic Resource Governance Under the BBNJ Agreement

Springer

Editor
Fran Humphries
Griffith Law School
Griffith University
Nathan, Queensland, Australia

ISSN 2523-3084 ISSN 2523-3092 (electronic)
Sustainable Development Goals Series
ISBN 978-3-031-72099-4 ISBN 978-3-031-72100-7 (eBook)
https://doi.org/10.1007/978-3-031-72100-7

This work was supported by the European Union's Horizon 2020 Research and Innovation Program, the European Union's Horizon Europe Program, Griffith University, UK Research and Innovation and the German Alliance of Scientific Organizations.

This Springer imprint is published by the registered company Springer Nature Switzerland AG
The registered company address is: Gewerbestrasse 11, 6330 Cham, Switzerland

This book is dedicated to marine biodiversity of areas beyond national jurisdiction.

Foreword

In the early hours of 4 March 2023, the gavel barely fell before the room erupted in jubilant applause. This was, without question, a triumph of multilateralism—two decades of effort and eight intense weeks of negotiation, culminating in a historic breakthrough. The new Agreement under the United Nations Convention on the Law of the Sea on the Conservation and Sustainable Use of Marine Biological Diversity of Areas beyond National Jurisdiction (BBNJ Agreement) marks a transformative chapter for ocean governance. From the Global South to the Global North, this verdict is unanimous.

When the President of the inter-governmental conference announced that the "*ship had reached the shore,*" I sat back, heaving a sigh of relief, as I felt a weight lift. My job was done. I had been responsible for facilitating negotiations on marine genetic resources, including questions on the sharing of benefits, a make-or-break factor for the agreement's success. Part II of the BBNJ Agreement—the last section to reach consensus—helped to secure this landmark accord at a critical moment.

The timing could not have been more significant. In the wake of the COVID-19 pandemic, with stark inequities exposed and amid the escalating crises of biodiversity loss, ocean degradation, and climate change, the need for unified action was undeniable. The BBNJ Agreement represents a profound statement of political will, an unequivocal choice to prioritize collective good over narrow national interests.

The Agreement has now been opened for signature until 20 September 2025. A preparatory commission has been established to prepare for entry into force and for the convening of the first meeting of the Conference of the Parties. Countries are conducting internal constitutional processes and other consultations to take the next step toward becoming Parties. In many ways, in spite of the Agreement, the true work has only just begun.

With 76 articles comprising a Preamble, twelve (12) Parts and two Annexes, the implementation of the BBNJ Agreement at the national and international levels will be a monumental task. And, implementation must be coherent, across all levels and scales. The Agreement is intended to be enduring. So how it is operationalized and what processes and arrangements are elaborated must similarly be future proofed. It is also expected to work in complementarity and coherence with other instruments and frameworks. It will need to be dynamic, nimble, anticipatory and responsive.

The contributors to *Decoding Marine Genetic Resource Governance under the BBNJ Agreement* recognize these complexities first hand. Their extensive involvement—from technical analyses in the preparatory phase to support for ongoing readiness efforts—underscores their commitment to the BBNJ Agreement's success. This book is a testament to their contributions and serves as a guide for the Agreement's future implementation.

Part II of the BBNJ Agreement introduces groundbreaking approaches to ocean governance. As detailed in the Commentary and Textual Analysis, its provisions on marine genetic resources bring together international legal and scientific practices, creating an innovative framework for fair and equitable benefit-sharing related to marine genetic resources in areas beyond national jurisdiction. Key elements include a new notification system, a Clearinghouse Mechanism, a standardized batch identifier for BBNJ, protocols for digital sequence information, integrated data management in ocean governance, and recognition of traditional knowledge associated with marine genetic resources.

These early chapters of *Decoding* provide a retrospective on the negotiation history of each article in Part II, offering insights into compromises reached and decisions made. They dissect these elements in depth, highlighting potential implications for national implementation, global application, and inter-institutional cooperation. The book's second part shifts the focus forward, examining the path from ratification to implementation.

The authors' insights are invaluable for governments, policymakers and practitioners, offering practical guidance and raising critical questions relevant to scientists, industry stakeholders, database managers and end users. Though grounded in pragmatism, *Decoding* keeps sight of the BBNJ Agreement's ambitious goals. Part II captures the spirit of equity that lies at the Agreement's core—a vision without which this achievement would not have been possible. Whether its provisions will be able to deliver on the objectives for fair and equitable benefit-sharing will ultimately depend on the dedication and actions of the Parties.

H.E. Mrs. Janine Coye Felson
Ambassador and Deputy Permanent Representative
for the Permanent Mission of Belize to the United Nations

Preface

This book provides practical guidance for understanding a new treaty adopted in June 2023 that will change the way biodiversity is governed in about two thirds of the oceans known as areas beyond national jurisdiction (ABNJ). The *Agreement under the United Nations Convention on the Law of the Sea on the Conservation and Sustainable Use of Marine Biological Diversity of Areas Beyond National Jurisdiction* (BBNJ Agreement) is an implementing agreement under the *United Nations Convention on the Law of the Sea* (UNCLOS). The treaty provisions in Part II on Marine Genetic Resource (MGR) governance were the most contentious part of negotiations, resulting in significant innovation and compromise. It has an ambitious framework with many details yet to be fleshed out after the treaty comes into force, including the infrastructure for the Conference of the Parties (CoP), Clearing House Mechanism and the Access and Benefit Sharing Committee. At the time of writing, there is no comprehensive guidance for stakeholders about what the treaty means for their current and future research and development (R&D) and commercialization of MGR of ABNJ, digital sequence information and associated traditional knowledge. This book aims to fill this gap in scholarship.

The idea for the book germinated at the fifth United Nations Intergovernmental Conference (IGC) of the treaty negotiations in New York when a group of scientists, lawyers, academics and policymakers met to discuss how the treaty might work in practice for marine bio-innovation and R&D by scientists and commercial end users. This conversation grew to the twenty-five authors of this book, located in Australia, Africa, South America, North America, Europe and the United Kingdom, offering diverse perspectives on research practices and expected implementation of treaty provisions in a range of cultural contexts. For five ICGs spanning 2018–2023, most of the authors were actively involved in the treaty negotiations as delegates, while some were part of the prior preparatory work and now in current implementation, with deep insights into the evolution of the final treaty provisions. Others have direct experience in collecting and using MGR from ABNJ for R&D purposes. During two workshops, two rounds of peer-review and ongoing feedback on each other's chapters, this book is a true collaboration with thoughtful and informed interpretation of treaty provisions from policy, legal and scientific perspectives.

Through a contextual doctrinal analysis, real-world examples and case studies, this book interprets and analyses the treaty text to offer practical considerations, guidelines and tools to assist scientists and commercial end users align their practices with the expected implementation of the treaty. It may also assist policymakers to consider options for implementing Party obligations under national laws. Part I of the book offers a historical and textual commentary on MGR governance provisions, including scope, definitions, infrastructure, notification and benefit-sharing schemes; monitoring and transparency procedures; and traditional knowledge requirements. Part II of this book provides guidance and insights into how the treaty might work in practice for challenging aspects of implementation—intellectual property, Party ratification, interlinkages with other access and benefit-sharing frameworks, identifiers and data management plans. It analyses a variety of scenarios of R&D pathways to offer practical considerations for scientists and commercial end users about how to align their current and future practices with the treaty framework. It ends with some insights into the transformative potential of the new treaty.

It is too early to accurately predict how the MGR obligations will be implemented by Parties. However, there are reputational and economic benefits for stakeholders to start aligning their practices with the known aspects of the BBNJ framework and to positively influence the pathway towards implementation.

Fran Humphries
Griffith University
Nathan, Queensland, Australia

Acknowledgements

The editor and co-authors of this collection warmly thank:

- our 17 expert peer reviewers during two rounds of chapter peer reviews for their thoughtful insights and suggestions for manuscript improvement;
- Elisa Morgera and Lydia Slobodian for their early feedback on the entire book manuscript;
- the University of Aberdeen for hosting our workshops for chapter collaboration and review;
- Marcel Jaspars for sharing terrific diagrams for figures throughout the chapters;
- our institutions, families and friends for supporting us in the creation of this work;
- our funders for generously covering the open-access fees so that this book is accessible to everyone. This work has received funding from:

  - the European Union's Horizon 2020 research and innovation programme under Grant Agreement no. 101000392 (MARBLES). This output reflects only the authors' view and the Research Executive Agency (REA) cannot be held responsible for any use that may be made of the information contained therein.
  - the European Union's Horizon Europe Programme under Grant Agreement No. 101082304 (BlueRemediomics). Views and opinions expressed are however those of the author(s) only and do not necessarily reflect those of the European Union or the European Research Executive Agency (REA). Neither the European Union nor the granting authority can be held responsible for them. UK partners are supported UK Research and Innovation (UKRI) under the UK Government's Horizon Europe funding guarantee Grant No. IFS 10057167 (University of Aberdeen).
  - Law Futures Centre and Griffith Law School of Griffith University, Australia; and
  - the ABS Science Hub project funded by the Alliance of Scientific Organizations (Allianz der Wissenschaftsorganisatioin) under a grant to the Leibniz Institute DSMZ German Collection of Microorganisms and Cell Cultures, Braunschweig, Germany.

The editor would like to thank all of her co-authors for their enthusiasm, passion and inspiration for our creation. Likewise, the co-authors would like to express their gratitude to the editor for her vision and hard work. The views and opinions in this book are those of the authors in their personal capacities and do not necessarily reflect those of their affiliated institutions.

Contents

Part II Treaty Implementation in Practice

Editor and Contributors

About the Editor

Fran Humphries (Ph.D., LLM, BA/LLB) has specialized in marine and biodiversity law and policy for over two decades in government, academia and consultancies. She is an Associate Professor at Griffith Law School, Griffith University, Australia, with a background in fisheries management and marine law. She has led large international research consultancies on access and benefit sharing of biological resources for governments, UN organizations, WorldFish. She was on the International Council of Environmental Law delegation for the Biodiversity Beyond National Jurisdiction (BBNJ) Agreement negotiations.

Contributors

Jeff A. Ardron Africa Oceans, The Nature Conservancy, Mombasa, Kenya

Todd Berry Law School, Monash University, Clayton, VIC, Australia

Arianna Broggiato European Commission, Brussels, Belgium

Abbe E. L. Brown School of Law, University of Aberdeen, Aberdeen, UK; World Commission On Environmental Law, Gland, Switzerland

Luciana Fernandes Coelho Stockholm Environment Institute, Stockholm, Sweden

Paul Dunshirn Department of Political Science, University of Vienna, Vienna, Austria

Hiroko Muraki Gottlieb Department of Organismic and Evolutionary Biology, Harvard University, Cambridge, MA, USA;
Elisabeth Haub School of Law at Pace University, White Plains, NY, USA

Tammy Horton Ocean BioGeosciences, National Oceanography Centre, Southampton, UK

Fran Humphries Griffith Law School, Griffith University, Nathan, QLD, Australia

Marcel Jaspars Department of Chemistry, Marine Biodiscovery Centre, University of Aberdeen, Aberdeen, UK

Daniel Kachelriess IUCN World Commission on Environmental Law, High Seas Alliance (Advisor), Vienna, Austria

Arne Langlet Department of Political Science, University of Vienna, Vienna, Austria

Jessica Lavelle Department of Environmental and Geographical Science, University of Cape Town, Rondebosch, South Africa

Charles Lawson Griffith Law School, Griffith University, Queensland, Australia

Christopher H. C. Lyal Department of Life Sciences, The Natural History Museum, London, UK

Elisa Morgera University of Strathclyde, Glasgow, UK

Chilenye Nwapi The Commonwealth Secretariat London, London, UK

Sergio Pena-Neira Universidad Mayor, Santiago, Chile

Muriel Rabone Department of Life Sciences, The Natural History Museum, London, UK

Amber H. Scholz Department of Science Policy and Internationalization, Leibniz Institute DSMZ-German Collection of Microorganisms and Cell Cultures, Braunschweig, Germany

Aysegul Sirakaya Department of Life Sciences, Natural History Museum, London, UK

Lydia Slobodian Environmental Law and Policy Program, Georgetown University Law Center, Washington D.C, USA

Alison Swaddling The Commonwealth Secretariat London, London, UK

Thomas Vanagt 3BIO, Brussels, Belgium

Rachel Wynberg Department of Environmental and Geographical Science, University of Cape Town, Rondebosch, South Africa

Abbreviations

ABMT	Area-Based Management Tool
ABNJ	Areas beyond national jurisdiction
ABS Committee	Access and Benefit Sharing Committee (art 15)
ABS	Access and Benefit Sharing
AHTEG	Ad hoc Technical Expert Group
AqGR	Aquatic Genetic Resources
ARIPO	African Regional Intellectual Property Organization
art	Article
ATA	Antarctic Treaty Area
ATCM	Antarctic Treaty Consultative Meeting
ATS	Antarctic Treaty System
AWNJ	Areas within national jurisdiction
BBNJ Agreement	Agreement under the United Nations Convention on the Law of the Sea on the Conservation and Sustainable Use of Marine Biological Diversity of Areas Beyond National Jurisdiction
BBNJ Identifier	BBNJ Standardized Batch Identifier
BBNJ treaty bodies	Conference of the Parties and its subsidiary bodies
BBNJ	Biodiversity beyond national jurisdiction
CB	Capacity Building
CBD	Convention on Biological Diversity
CBTMT Committee	Capacity Building and Transfer of Marine Technology Committee (art 46)
CBTMT	Capacity Building and Transfer of Marine Technology
CCAMLR	Convention on the Conservation of Antarctic Marine Living Resources
CGIAR	Consultative Group on International Agricultural Research
CHM	Clearing House Mechanism (art 51)
CNA	Competent National Authority (Nagoya Protocol)
CoP	Conference of the Parties
CoP 15	15th Conference of the Parties to the CBD
DMP	Data Management Plan
DNA	Deoxyribonucleic acid
DOI	Digital object identifier

DSI	Digital Sequence Information
DTA	Data Transfer Agreement
EIA	Environmental Impact Assessment
ESD	Ecologically sustainable development
FAO	Food and Agriculture Organization of the United Nations
FPIC	Free Prior and Informed Consent
FTA	Free Trade Agreement
G77/China	Group of 77 and China
GBF	Kunming-Montreal Global Biodiversity Framework
GEF	Global Environment Facility
GLIS	Global Information System (Plant Treaty)
GRFA	Genetic Resources for Food and Agriculture
ICCs	Indigenous Cultural Communities
ICESCR	International Covenant on Economic, Social and Cultural Rights
IGC	Intergovernmental Committee
INSDC	International Nucleotide Sequence Database Collaboration
IP	Intellectual Property
IPLC	Indigenous Peoples and local communities
IRCC	Internationally Recognized Certificate of Compliance
ISA	International Seabed Authority
ITPGRFA	International Treaty on Plant Genetic Resources for Food and Agriculture
IUCN	International Union for Conservation of Nature
IVTM	Influenza Virus Traceability Mechanism
LDC	Least Developed Countries
Madrid Protocol	Protocol on Environmental Protection to the Antarctic Treaty
MAT	Mutually Agreed Terms
MDG	Millennium Development Goals
MGR	Marine Genetic Resources
MPA	Marine Protected Area
MSR	Marine Scientific Research
MTA	Material Transfer Agreement
Nagoya Protocol	Nagoya Protocol on Access to Genetic Resources and the Fair and Equitable Sharing of Benefits arising from their Utilization to the Convention on Biological Diversity
NFP	National Focal Point (Nagoya Protocol)
PABS System	WHO Pathogenic Access and Benefit-sharing System
PIC	Prior Informed Consent
PIP Framework	Pandemic Influenza Preparedness Framework for Sharing of Influenza Viruses and Access to Vaccines and other Benefits

Plant Treaty	International Treaty on Plant Genetic Resources for Food and Agriculture
R&D	Research and development
RNA	Ribonucleic acid
SDG	Sustainable Development Goals (UNEP)
SMTA	Standard Material Transfer Agreement
STB	Scientific and Technical Body (art 49)
TK	Traditional Knowledge
TMT	Transfer of Marine Technology
UN	United Nations
UNCLOS	United Nations Convention on the Law of the Sea
UNDRIP	United Nations Declaration on the Rights of Indigenous Peoples
UNEP	United Nations Environment Program
UNGA	United Nations General Assembly
WHO	World Health Organization
WIPO	World Intellectual Property Organization
WTO	World Trade Organization

List of Figures

List of Tables

Bridging Divides: The Evolution of Marine Genetic Resource Governance Beyond National Jurisdiction

1

Fran Humphries, Todd Berry, and Hiroko Muraki Gottlieb

Abstract

Part II of *The Agreement under the United Nations Convention on the Law of the Sea on the Conservation and Sustainable Use of Marine Biological Diversity of Areas Beyond National Jurisdiction* (BBNJ Agreement) is a story of innovation, ambition, and compromise for a framework of information and benefit sharing concerning marine genetic resources of areas beyond national jurisdiction and associated digital sequence information and Traditional Knowledge. Understanding the scope, purpose and proposed implementation of the new treaty requires reflection on the (over) twenty years of preparation leading up to the final treaty text. This introductory chapter for the edited collection 'Decoding Marine Genetic Resource Governance under the BBNJ Agreement' analyzes the need for knowledge on marine biodiversity and genetic resources, the jurisdictional context and gaps in ocean governance that the treaty aimed to fill, the preparatory work leading to the negotiations, and the key areas of convergence and divergence during the intergovernmental conference treaty negotiations. It outlines how subsequent chapters in this collection build on this context, by analyzing and interpreting the obligations under the Part II framework and offering practical considerations for its implementation under national law and good scientific practice.

Keywords

BBNJ agreement · High seas · Biodiversity · Marine genetic resources · Digital sequence information · Traditional knowledge · Negotiation history · UNCLOS · Ocean governance · Conservation and benefit sharing

F. Humphries (✉)
Griffith Law School, Griffith University, Nathan, QLD, Australia
e-mail: fran.humphries@griffith.edu.au

T. Berry
Law School, Monash University, Clayton, VIC, Australia

H. M. Gottlieb
Department of Organismic and Evolutionary Biology, Harvard University, Cambridge, MA, USA

1.1 Introduction

In June 2023, countries adopted the historic *Agreement under the United Nations Convention on the Law of the Sea on the Conservation and Sustainable Use of Marine Biological Diversity of Areas Beyond National Jurisdiction* (BBNJ

F. Humphries (ed.), *Decoding Marine Genetic Resource Governance Under the BBNJ Agreement*, Sustainable Development Goals Series, https://doi.org/10.1007/978-3-031-72100-7_1

Agreement) by consensus. Part II provisions on the sharing of benefits from the use of marine genetic resources (MGR) of areas beyond national jurisdiction (ABNJ) were the most contentious part of the negotiations, with initially stark divides on questions ranging from 'if, how, and what type of benefits (monetary or non-monetary) would be shared', to the traceability system needed to achieve it and to the underlying principles of international law (e.g., see UNGA, 2018a). The resulting text of the BBNJ Agreement, reflecting significant compromises from all sides, establishes the key rules for activities concerning MGR of ABNJ, but leaves many detailed provisions to be fleshed out over the coming years at both international and national levels. Several issues are still to be resolved at the conceptual level, such as application of the principles of due regard of state interests and common heritage of humankind (see Marciniak (2017), as well as at a practical level, such as the modalities for notification and sharing of monetary benefits for the physical MGR and digital sequence information (DSI) on MGR. At the time of writing, there are 105 signatories to the BBNJ Agreement with eight Parties.[1] The BBNJ Agreement, however, needs 60 Parties before it comes into force 120 days later. Even before the BBNJ Agreement commences with legal effect, a state that is a signatory to the BBNJ Agreement is obliged to refrain from acts that would defeat its object and purpose (*Vienna Convention on the Law of Treaties* art 18).

Achieving conservation and sustainable use of marine biodiversity in ABNJ, and fair and equitable sharing of benefit from the use of MGR are crucial for meeting a range of the United Nations' Sustainable Development Goals (SDGs). The 2030 Agenda for Sustainable Development was a plan of action for people, planet, and prosperity, outlining 17 SDGs, many of which are relevant to the objectives of the BBNJ Agreement (UNGA, 2015). The most directly applicable SDG, Goal 14, is to 'conserve and sustainably use the oceans, seas and marine resources for sustainable development.' Actions include reducing marine pollution, strengthening marine ecosystems, minimizing the impacts of ocean acidification, end over fishing, and conserving at least 10% of coastal/marine areas by 2020, increase economic benefits to small island developing States and least developed countries from sustainable use of marine resources, and enhance the conservation and sustainable use of oceans and their resources by implementing what was to become the BBNJ Agreement (UNGA, 2015, p. 23–4). According to Blasiak et al., (2023), SDG14 receives the least development funding of all the SDGs (p. 483). In response to the continued alarming loss of biodiversity, the United Nations General Assembly (UNGA) adopted the 2022 Kunming-Montreal Global Biodiversity Framework (GBF). The GBF supports implementation of the three objectives of the *Convention on Biological Diversity* (CBD) (conservation, sustainable use, and fair and equitable sharing of biological resources—art 1) and supports progress toward the SDGs. GBF Goal C aims to see a substantial increase by 2050 in benefits from the utilization of genetic resources and DSI on genetic resources and of Traditional Knowledge associated with genetic resources, while ensuring Traditional Knowledge is appropriately protected, 'thereby contributing to the conservation and sustainable use of biodiversity' (UNEP, 2022a p. 9). The United Nations General Assembly has urged Parties to the CBD to ensure coherence and complementarity of the GBF with other international frameworks (UNGA Res 78/155 para 13), such as the BBNJ Agreement.

The BBNJ Agreement is groundbreaking in its geographical and jurisdictional coverage. It covers biodiversity in the majority of the world's ocean where sovereignty and sovereign rights over waters and resources are not recognized under the *United Nations Convention on the Law of the Sea* (UNCLOS). This book uses the shortened term 'BBNJ Agreement' because at the time of writing, the correct use of

[1] https://treaties.un.org/pages/ViewDetails.aspx?src=TREATY&mtdsg_no=XXI-10&chapter=21&clang=_en.

a shortened version is not settled in the United Nations' documents and some of the other titles, including High Seas Treaty, may misrepresent the scope of the treaty for biodiversity in the high seas water column and the deep seabed below (Mendenhall and Bateh, 2024).

The BBNJ Agreement objective is to ensure the conservation and sustainable use of marine biological diversity of ABNJ for the present and in the long term, through effective implementation of the relevant provisions on the *United Nations Convention on the Law of the Sea* (UNCLOS) and further international cooperation and coordination (art 2). It covers four key elements and crosscutting issues. The four elements are:

1. marine genetic resources, including the fair and equitable sharing of benefits (Part II);
2. measures, such as area-based management tools, including marine protected areas (Part III);
3. environmental impact assessments (Part IV); and
4. capacity building and the transfer of marine technology (Part V).

Crosscutting issues, which represent matters that are generally relevant for the above four substantive elements, include:

- the relationship between the BBNJ Agreement, UNCLOS, and other global, regional, subregional, and sectorial bodies (art 5);
- institutional arrangements, including decision-making and advisory bodies and committees (art 15 and Part IV);
- finance and funding arrangements (Part VII);
- implementation and compliance (Part VIII); and
- settlement of disputes (Part IX).

Part II of the BBNJ Agreement, which is the focus of this book, has four objectives (art 9): first, the fair and equitable sharing of benefits arising from activities with respect to MGR and DSI of ABNJ for the conservation and sustainable use of marine biological diversity of ABNJ; second, the building and development of the capacity of Parties to carry out these activities, particularly low- and middle-income countries; third, the generation of knowledge, scientific understanding, and technological innovation including through the development and conduct of marine scientific research; and fourth, the development and transfer of marine technology, including inter alia, information, and data, provided in a user-friendly format, on marine sciences and related marine operations and services (art 1(10)). The MGR framework includes, among other things, new notification, benefit sharing and transparency mechanisms for MGR and DSI, and access obligations for Traditional Knowledge associated with MGR in ABNJ.

At the time of writing, there is no comprehensive guidance for stakeholders about what the BBNJ Agreement means for their current and future research, development, and commercialization activities with MGR, associated DSI, and relevant Traditional Knowledge and how they can begin to think about how they can align their practices within the treaty framework. Stakeholders include users of MGR, DSI, and Traditional Knowledge and holders of Traditional Knowledge. There may be reputation and economic benefits for stakeholders to start exploring how their practices may align with the known aspects of the skeleton framework of the BBNJ Agreement and to continue to modify their practices as international and national communities begin to operationalize the MGR requirements. Through a contextual doctrinal legal analysis and case studies, this book interprets and analyzes treaty text relevant to MGR governance and offers practical considerations, guidelines, and tools for stakeholders to understand how the changes may affect them.

It is too early to accurately predict how all the treaty obligations will be implemented by the Parties to the BBNJ Agreement. The BBNJ Agreement bodies and committees including the Conference of the Parties (COP) can only come into being after the BBNJ Agreement enters into force and there are many policies and procedures that need to be agreed upon before Part II can be fully operationalized. While the travaux préparatoires can offer glimpses of negotiator intent about the meaning behind provisions, scope, and framework at the time of negotiation, subsequent state practice and COP decisions may build on and further clarify the original intention. In the meantime, this book draws from the experience and research of authors who were either involved or who are practitioners in the field of MGR. The target audience of this book are policy makers, scientists and commercial end users of MGR, DSI and Traditional Knowledge.

The aim of this chapter is to provide the historical and contemporary context for Part II of the BBNJ Agreement to aid interpretation and analysis in subsequent chapters of this book. It highlights the environmental and knowledge context for biodiversity in ABNJ (Sect. 1.2) and the jurisdictional context and gaps in ocean governance that underlie the need for the framework (Sect. 1.3). It outlines some economic, social, and cultural context for Part II, including social justice issues for fairly and equitably sharing the benefits from the use of MGR of ABNJ and associated DSI. Sections 1.4 and 1.5 outline the preparatory work for the BBNJ Agreement to address gaps in ocean governance and the intergovernmental committee (IGC) meetings, with context about key issues of convergence and divergence during BBNJ Agreement negotiations. Section 1.6 outlines the structure of this book, outlining how each of the chapters contribute to a comprehensive understanding of the BBNJ Agreement.

1.2 Knowledge on Marine Biodiversity and MGR of ABNJ

Marine biological diversity in ABNJ is the centrepin of the BBNJ Agreement framework. ABNJ contains 90% of the total biomass of the ocean (Crespo et al., 2020). This includes a wide range of ecological processes, dynamics, and ecosystems rich in biodiversity, from viruses and bacteria to the largest animals that have ever existed (Crespo et al., 2020; Rogers et al., 2021, 2023). It is estimated that there are '2.2 million (range 0.3–10 million) eukaryotic marine species' that 'probably exist in the ocean, of which 230,000 are confirmed' (Blasiak et al., 2020). Despite this, ABNJ remains the least studied biodiversity areas on the planet (Gonçalves, 2023) and significant knowledge gaps remain (Morgera, 2022). Roughly 91% of marine species remain undescribed while 80% of the ocean floor is still unmapped and unexplored (Gonçalves, 2023). The BBNJ Agreement itself does not define the term 'marine biological biodiversity.' However, drawing from the CBD definition of 'biological diversity,' 'marine biological diversity' (CBD art 2) may be understood as 'the variability among living organisms from marine ecosystems and the ecological complexities of which they are part, including diversity within species, between species, and of ecosystems' (Roach, 2021, p. 82).

Conserving and sustainably using marine biological diversity in ABNJ is essential for maintaining the balance of the Earth's ecosystem (Qureshi, 2018). Policymakers have realized over the past decades that urgent action is required if humanity wants to continue enjoying the benefits of the ocean (Xiao, 2020). Yet, the loss of global biodiversity is accelerating at an alarming rate (Hoel, 2021). There are an estimated eight million species of animals and plants, with about one million of these facing the risk of extinction over coming decades (Lee, 2021). For marine biodiversity, there has been an estimated decline in marine species of 35% since 1970 (Díaz, 2019), with biodiversity declines from coastal waters to the deep sea (Díaz, 2019, Lee, 2021). In 2008, an estimated 40% of ocean areas had been 'strongly affected by multiple drivers' (Díaz, 2019, p. 24). In 2014, an estimated 66% of the world's ocean was experiencing increasing cumulative impacts as a consequence (Díaz, 2019) and these rapid global changes are expected to continue beyond 2050. The human impacts upon the ocean are 'indisputable', and some of the consequences are only now beginning to be understood (Gonçalves, 2023; Harden-Davies, 2021).

The main human drivers impacting marine biodiversity in ABNJ are pollution, overfishing, oil exploratory activities, and bottom trawling (Qureshi, 2018). Researchers argue that of these, overfishing had the biggest impact upon marine biodiversity in the past 50 years (Díaz, 2019, Barnes, 2022). Globally, 90% of fisheries are either fully exploited or overexploited with a 2018 major study (Freestone, 2019) suggesting that roughly only 13% of the ocean can be considered to have an 'intact ecosystem with low impact from human pressure' (Gonçalves, 2023, p. 132). 30% of fish stocks are caught beyond sustainable limits with 61% considered 'fully fished' (Long, 2018, p. 288). An estimated 62% of the total number of all fish caught in ABNJ was done by only 10 States (Cheung et al., 2017).

Other challenges facing marine biodiversity in ABNJ include ocean plastic pollution with estimates suggesting that the amount has increased tenfold since 1980 (Díaz, 2019). For example, an estimated 90% of Fulmar sea birds have plastic in their stomachs and on average there are 712 items of litter found for every 100 miles of the Atlantic Coast (Long, 2018). 80% of the world's wastewater is discharged back into the environment without treatment, and '300–400 million tons of heavy metals, solvents, toxic sludge, and other wastes from industrial facilities are dumped into the world's waters each year' (Díaz, 2019, p. 28). Moreover, deep seabed mining, if it begins, will pose a substantial threat to marine biodiversity due to the destruction of local and surrounding habitats (Gjerde, 2006). There are concerns that deep seabed mining will expand its reach and could expand into the Arctic and Antarctic regions as the ice begins to melt (Díaz, 2019).

Other human activities, such as pollution, and climate change impacts, including ocean acidification, directly threaten the survival of marine biodiversity (Qureshi, 2018). Ocean acidification is a result of increased carbon dioxide levels (Díaz, 2019) and is a major threat to coral and Artic Ocean ecosystem (Scott, 2018). Climate models demonstrate a decline in global ocean oxygen and predict continuing and accelerating ocean deoxygenation (Oschlies et al., 2018). Marine life in both hemispheres is expected to move poleward due to the warming of the oceans (Díaz, 2019). This does not, however, imply that there will be an increase in marine biodiversity in the polar regions (Díaz, 2019). Rather, declines are expected for marine biodiversity located in 'boreal, subpolar, and polar regions' due to warming seas, ice retreating, and ocean acidification (Díaz, 2019, p. 16). The GBF includes a target to minimize the impact of climate change and ocean acidification on biodiversity and increasing its resilience (GBF target 8). The UNFCCC's Glasgow Climate Pact in 2021 recognized the essential role of the ocean in climate change mitigation and adaptation policies (Lennan & Morgera, 2022).

Understanding the properties, benefits, conservation, and sustainable use of MGR of ABNJ through science and Traditional Knowledge is essential for biological diversity law and policy. One of the key drivers of Part II outlined in Sect. 1.3 below concerns understanding the scientific and commercial value or uses of MGR in ABNJ and the expected fair and equitable sharing of benefits from these uses. From a scientific perspective, MGR are 'the genetic material present in all marine life' (Rogers et al., 2021, p. 4). The BBNJ Agreement defines MGR as 'any material of marine plant, animal, microbial or other origin containing functional units of heredity of actual or potential value' (art 1(8)). Scientists are keen to explore ABNJ for MGR for a range of commercial and non-commercial reasons (Blasiak et al., 2022; Carroll et al., 2024; Rogers et al., 2021; Sigwart et al., 2021). The High Seas and deep seabed are home to extremophiles: creatures that have adapted to live in extreme temperatures, pressure, salinity, and darkness (Gameiro, 2023). The unique genetic characteristics that allow them to survive in extreme conditions could be useful for development of medicines, cosmetics, and food (Jaspars & Brown, 2021). Some examples of scientific interest in MGR include wound healing using jellyfish collagen; drug development using marine sponges; and soft robotics development inspired by seastars (Harden-Davies, 2021). The use of MGR in marine biotechnology can support conservation efforts, for example, ecosystem monitoring (where collection and sequencing of MGR can provide baseline data for taxonomy and conservation activities), bioremediation, and ecosystem adaptation strategies, such as bionic corals using gene editing techniques (Blasiak et al., 2023).

Estimating the actual and potential value of MGR for research and development (R&D) and commercialization is important for ascertaining the potential benefits that might accrue from their use. Sources of data for these estimates include scientific literature (e.g., Carroll et al., 2024), sequence databases (e.g., Scholz et al., 2021), and patent databases (e.g., Oldham et al., 2014, etc.). However, the gaps in data on the commercial application and value of MGR of ABNJ are well documented (Blasiak et al., 2018; Sigwart et al., 2021). Challenges for accurate data include an under-reporting of origin or source of a specific genetic resource in sequence and patent databases (Blasiak et al., 2019) and the nature of the R&D process where one MGR product or process may incorporate a variety of genetic material from several jurisdictional areas (Rogers et al., 2021). Recent international developments may slowly address the gaps in data on origin, such as database and publication requirements to specify provenance of genomic data and materials (Huang & Qin, 2024) and the 2024 *WIPO Treaty on Intellectual Property, Genetic Resources and Associated Traditional Knowledge*, obliging Parties to require patent applications in their jurisdiction to disclose the origin or source of genetic resources or Traditional Knowledge on which a claimed invention is based (WIPO, 2024, art 3). The notification and information sharing provisions of the BBNJ Agreement may produce data that continue to build a more complete picture of the actual and potential value of MGR, both commercial and non-commercial.

1.3 Jurisdictional Context and Gaps in Ocean Governance

The BBNJ Agreement applies to biodiversity outside of any country's boundaries, known as ABNJ (art 3). ABNJ refers to two maritime zones—the High Seas and the Area. The High Seas as defined by UNCLOS are all areas of water, including the water column, that are not within national jurisdiction (UNCLOS art 86, Parts VII, XI). Geographically, ABNJ cover roughly 70% of the world's surface (Battaglia, 2023); jurisdictionally, ABNJ covers roughly 60% of the ocean's surface (Elferink et al., 2022; Rogers et al., 2021) and seabed (Mengerink, 2018) and between 73% (Rogers et al., 2021) and 95% (Jarvis & Young, 2023; Payne, 2020) of the volume of the world's ocean. ABNJ includes all the seabeds, ocean floor, and subsoil that are not within national jurisdiction (UNCLOS art 1(1)). Unlike the High Seas that begin at the end of a State's 200 nm Exclusive Economic Zone (EEZ), the starting point for the Area is more complex (UNCLOS art 76). The Area begins where a coastal State's continental shelf ends and includes all seabed, ocean floor, and subsoil beyond such point (UNCLOS art 89; Korkut & Fowler, 2019).

This BBNJ Agreement will be the third implementing agreement under UNCLOS. The first implementing agreement is *The Agreement Relating to the Implementation of Part XI of the United Nations Convention on the Law of the Sea of 10 December 1982* (Part IX Agreement). The Part IX Agreement provides specific regulations on the conservation and sustainable use of the seabed, ocean floor, and subsoil of the Area as a space whereby all its resources have been declared to belong to the common heritage of humankind (UNCLOS art 136). The Authority, now known as the International Seabed Authority (ISA) (UNCLOS art 1(2)), has the legal power under UNCLOS to regulate this space (UNCLOS art 137), including administrating mineral resources in the Area

(UNCLOS art 157(1)). The BBNJ Agreement is concerned with marine biological diversity, not mineral resources (art 2). In keeping with the 'not undermining' principle under Article 5, the BBNJ Agreement does not impact the ISA's legal authority over the governance of mineral resources located in the Area.

The second implementing instrument is the United Nations *Agreement for the Implementation of the Provisions of the Convention relating to the Conservation and Management of Straddling Fish Stocks and Highly Migratory Fish Stocks* (UNFSA), which entered into force in 1995. The UNFSA provides regulations on key fisheries and includes obligations for States 'not to cause pollution and other harm to marine life and resources and the marine environment' (Qureshi, 2018, p. 845). Its general principles are to 'conserve and manage straddling fish stocks and highly migratory fish stocks' (UNFSA art 5). The framework relies on the cooperation of States to act in accordance with the convention when fishing on the High Seas (UNFSA art 5). The mechanism for this stipulates that States ought to 'pursue cooperation in relation to straddling fish stocks and highly migratory fish stocks either directly or through appropriate subregional or regional fisheries management organizations or arrangements' (UNFSA art 8). UNFSA lists out explicitly what is expected by these subregional and regional organizations or arrangements (UNFSA art 10). It recognizes States' rights to fish on the High Seas with certain conditions (UNCLOS art 116). In doing so, it created obligations upon States to take measures for the 'conservation of living resources of the high seas' (UNCLOS arts 116, 117, 118, 119 and 120), such as creating Regional Fisheries Management Organisations (RFMOs) (UNCLOS art 118). RMFOs provide governance measures on fisheries located in some parts of ABNJ (Russell & VanderZwaag, 2010).

Part II of the BBNJ Agreement covers activities with respect to MGR and associated DSI but does not apply to fishing regulated under relevant international law, fishing-related activities or fish taken in such activities (unless it is subject to R&D on its genetic and/or biochemical composition, including through the application of biotechnology) (art 10(2)). The BBNJ Agreement must be interpreted and applied in a manner that does 'not undermine' relevant legal instruments and frameworks (art 5(2)), including the UNFSA, meaning biological resources from fish collected in ABNJ may be regulated under both the BBNJ Agreement and UNFSA, depending on the activity and purpose of use. The BBNJ Agreement text does not define or provide an explanation of what is meant by 'not undermine' (see Humphries, 2025).

There is a range of international fora that regulate different activities in ABNJ, producing a level of complexity for regulation and governance of this jurisdictional area. Ardron and Warner (2015) demonstrate the range of bodies and conventions that have an impact on governance of ocean biological resources as at 2015, and these are increasing as new agreements are reached. Table 1.1 includes a non-inclusive list of agreements that apply to living resources in ABNJ within different contexts. Each agreement has its own scope of regulatory authority and subject matter, which has led to a patchwork of international policies on the conservation and sustainable use of marine biodiversity in ABNJ (Dalaker, 2022; Freestone, 2019; Qureshi, 2018). Additional treaties that relate to biodiversity in ABNJ may enter into effect in the future, such as the proposed Plastic Treaty—UNEP, 2022a; therefore, legal complexity is likely to increase in the future.

The negotiating countries identified several gaps in ABNJ governance that the BBNJ Agreement has the potential fill. UNCLOS includes obligations upon States 'to protect and preserve the marine environment' (UNCLOS art 192) and not to pollute it (UNCLOS 194). However, there were concerns that existing obligations under UNCLOS did not provide a coordinated or effective framework for the conservation and sustainable use of marine biodiversity in ABNJ (Ardron et al., 2014; Druel & Gjerde, 2014; Nishimura, 2018; Tiller et al., 2019). There was no institutional mechanism or processes to implement area-based marine

Table 1.1 Examples of existing international agreements relating to marine life in ABNJ

Agreements	Geographical jurisdiction	Subject matter
International convention for the regulating of whaling (1946) (whaling convention)	All waters in which whaling is engaged by factory ships, land stations, and whale catchers (whaling convention art 1(2))	Conservation of whaling stocks and orderly development of the whaling industry (whaling convention preamble)
Convention on the conservation of antarctic marine living resources (1980) (CCAMLR)	'Area south of 60° South latitude' and 'the area between that latitude and the Antarctic Convergence' (CCAMLR art 1(1))	Antarctic marine living resources (CCAMLR art 1(1))
Convention for the conservation of salmon in the North Atlantic Ocean (1982) (NASCO)	'Salmon stocks which migrate beyond areas of fisheries jurisdiction of coastal states of the Atlantic Ocean north of 36° N latitude throughout their migratory range' (NASCO art 1(1))	Salmon stocks (NASCO art 1(1))
The conservation and management of straddling fish stocks and highly migratory fish stocks (1994) (UNFSA)	ABNJ, except articles 6 and 7 also apply to AWNJ (UNFSA art 3(1))	The 'conservation and sustainable use of straddling fish stocks and highly migratory fish stocks' (UNFSA art 2)
Convention for the conservation of anadromous stocks in the North Pacific Ocean (1992) (NPAFC)	North Pacific Ocean and its adjacent seas, north of 33° North Latitude beyond 200 nm from the territorial sea baseline, scientific purposes can extend south of this (NPAFC art 1)	Fishing of anadromous fish (NPAFC art 3(1)), exception for scientific research purposes (NPAFC art 3(2))
Agreement on cooperation in research, conservation and management of marine mammals in the North Atlantic (1992) (NAMCO)	North Atlantic (NAMCO art 2)	Conservation, rational management, and study of marine mammals (NAMCO art 2)
Convention for the protection of the marine environment of the north-east Atlantic (1992) (OSPAR)	Land and sea (OSPAR arts 2(4) and 3)	Prevent and eliminate pollution, protect maritime areas, safeguard human health, conserve marine ecosystems, and restore marine areas (OSPAR art 2(1)(a))
Convention for the conservation of southern Bluefin Tuna (1993) (CCSBT)	Any operation at sea (CCSBT art 2(b))	Southern bluefin tuna (CCSBT art 1) and 'ecologically related species' (CCSBT arts 2(a) and 5)
Convention for the conservation and management of highly migratory fish stocks in the western and central pacific ocean (2000) (WCPFC)	'All waters of the Pacific Ocean bounded to the south and to the east by' particular coordinates (WCPFC art 3(1))	All highly migratory fish stocks of the western and central Pacific Ocean (WCPFC preamble), except sauries, within the Convention Area (WCPFC art 3(3))
Agreement on the conservation of albatrosses and petrels (2001) (ACAP)	Areas where albatrosses and petrels are located (ACAP art 3)	Species of albatrosses and petrels set out in Annex 1 of the agreement (ACAP art 1(1))
Convention on the conservation and management of the high seas fishery resources of the south pacific ocean (2009) (SPRFMO)	Waters of the pacific ocean beyond areas of national jurisdiction (SPRFMO art 5)	Long-term conservation and sustainable use of fishery resources and safeguarding marine ecosystems where these resources occur (SPRFMO art 2)

management tools, including marine protected areas in ABNJ and no standardized methodologies to conduct and report environmental impact assessments in ABNJ, which would prevent, mitigate, and manage significant adverse impacts (Gonçalves, 2023; Nguyen, 2022).

As technologies and the High Seas have become increasingly accessible in recent years, policymakers turned their minds to regulating bioprospecting in ABNJ (UNGA, 2005). UNCLOS includes provisions for marine scientific research (UNCLOS Part XIII), and capacity building and technology transfers to assist geographically disadvantage States (UNCLOS art 266) and developing States (UNCLOS arts 144, 266, 276). It does not however offer any guidance in relation to regulating MGR, bioprospecting activities (Deasy, 2023), or the fair and equitable sharing of benefits from the use of MGR of ABNJ (Honkonen et al., 2021; Harden-Davies et al., 2022; Nishimura, 2018). Bioprospecting is, generally, 'searching for biochemical and genetic information in nature' that 'can lead to new products with commercial value' (Payne, 2020, p. 346). At the time of the adoption of the BBNJ Agreement, there were few commercialized products associated with biological materials of ABNJ (Deasy, 2023). While there may be potential for high profits (Deasy, 2023), estimates vary widely (Mendenhall et al., 2019). Bioprospecting, however, requires enormous financial and technological resources to provide access to remote areas in ABNJ for sampling and laboratory equipment and human expertise within national jurisdiction to study and develop marketable products (Rogers et al. 2021). It costs approximately US$1 billion to develop a drug for market, and all drugs based on marine organisms on the market were developed by companies in Europe, Japan, and North America (Blasiak et al., 2023). Many countries argued that the BBNJ Agreement needs to address inequality arising for those states without the technological and financial resources to exploit MGR (de la Concepcion, 2024; Vanagt et al., 2019).

These types of equity issues around benefit sharing were addressed in other international fora concerning biological resources through the concept of 'access and benefit sharing' (ABS). ABS was originally designed as an economic incentive to conserve and sustainably use biodiversity *within* national jurisdiction after UNCLOS was adopted in the 1980s (Lawson, 2012; Sonesson et al., 2023). The concept was included in the 1992 *Convention on Biological Diversity* (CBD) and its implementing agreement the 2010 *Nagoya Protocol on Access to Genetic Resources and the Fair and Equitable Sharing of Benefits Arising from their Utilization to the Convention on Biological Diversity* (Nagoya Protocol). The CBD recognized the sovereign rights of States over their natural resources and the authority for national governments to determine access to genetic resources (CBD art 15(1)). Each Contracting Party to the CBD is required to endeavor to create conditions to facilitate access to genetic resources for environmentally sound uses by other contracting Parties (CBD art 15(2)). Access must be subject to the prior informed consent of the provider Party, and where granted, access must be on mutually agreed terms (CBD arts 15(4), 15(5)). Benefits from the 'utilization' and commercialization of the genetic resources and Traditional Knowledge associated with the genetic resources must be shared in a 'fair and equitable way' with the provider Party that is the country of origin or who acquired them in accordance with the CBD (Nagoya Protocol art 5). These bilateral agreements only extend to biological resources within national jurisdiction or to activities or processes within or beyond the limits of national jurisdiction that are carried out under a state's jurisdiction or control regardless of where their effects occur (CBD art 4).

There are a number of other frameworks that deal with genetic resources, which have benefit sharing procedures and mechanisms for specific purposes in the fields of plant genetic resources and H5N1 virus genetic resources.[2] Extending benefit sharing mechanisms to virus genetic resources with pandemic potential and a multilateral mechanism for DSI are under discussion, and the latter of which may have a significant effect on how the BBNJ Agreement manages DSI on MGR of ABNJ (UNEP, 2022b). As this book outlines in several chapters, the BBNJ Agreement sets up a skeleton framework for fair and equitable benefit sharing from the use of MGR and DSI (art 14) but whether it achieves its social and economic justice aims will depend on how Parties implement their obligations.

The lacuna in global policy on fair and equitable benefit sharing from the use of MGR of ABNJ was a key priority of States who did not have the technology or resources to exploit them and who saw it as an economic means to contribute to the conservation and sustainable use of marine biodiversity in areas beyond national jurisdiction (UNGA, 2005; Nishimura, 2018). Treaty aims to address the inequalities caused by the dominance of a disproportionately small number of organizations in developed countries (mostly US, Japan, and Germany) benefiting from the commercialization of MGR of ABNJ (Blasiak et al., 2022). Ministers of the Group of 77 and China argued that the principle of common heritage of humankind should 'guide and underpin' the new legal regime for MGR because it 'would allow all countries to benefit from the potential that marine biodiversity represents in terms of global food security and economic prosperity, and to address the challenges of conservation and sustainable use of MGRs' of ABNJ (UNGA Resolution A/72/511 Annex to the letter (2017) Para. 159). Analysis of the common heritage of humankind principle is beyond the scope of this chapter (see Vadrot et al., 2022), but it was central to the ideological divide throughout negotiations between those countries that argued that freedom of the High Seas (UNCLOS art 87) does not impact benefit sharing and the countries that argued that the common heritage of mankind principle requires mandatory benefit sharing, including monetary benefits (see Sect. 1.5 below). Article 87 provides that the High Seas are open to all States, but with certain restrictions. For example, the freedom of the High Seas comprises among other things freedom of scientific research, but these freedoms are to be exercised with due regard for the interests of other States (UNCLOS art 87). The article has generally been interpreted by some States to mean that access to MGR in ABNJ should remain open and unimpeded by rules and procedures (Mendenhall et al., 2019)—essentially, a first-in-first served approach.

UNCLOS does not have explicit provisions about the protection and facilitation of Indigenous Peoples and local communities' (IPLCs) rights concerning Traditional Knowledge of the oceans and its biodiversity. There is a lack of research and international guidance on how to take into account both scientific and Traditional Knowledge of IPLCs when making rules and decisions about the conservation and sustainable use of marine biodiversity in ABNJ (Mulalap et al., 2020). The BBNJ Agreement recognizes the importance of the use of best available science and the use of relevant Traditional Knowledge, while protecting the rights of IPLCs that hold such knowledge (art 7). Specifically, the text stipulates that Traditional Knowledge of IPLCs associated with MGR in ABNJ can only be accessed with 'the free, prior, and informed consent or approval and involvement of these Indigenous Peoples and local communities' (art 13). The general principles of the BBNJ Agreement include respecting, promoting, and considering the rights of IPLCs when taking action to address the conservation and sustainable use of marine biological diversity in ABNJ (art 7).

1.4 Preparatory Work Leading to the Negotiations

Discussions and negotiations at the United Nations on the BBNJ Agreement, from start to the adoption by consensus, lasted over 20 years.

[2] *International Treaty on Plant Genetic Resources for Food and Agriculture*, 3 November 2001, 2400 U.N.T.S. 303; *Pandemic influenza preparedness framework for the sharing of influenza viruses and access to vaccines and other benefits.* Geneva: World Health Organization; 2011, https://www.who.int/publications/i/item/9789240024854.

The groundwork for the BBNJ Agreement at the UN began in 2004 with the UN General Assembly adopting a resolution, which established an Ad Hoc Open-ended Informal Working Group (Working Group) to study issues associated with the conservation and sustainable use of marine biodiversity beyond areas of national jurisdiction (UNGA Res. 59/24, 2005). A prominent debate during this time was how to balance High Seas freedoms, the duty to protect and preserve the marine environment, and the application of the common heritage of humankind to MGR in ABNJ (Wright et al., 2016). The following year, the UN General Assembly adopted a resolution, which included a recommendation for the Member States to engage in discussions on the conservation and sustainable use of marine biodiversity in the Working Group (UNGA Res. 60/30, 2006). From 2006 through 2010, the Working Group was tasked with determining possible substantive areas to be covered by a new governance on marine biological diversity of ABNJ. In 2010, the General Assembly adopted a resolution that called for reconvening the Working Group the following year, emphasizing the need for continued discussions on MGR, marine protected areas, and environmental impact assessment processes (UNGA Res. 65/37, 2021).

A breakthrough happened in 2011 during the fourth session of the Working Group. A small group within the Working Group provided recommendations, which was adopted by consensus, on initiating a process on the legal framework for the conservation and sustainable use of marine biodiversity beyond areas of national jurisdiction (Earth Negotiations Bulletin, 2011). To that end, the process would identify gaps and ways forward, including through the implementation of existing instruments and the possible development of a multilateral agreement under UNCLOS (Earth Negotiations Bulletin, 2011). The recommendations also included a 'package' of issues to be addressed 'as a whole' in this process, namely MGR, including questions on benefit sharing; measures, such as area-based management tools, including marine protected areas and environmental impact assessments; capacity building, and the transfer of marine technology. The four elements became known as the '2011 Package' among the negotiators. The UN General Assembly adopted a resolution at the end of 2011 that requested the Secretary-General to convene a meeting of the Working Group the following year to make progress on all issues under examination within the Working Group and to provide recommendations to the General Assembly at its sixty-seventh session. However, the subsequent sessions in 2012 did not lead to starting an intergovernmental conference to negotiate a treaty.

In 2014, the UN General Assembly adopted a resolution with a mandate for the Member States to discuss the scope, parameters, and feasibility of a possible new treaty on marine biodiversity in ABNJ under UNCLOS (UNGA Res. 68/70, 2014). Pursuant to the resolution, three sessions of the Working Group took place in April 2014, June 2014, and January 2015. In the third session, the Member States reached a consensus after long and intense negotiations on a decision to be taken at the 69th session of the UN General Assembly to develop a new treaty on marine biodiversity of ABNJ under UNCLOS (Earth Negotiations Bulletin, 2015). The Member States also agreed by consensus about the negotiation process, which would include four sessions of Preparatory Committee meetings that would determine a set of recommendations on the elements of a draft text of a treaty that would be submitted to the UN General Assembly in 2017. Based on the recommendations, the General Assembly at its 72nd session would decide whether to convene an intergovernmental conference (Earth Negotiations Bulletin, 2015).

The Preparatory Committee convened for four sessions (two sessions each in 2016 and 2017) pursuant to UN General Assembly resolution 69/292, with each session lasting two weeks. It is worth noting that all the sessions were open to registered observers with certain rights to make statements on the floor. After two years of discussions, the Preparatory Committee issued a report (31 July 2017) that provided recommendations to the UN General Assembly for its consideration, with a view to the development of a draft text of an international legally binding instrument under UNCLOS on the conservation and sustainable use of marine biological diversity of areas beyond national jurisdiction (UNGA, 2017). There are two sections of the Preparatory Committee report that provided some of the key ideas (i.e., not all ideas were included). Section A covers non-exclusive elements that generated convergence among most delegations. Section B provides the main issues on which there was divergence of views. As the chair of the Preparatory Committee emphasized throughout the discussions, both sections were deemed to be without prejudice to the positions of Member States and the entire report was issued as a reference document. The Preparatory Committee also recommended that the UN General Assembly decide, as soon as possible, to convene an intergovernmental conference to consider the recommendations of the Preparatory Committee's report on the elements and to elaborate the text of a treaty.

1.5 The BBNJ Agreement Negotiations on Marine Genetic Resources

On 24 December 2017, the UN General Assembly adopted a resolution to convene an intergovernmental conference (IGC) from 2018 through 2020 for the Member States to negotiate an international legally binding instrument under UNCLOS on the conservation and sustainable use of marine biological diversity of areas beyond national jurisdiction (UNGA Res. 72/249, 2018). All four elements of the BBNJ Agreement needed agreement before adoption, and Part II of the BBNJ Agreement was the hardest of the treaty elements to reach compromise (de la Concepción, 2024). Most of the negotiating sessions were highly transparent with a large number of accredited observers involved, including intergovernmental organization and non-governmental organizations, who made interventions and contributed a range of scientific and policy expertise.

While there was much good will and compromise between delegations fostered by the President Ambassador Lee and Part II Facilitator Ambassador Felson during the negotiations on the MGR provisions, there were key areas of divergence throughout the negotiations. These include:

- the extent and modalities of benefit sharing, including what is shared and who it is shared with;
- whether to include DSI as the subject matter of benefit sharing and if so, how;
- the extent of monitoring and compliance infrastructure;
- the relationship between intellectual property and the instrument; and
- principles guiding benefit sharing, including principles of the common heritage of humankind and freedom of the high seas (facilitator reports, see Table 1.2).

The original program of work included four IGCs but due to lack of agreement, the UN funded a fifth IGC. Table 1.2 provides a summary of the text versions discussed at each of the IGCs, with links to the President and Facilitator reports for the relevant IGC.

At the first intergovernmental conference in September 2018 (IGC1), states considered the President's Aid to Discussions including a non-exhaustive list of issues, questions, and options (UNGA, 2018b). Each substantive element and crosscutting issues were presided by a Facilitator, who led the discussions and reported back to the President of the IGC. In terms of MGR, there was clear divergence on all of the key access, benefit sharing, and monitoring issues and special rights of coastal states (UNGA, 2018a, p. 25). While there was some convergence on the need for meaningful benefit sharing without disproportionate burdens on fostering science and the private sector, developed countries did not support any type of monetary benefit sharing in the BBNJ

Table 1.2 Texts of president's aid documents and draft and final treaty texts with key MGR ICG reports

ICG	IGC date	Document or text discussed during the IGC	Text date	President and facilitator reports
1	4–17 September 2018	President's aid to discussions UNGA, 2018b	25 June 2018	UNGA, 2018a 20 September 2018
2	25 March-5 April 2019	President's aid to negotiations UNGA, 2018c	3 December 2018	UNGA, 2019a 18 April 2019
3	19–30 August 2019	Draft text of an agreement under the United Nations Convention on the Law of the Sea on the conservation and sustainable use of marine biological diversity of areas beyond national jurisdiction UNGA, 2019b	17 May 2019	UNGA, 2019d 13 September 2019
Virtual intersessional work—UN decision 74/543 to postpone work during COVID-19				
4	7–18 March 2022	Revised draft text of an agreement […] UNGA, 2019c	18 November 2019	UNGA, 2022a 14 April 2022
5	15–26 August 2022	Further revised draft text of an agreement […] UNGA, 2022b	1 June 2022	UNGA, 2022c 14 September 2022
		Refreshed draft text of an agreement […] UNGA, 2022d		Outcomes of the small working groups A/CONF.232/2023/INF.2 1 February 2023
		Further refreshed draft text of an agreement […] A/CONF.232/2022/CRP.13		
5#	20 February-3 March 2023	Further refreshed draft text of an agreement […] UNGA, 2022e	12 December 2022	
		Updated draft text of an agreement […] A/CONF.232/2023/CRP.1	25 February 2023	
		Draft agreement […] UNGA, 2023a	3 May 2023	
		Draft agreement […] UNGA, 2023b	6 June 2023	
5^	19–20 June 2023	Agreement under the United Nations Convention on the Law of the Sea on the conservation and sustainable use of marine biological diversity of areas beyond national jurisdiction UNGA, 2023c (final adopted text)	19 June 2023	UNGA, 2023d 30 June 2023

* ICG means intergovernmental conference. # Means resumed session. ^ Means further resumed session. […] means the remainder of the text name common to each document 'under the United Nations Convention on the Law of the Sea on the conservation and sustainable use of marine biological diversity of areas beyond national jurisdiction'

Agreement, nor the inclusion of DSI as subject matter of access or benefit sharing provisions (Tiller et al., 2019). Nevertheless, the mood of delegates suggested an air of optimism as they worked toward finding common ground (Tiller et al., 2019).

The President introduced a 'President's Aid to Negotiations' (UNGA, 2018c) for discussion at IGC2 (March–April 2019), with a range of suggested options for various provisions explored at IGC1 and the preparatory work (UNGA, 2018a). While there was some convergence on the objectives of benefit sharing, namely that it should contribute to the conservation and sustainable use of MGR and build the capacity of developing countries to access and use them, most of the options were left for the subsequent IGC without agreement (UNGA, 2019a). Mendenhall et al., (2019) noted that the nature and breadth of disagreements raised at IGC2 had not significantly altered since the Preparatory Committee phase (Mendenhall et al., 2019). Debate further entrenched polarized positions generalized as: (a) those countries (e.g., G77 and China) that supported the common heritage principle, ambitious goals, and institutional arrangements; and (b) developed countries that supported the freedom of the High Seas, narrower goals, and limited institutional arrangements (Mendenhall et al., 2019). The latter tended to be countries most likely to engage in research activities in the High Seas or utilizing the materials and DSI from the high seas and who were cautious about adding regulatory burdens while the former (seeking additional regulation) lacked capacity to undertake the research and development activities (De Santo et al., 2020). Several countries and regions, such as Canada, EU, and Australia, advocated for middle positions between the polarized stands (De Santo et al., 2020).

The first (zero) draft text emerging from IGC2 (UNGA, 2019b) was introduced at IGC3 (August 2019) for discussion. Commentators observed that negotiations reverted to the positions on common heritage of humankind and freedom of the High Seas principles despite the suggestion for less polarized alternatives (De Santo et al., 2020). However, by the end of IGC3, the Facilitator recognized progress in moving away from conceptual discussions toward identifying textual solutions (UNGA, 2019d). Whether or not to include intellectual property provisions was a sticking point (De Santo et al., 2020), with options including disclosure provisions and restrictions on patent applications and claims that would have significantly altered the status quo under existing international frameworks and would have impacted national legislation (UNGA, 2019b). The outcome of IGC3 was captured in the draft text of 18 November 2019, with a pared down version of the draft intellectual property article. Further, all draft articles on intellectual property were in brackets with no agreement in sight.

The COVID pandemic led to unanticipated delays for hosting IGC4, originally expected in March 2020 (see Vadrot et al., 2021). In the interim, the President of the IGC hosted informal intersessional virtual meetings to help the delegations to further understand various positions with encouragement for the States to find common ground on various open questions. The President made clear that the informal intersessional virtual meetings were not formal negotiations. For that reason, during this period, the draft text did not change.

Momentum toward convergence on MGR issues was slow but with some progress made at IGC4. Countries negotiated on the November 2019 draft text seemingly clung to their diverging positions (UNGA, 2022a), but with proposed language that indicated some flexibility (Mendenhall et al., 2022). There was some movement toward the development of a notification system relating to collection of MGR in ABNJ, but views still significantly divided on intellectual property, benefit sharing, and monitoring systems (UNGA, 2022a). The effects of the pandemic resulted in limited in-person participation in the conference room to two people per state delegation with no observers (although restrictions were eased in the second week). Stretched capacity for meaningful and informed negotiations and an observed lack of political will for progress including robust new institutions (Mendenhall et al., 2022; Tiller et al., 2023) contributed to even greater divergence on DSI and monetary benefit sharing in the outcome draft text of 12 December 2022 (de la Concepción, 2024). The IGC concluded without agreement on key issues in Part II and the UN approved funding for one more IGC to conclude negotiations in August 2022.

The fifth IGC extended to two substantive sessions. During the August 2022 session, there was extensive transparency and access by observers (Tiller et al., 2023), but there were challenges for delegations to stay on top of coordination and consensus building because of the small group formats, long hours, and lack of access to translators (Tiller et al., 2023). Substantial progress was made on the notification system to govern MGR of ABNJ and agreement that there be monetary and non-monetary benefits, but continued lack of agreement on intellectual property provisions, modalities for monetary benefit sharing, and the inclusion of DSI as subject matter remained (UNGA, 2022c). There was a notable resurgence of the common heritage of humankind principle (Tiller et al., 2023), and the revised draft text (publicly released on 12 December 2022) (UNGA, 2022e) failed to achieve the expected outcome for ambitious and practical benefit sharing arrangements desired by G77 and China (de la Concepción, 2024).

With no agreement in sight, the fifth session was suspended and resumed in February through March of 2023 (Resumed IGC5) (UNGA Res. 77/248, 2023). The text of 12 December 2022 was expected to be the starting point in the negotiations on MGR, but this was sidestepped as G77 and China proposed an ambitious framework on the first day, which became the basis for additional options in the 25 February 2023 draft text. For the first time in the treaty text, DSI became the subject matter of the utilization notification and benefit sharing arising from commercialization and part of the other information sharing arrangements under the notification and transparency measures. A range of factors led to a breakthrough in the DSI issue, including the strong and unified position of G77 and China (de la Concepción, 2024) and the Conference of the Parties (COP) to the CBD decision to introduce a DSI benefit sharing multilateral mechanism (UNEP 2022c).

There were further numerous breakthroughs in the modalities of benefit sharing, including royalties, milestones, and a tiered fee based on aggregate level of activities within a Party's jurisdiction (Broggiato et al., 2025). The G77 and China proposed a BBNJ Batch Identifier mechanism as a transparency measure that enables the tagging of MGR and DSI and, together with other reporting requirements, would contribute to estimating the aggregate level of MGR-related activities for possible future rules for calculating monetary benefit sharing contributions under the tiered fee model (de la Concepción, 2024; Oldham et al., 2023). The draft intellectual property provision disappeared as a stand-alone provision (Brown, 2025) and a final compromise to the polarizing ideological divide came in the final hours of negotiations when both the principles of the common heritage of humankind and the freedom of marine scientific research (together with other freedoms of the High Seas) were placed together in the general principles section outside Part II (art 7). Reaching agreement on Part II BBNJ Agreement was a remarkable achievement, but there are many gaps in interpretation, policy, and procedure that will need to be addressed over the coming years, as this book examines.

On the night of 4 March 2023, the draft final text of the BBNJ Agreement was adopted by consensus, with the President of the negotiations recognizing that the hard work toward implementation was only beginning (Tiller & Mendenhall, 2023). The IGC established an open-ended informal working group to ensure the uniformity of terminology throughout the draft final text in English and harmonizing the versions in the six official languages of the United Nations (English, Arabic, French, Spanish, Chinese, Russian). At the further resumed fifth IGC, the BBNJ Agreement was officially adopted by consensus on 19 June 2023 (UNGA, 2023d). To support the early implementation of the BBNJ Agreement, the President of the negotiations sought support from the UN General Assembly for the creation of a Preparatory Commission and an interim secretariat to promote better understanding and ratification of the BBNJ Agreement (UNGA Res. A/78/272, 2024).

1.6 Structure of This Book

The twenty-five authors of this edited collection from a range of continents have extensive experience in marine research, law, and policy. They are part of a multidisciplinary network of researchers, practitioners, and delegates for the intergovernmental conference (ICG) negotiations of the BBNJ Agreement but the views they express are their own and not representative of their institutions or countries.

The peer-reviewed book is divided into two parts (Table 1.3). Part 1 is a commentary that includes interpretation of the main articles in Part II of the BBNJ Agreement, including an analysis of the history and evolution of the provision (where relevant), key innovations, gaps in policy that need clarification by the BBNJ

Table 1.3 Topics and articles analyzed in this book

Chapters	Topic and treaty articles (if relevant)	Title
1	Negotiating history	Bridging divides: the evolution of marine genetic resource governance beyond national jurisdiction
Part I: treaty interpretation and analysis		
2	Institutional arrangements (Part VI, art 15)	BBNJ agreement: a new infrastructure to foster benefit sharing of marine genetic resources
3	Definitions and scope (arts 1, 3, 4, 5, 6, 7, 8, 9, 10, 11)	Marine genetic resources beyond national jurisdiction: the expansive scope of the BBNJ agreement
4	Objectives, principles, preamble (arts 2, 7, 9)	Understanding the preamble, objectives and principles of the BBNJ agreement: a focus on the fair and equitable sharing of benefits of marine genetic resources
5	Activities and notification (arts 11, 12)	The novel notification information system for marine genetic resources under the BBNJ agreement
6	Benefit sharing (art 14)	Monetary and non-monetary benefit sharing under the BBNJ agreement
7	Monitoring and transparency (art 16)	Monitoring and transparency aspects of MGR-utilization under the BBNJ agreement
8	Traditional knowledge (art 13)	Traditional knowledge associated with marine genetic resources in areas beyond national jurisdiction
Part II: treaty implementation in practice		
9	Intellectual property	The place of intellectual property under the BBNJ agreement
10	Ratification	Considerations concerning state ratification of the BBNJ agreement
11	Relationships with ABS frameworks	Marine genetic resources and digital sequence information under the BBNJ agreement—interlinkages with other access and benefit-sharing frameworks
12	BBNJ identifier and data management plans	Data management and the 'BBNJ standardized batch identifier' under the BBNJ agreement
13	Benefit sharing in practice (art 14)	Benefit sharing under the BBNJ agreement in practice
14	Practical guidance for MGR, DSI and traditional knowledge users	BBNJ agreement: considerations for scientists and commercial end users of MGR at research, development and commercialisation stages
15	Where to from here	Conclusions: equity, sustainability and transformation under the BBNJ agreement

Agreement bodies or committees, and considerations for practical implementation. Part 2 draws from and builds on Part 1 by focusing on practical considerations for treaty implementation. It critically analyzes key areas of the BBNJ Agreement (including ratification, interlinkages with other ABS frameworks, the BBNJ Identifier, benefit sharing, and guidelines for scientists and commercial end users) to suggest the various ways that obligations may be put into practice by policymakers and the role of stakeholders, including scientists, commercial entities, and IPLCs in implementation.

Part 1 Interpretation and Commentary

The aim of this chapter (Humphries et al., 2025a) was to set out the environmental, economic, social, cultural, jurisdictional, and legal context for the development of the BBNJ Agreement and the factual background to the preparatory work and negotiations on Part II. It outlined how chapters in this book offer practical options for MGR and Traditional Knowledge Holders, policymakers, and academic and research institutions to consider the practical application of the elements of the MGR governance scheme in the BBNJ Agreement.

Chapter 2 (Muraki Gottlieb et al., 2025a) analyzes the BBNJ Agreement bodies, organs, and committees that are integral to the Part II MGR framework. These include the Conference of the Parties, Secretariat, Scientific and Technical Body, the Clearing House Mechanism, and subject matter committees—the Access and Benefit Sharing Committee, Finance Committee, Capacity Building and Transfer of Marine Technology Committee, and Implementation and Compliance Committee. The analysis emphasizes the novelty of the BBNJ Agreement institutions compared with other UNCLOS institutions and reflects on their role for transparency and monitoring. It highlights areas where further clarity is needed about the roles and functions of these institutional arrangements for stakeholders.

Chapter 3 (Humphries, 2025) interprets the scope and key definitions in Part II of the BBNJ Agreement. The purpose of this chapter is: (a) to aid practitioners and policymakers' understanding of the suggested rationale and intent underlying the provisions; (b) to set out the definitions and scope for cross referencing in subsequent chapters in this book; and (c) to outline the gaps in interpretation and provide practical considerations for how these elements might be implemented in practice by policymakers. It analyzes article 1 definitions including 'marine genetic resources,' 'areas beyond national jurisdiction,' 'collection in situ,' 'utilization of marine genetic resources,' and 'biotechnology.' It analyzes key undefined terms that are critical to the framework including 'digital sequence information,' 'samples,' 'traditional knowledge associated with marine genetic resources,' and 'access.' It interprets geographical scope (article 3), activity, and temporal scope (articles 4, 10, and 11).

Chapter 4 (Muraki Gottlieb et al., 2025b) outlines the significance of the BBNJ Agreement's general principles (article 7), objectives (articles 3 and 9), and the preamble that are relevant to interpretation of Part II MGR framework. The analysis of general principles includes the principles of common heritage of humankind, freedom of the High Seas, equity, and fair and equitable sharing of benefits, the use of best available science and scientific information and principles concerning Traditional Knowledge, and the recognition of special interests of certain states. It analyzes the general objective of the conservation and sustainable use of marine biological diversity of areas beyond national jurisdiction, for the present and in the long term and more specific objectives for Part II including those relating to benefit sharing, capacity building, and Traditional Knowledge.

The next three chapters analyze the core of the Part II MGR multilateral system—notification, benefit sharing, and monitoring systems. Chapter 5 (Humphries et al., 2025b) interprets and analyzes the notification system under article 12 of the BBNJ Agreement. Chapter 6 (Broggiato et al., 2025) interprets and analyzes the benefit sharing system under article 14 of the BBNJ Agreement. Chapter 7 (Langlet et al., 2025) interprets and analyzes the monitoring/

transparency system under article 16 of the BBNJ Agreement. These chapters follow a consistent format with a brief overview of the key elements of the systems before providing historical context and a detailed interpretation of the intent of these elements. The chapters analyze the innovative elements of these articles, how they relate to other parts of the BBNJ Agreement, highlight potential gaps in policy, and suggest how these systems might be implemented in practice.

Chapter 8 (Pena-Neira & Coelho, 2025) interprets and analyzes article 13 concerning the utilization of Traditional Knowledge associated with MGR in ABNJ. It traces the evolution of the obligations, including the requirement for prior informed consent and mutually agreed terms and how the Traditional Knowledge obligation may work in practice with the other multilateral notification, benefit sharing, and monitoring systems in the MGR framework.

Part 2 Practical Considerations for Treaty Implementation

Part 2 of the book commences with an analysis of the role of intellectual property under the BBNJ Agreement. While the final text did not contain the former draft article on intellectual property, Chap. 9 (Brown, 2025) outlines the history of the negotiations relating to intellectual property, analyzes its relationship with the BBNJ Agreement, and suggests paths for implementation of the final treaty text with respect to disclosure of origin under patent law and consistency with intellectual property frameworks more widely.

Chapter 10 (Ardron et al., 2025) presents issues that States and regional economic integration organizations may wish to consider before and after ratifying the BBNJ Agreement. Regarding the adoption of national legislation necessary to implement the Agreement, the chapter briefly examines the behaviors of Parties to the other two previously ratified implementing agreements to UNCLOS. It concludes with some ratification lessons learnt from other treaties outside of the UNCLOS framework, which are relevant to MGR of ABNJ. The primary audience for this chapter is policymakers but its insights are important for users of MGR and Traditional Knowledge in understanding the process of giving legal effect to the provisions under national law.

The purpose of Chap. 11 (Kachelriess et al., 2025) is to highlight several of the other instruments and bodies that will be relevant for the future implementation of the MGR and related benefit sharing provisions of the BBNJ Agreement and how these might co-exist with the BBNJ Agreement's multilateral system. These include the World Health Organization's Intellectual Property agreements, the International Seabed Authority's deep sea regime, and key access and benefit sharing agreements—Convention on Biological Diversity, the Nagoya Protocol, the forthcoming Digital Sequence Information multilateral mechanism, and agreements under the United Nations Food and Agriculture Organization and World Health Organization. The chapter highlights some challenges and opportunities for co-existence between these regimes, which in many cases share the same subject matter from different jurisdictional areas.

Chapter 12 (Lawson et al., 2025) focuses on two innovations in the BBNJ Agreement—data management plans and the 'BBNJ Standardized Batch Identifier' (BBNJ Identifier). It critically analyzes these innovations and builds on the interpretation of information sharing requirements outlined in Chap. 5 (notification obligations), Chap. 6 (benefit sharing obligations), and Chap. 7 (monitoring and transparency obligations). It draws from lessons learned for the data ecosystem from other agreements outlined in Chap. 11. The chapter offers practical insights into how the BBNJ Identifier system could be operationalized from an international governance viewpoint.

Chapter 13 (Lavelle & Wynberg, 2025) considers how the implementation of the benefit sharing system under article 14 may work in practice within the unique context of the BBNJ Agreement, focusing on benefits for the conservation and sustainable use of marine biodiversity. It provides examples of monetary and non-monetary benefit sharing and how these might be applied in the treaty context.

Chapter 14 (Rabone et al., 2025) provides insights into how the BBNJ Agreement may be applied in practice using different scenarios to exemplify various ways in which MGR and DSI may be collected and used. It draws from realistic examples based on existing use of MGR within national jurisdiction and in ABNJ. These scenarios are based on the requirements in the BBNJ Agreement, not based on a particular Party's legislation or policies. Further, it identifies and considers how practitioners might incorporate into their practices and processes elements of obligations that are expected to flow through to national law. This chapter draws from all other chapters in the book to provide a visual and practical guide for scientists, businesses, funders, repositories, databases, and other relevant entities.

Chapter 15 (Morgera, 2025) offers some concluding perspectives on the treaty from an equity and innovation perspective. It ties together the analysis of previous chapters in this collection and reflects on how Part II of the BBNJ Agreement might promote transformative change in ocean science, equity, and governance.

1.7 Conclusion

The BBNJ Agreement was an extraordinary achievement given the long-standing diverging views between negotiating groups about how to fill the gaps in ocean governance in ABNJ generally and benefit sharing from MGR specifically. The MGR framework includes, among other things, new notification, benefit sharing and transparency mechanisms for MGR and DSI, and access obligations for Traditional Knowledge associated with MGR in ABNJ. It offers new procedural mechanisms, such as the BBNJ Identifier and Data Management Plans for the purpose of information sharing, transparency, notification, and benefit sharing outcomes. It was a massive global effort to address governance gaps for marine biodiversity in 60% of the world's oceans, while at the same time continuing the long pathway of achieving fairness and equity between technologically rich and technologically poor countries in the distribution of benefits from the use of marine genetic resources of ABNJ.

As with all treaty negotiations, the diverging interests of countries and the political process make it difficult to glean 'the intent' of negotiators when drafting the text. The resulting wording is a product of careful political compromise and 'constructive ambiguity.' Constructive ambiguity is the idea that ambiguously worded text can offer space for advancing the interests of parties in disagreement, based on the assumption that tackling them in an unambiguous way would cause a breakdown in negotiations (Friedman, 2017). Part II of the BBNJ Agreement has a history of entrenched positions of negotiating groups at the IGCs that carried through from the preparatory commission work and is likely to continue into the preparatory commission and COP work when the agreement is adopted. Key areas of divergence, including governance of DSI, modalities for benefit sharing, and the role of intellectual property, were skillfully addressed right up until the marathon last negotiating session with language enabling future Parties to interpret their obligations in a way that meets their interests and unique circumstances. For most provisions, there is no single comprehensive interpretation of text wording, which will be further developed through state practice and COP decisions for many years to come.

The aim of this book is to offer insights into how treaty provisions might be interpreted (often in multiple ways), considerations for implementing obligations, and suggestions about how stakeholders could align their R&D, notification, benefit sharing, and reporting practices with these interpretations. Treaty obligations are on the Parties (not stakeholders), and they may adopt differing policies and procedures based on their interpretations of the obligations. Likewise, stakeholders will need to align their practices to meet the obligations in the countries where they are undertaking their notification and utilization activities. These practices are likely to diverge unless there is strong guidance by BBNJ Agreement bodies about cooperation and where possible, consistency, of implementation approaches between Parties. In the meantime, however, this book demonstrates that there are existing scientific and Traditional Knowledge practices enhanced by the BBNJ Agreement framework that can be developed and adopted by research institutions in anticipation of implementation of the treaty.

References

Ardron, J., et al. (2014). The sustainable use and conservation of biodiversity in ABNJ: What can be achieved using existing international agreements? *Marine Policy, 49*, 98–108.

Ardron, J., Kachelriess, D., Lyal, C., Nwapi, C., Rabone, M., Sirakaya, A., & Swaddling, A. (2025). Considerations concerning stateratification of the BBNJ Agreement. In F. Humphries (Ed.), *Decoding marine genetic resource governance under the BBNJ Agreement*. Springer.

Ardron, J. A., & Warner, R. (2015). International marine governance and protection of biodiversity. In H. D. Smith, J. L. Suárez de Vivero, & T. S. Agardy (Eds.), *Routledge handbook of ocean resources and management* (pp. 55–72). Routledge.

Barnes, R. (2022). The construction of ocean space in areas beyond national jurisdiction. In R. Churchill & A. O. Elferink (Eds.), *International law and marine areas beyond national jurisdiction* (pp. 275–315). Brill Nijhoff.

Battaglia, F. (2023). Climate change and the ocean: The disruption of the coral reef. In M. G. Garcia & A. Cortês (Eds.), *Blue planet law* (pp. 121–130). Springer.

Blasiak, R., Jouffray, J. B., Amon, D. J., et al. (2023). Making marine biotechnology work for people and nature. *National Ecology Evolution, 7*, 482–485. https://doi.org/10.1038/s41559-022-01976-9

Blasiak, R., Jouffray, J. B., Amon, D. J., Moberg, F., Claudet, J., et al. (2022). A forgotten element of the blue economy: Marine biomimetics and inspiration from the deep sea. *PNAS Nexus, 1*(4), pgac196.

Blasiak, R., Jouffray, J. B., Wabnitz, C. C., & Österblom, H. (2019). Scientists should disclose origin in marine gene patents. *Trends in Ecology and Evolution, 34*(5), 392–395.

Blasiak, R., Jouffray, J. B., Wabnitz, C., Sundström, E., & Österblom, H. (2018). Corporate control and global governance of marine genetic resources. *Science Advances, 4*, 5237.

Blasiak, R., Wynberg, R., Grorud-Colvert, K., Thambisetty, S., Bandarra, N. M., Canário, A. V. M., da Silva, J., Duarte, C. M., Jaspars, M., Rogers, A., Sink, K., & Wabnitz, C. C. C. (2020). The ocean genome and future prospects for conservation and equity. *Nature Sustainability, 3*, 588–596.

Broggiato, A., Dunshirn, P., Jaspars, M., & Pena-Neira, S. (2025). Monetary and non-monetary benefit sharing under the BBNJ Agreement. In F. Humphries (Ed.), *Decoding governance of marine genetic resources under the BBNJ Agreement*. Springer.

Brown, A. E. L. (2025). The place of intellectual property under the BBNJ Agreement. In F. Humphries (Ed.), *Decoding governance of marine genetic resources under the BBNJ Agreement*. Springer.

Carroll, A. R., Copp, B. R., Grkovic, T., Keyzers, R. A., & Prinsep, M. R. (2024). Marine natural products. *Natural Product Reports, 41*(2), 162–207.

Cheung, W., Jones, M., Lam, V., Miller, D., Ota, Y., The, L., & Sumaila, U. (2017). Transform high seas management to build climate resilience in marine seafood supply. *Fish and Fisheries, 18*, 254–263.

Crespo, G. O., Mossop, J., Dunn, D., Gjerde, K., Hazen, E., Reygondeaue, G., Warner, R., Tittensor, D., & Halpina, P. (2020). Beyond static spatial management: Scientific and legal considerations for dynamic management in the high seas. *Marine Policy, 122*, 104102.

Dalaker, K. E. (2022). Imagining a polycentric approach to institutional governance for marine areas beyond national jurisdiction. In R. Churchill & A. O. Elferink (Eds.), *International law and marine areas beyond national jurisdiction* (pp. 353–391). Brill Nijhoff.

de la Concepción, R. T. (2024). Negotiating fair and equitable sharing of benefits in the BBNJ agreement: Role of the Group of 77 and China. *Marine Policy, 163*, 106085.

De Santo, E., Mendenhall, E., Nyman, E., & Tiller, R. (2020). Stuck in the middle with you (and not much time left): The third intergovernmental conference on biodiversity beyond national jurisdiction. *Marine Policy, 117*, 103957.

Deasy, K. (2023). What we know about the new treaty. *NPJ Ocean Sustainability, 2*(7), 1–3.

Díaz, H., et al. (2019). *Summary for policymakers of the global assessment report on biodiversity and ecosystem services of the intergovernmental science-policy platform on biodiversity and ecosystem services*. https://zenodo.org/records/3553579

Druel, E., & Gjerde, K. M. (2014). Sustaining marine life beyond boundaries: Options for an implementing agreement for marine biodiversity beyond national jurisdiction under the united nations convention on the law of the sea. *Marine Policy, 49*, 90–97.

Elferink, A. O., Lucia, V. D., & Nguyen, L. N. (2022). Areas beyond national jurisdiction—looking at and beyond the BBNJ process. In R. Churchill & A. O. Elferink (Eds.), *International law and marine areas beyond national jurisdiction* (pp. 1–13). Brill Nijhoff.

Freestone, D. (2019). The UN process to develop an international legally binding instrument under the 1982 law of the sea convention: Issues and challenges. In D. Freestone (Ed.), *Conserving biodiversity in areas beyond national jurisdiction* (pp. 3–48). Brill Nijhoff.

Friedman, E. (2017). Evasion strategies in international documents: When 'constructive ambiguity' leads to oppositional interpretation. *Critical Discourse Studies, 14*(4), 385–401.

Gameiro, M. I. (2023). Law and marine genetic resources. In M. G. Garcia & A. Cortês (Eds.), *Blue planet law* (pp. 227–236). Springer.

Gjerde, K. M. (2006). Ecosystems and biodiversity in deep waters and high seas (No. 178). UNEP/ Earthprint.

Gonçalves, E. J. (2023). Marine protected areas as tools for ocean sustainability. In M. G. Garcia & A. Cortês (Eds.), *Blue planet law* (pp. 131–144). Springer.

Harden-Davies, H. (2021). Marine technology transfer. In M. H. Nordquist & J. N. Moore (Eds.), *Marine biodiversity of areas beyond national jurisdiction* (pp. 231–240). Brill Nijhoff.

Harden-Davies, H., Amon, D. J., Chung, T. R., Gobin, J., Hanich, Q., Hassanali, K., et al. (2022). How can a new UN ocean treaty change the course of capacity building? *Aquatic Conservation: Marine and Freshwater Ecosystems, 32*(5), 907–912.

Hoel, A. H. (2021). The BBNJ process and capacity building. In M. H. Nordquist & J. N. Moore (Eds.), *Marine biodiversity of areas beyond national jurisdiction* (pp. 213–230). Brill Nijhoff.

Honkonen, T., Kulovesi, K., Morgera, E., Recio, M. E., & van Asselt, H. (2021). The Siena negotiations: A multilateral simulation exercise: An international legally binding instrument on the conservation and sustainable use of marine biological diversity of areas beyond national jurisdiction. *Global Environmental Law Annual, 12*, 179.

Huang, H., & Qin, J. (2024). Metadata functional requirements for genomic data practice and curation. *Information Research, 29*(2), 3–29.

Humphries, F. (2025). Marine genetic resources beyond national jurisdiction: the expansive scope of the BBNJ Agreement. In F. Humphries (Ed.), *Decoding governance of marine genetic resources under the BBNJ Agreement.* Springer.

Humphries, F., Berry, T., & Muraki Gottlieb, H. (2025a). Bridging divides: The evolution of marine genetic resources governance beyond national jurisdiction. In F. Humphries (Ed.), *Decoding governance of marine genetic resources under the BBNJ Agreement.* Springer.

Humphries, F., Jaspars, M., Lavelle, J., & Kachelriess, D. (2025b). The novel notification information system for marine genetic resources under the BBNJ Agreement. In F. Humphries (Ed.), *Decoding governance of marine genetic resources under the BBNJ Agreement.* Springer.

IISD. (2011). *Earth Negotiations Bulletin, 25*(70).

IISD. (2015). *Earth Negotiations Bulletin, 25*(94).

Jarvis, R., & Young, T. (2023). Pressing questions for science, policy, and governance in the high seas. *Environmental Science and Policy, 139*, 177–184.

Jaspars, M., & Brown, A. E. L. (2021). Benefit sharing. In M. H. Nordquist & J. N. Moore (Eds.), *Marine biodiversity of areas beyond national jurisdiction* (pp. 97–130). Brill Nijhoff.

Kachelriess, D., Dunshirn, P., Langlet, A., Brown, A. E. L., & Scholz, A.H. (2025). Marine genetic resources and digital sequence information under the BBNJ Agreement: Interlinkages with other access and benefit sharing frameworks. In F. Humphries (Ed.), *Decoding governance of marine genetic resources under the BBNJ Agreement.* Springer.

Korkut, E., & Fowler, L. B. (2019). Melting ice and deep waters: The United States and deep seabed mining in the arctic. *Natural Resources and Environment, 34*(2), 27–30.

Langlet, A., Dunshirn, P., Jaspars, M., Humphries, F., & Kachelriess, D. (2025). Monitoring and transparency aspects of MGR-utilization under the BBNJ Agreement. In F. Humphries (Ed.), *Decoding governance of marine genetic resources under the BBNJ Agreement.* Springer.

Lavelle, J., & Wynberg, R. (2025). Benefit sharing under the BBNJ Agreement in practice. In F. Humphries (Ed.), *Decoding governance of marine genetic resources under the BBNJ Agreement.* Springer

Lawson, C. (2012). *Regulating genetic resources: Access and benefit sharing in international law.* Edward Elgar Publishing.

Lawson, C., Humphries, F., Jaspars, M., & Rabone, M. (2025). Data management and the 'BBNJ standardized batch identifier' under the BBNJ Agreement. In F. Humphries (Ed.), *Decoding governance of marine genetic resources under the BBNJ Agreement.* Springer.

Lee, R. (2021). The journey to realisation. In M. H. Nordquist & J. N. Moore (Eds.), *Marine biodiversity of areas beyond national jurisdiction* (pp. 3–6). Brill Nijhoff.

Lennan, M., & Morgera, E. (2022). The glasgow climate conference (COP26). *The International Journal of Marine and Coastal Law, 37*(1), 137–151.

Long, R. (2018). Regulating and managing fisheries resources: Five decades of triumph and failure in the European Union. In H. N. Scheiber, N. Oral, & M. Kwon (Eds.), *Ocean law debates* (pp. 287–328). Brill Nijhoff.

Marciniak, K. J. (2017). Marine genetic resources: Do they form part of the common heritage of mankind principle? In L. Martin, C. Salonidis, & C. Hioureas (Eds.), *Natural resources and the law of the sea: Exploration, allocation, exploitation of natural resources in areas under national jurisdiction and beyond* (p. 373). Juris Publishing.

Mendenhall, E. & Bateh, F. (2024). 'High Seas Treaty' name is inaccurate and should centre biodiversity(Commentary)'. Mongabay News, 19 February 2024. https://news.mongabay.com/2024/02/high-seas-treaty-name-isinaccurate-and-should-center-biodiversity-commentary/

Mendenhall, E., De Santo, E., Jankila, M., Nyman, E., & Tiller, R. (2022). Direction, not detail: Progress towards consensus at the fourth intergovernmental conference on biodiversity beyond national jurisdiction. *Marine Policy, 146*, 105309.

Mendenhall, E., De Santo, E., Nyman, E., & Tiller, R. (2019). A soft treaty, hard to reach: The second intergovernmental conference for biodiversity beyond national jurisdiction. *Marine Policy, 108*, 103664.

Mengerink, K. (2018). Defining "serious harm" and "harmful effects" for deep seabed mining. In H. N. Scheiber, N. Oral, & M. Kwon (Eds.), *Ocean law debates* (pp. 447–477). Brill Nijhoff.

Morgera, E. (2025). Conclusions: Equity, Sustainability, and Transformation under the BBNJ Agreement. In F. Humphries (Ed.), *Decoding governance of marine genetic resources under the BBNJ Agreement.* Springer.

Morgera, E. (2022). The relevance of the human right to science for the conservation and sustainable use of marine. In R. Churchill & A. O. Elferink (Eds.), *International law and marine areas beyond national jurisdiction* (pp. 242–274). Brill Nijhoff.

Mulalap, C. Y., Frere, T., Huffer, E., Hviding, E., Paul, K., Smith, A., & Vierros, M. K. (2020). Traditional knowledge and the BBNJ instrument. *Marine Policy, 122*, 104103.

Muraki Gottlieb, H., Ardron, J., & Brown, A. E. L. (2025a). BBNJ Agreement: A new infrastructure to foster benefit sharing of marine genetic resources. In F. Humphries (Ed.), *Decoding governance of marine genetic resources under the BBNJ Agreement.* Springer.

Muraki Gottlieb, H., Kachelriess, D. & Slobodian. L. (2025b). Understanding the preamble, principles and objectives of the BBNJ Agreement: A focus on the fair and equitable sharing of benefits of marine genetic resources. In F. Humphries (Ed.), *Decoding governance of marine genetic resources under the BBNJ Agreement.* Springer.

Nguyen, L. G. (2022). Principled challenges. In R. Churchill & A. O. Elferink (Eds.), *International law and marine areas beyond national jurisdiction* (pp. 114–147). Brill Nijhoff.

Nishimura, T. (2018). The legal framework for marine genetic resources: The relationship between the UNCLOS and the convention on biological diversity regime. *Asian Business Lawyer, 21*, 63–80.

Oldham, P., Hall, S., Barnes, C., Oldham, C., Cutter, M., Burns, N., & Kindness, L. (2014). *Valuing the deep: Marine genetic resource in areas beyond national jurisdiction Defra contract MB0128—a review of current knowledge regarding marine genetic resources and their current and projected economic value to the UK economy*. Final Report Version One. DEFRA.

Oldham, P., Chiarolla, C., & Thambisetty, S. (2023). Digital sequence information in the UN high seas treaty: Insights from the global biodiversity framework-related decisions. In *LSE Law-Policy Briefing Paper*.

Oschlies, A., Brandt, P., Stramma, L., & Schmidtko, S. (2018). Drivers and mechanisms of ocean deoxygenation. *Nature Geoscience, 11*(7), 467–473.

Payne, C. R. (2020). New law for the high seas. *Berkeley Journal of International Law, 37*(2), 345–368.

Pena-Neira, S., & Coelho, L.F. (2025). Traditional knowledge associated with genetic resources in areas beyond national jurisdiction. In F. Humphries (Ed.), *Decoding governance of marine genetic resources under the BBNJ Agreement*. Springer.

Qureshi, W. A. (2018). Marine biodiversity conservation: The international legal framework and challenges. *Houston Journal of International Law, 40*(3), 845–936.

Rabone, M., Horton, T., Humphries, F., Lyal, C., Muraki Gottlieb, H., Scholz, A. H., Vanagt, T., & Jaspars, M. (2025). BBNJ Agreement: Considerations for scientists and commercial end users of MGR at research, development and commercialisation stages. In F. Humphries (Ed.), *Decoding governance of marine genetic resources under the BBNJ Agreement.* Springer.

Roach, J. A. (2021). BBNJ treaty negotiations 2019. In M. H. Nordquist & J. N. Moore (Eds.), *Marine biodiversity of areas beyond national jurisdiction* (pp. 25–89). Brill Nijhoff.

Rogers, A. D., Appiah-Madson, H., Ardron, J. A., Bax, N. J., Bhadury, P., Brandt, A., Buttigieg, P. L., De Clerck, O., Delgado, C., Distel, D. L., & Glover, A. (2023). Accelerating ocean species discovery and laying the foundations for the future of marine biodiversity research and monitoring. *Frontiers in Marine Science, 10*, 1224471.

Rogers, A. D., Baco, A., Escobar-Briones, E., Currie, D., Gjerde, K., Gobin, J., Jaspars, M., Levin, L., Linse, K., Rabone, M., & Ramirez-Llodra, E. (2021). Marine genetic resources in areas beyond national jurisdiction: Promoting marine scientific research and enabling equitable benefit sharing. *Frontiers in Marine Science, 8*, 1.

Russell, D. A., & VanderZwaag, D. L. (2010). *Recasting transboundary fisheries management arrangements in light of sustainability principles*. Martinus Nijhoff Publishers.

Scholz, A. H., Lange, M., Habekost, P., Oldham, P., Cancio, I., Cochrane, G., & Freitag, J. (2021). Myth-busting theprovider-user relationship for digital sequence information. *GigaScience, 10*(12), giab085.

Scott, K. N. (2018). Not an intractable challenge geoengineering MSR. In M. H. Nordquist & J. N. Moore (Eds.), *Marine biodiversity of areas beyond national jurisdiction* (pp. 189–212). Brill Nijhoff.

Sigwart, J. D., Blasiak, R., Jaspars, M., Jouffray, J. B., & Tasdemir, D. (2021). Unlocking the potential of marine biodiscovery. *Natural Product Reports, 38*(7), 1235–1242.

Sonesson, A. K., Hallerman, E., Humphries, F., Hilsdorf, A. W. S., Leskien, D., Rosendal, K., Bartley, D., Hu, X., Garcia Gomez, R., & Mair, G. C. (2023). Sustainable management and improvement of genetic resources for aquaculture. *Journal of the World Aquaculture Society, 54*, 364–396.

Tiller, R., De Santo, E., Mendenhall, E., & Nyman, E. (2019). The once and future treaty: Towards a new regime for biodiversity in areas beyond national jurisdiction. *Maine Policy, 99*, 239–242.

Tiller, R., & Mendenhall, E. (2023). And so it begins: The adoption of the 'biodiversity beyond national jurisdiction' treaty. *Marine Policy, 157*, 105836.

Tiller, R., Mendenhall, E., De Santo, E., & Nyman, E. (2023). Shake it off: Negotiations suspended, but hope simmering, after a lack of consensus at the fifth intergovernmental conference on biodiversity beyond national jurisdiction. *Marine Policy, 148*, 105457.

UNEP (2022a) Resolution 5/14 Resolution adopted by the United Nations Environment Assembly on 2 March 2022. *End plastic pollution: Towards an international legally binding instrument*. UNEP/PP/OEWG/1/INF/1. 10 May 2022.

UNEP. (2022b). Decision adopted by the conference of the parties to the convention on biological diversity, 15/4 Kunming-Montreal global biodiversity framework. In *Conference of the parties to the convention on biological diversity*. CBD/COP/DEC/15/4. 19 Dec 2022.

UNEP. (2022c). Decision adopted by the conference of the parties to the convention on biological diversity, 15/9 digital sequence information on genetic resources. In *Conference of the parties to the convention on biological diversity*. CBD/COP/DEC/15/9. 19 Dec 2022.

UNGA Res. 59/24 Resolution adopted by the General Assembly on 17 November 2004: Oceans and the law of the sea. A/RES/59/24. 4 Feb 2005.

UNGA. (2005). *Oceans and the law of the sea: Report of the secretary-general*. A/60/63/Add.1. 15 July 2005.

UNGA Res. 60/30 Resolution adopted by the General Assembly on 29 November 2005: Oceans and the law of the sea. A/RES/60/30. 8 March 2006.

UNGA Res. 68/70 Resolution adopted by the General Assembly on 9 December 2013: Oceans and the law of the sea. A/RES/68/70. 27 Feb 2014.

UNGA. (2015). *Transforming our world: The 2030 agenda for sustainable development*. A/RES/70/1.

UNGA (2017) Report of the preparatory committee established by general assembly resolution 69/292: Development of an international legally binding instrument under the United Nations convention on the Law of the Sea on the conservation and sustainable use of marine biological diversity of areas beyond national jurisdiction, A/AC.287/2017/PC.4/2

UNGA Res. 72/249 Resolution adopted by the General Assembly on 24 December 2017: international legally binding instrument under the United Nations Convention on the Law of the Sea on the conservation and sustainable use of marine biological diversity of areas beyond national jurisdiction. Seventy-second session. A/RES/72/249. 19 Jan 2018.

UNGA. (2018a). Statement by the president of the conference at the closing of the first session. In *Intergovernmental conference on an international legally binding instrument under the United Nations Convention on the Law of the Sea on the conservation and sustainable use of marine biological diversity of areas beyond national jurisdiction*. First session. A/CONF.232/2018/7. 20 Sept 2018.

UNGA. (2018b). President's aid to discussions. In *Intergovernmental conference on an international legally binding instrument under the United Nations convention on the law of the Sea on the conservation and sustainable use of marine biological diversity of areas beyond national jurisdiction*. First session. A/CONF.232/2018/3. 25 June 2018.

UNGA. (2018c). President's aid to negotiations. In *Intergovernmental conference on an international legally binding instrument under the United Nations Convention on the Law of the Sea on the conservation and sustainable use of marine biological diversity of areas beyond national jurisdiction*. Second session. A/CONF.232/2019/1*. 3 Dec 2018.

UNGA. (2019a). Statement by the President of the conference at the closing of the second session. In *Intergovernmental conference on an international legally binding instrument under the United Nations Convention on the Law of the Sea on the conservation and sustainable use of marine biological diversity of areas beyond national jurisdiction*. Second session. A/CONF.232/2019/5. 18 April 2019.

UNGA. (2019b). Draft text of an agreement under the United Nations Convention on the Law of the Sea on the conservation and sustainable use of marine biological diversity of areas beyond national jurisdiction: Note by the President. In *Intergovernmental conference on an international legally binding instrument under the United Nations Convention on the Law of the Sea on the conservation and sustainable use of marine biological diversity of areas beyond national jurisdiction*. Third session. A/CONF.232/2019/6. 17 May 2019.

UNGA. (2019c). Revised draft text of an agreement under the United Nations Convention on the Law of the Sea on the conservation and sustainable use of marine biological diversity of areas beyond national jurisdiction: Note by the President. In *Intergovernmental conference on an international legally binding instrument under the United Nations Convention on the Law of the Sea on the conservation and sustainable use of marine biological diversity of areas beyond national jurisdiction*. Fourth session. A/CONF.232/2020/3. 18 Nov 2019.

UNGA. (2019d). Statement by the President of the conference at the closing of the third session. In *Intergovernmental conference on an international legally binding instrument under the United Nations Convention on the Law of the Sea on the conservation and sustainable use of marine biological diversity of areas beyond national jurisdiction*. Third session. A/CONF.232/2019/10*. 13 Sept 2019.

UNGA Res. 65/37 Resolution adopted by the General Assembly on 7 December 2010: Oceans and the law of the sea. A/RES/65/37.17 March 2021.

UNGA. (2022a). Report of the intergovernmental conference on an internationally legally binding instrument under the United Nations Convention on the Law of the Sea on the conservation and sustainable use of marine biological diversity of areas beyond national jurisdiction. In *Intergovernmental conference on an international legally binding instrument under the United Nations Convention on the Law of the Sea on the conservation and sustainable use of marine biological diversity of areas beyond national jurisdiction*. Fourth session. A/CONF.232/2022/4. 14 April 2022.

UNGA. (2022b). Further revised draft text of an agreement under the United Nations Convention on the Law of the Sea on the conservation and sustainable use of marine biological diversity of areas beyond national jurisdiction: Note by the President. In *Intergovernmental conference on an international legally binding instrument under the United Nations Convention on the Law of the Sea on the conservation and sustainable use of marine biological diversity of areas beyond national jurisdiction*. Fifth session. A/CONF.232/2022/5. 1 June 2022.

UNGA. (2022c). Statement by the President of the conference issues after the suspension of the fifth session. In *Intergovernmental conference on an international legally binding instrument under the United Nations Convention on the Law of the Sea on the conservation and sustainable use of marine biological diversity of areas beyond national jurisdiction*. Fifth session. A/CONF.232/2022/9. 14 Sept 2022.

UNGA. (2022d). Refreshed draft text of an agreement under the United Nations Convention on the Law of the Sea on the conservation and sustainable use of marine biological diversity of areas beyond national jurisdiction: Note by the President. In *Intergovernmental conference on an international legally binding instrument under the United Nations Convention on the Law of the Sea on the conservation and sustainable use of marine biological diversity of areas beyond national jurisdiction*. Fifth session. A/CONF.232/2022/CRP.12.

UNGA. (2022e). Further refreshed draft text of an agreement under the United Nations Convention on the Law of the Sea on the conservation and sustainable use of marine biological diversity of areas beyond national jurisdiction: Note by the President. In *Intergovernmental conference on an international legally binding instrument under the United Nations Convention on the Law of the Sea on the conservation and sustainable use of marine biological diversity of areas beyond national jurisdiction*. Resumed fifth session. A/CONF.232/2023/2. 12 Dec 2022.

UNGA Res. 78/155 Resolution adopted by the General Assembly on 19 December 2023: Implementation of the Convention on Biological Diversity and its contribution to sustainable development. A/RES/78/155. 21 Dec 2023.

UNGA. (2023a). Draft agreement under the United Nations Convention on the Law of the Sea on the conservation and sustainable use of marine biological diversity of areas beyond national jurisdiction. In *Intergovernmental conference on an international legally binding instrument under the United Nations Convention on the Law of the Sea on the conservation and sustainable use of marine biological diversity of areas beyond national jurisdiction*. Further resumed fifth session. A/CONF.232/2023/CRP.2/Rev.2. 3 May 2023.

UNGA. (2023b). Draft agreement under the United Nations Convention on the Law of the Sea on the conservation and sustainable use of marine biological diversity of areas beyond national jurisdiction. In *Intergovernmental conference on an international legally binding instrument under the United Nations Convention on the Law of the Sea on the conservation and sustainable use of marine biological diversity of areas beyond national jurisdiction*. Further resumed fifth session. A/CONF.232/2023/CRP.3. 6 June 2023.

UNGA. (2023c). Agreement under the United Nations Convention on the Law of the Sea on the conservation and sustainable use of marine biological diversity of areas beyond national jurisdiction. In *Intergovernmental conference on an international legally binding instrument under the United Nations Convention on the Law of the Sea on the conservation and sustainable use of marine biological diversity of areas beyond national jurisdiction*. Further resumed fifth session. A/CONF.232/2023/4*.19 June 2023.

UNGA. (2023d). Report of the intergovernmental conference on an internationally legally binding instrument under the United Nations Convention on the Law of the Sea on the conservation and sustainable use of marine biological diversity of areas beyond national jurisdiction at its fifth session. In *Intergovernmental conference on an international legally binding instrument under the United Nations Convention on the Law of the Sea on the conservation and sustainable use of marine biological diversity of areas beyond national jurisdiction*. Fourth session. A/CONF.232/2023/5. 30 June 2023.

UNGA Res. 78/272 of 24 April 2024 (currently available as A/78/L.41 Singapore Draft Resolution: Agreement under the United Nations Convention on the Law of the Sea on the Conservation and Sustainable use of Marine Biological Diversity of Areas beyond National Jurisdiction, seventy-eight session). 6 Feb 2024.

Vadrot, A. B., Langlet, A., Tessnow-von Wysocki, I., Tolochko, P., Brogat, E., & Ruiz-Rodríguez, S. C. (2021). Marine biodiversity negotiations during COVID-19: A new role for digital diplomacy? *Global Environmental Politics, 21*(3), 169–186.

Vadrot, A. B. M., Langlet, A., & Wysocki, I. T. (2022). Who owns marine biodiversity? Contesting the world order through the 'common heritage of humankind' principle. *Environmental Politics, 31*(2), 226–250.

Vanagt, T., Broggiato, A., Lallier, L. E., Jaspars, M., Burton, G., & Muyldermans, D. (2019). *Mare Geneticum*: Towards an implementing agreement for marine genetic resources in international waters. In D. Freestone (Ed.), *Conserving biodiversity in areas beyond national jurisdiction* (pp. 267–297). Brill Nijhoff.

WIPO. (2024). Diplomatic conference to conclude an international legal instrument relating to intellectual property, genetic resources and traditional knowledge associated with genetic resources. In *WIPO treaty on intellectual property, genetic resources and associated traditional knowledge.* GRATK/DC/7.

Wright, G., Rochette, J., Druel, E., & Gjerde, K. (2016). *The long and winding road continues: Towards a new agreement on high seas governance*. IDDRI.

Xiao, T. (2020). Area-based management tools in areas beyond national jurisdiction: Main debates and proposed solutions. *China Oceans Law Review, 1*, 112–129.

Fran Humphries has specialized in marine and biodiversity law and policy for over two decades in government, academia and consultancies. She is an associate professor at Griffith Law School, Griffith University, Australia, with a background in fisheries management and marine law. She has led large international research consultancies on access and benefit sharing of biological resources for governments, UN organizations and other institutions. She was on the International Council of Environmental Law delegation for the Biodiversity Beyond National Jurisdiction (BBNJ) Agreement negotiations.

Todd Berry is a lecturer at the Monash University Faculty of Law. He holds a Doctor of Philosophy from the Griffith Law School. His research interests include sovereignty and international law.

Hiroko Muraki Gottlieb is a licenced lawyer in the USA with a diversified career in global businesses, intergovernmental organizations, non-governmental organizations and academic institutions. She brings a unique blend of expertise in climate change and biodiversity conservation strategy, environmental law (domestic and international), policymaking, regulatory compliance and stakeholder engagement. She is the representative for the Ocean and led the International Council of Environmental Law to the BBNJ Agreement negotiations as the Head of Delegation and holds appointments with the Department of Organismic and Evolutionary Biology at Harvard University and Elisabeth Haub School of Law at Pace University.

Part I
Treaty Interpretation and Analysis

BBNJ Agreement: A New Infrastructure to Foster Benefit Sharing of Marine Genetic Resources

2

Hiroko Muraki Gottlieb, Jeff A. Ardron, and Abbe E. L. Brown

Abstract

An effective implementation of the Part II of the Agreement under the United Nations Convention on the Law of the Sea on the Conservation and Sustainable Use of Marine Biological Diversity of Areas Beyond National Jurisdiction (BBNJ Agreement) will require understanding the relevance of the infrastructure in the treaty text. The Agreement's Part II provides a new framework of information and benefit sharing of marine genetic resources of areas beyond national jurisdiction and associated Digital Sequence Information and Traditional Knowledge. This chapter for the edited collection "Decoding Marine Genetic Resource Governance under the BBNJ Agreement" explores the infrastructure of the Agreement. The chapter views infrastructure of an implementing agreement to be robust when there is a Conference of the Parties (COP) with a decision-making function, a Secretariat and various subsidiary bodies (e.g., Scientific and Technical Body) that provide support for the COP to make informed decisions in an objective manner. A strong infrastructure also facilitates transparency through an open-access information platform and inclusivity provisions, which allow a wide range of stakeholders to be timely informed and facilitate active participation in meetings. Further, a strong infrastructure builds in functions to review the progress of implementation so that any gaps can be addressed by the COP or the Parties, with the support of the subsidiary bodies. These aspects are essential to ensure that with input from various stakeholders, the COP can make timely and well-informed decisions to enhance the health of the two-thirds of the global ocean.

Keywords

BBNJ agreement · High seas · Biodiversity · Conservation · Marine genetic resources · Digital sequence information · Institutional arrangements · Transparency · Fish stocks agreement · Deep seabed mining · UNCLOS · Ocean governance

H. M. Gottlieb (✉)
Department of Organismic and Evolutionary Biology, Harvard University, Cambridge, MA, USA
e-mail: hmurakigottlieb@fas.harvard.edu

J. A. Ardron
Africa Oceans, The Nature Conservancy, Mombasa, Kenya

A. E. L. Brown
School of Law, University of Aberdeen, Aberdeen, UK

F. Humphries (ed.), *Decoding Marine Genetic Resource Governance Under the BBNJ Agreement*, Sustainable Development Goals Series, https://doi.org/10.1007/978-3-031-72100-7_2

2.1 Introduction

The most recent implementing agreement to govern nearly two-thirds of the global ocean, *the Agreement under the United Nations Convention on the Law of the Sea on the Conservation and Sustainable Use of Marine Biological Diversity of Areas Beyond National Jurisdiction* (2023) (BBNJ Agreement), was adopted on 19 June 2023. Nearly 200 of the UN Member States reached this historic achievement by consensus, after over 20 years of studies and subsequent negotiations (Humphries et al., 2025). The main objective of the BBNJ Agreement "…is to ensure the conservation and sustainable use of marine biological diversity of areas beyond national jurisdiction, for the present and in the long term, through effective implementation of the relevant provisions of the United Nations Convention on the Law of the Sea (UNCLOS) and further international cooperation and coordination" (BBNJ Agreement, art. 2). The BBNJ Agreement aims to achieve such overall objective with requirements on four substantive elements: (1) marine genetic resources (MGRs), including the fair and equitable sharing of benefits, (2) measures, such as area-based management tools, including marine protected areas (ABMTs, including MPAs), (3) environmental impact assessments (EIAs), and (4) capacity building and the transfer of marine technology (CB/TMT) (BBNJ Agreement, Part II, III, IV, and V). This new implementing agreement also incorporates a unified vision for the Parties through the preamble, general principles and approaches, and subject matter-based objectives (Muraki Gottlieb et al., 2025).

The BBNJ Agreement is innovative because it incorporates new provisions to govern MGRs of areas beyond national jurisdiction (ABNJ) and associated digital sequence information (DSI) alongside the growing desire of the countries to address urgent conservation challenges that we face today and for the long run while considering economic, social, and cultural aspects (BBNJ Agreement, Preamble). Also, there is an emphasis on addressing equity issues through capacity building and the transfer of marine technology, which is reflected throughout the text. Finally, with the support of the scientific community, the negotiators were able to consider the best scientific practices to avoid unintended consequences of hindering scientific development and collaboration, with regard to the collection and peaceful use of MGRs (Rabone et al., 2025).

This chapter begins with how the negotiators designed the BBNJ Agreement's infrastructure by reflecting on some of the lessons learned from the two previous UNCLOS implementing agreements and other multilateral environmental agreements. In the BBNJ Agreement, there is a focus on transparency and inclusivity as the bedrock of the implementing provisions. To that end, the Agreement's infrastructure ensures that the COP can conduct regular meetings with Parties and observers, monitor progress, and review the effectiveness of implementation.

This chapter will then provide the foundational information on the BBNJ Agreement's infrastructure and some of the challenges and opportunities, focused on the benefit sharing of MGRs and DSI (MGR Framework). [A detailed exploration of the MGR Framework and its practical implications are available in Chap. 14 of this book (Rabone et al., 2025)]. Further, this chapter will offer some considerations on transparency and inclusivity that could lead to avoiding unintended consequences in implementing the MGR Framework provisions. Some of the enablers and specific examples of how the active involvement of a wide range of stakeholders (including the scientific community and more generally civil society) are explored.

Finally, this chapter views infrastructure of an implementing agreement to be robust when there is a COP with a decision-making function, a Secretariat, and various subsidiary bodies that provide support for the COP to make informed decisions. A strong infrastructure also facilitates transparency through an open-access information platform and inclusivity provisions, which allow a wide range of stakeholders to be timely informed and facilitate active participation in meetings. These aspects are essential to ensure that with input from various stakeholders, the COP can make objective and well-informed decisions. Finally, a strong infrastructure builds in functions to review the progress of implementation so that any gaps can be addressed by the COP or the Parties, with the support of the subsidiary bodies.

2.1.1 Lessons Learned from Two Previous UNCLOS Implementation Agreements

The infrastructure of a multilateral agreement is arguably the most important aspect of implementing requirements, especially when seeking to include significantly divergent interests of countries. UNCLOS, the agreement which established the framework for governance the ocean, has implementing agreements operationalizing certain aspects of UNCLOS' requirements. An implementing agreement can provide institutional arrangements, such as the COP, and include more granular requirements that facilitate a better implementation of UNCLOS' provisions.

Since UNCLOS entered into force on Nov. 1, 1994, two implementing agreements have followed. First is the Agreement relating to the implementation of Part XI of the United Nations Convention on the Law of the Sea of 10 December 1982 (Part XI Agreement) adopted on 28 July 1994 (A/RES/48/263). The objective of the Part XI Agreement is to better implement the requirements of UNCLOS associated with the governance of the exploration and exploitation of deep seabed minerals in the Area. The Area covers "the deep seabed and ocean floor and subsoil thereof, beyond the limits of national jurisdiction" (UNCLOS, art. 1.1(a)). The second implementing agreement is the *Agreement for the Implementation of the Provisions of the United Nations Convention on the Law of the Sea of 10 December 1982 Relating to the Conservation and Management of Straddling Fish Stocks and Highly Migratory Fish Stocks* (Fish Stocks Agreement) (A/CONF.164/37). Entered into force on 28 July 1996, the Fish Stocks Agreement operationalizes Parts V and VII of UNCLOS and mandates that the target fish stocks governance must be based on the best available science and the precautionary approach (Fish Stocks Agreement, art. 5(b) and 5(c)). Further, the Fish Stocks Agreement requires countries to cooperate to promote optimum utilization of fisheries resources both within and beyond the exclusive economic zone (Fish Stocks Agreement, art. 5(a)).

This section briefly reviews and compares the infrastructure of the Fish Stocks Agreement and the Part XI Agreement with the BBNJ Agreement's infrastructure. The differences reflect the resolve of the negotiators of the BBNJ Agreement to create an infrastructure that could improve some of the challenges of the other two implementing agreements. Part XI Agreement that has an infrastructure with a decision-making function but lacks transparency and inclusivity. The Fish Stocks Agreement does not have a decision-making infrastructure function and, therefore, makes it difficult for a decision to be made at the global level, but arguably provides somewhat more transparency than the Part XI Agreement (Ardron, 2018).

The analysis of this section concludes that without a decision-making body with supporting subsidiary bodies and a foundation of transparency, an infrastructure of an agreement will lack strengths and compromises effective implementation of UNCLOS. Based on the lessons learned from the Part XI Agreement and the Fish Stocks Agreement, the new infrastructure of the BBNJ Agreement reflects the desire for improved access for a wide range of stakeholders on critical matters of BBNJ governance. Such conviction continues the spirit of transparency and inclusivity, the two concepts which were paramount during the negotiations of the BBNJ Agreement (Humphries et al., 2025).

2.1.1.1 Fish Stocks Agreement

The Fish Stocks Agreement relies on informal State Parties consultations and a review conference led by the UN Secretary-General. Countries provide recommendations through the UN General Assembly resolutions, which are not legally binding (UN Charter, art. 10). Rather, implementation decisions of the Fish Stocks Agreement are made at the Regional Fisheries Management Organisations (RFMOs) that have their own mandates, including specific geographical coverage. Such infrastructure is due to the fact that, "[a]t the heart of UNFSA is a commitment to decentralization, namely through an accompanying regime of regional fisheries management organizations/arrangements (RFMO/As), which were delegated the responsibility for management, monitoring and compliance issues within their respective regions" (Blasiak and Yagi, 2016).

The BBNJ Agreement is more like the Part XI Agreement's infrastructure discussed below (Sect. 2.1.1.2). The Agreement has global institutional arrangements, from the COP (which is the main decision-making body except for certain matters regarding environmental impact assessments) to subsidiary bodies (BBNJ Agreement, art. 47 and Part VI.) The BBNJ Agreement also built in flexibility for the COP to create additional subsidiary bodies, should they agree to do so to fill any gap that hinders informed decision to be made. Such strong global infrastructure is absent from the Fish Stocks Agreement. To date, the Fish Stocks Agreement, as implemented through RFMOs, has been marginally more transparency than the Part XI Agreement implemented through the International Seabed Authority (ISA) (Ardron, 2018). However, the lack of a global decision-making body mandate has created regional inconsistencies and variability in managing the straddling fish stocks (Blasiak and Yagi, 2016).

A review conference has been an important aspect of the Fish Stocks Agreement, where critical shortcomings of the implementing agreements had been identified. For example, in the opening statements at the 2023 Fish Stocks Agreement resumed Review Conference, on behalf of the Secretary-General, the Assistant Secretary-General for Legal Affairs (representing the Under-Secretary-General for Legal Affairs and United Nations Legal Counsel) stated, "…although the overall status of straddling fish stocks and highly migratory fish stocks continued to deteriorate, the recommendations adopted in 2006, 2010 and 2016 had a considerable impact on the practices of States and regional fisheries management organizations and arrangements, and had provided the impetus for many international efforts" (A/CONF.210/2023/6).

A review conference is one of the functions of the COP in the BBNJ Agreement. The relevance of such conference is discussed in Sect. 2.2.3.1 below.

2.1.1.2 Part XI Agreement

The Part XI Agreement that governs deep seabed mining is implemented through the ISA that has a decision-making function through its executive organ, the Council, as well as the Assembly (similar to a COP). A subsidiary body, the Legal and Technical Commission (LTC), provides a wide range of recommendations regarding, inter alia, contracts, environmental impact assessments, standards, and guidelines. There are some similarities but also significant differences between the functions of the LTC and the STB of the BBNJ Agreement. Table 2.1 compares the differences between the LTC and the STB in terms of their roles, how the meetings are conducted, information disseminated, and the rights granted to the observers to participate in the meetings.

As summarized in Table 2.1, there are differences between how the BBNJ Agreement's STB and the Part XI Agreement's LTC address transparency and inclusivity. The BBNJ Agreement's COP, the main decision-making body, must "promote transparency in decision-making processes and other activities carried out under the Agreement" (BBNJ Agreement, art. 48.1). While the details will be decided by the COP through the rules of procedure and its decisions, the meetings of the COP and its subsidiary bodies (which includes the STB) must be open to observers and the COP's decision must be published and maintained (BBNJ Agreement, art. 48.3). The COP must promote transparency in implementing the BBNJ Agreement, and actions to do so include public dissemination of information, facilitation of participation, and consultation with a wide range of stakeholders (BBNJ Agreement, art. 48.3). The transparency article

Table 2.1 A comparison between Part XI Agreement and BBNJ Agreement on information dissemination and observer participation

	Scientific and Technical Body	Legal and Technical Commission
Role	Provides scientific and technical advice to the COP, taking into account the multidisciplinary expertise. (BBNJ Agreement, Art. 49.4)	Work includes, inter alia, the review of applications for plans of work, supervision of exploration or mining activities (including review of annual reports submitted by contractors), development of environmental management plans, assessment of the environmental implications of activities in the Area, formulation and keeping under review the rules, regulations and procedures in relation to activities in the Area, and makes recommendations to the Council on all matters relating to exploration and exploitation of non-living marine resources (such as polymetallic nodules, polymetallic sulfides, and cobalt-rich ferromanganese crusts). (UNCLOS, art. 165)
Meetings	All meetings are open to observers participating in accordance with the rules of procedure unless otherwise decided by the COP (BBNJ Agreement, art. 48.2). The COP facilitates the participations of, and consultation with a wide range of stakeholders (e.g., relevant global, regional, subregional, and sectoral bodies, Indigenous Peoples and local communities with relevant traditional knowledge, the scientific community, civil society, and other relevant stakeholders) (BBNJ Agreement, art. 48.3)	All meetings are held in private, unless the LTC decides otherwise. The Commission must consider the desirability of holding open meetings when issues of general interest to members of the Authority, which do not involve the discussion of confidential information, are being discussed. (ISBA/6/C/9, Rule 6)
Dissemination of information	Public dissemination of information to a wide range of stakeholders (e.g., relevant global, regional, subregional, and sectoral bodies, Indigenous Peoples and local communities with relevant traditional knowledge, the scientific community, civil society, and other relevant stakeholders) (BBNJ Agreement, art. 48.2 and 48.3)	The Secretary-General must notify the members of the Commission and the members of the Authority as early as possible of the date and duration of each session and must seek confirmation of their attendance. (ISBA/6/C/9, Rule 4)

includes the requirement to provide timely access to all relevant information. Further, non-Parties have the right to request to participate as observers, and the modalities for such participation must "not be unduly restrictive." (BBNJ Agreement, art. 48.3).

The LTC in the Part XI Agreement follows a different approach. Unlike the BBNJ Agreement, all meetings are held in private, unless the LTC decides otherwise. Further, the LTC must consider the desirability of holding open meetings when issues of general interest to members of the ISA, which do not involve the discussion of confidential information, are being discussed (ISBA/6/C/9, Rule 6). The Secretary-General of the ISA must provide notification of the meetings, but only to the members of the LTC and the members of the ISA, "as early as possible" (ISBA/6/C/9, Rule 4). The strict limitation on those with the right to participate and the qualifier of "as early as possible" gives a significant room for the ISA's Secretary-General to control when to send out notice of the meetings.

Further, the opacity of the decision-making on a wide range of important matters [e.g., the review of applications for plans of work, supervision of exploration or mining activities, development of environmental management plans, assessment of the environmental implications of activities in the Area, formulation and keeping under review the rules, regulations, and procedures in relation to activities in the Area makes recommendations to the LTC on all matters relating to exploration and exploitation of non-living marine resources (UNCLOS, art. 165) gives the LTC power, usually without input from stakeholders. In addition, the ISA Council has very limited ability to go against an LTC recommendation, thus giving the LTC almost absolute power, behind closed doors. To address such concern, in the 2017 review of the ISA, it was recommended that the LTC should be encouraged to hold more open meetings to allow for greater transparency in its work (ISBA/23/A/3, recommendation 16). As of July 2024, however, no open meetings have been held since that recommendation was made.

2.1.2 Concepts Borrowed from Other Multilateral Environmental Agreements

The BBNJ Agreement also reflects consideration of the infrastructure of the so-called Rio Agreements since the 1992 United Nations Conference on Environment and Development (UNCED), in Rio de Janeiro (Rio Earth Summit), was the venue for adoption of certain multilateral environmental agreements. For example, the UN Framework Convention on Climate Change (UNFCCC) adopted at the Rio Earth Summit established a COP, a multidisciplinary Subsidiary Body for Scientific and Technological Advice, and a Subsidiary Committee for Implementation (UNFCCC, art. 7, 9, and 10). The Paris Agreement, under the UNFCCC, established the Technology Framework and has a focus on transparency (Paris Agreement, art. 2, 10, and 13). Another example of the Rio Agreements is the Convention on Biological Diversity (CBD), which has a COP, a Subsidiary Body on Scientific, Technical and Technological Advice, and refers to the COP to establish a Clearing-House Mechanism to promote and facilitate technical and scientific cooperation (CBD, art. 23, 25, and 18(3)). The CBD's Nagoya Protocol also has a COP, a Clearing House, and an information-sharing mechanism (Nagoya Protocol, art. 26, 10, and 14).

While the UNFCCC and UNCLOS both entered into force in 1994, the difference in the infrastructure represents the fact that UNCLOS was negotiated in the 1970s. The countries that participated in the BBNJ Agreement negotiations are Parties to the above-discussed multilateral environmental agreements. The experience in negotiating and participating in the subsequent meetings of these other decision-making bodies and the subsidiary bodies allowed the countries to have a good understanding of the pros and cons of different types of infrastructure. Such knowledge and experience provided invaluable insights that are reflected in the design of the BBNJ Agreement's infrastructure.

2.1.3 MGR Framework and the Implications of the BBNJ Agreement's Infrastructure

The BBNJ Agreement's infrastructure has great potential to foster benefit sharing through the MGR Framework without unintended consequences. The Agreement is indeed progressive and could create higher standards of transparency and inclusivity for future multilateral environmental agreements. The infrastructure, by design, provides ample support for the COP to make informed decisions by input from expert committee members and potentially from a wide range of stakeholders, in light of the infrastructure's focus on transparency and inclusivity.

While the basic infrastructure design is laid out in the BBNJ Agreement's text, there are further details that will need to be decided by the COP to support effective implementation of the MGR Framework. The table below provides some of the questions to be resolved by the COP with a cross-reference to the relevant sections in this chapter for ease of reference (Table 2.2).

Table 2.2 Infrastructure questions relevant for MGR framework

Infrastructure questions	Relevant sections in this chapter
Rules of procedures for the COP	Section 2.3.1
Terms of reference, modalities for the operation, and rules of procedure for the subsidiary bodies	• STB (Sect. 2.3.2) • Access and benefit sharing (ABS) committee (Sect. 2.4.1) • Capacity building and transfer of marine technology (CB/TMT) committee (Sect. 2.4.2) • Finance committee (Sect. 2.4.3) • Implementation and compliance (IC) committee (Sect. 2.4.4)
Selection process of the members of subsidiary bodies	• STB (Sect. 2.3.2) • ABS committee (Sect. 2.4.1) • CB/TMT committee (Sect. 2.4.2) • Finance committee (Sect. 2.4.3.1) • IC committee (Sect. 2.4.4)
Modalities for the operation of the CIHM	Section 2.5
Arrangements for the functioning of the Secretariat	Section 2.3.3
Modalities and mechanisms to enhance cooperation with relevant legal instruments and frameworks and relevant global, regional, subregional, and sectoral bodies (LIFBs)	• COP (Sect. 2.3.1) • Secretariat (Sect. 2.3.3) • ABS committee (Sect. 2.4.1) • CB/TMT committee (Sect. 2.4.2) • Finance committee (Sect. 2.4.3) • CIHM (Sect. 2.5)
Financial rules governing the funding of the COP, Secretariat, and subsidiary bodies	• COP (Sect. 2.3.1) • STB (Sect. 2.3.2) • ABS committee (Sect. 2.4.1) • CB/TMT committee (Sect. 2.4.2) • Finance committee (Sect. 2.3.3) • IC committee (Sect. 2.4.4)

2.2 Key Infrastructure Aspects Related to the MGR Framework

2.2.1 Infrastructure Before Entry into Force

Before considering the infrastructure after entry into force, it is helpful to reflect on what will happen beforehand. The Agreement will enter into force 120 days after the 60th ratified instrument is deposited with the UN Secretary-General (BBNJ Agreement, art. 68.1). The BBNJ Agreement's infrastructure will be in place after the first COP, which will take place within one year of entry into force (BBNJ Agreement, art. 47.2). The UN Secretary-General, through the UN Division on Ocean Affairs and Law of the Sea (DOALOS), will be the interim Secretariat since the seat and the functioning of the permanent Secretariat will be decided at the first COP (BBNJ Agreement, arts. 50.2 and 50.1). The UN General Assembly adopted a resolution on 1 August 2023 that approved that UN Secretary-General to perform the depositary functions and to serve in the capacity of a Secretariat through DOALOS, until a permanent Secretariat begins its functions (A/RES/77/321).

In anticipation of the first COP, the President of the Intergovernmental Conference, Ms. Rena Lee, issued a letter to the President of the General Assembly in which she stated, "...a preparatory process, to be established under the auspices of the General Assembly, to carry out the required preparations for that first meeting and to provide guidance on the work of the interim Secretariat until the convening of that meeting would be very helpful" (A/77/945). In response, DOALOS provided secretariat services, including an information document summarizing various modalities elements (A/AC.296/2024/3). The organizational meeting to prepare for the Preparatory Commission (PrepCom) meetings took place at the UNHQ in June 2024 (A/RES/78/272). At the meeting, certain matters, such as the dates of the PrepCom meetings and some of the program of work, were decided (Chasek, 2024).

2.2.1.1 Implications for the MGR Framework

At the PrepCom organizational meeting in June 2024, two co-chairs to lead the PrepCom were appointed (Chasek, 2024). One of the co-chairs elected to the position, Ambassador Janine Coye-Felson of Belize, is well versed in MGRs. She was the facilitator for MGRs during the Preparatory Committee meetings and during the Intergovernmental Conference (Humphries et al., 2025). The years of preparatory work and the intergovernmental conference involved leading discussions and negotiations on the MGR Framework, which is very technical and often fraught with differences in countries' positions, including on legal principles (Muraki Gottlieb et al., 2025). Having a co-chair with the expertise and experience on MGRs will allow her to effectively lead the complex discussions on the details of the BBNJ Agreement's infrastructure that could impact the implementation of the MGR Framework.

2.2.2 BBNJ Agreement's Infrastructure

This section introduces the foundational information on the infrastructure of the BBNJ Agreement. Chart 1 shows the organization of the BBNJ Agreement's infrastructure. It shows COP as the main decision-making body. Reporting to the COP are the following functions: the Secretariat, the STB, and the subsidiary bodies (i.e., Access and Benefit Sharing Committee, Finance Committee, Capacity Building and Transfer of Marine Technology Committee, and the Implementation and Compliance Committee). The Secretariat will manage the Clearing-House Mechanism. The entire infrastructure is based on, inter alia, the principles of transparency and inclusivity. How they are operationalized in the text of the BBNJ Agreement is further discussed in each of the sections below.

2.2.3 Institutional Arrangements and Other Subsidiary Bodies

The BBNJ Agreement's infrastructure includes the COP and certain subsidiary bodies identified as "institutional arrangements" (BBNJ Agreement, Part VI). The Agreement also established subject matter specific committees, such as the Access and Benefit Sharing (ABS) Committee (BBNJ Agreement, art. 15.1). Institutional arrangements also include the procedural principle of transparency, and the Clearing-House Mechanism (ClHM), an "open-access platform" to enable exchange of information, knowledge, and data relevant to the Agreement. Each of the functions are discussed below, followed by a commentary on the key aspects of the infrastructure as it relates to the MGR Framework. Many of the key aspects related to the MGR Framework are also important for other substantive elements [e.g., area-based management tools, (ABMTs), including marine protected areas (MPAs), environmental impact assessments (EIAs), Capacity Building and the Transfer of Marine Technology (CB/TMT)] but a detailed discussion of such relevance are beyond the scope of this chapter.

2.2.3.1 Conference of the Parties

Introduction to the COP

The COP will consist of the Parties to the BBNJ Agreement. It is the Agreement's decision-making body, except for certain matters associated with conducting, deciding on the outcome thereof, monitoring, reporting, and review of environmental impact assessments (BBNJ Agreement, Part IV). The COP has a wide range of tasks, from adoption of decisions and recommendations associated with the BBNJ Agreement to deciding on the budget (BBNJ Agreement, arts. 47.5 and 47.6). It also has the power to establish new subsidiary bodies as deemed necessary for the implementation of the Agreement. The COP also has facilitative functions, including providing relevant information and promoting appropriate processes, cooperation, and coordination with and among relevant legal instruments and frameworks and relevant global, regional, subregional, and sectoral bodies, with a view to promoting coherence among efforts toward the conservation and sustainable use of marine biological diversity of areas beyond national jurisdiction [BBNJ Agreement, art. 47.6(c)]. Further, the COP could decide to hold extraordinary meetings outside of the regular meetings according based on the rules of procedures that will be established by the COP at its first meeting (BBNJ Agreement, art.47.2).

Some of the challenges the COP may face will begin at its first meeting. The volume and complexity of matters that the BBNJ Agreement mandates the COP to decide *by consensus* (matters besides those listed below can go to a vote, as discussed below) are significant. It may take significant efforts to achieve consensus on these five matters that will set the rules for decades to come. Each item has granular aspects that the COP will need to unpack in detail:

1. adopt rules of procedure for itself and its subsidiary bodies (i.e., Scientific and Technical Body, Access and Benefit Sharing Committee, capacity building and transfer of marine technology committee, and Implementation and Compliance Committee) (BBNJ Agreement, art. 47.2);
2. financial rules on funding itself and the Secretariat as well as the subsidiary bodies, and thereafter, rules of procedure and financial rules for any additional subsidiary body that it may decide to establish (BBNJ Agreement, art. 47.2);
3. determine the terms of reference and modalities for the operation of some of its subsidiary bodies (i.e., capacity building and transfer of marine technology committee, Scientific and Technical Body and Implementation and Compliance Committee) (BBNJ Agreement, arts. 46.2, 49.2 and 55.3);
4. determine the seat and the functions of the Secretariat (BBNJ Agreement, art. 50.1); and
5. enter into an agreement with the Global Environment Facility (GEF) to give effect to relevant funding provisions (BBNJ Agreement, art. 52.10).

The BBNJ Agreement also took good governance into consideration in the COP's decision-making. While reaching a consensus is important to ensure that there is broad acceptance by the Parties of the COP's decision, there will be times when a Party or a regional economic integration organization may strongly object. In that case, rather than the COP being unable to make an essential decision, it would be better to have an option to vote. Such consideration is reflected in the BBNJ Agreement, where the COP, after exhausting every effort to adopt decisions and recommendations by consensus (unless otherwise stated in the Agreement), can take the following voting options:

1. For questions on substance, two-thirds of the majority of the Parties present and voting (BBNJ Agreement, art. 47.5).
2. For questions on procedure, majority of the Parties present and voting (BBNJ Agreement, art. 47.5).
3. For adoption a budget, three-fourths majority of the Parties present and voting. (BBNJ Agreement, art. 47.6(e)).

The COP also has an option to obtain an advisory opinion from the International Tribunal for the Law of the Sea (ITLOS) on a legal question on the conformity of a proposal before the COP on any matter *within its competence* (BBNJ Agreement, art. 47.7). The focus on competence is included because some negotiating countries wanted to ensure that certain matters about which there was concern about the BBNJ Agreement undermining the mandates of existing organizations (e.g., regional fisheries management organizations, questions on establishing extended continental shelf, territorial disputes, etc.) would be carved out. Specifically, the BBNJ Agreement states that an advisory opinion "must not be sought on a matter within the competence of other global, regional, subregional or sectoral bodies, or on a matter that necessarily involves the concurrent consideration of any dispute concerning sovereignty or other rights over continental or insular land territory or a claim thereto or the legal status of an areas within national jurisdiction" (BBNJ Agreement, art. 47.7). Clearly, the language shows that the limited scope of the COP's right to obtain an advisory opinion shows a balancing act among the negotiators that wanted to include an operative paragraph on an advisory opinion and those that were weary that the venue would be used to interfere with their political aspirations.

Finally, the COP is required to host a review conference within five years of the Agreement's entry into force (BBNJ Agreement, art. 47.8). At the meeting, the COP is to "assess and review the adequacy and effectiveness of the provisions" of the Treaty "and, if necessary, propose means of strengthening the implementation of those provisions" to meet the general objective of the Treaty (BBNJ Agreement, art. 47.8). A condition that may weaken the BBNJ Agreement's effectiveness of the review conference is the term, "if necessary" in art. 47.8 quoted above. Also, the timeliness of identifying and addressing the issues will depend on the interval of the review conference that will be determined by the COP. However, the BBNJ Agreement's requirement to promote transparency (BBNJ Agreement, art. 48) in its meetings has a great potential to act as a balancing factor if there is a wide range of stakeholders' active engagement at the review conference.

COP and the MGR Framework

Since the COP is the main decision-making body, its mandates and functions can impact the implementation of the MGR Framework in various ways:

(a) Preparation for and decisions made at the first COP meeting. The COP will need to consider the following elements in the context of their impacts on the MGR Framework;
   a. rules of procedure for itself and its subsidiary bodies (BBNJ Agreement, art. 47.2);
   b. financial rules on funding itself and the Secretariat as well as the subsidiary bodies, and thereafter, rules of procedure and financial rules for any additional subsidiary body that it may decide to establish (BBNJ Agreement, art. 47.2); and
   c. terms of reference and modalities for the operation of some of its subsidiary bodies (i.e., capacity building and transfer of marine technology committee, Scientific and Technical Body and Implementation and Compliance Committee) (BBNJ Agreement, arts. 46.2, 49.2, and 55.3).

(b) Review conference. The review conference will shed light on the progress in terms of monetary and non-monetary benefits and how the gaps can be addressed in a systematic and at regular intervals. Such conference may provide additional opportunities (besides the biannual reviews in art. 14.10) for the developed and developing countries to explore how best to foster greater compliance with the MGR Framework.

It is worth noting that the advisory opinion provision is unlikely but may be brought forward for the MGR provisions on DSI. AS discussed above, the COP also has an option to obtain an advisory opinion from the ITLOS on a legal question on the conformity of a proposal before the COP on any matter *within its competence* (BBNJ Agreement, art. 47.7). While the definition or the scope of DSI is a subject of negotiations in other multilateral environmental agreements, such as the CBD, the negotiators included provisions on DSI throughout the

BBNJ Agreement. For that reason, there may be an argument that DSI is within the competence of the BBNJ Agreement's COP, even though discussions on the definition of DSI have been the subject of many years of debate, for example, at the CBD.

Further, the COP's voting rule on its decisions on monetary benefit sharing on MGRs and associated DSI is in line with its decision on budgets, where if all efforts to reach consensus are exhausted, the COP is to adopt a decision by three-fourths majority of the Parties present and voting [BBNJ Agreement, art. 47.6(e) and 14.7]. The supermajority provision presents an opportunity for the developing countries on the COP's decision on monetary benefit sharing. While the UN Statistical Division notes that there is no definition of "developing" or "developed" categorization in the UN system or in the BBNJ Agreement, the Group of 77 plus China has a membership of 134 countries (out of 193 UN Member States), which was quite firm on their position on mandatory monetary benefit sharing (Humphries et al., 2025). It is possible that with the participation of all the Group of 77 plus China members and possibly other non-member developing countries, three-quarters majority could be reached, with a favorable monetary benefit terms for the developing country Parties.

2.2.3.2 Scientific and Technical Body (STB)

The BBNJ Agreement established the STB (BBNJ Agreement, art. 49.1). The STB, under the authority of the COP, will provide scientific and technical advice to the COP, including reports on its work (BBNJ Agreement, art. 49.4). To that end, the STB will consider the multidisciplinary expertise of the STB members and may also draw on "appropriate advice emanating from relevant legal instruments and frameworks and relevant global, regional, subregional and sectoral bodies, as well as from other scientists and experts" (BBNJ, art. 49.4). The Agreement takes a broad view of multidisciplinary expertise, which includes "relevant scientific and technical expertise and expertise in relevant traditional knowledge of Indigenous Peoples and local communities, gender balance and equitable geographical representation" (BBNJ Agreement, art. 47.1). Experts will serve in their expert capacity (i.e., not as a representative of a Party or an organization) "and in the best interests of the Agreement." (BBNJ Agreement, art. 49.2).

The terms of reference that the COP will adopt will influence the effectiveness of the STB. For example, Gaebel et al. conducted a study and identified "eight desirable characteristics" of the STB based on interviews of BBNJ stakeholders. The characteristics are as follows: synergistic (works with and adds to the current governance system), proficient (members are qualified and advice is pertinent), de-politicized (process and actors are free from political influences), transparent (open access to the process, information, and outcomes), influential (advice is used in the decision-making process), dynamic (flexible and adaptable to meet changing needs), inclusive (equitable access to and participation in the STB), and multidisciplinary (incorporates diverse knowledge systems) (Gaebel et al., 2024, Fig. 2.1). The authors discussed that there are synergies and trade-offs that will need to be carefully considered for the design of the STB to achieve its objective.

MGR Framework and the STB

The BBNJ Agreement mandates the ABS Committee (discussed below in Sect. 2.2.4.1) to provide recommendations to COP on the following, which will have scientific and technical aspects:

- guidelines or a code of conduct for activities with respect to MGRs and digital sequence information on MGRs of ABNJ (BBNJ Agreement, art. 15.3(a)); and
- measures to implement decisions taken in accordance with the requirements on MGRs of ABNJ and associated digital sequence information (BBNJ Agreement, art. 15.3(b)).

That said, there will be an overlap on the scientific and technical (e.g., traditional knowledge associated with MGRs) aspects, which

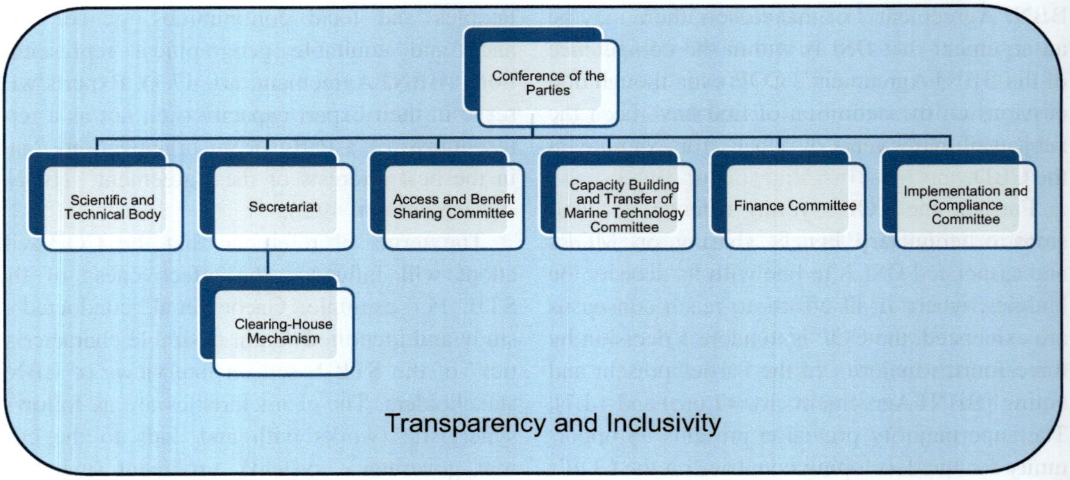

Fig. 2.1 Organization of BBNJ Agreement's infrastructure. All the functions are based on the foundation of transparency and inclusivity, which are discussed in Sect. 2.3 of this chapter

will present great opportunities for collaboration. Further, having the expertise of the STB will allow efficient use of resources by not staffing the STB and the ABS Committee with expertise on the same disciplines. Therefore, the COP may clarify in the terms of reference for both the STB and the ABS Committee that they are to closely collaborate and for the STB's experts to provide input on the work of the ABS Committee, since there are no cooperation or collaboration mandate among the subsidiary bodies in the BBNJ Agreement. While not in the BBNJ Agreement text, the STB can also collaborate with the COP and the Secretariat on the ClHM, which will be key in the transparency and inclusivity aspects of the MGR Framework.

The STB could also provide input to other subsidiary bodies on the members' relevant expertise on MGR Framework:

- CB/TMT Committee since both the MGR Framework and the capacity building and the transfer of marine technology provisions focus on MGRs and associated DSI;
- Implementation and Compliance (IC) Committee on technical aspects with the MGR Framework to foster Implementation and Compliance; and
- Finance Committee on the MGR Framework since the financial mechanism includes a special fund that will be funded from monetary benefit sharing (BBNJ Agreement, art. 52.4) and provide insights on financial resources to support the implementation of the Agreement (BBNJ Agreement, art. 52.7).

Given its wide range of duties, it is imperative that the STB is adequately staffed and funded. One way that the COP may consider maximizing the STB's resources is to allow the STB to create a pool of experts who may not be members but could be called upon for specific matters on the MGR Framework. There are certainly precedents in the consideration for the use of pool of experts, including for the Global World Ocean Assessment (A/68/82) and in various multilateral environmental agreements, such as the CBD's Subsidiary Body on Scientific, Technical and Technological Advice.

2.2.3.3 Secretariat

The BBNJ Agreement established a Secretariat, which will provide administrative and logistical support to the COP and its subsidiary bodies. Such services will range from arranging and servicing the meetings of the COP to circulating information relating to the implementation of the Treaty "in a timely manner" (BBNJ Agreement, art. 50.4). The Secretariat is required to make decisions of the COP publicly available and transmit to all Parties as well as to relevant legal instruments and frameworks and relevant global, regional, subregional, and sectoral bodies (LIFBs) (BBNJ Agreement, art. 50.4(c)).

The Secretariat and the host State may enter into a headquarters agreement. The BBNJ Agreement mandates that the "Secretariat shall enjoy legal capacity in the territory of the host State and be granted such privileges and immunities by the host State as are necessary for the exercise of its functions" (BBNJ Agreement, art. 50.3). A grant for privileges and immunities is consistent with the *Convention on the Privileges and Immunities of the United Nations*, adopted by the UN General Assembly on 13 February 1946 (UNGA Res. 22A(I)) (General Convention). The General Convention contains clauses that will be important for the Secretariat, such as the clause on "juridical personality" that grants the rights "to contract, to acquire and dispose of immovable and movable property, and to institute legal proceedings" (General Convention, art. I Sect. 2.1). The Convention also has provisions on property funds and assets, which includes tax exemption (General Convention, art. II). Further, the Convention includes articles on communications (General Convention, art. III), privileges and immunities including arrests (General Convention, art. IV), officials' privileges and immunities, including exemption from national service (General Convention, art. V), privileges and immunities for experts' on missions for the UN (General Convention, art. VI), UN laissez-passer, which is regarding travel documents (General Convention, art. VII), and settlement of disputes (General Convention, art. VIII). The General Convention has had impact on headquarters or seat arrangements (Reinisch, 2009). Therefore, the headquarters agreement is likely to follow the precedence set by other arrangement.

DOALOS, the interim Secretariat, is part of the UN's Office of Legal Affairs. It has a wide area of expertise, including on the Fish Stocks Agreement and the Part XI Agreement. DOALOS has a broad mandate, which includes maintaining a comprehensive information system, websites, and a research library containing materials on ocean affairs and the law of the sea (ST/SGB/2021/1). As the Secretariat, DOALOS oversaw discussions and negotiations for almost 20 years leading up to the adoption of the BBNJ Agreement. All the meetings took place and will continue to take place at the UN Headquarters (UNHQ) in New York City until the permanent Secretariat starts to function. There are benefits for DOALOS to continue acting in the capacity of the interim Secretariat due to its vast experience and the existing UNHQ conference rooms and services. The conference services require various resources, including conference rooms that can provide space for nearly 200 countries and observers, and interpreters to accommodate six official UN languages (i.e., Arabic, Chinese, English, French, Spanish, Russian). DOALOS is highly experienced in facilitating meetings, including providing key documents necessary, creating, and maintaining dedicated websites or via e-mail correspondence.

MGR Framework and the Secretariat

The Secretariat will also manage the Clearing-House Mechanism (BBNJ Agreement, art. 51.4) and facilitate cooperation and coordination with the other Secretariats of LIFBs as may be required, subject to approval by the COP (BBNJ Agreement, art 50.4(c)). In managing the Clearing-House Mechanism, the Secretariat must do so "without prejudice to possible cooperation with other relevant legal instruments and frameworks and relevant global, regional, subregional, and sectoral bodies as determined by the COP." Such existing organizations will include the Intergovernmental Oceanographic Commission of the United Nations Educational, Scientific and Cultural Organization, the International Seabed Authority, the International Maritime Organization, and the Food and Agriculture Organization of the United Nations (BBNJ Agreement, art. 51.4). The listed organizations have experience in creating certain databases and websites that could be useful in the creation of and management of the BBNJ Agreement's Clearing-House Mechanism.

As discussed below in Sect. 2.2.4.1, one area where the Secretariat may find synergies in terms of MGRs is the ABS Committee's mandate to make recommendations on the matters in relation to the Clearing-House Mechanism (BBNJ Agreement, art. 50.4(c)). Since the Secretariat will be managing the Clearing-House Mechanism, there would be benefits for the ABS Committee to collaborate with the Secretariat to ensure coherence. To that end, as discussed above, collaboration with the STB and other subsidiary bodies will be important to ensure coherence.

As for the MOU between the COP and the permanent seat of the Secretariat discussed above, there could be implications for the ClHM should there be infrastructure or resources that will be needed to be established at the new seat. Such implications may include considerations on taxation.

2.2.4 Subject Matter Committees

The BBNJ Agreement established committees on certain subject matters. A summary of each committee is provided below. The Access and Benefit Sharing (ABS) Committee is the main subject matter committee for the MGR Framework. Therefore, the implications of the MGR Framework will not be separately discussed. For the other three subject matter committees, an introduction will be provided, followed by implications for the MGR Framework.

2.2.4.1 MGR Framework and the Access and Benefit Sharing (ABS) Committee

The objective of the ABS Committee is to establish guidelines for benefit sharing on MGRs of ABNJ and associated digital sequence information (BBNJ Agreement, art. 15.1). The ABS Committee is tasked with providing transparency and ensuring a fair and equitable sharing of both monetary and non-monetary benefits (clauses on monetary and non-monetary benefit sharing is in BBNJ Agreement, art. 14) (BBNJ Agreement, art. 15.1). The ABS Committee will be composed of 15 members "possessing appropriate qualification in related fields, so as to ensure the effective exercise of the functions of the committee" (BBNJ Agreement, art. 15.2). The Parties will nominate members and they will be elected by the COP. In doing so, the COP must consider "gender balance and equitable geographical distribution and providing representation on the committee from developing States" (BBNJ Agreement, art. 15.2). In terms of "developing countries" the clause includes least developed countries, Small Island Developing States and landlocked developing countries. The terms of reference and modalities for the operation of the ABS Committee will be determined by the COP (BBNJ Agreement, art. 15.2).

The ABS Committee will make recommendations to the COP on the following (BBNJ Agreement, art 15.3):

- guidelines or a code of conduct for activities with respect to MGRs and digital sequence information on MGRs of ABNJ;
- measures to implement decisions taken in accordance with the requirements on MGRs of ABNJ and associated digital sequence information;
- rates or mechanisms for the sharing of monetary benefits;
- matters in relation to the ClHM;
- matters in relation to the financial mechanism (BBNJ Agreement, art 52); and
- any other matters that the COP may request.

Through the ClHM, each Party is required to provide to the ABS Committee the following information (BBNJ Agreement, art. 15.4):

- legislative, administrative, and policy measures on access and benefit sharing;
- contact details and other relevant information on national focal points; and
- other information required pursuant to the decisions taken by the COP.

The ABS Committee may consult and facilitate the exchange of information with relevant legal instruments and frameworks and relevant global, regional, subregional, and sectoral bodies on activities under its mandate, including benefit sharing, the use of DSI on MGRs, best practices, tools and methodologies, data governance, and lessons learned (BBNJ Agreement, art. 15.5). The ABS Committee may make recommendations to the COP in relation to such information and this could include points in relation to different forms of intellectual property (IP) licensing, as considered in Chap. 9 of this book (Brown, 2025).

The COP is required to determine the terms of reference and modalities for the ABS Committee and the Finance Committee, but there is no specific text that mandates the COP to do so at its first meeting (BBNJ Agreement, art. 15.2 and 52.14). That said, the two committees will play critical roles in terms of financing the implementation of the Agreement. Further, as discussed above, close collaboration with the STB will be essential to maximize resources and to ensure coherence in terms of scientific and technical information associated with the fair and equitable sharing of benefits of MGRs and DSI. Since the BBNJ Agreement requires that the COP determine the terms of reference for the STB at its first meeting, coherence can be ensured if the two committees' details will also be decided at the first COP meeting.

2.2.4.2 Capacity Building and Transfer of Marine Technology (CB/TMT) Committee

The BBNJ Agreement established the CB/TMT Committee (BBNJ Agreement, art. 46.1). Unlike the ABS Committee discussed above, the Agreement does not specify the number of CB/TMT members, but the members are to possess "appropriate qualification in related fields, so as to ensure the effective exercise of the functions of the committee" (BBNJ Agreement, art. 46.2). The Parties will nominate members and they will be elected by the COP, "taking into account gender balance and equitable geographical distribution and providing representation on the committee from developing States" (BBNJ Agreement, art. 46.2). As with the ABS Committee, "developing countries" in this clause includes least developed countries, Small Island Developing States, and landlocked developing countries. The terms of reference and modalities for the operation of the CB/TMT Committee will be determined by the COP at its first meeting (BBNJ Agreement, art. 46.2).

One of the specific work items for the CB/TMT Committee is to submit reports and recommendations for the consideration and action of the COP as appropriate (BBNJ Agreement, art. 46.3). Such requirement will foster the requirements for transparency (BBNJ Agreement, art. 48) and inclusivity, since a wide range of stakeholders will be able to have access to the work and the recommendations of the CB/TMT Committee.

MGR Framework and the CB/TMT Committee

While the substantive element, CB/TMT, has its stand-alone section in Part V of the BBNJ Agreement, CB/TMT is embedded throughout the Agreement, including in the MGR Framework (BBNJ Agreement, Part II). There are also specific references to matters of the MGR Framework in Part V of the BBNJ Agreement. Therefore, even though the CB/TMT Committee's mandate has no specific reference to MGRs Framework, the ABS Committee and the CB/TMT Committee would benefit from close cooperation to ensure integration and coherence across their work. Such partnership will create positive synergies and has the potential to maximize resources that the COP will provide to both committees.

There is also scope for this collaboration to include matters on IP licensing and how to best manage technology transfer that may need consideration on IP rights, including trade secrets, held by a private entity; and the IP owner may wish to rely on their rights to prevent others dealing with the technology (Brown, 2025). Such consideration is separate from the tangential reference to notification requirement on patents in the MGR Framework (BBNJ Agreement, art. 12.8(a)).

2.2.4.3 Finance Committee

The BBNJ Agreement mandates the COP to establish a Finance Committee (BBNJ Agreement, art. 52.14) as part of its "financial resources and mechanism" (BBNJ Agreement, Part II). Similar to the CB/TMT Committee discussed above, the BBNJ Agreement does not specify the number of Finance Committee members, but the members are to possess "appropriate qualification and expertise, taking into account gender balance and equitable geographical distribution" (BBNJ Agreement, art. 52.14). The terms of reference and modalities for the operation of the Finance Committee will be determined by the COP, but the Agreement does not specifically state that a decision will be made at its first meeting (BBNJ Agreement, art. 52.14). A delay in establishing the Finance Committee could significantly compromise the effectiveness of the financial mechanism given its function discussed below.

The Finance Committee must periodically report and make recommendations on the identification and mobilization of funds under the BBNJ Agreement's financial mechanism (BBNJ Agreement, art. 52.14). It will collect information and report on funding under other mechanisms and instruments contributing directly or indirectly to the achievement of the objectives of the Agreement (BBNJ Agreement, art. 52.14). The Finance Committee will also consider the following: (1) the assessment of the needs of the Parties, in particular developing Party countries; (2) the availability and timely disbursement of funds; (3) the transparency of decision-making and management processes concerning fundraising and allocations; and (4) the accountability of the recipient developing Party countries with respect to the agreed use of funds (BBNJ Agreement, art. 52.14(a)–(d)). The COP is required to consider the Finance Committee's reports and recommendations "and take appropriate action" (BBNJ Agreement, art. 52.15). What an "appropriate action" may consist of will be up to the COP's decision. For that reason, robust implementation of the concept of transparency in the BBNJ Agreement's art. 48 and inclusivity will be imperative.

MGR Framework and the Finance Committee

One of the funding sources of the BBNJ Agreement's financial mechanism is the special fund that will be provided from the monetary benefit sharing of MGRs of ABNJ (BBNJ Agreement, art. 52.14 (b)(ii)). Other sources of funding are as follows: (1) a voluntary trust fund established by the COP to facilitate the participation of representatives of developing States Parties, and a special fund that will be funded through the following: (a) Annual contributions by the COP, (b) additional voluntary contributions from Parties and private entities that wish to fund conservation and sustainable use of BBNJ, and (c) the GEF trust fund (BBNJ Agreement, art. 52.4 (a)–(c)). The modalities of the various funding mechanisms, including financial rules governing the funding of the COP, will be decided by the COP. Further, operationalization of other provisions of financial resource, including the scale of assessed contributions, will also be decided by the COP.

The ABS Committee is tasked with providing transparency and ensuring a fair and equitable sharing of both non-monetary and monetary benefits from the utilization of MGRs of ABNJ and associated digital sequence information (BBNJ Agreement, art. 14 and 15.1). Since the Finance Committee also has reporting obligations associated with funding, the two committees would benefit from sharing information before providing the required reports to the COP. This is particularly important since the BBNJ Agreement states that the financial mechanism "should seek to ensure that duplication is avoided, and complementarity and coherence promoted, among the utilization of the funds within the mechanism" (BBNJ Agreement, art. 52.7).

2.2.4.4 Implementation and Compliance Committee (IC Committee)

The BBNJ Agreement established the IC Committee (BBNJ Agreement, art. 53). The IC Committee's membership will consist of those with "appropriate qualifications and experience," and the Parties will nominate the members (BBNJ Agreement, art. 55.2). The COP will elect the members with "due consideration given to gender balance and equitable geographical representation" (BBNJ Agreement, art. 55.2). The IC Committee will operate under the modalities and rules of procedure, which the COP will adopt at its first meeting (BBNJ Agreement, art. 55.3).

While the title of the IC Committee may give the perception that it is an enforcement mechanism, the BBNJ Agreement is clear that the IC Committee will be "facilitative in nature and function in a transparent, non-adversarial and non-punitive manner" (BBNJ Agreement, art. 55.1). To that end, the IC Committee will "consider issues of Implementation and Compliance at the individual and systemic levels, inter alia, and report periodically and make recommendations" (BBNJ Agreement, art. 55.3). One of the challenges that the IC Committee may encounter is ensuring that its recommendations are "appropriate, while cognizant of respective national circumstances" (BBNJ Agreement, art. 55.3). Such qualifiers may weaken the role of the IC Committee, and therefore, the overall effectiveness of Implementation and Compliance requirements in the Agreement. This is particularly the case because each Party must take "the necessary legislative, administrative or policy measures, *as appropriate*, to ensure" (emphasis added) the BBNJ Agreement's implementation. The repetition of the terminology "appropriate" emphasizes the point that some Parties may position themselves to implement weak domestic requirements.

MGR Framework and the IC Committee

As discussed above in Sect. 2.2.4.1, the ABS Committee also has tasks associated with implementation of the MGR requirements; specifically, the committee is to establish:

- guidelines or a code of conduct for activities with respect to MGRs and digital sequence information on MGRs of ABNJ; and
- measures to implement decisions taken in accordance with the requirements on MGRs of ABNJ and associated digital sequence information (BBNJ Agreement, art. 15.3(b)).

The IC Committee and the ABS Committee could share data, information, and knowledge to ensure that there is coherence in their approaches. It is important to note that there is a hierarchy in the useful nature of the three types of benefits. Data provides raw information that does not include analysis; therefore, unless there is a way to use the data, they will provide little benefit. Information could range from dates to the result of research, and the level of usefulness will depend on the content. Knowledge that could be transferred through extensive education and training, on the other hand, could be the key to unlocking the greatest scope of benefits associated with data and information.

2.2.5 Clearing-House Mechanism (ClHM)

The BBNJ Agreement established the ClHM (BBNJ Agreement, art. 51.1). The ClHM will be primarily "an open-access platform" and serve as a centralized platform to facilitate transparency and international cooperation and collaboration (BBNJ Agreement, art. 51.2). Uncertainty to what "open access" means and possible intersections with IP rights are explored in Chap. 9 of this book (Brown, 2025). The ClHM will enable Parties to access, provide, and disseminate information on the four substantive elements of the BBNJ Agreement: (1) MGRs, including the fair and equitable sharing of benefits, (2) establishment and implementation of area-based management tools, including marine protected areas, (3) environmental impact assessment; and (4) capacity building and the transfer of marine technology (BBNJ Agreement, art. 51.3).

Note that the information associated with CB/TMT is intended to be robust, including requests for CB/TMT and "opportunities with respect thereto, including research collaboration and training opportunities, information on sources and availability of technological information and data for the transfer of marine technology, opportunities for facilitated access to marine technology and the availability of funding" (BBNJ Agreement art 51.3 (a)(iv)). Such language in the BBNJ Agreement shows the determination of the developing country negotiators to ensure that the current gap in CB/TMT is adequately filled in a sustained way. In terms of capacity building, the ClHM will provide other functions, such as facilitating the matching of capacity building needs and transfer of marine technology with those interested in providing as donors (BBNJ Agreement art 51.3(b)). In that regard, the BBNJ Agreement identifies a broad range of stakeholders, including governmental, non-governmental, or private entities that may be interested in transfer of marine technology and facilitating access to related know-how and expertise (BBNJ Agreement art 51.3(b)).

The BBNJ Agreement has specific considerations associated with the management of the ClHM for Small Island Developing States (SIDS) (BBNJ Agreement art 51.5). Specifically, the Agreement states that "[i]n the management of the Clearing-House Mechanism, full recognition shall be given to the special requirements of developing States Parties, as well as the special circumstances of Small Island Developing States Parties, and their access to the mechanism shall be facilitated to enable those States to utilize it without undue obstacles or administrative burdens. Information shall be included on activities to promote information-sharing, awareness-raising and dissemination in and with those States, as well as to provide specific programmes for those States" (BBNJ Agreement, art 51.5). The concept of the "special circumstances of Small Island Developing States" was first introduced at the UN Conference on Environment and Development in June 1992 in Rio de Janeiro (A/RES/44/228). The report from the conference states, "[S]mall Island Developing States, and islands supporting small communities are a special case both for environment and development. They are ecologically fragile and vulnerable. Their small size, limited resources, geographic dispersion and isolation from markets, place them at a disadvantage economically and prevent economies of scale" (A/Conf.151/26/Rev.1 (Vol. I) Chapter 17.123). The specific reference of the developing countries, especially the SIDS, would require that the ClHM is not overly burdensome to use for countries with weak IT infrastructure.

The BBNJ Agreement also considers issues associated with maintaining confidentiality. The confidentiality clause ensures that the Agreement would not require "sharing of information that is protected from disclosure under the domestic law of a Party or other applicable law" (BBNJ Agreement, art. 51.6). Such provision is necessary given the strong emphasis on transparency throughout the BBNJ Agreement. That said, there could be an abuse of such carve-out if there is a broad reading of the confidentiality clause that favors blocking information from disclosure.

The architecture of the ClHM and its functions were discussed during the IGC, but the negotiators ultimately decided to not include the details of the ClHM (Humphries et al., 2025). Since the timing of the establishment of the ClHM is not specified in the BBNJ Agreement, there could be a substantial lag between the Agreement's entry into force and the operationalization of the ClHM. Such delay would cause significant issues for the transparency and inclusivity provisions in the Agreement.

As discussed above, the BBNJ Agreement's Secretariat will manage the ClHM with possible cooperation with other relevant legal instruments and frameworks and relevant global and regional organizations determined by the COP (BBNJ Agreement, art. 51.4). Without the ClHM, the Secretariat would not have the proper capabilities to execute its obligations. Should it not be possible to have a fully functioning ClHM, there may be an option to create a temporary ClHM with minimal capabilities, with the potential to rapidly build up to a full ClHM. Another option may be to start to consider the design and functions of the ClHM prior to entry into force and even conduct pilot studies to determine the best information technology infrastructure.

2.2.5.1 MGR Framework and the ClHM

The MGR Framework is heavily based on the use of the ClHM. For example, monitoring and transparency with regard to benefit sharing must be achieved through notification to the ClHM (BBNJ Agreement, art. 16.1). The ABS Committee must prepare a report based on the information received through the ClHM (BBNJ Agreement, art. 16.3). All the notification requirements on MGRs and associated DSI must be made using the ClHM (BBNJ Agreement, art. 12). It will also provide links to relevant global, regional, subregional, national, and sectoral Clearing-House Mechanisms and other gene banks, repositories, and databases, including those pertaining to relevant traditional knowledge of Indigenous Peoples and local communities (BBNJ Agreement, art. 51.3). Without the existence and effective functioning of the ClHM, the MGR Framework's requirements will be unmet.

While the BBNJ Agreement "established" the ClHM, it only exists as a concept. The COP will need to determine the specific modalities for the operation of the ClHM (BBNJ Agreement, art. 51.2). Such modalities will include complex issues, such as the type, architecture, and functionalities of the platform. An aspect that will be of particular interest for the scientific community will be the process for generating the "BBNJ" standardized batch identifier that is part of the MGR Framework, which has various challenges and opportunities as discussed in Chap. 14 of this book (Rabone et al., 2025). To that end, partnerships with the private sector with advanced information technology expertise and databases that have wide uses (e.g., International Nucleotide Sequence Database Collaboration, etc.) could lead to futureproofing the ClHM.

Another aspect that the COP will need to decide will be on how the ClHM will manage traditional knowledge of Indigenous Peoples and local communities that is part of the MGR Framework. While part of the MGR Framework, the BBNJ Agreement makes it optional for the use of ClHM to access traditional knowledge (BBNJ Agreement, art. 13). An exploration of various matters associated with traditional knowledge in the MGR Framework is available in Chap. 8 of this book. (Pena-Neira and Coelho, 2025).

2.3 MGR Framework and Transparency and Inclusivity

2.3.1 Transparency as Operationalized in the BBNJ Agreement

The principle of transparency crosscuts throughout the BBNJ Agreement's infrastructure. For example, the COP must "promote transparency in decision-making processes and other activities carried out" pursuant to the Agreement (BBNJ Agreement, art. 48.1). The Agreement also mandates that the rules and procedures that will be adopted by the COP require that the COP provide to the non-Party observers "timely access" to all relevant information and its decisions must be published and maintained as public record (BBNJ Agreement, art. 48.1 and 48.4). Since the word "timely" is not defined, there may be questions as to how much time the Secretariat would have to release information to the observers. Further, "relevant information" may call into question what may be relevant and for whom. For such reasons, development of agreed-upon procedures will be key for successfully implementing the BBNJ Agreement's transparency requirements.

2.3.2 Inclusivity

Inclusivity is included alongside the transparency provisions of the BBNJ Agreement, consistent with international best practices (Ardron et al., 2023). For example, all meetings of the COP and its subsidiary bodies must be open to non-Party observers participating in accordance with the rules and procedures adopted by the COP, unless otherwise decided by the COP (BBNJ Agreement, art. 48.2). The BBNJ Agreement identifies a wide range of stakeholders, including non-Parties, Indigenous Peoples, and local communities, the scientific community and "other relevant stakeholders" with "an interest in matters pertaining" to the COP and of its subsidiary bodies, to participate as observers (BBNJ Agreement, art. 48.3). The COP's rules and procedures regarding meetings must provide modalities for such participation, and they must not be unduly restrictive (BBNJ Agreement, art. 48.4). The COP also has the requirement to facilitate the participation of and consultation with the identified stakeholders (BBNJ Agreement, art. 48.3). The strong emphasis on inclusive meetings reflects the long history of observer participation in the meetings and negotiations leading up to the adoption of the BBNJ Agreement in June of 2023 (Humphries et al., 2025). While the power of the COP to close the doors for observer participation may hinder robust inclusive participation, they may also allow for difficult deadlocks to be broken. However, the transparency provisions in the BBNJ Agreement generally support a case for the COP to make inclusive meetings the norm.

2.3.3 Open Communication Among the Subsidiary Bodies

The BBNJ Agreement's focus on cooperation (BBNJ Agreement, art. 8.3) and coherence (BBNJ Agreement, Preamble) could give the COP direction to look for positive synergies among the subject matter committees (i.e., ABS Committee, financial committee, CI Committee, and CB/TMT Committee), the Secretariat, and the STB. Key themes could include regard for state obligations under other relevant areas of law, such as intellectual property in respect of technology transfer, as noted above, and also regard to science, including indigenous and traditional knowledge (Collective Statement from Edinburgh High Seas Treaty Symposium, 2023; *see also*, Harden-Davies, 2024). The substantive committees, the STB, and the Secretariat could cooperate in several ways, including written correspondence as well as in-person informal meetings leading up to the preparatory meetings, preparatory commission meetings, informal meetings post-preparatory commission meeting, and the first COP meeting.

While some information about the role of the ABS Committee is available as discussed above, the terms of reference and modalities of the BBNJ Agreement's subject matter committees are yet to be decided. Furthermore, the extent to which the ABS Committee will be directed to consult or collaborate is unclear (i.e., with other substantive committees or with the subsidiary bodies under the institutional arrangements, the STB, and the Secretariat). However, given the crosscutting interests of the Access and Benefit Sharing Committee, one can assume that some cooperation with other committees will be seen as beneficial.

2.3.4 Timing of the ClHM Availability

While the BBNJ Agreement describes overall functions of the information and data platform, its specific timing and modalities are to be determined by the COP (BBNJ Agreement, art. 51.2). During the negotiations, there was a conscious effort by the negotiators not to specify the modalities of the ClHM because of the importance to ensure that the BBNJ Agreement would not create an IT system that will not in short duration become outdated. However, the ClHM is at the heart of the Agreement's mandate to ensure transparency. Rather than for the Parties to wait until the first COP, which will take place "no later than one year after the entry into force of this agreement," (BBNJ Agreement, art. 47.2) the Parties could explore various options during informal meetings leading up to the preparatory meetings and at the preparatory meetings. For such meetings, engaging with the private sector and data experts, with input from developing States, especially, SIDS, may provide options that are fit for purpose.

2.3.5 Various Types of Meetings to Foster Transparency

One benefit that the countries and observers have is the experience of hosting and participating in informal and UN led meetings from the very first meeting to discuss the importance of BBNJ in 2004 and to the finish line when the BBNJ Agreement was adopted in June 2023. Broadly speaking, there are two types of meetings: (1) Meetings hosted by the UN and (2) Informal meetings hosted by countries or observers. A detailed history of UN led meetings is available in Chap. 1 (Humphries et al., 2025).

One can also look at how other multilateral environmental agreement meetings are held. For example, at the Convention on Biological Diversity's ad hoc technical expert group are "encouraged to use innovative means of communication and to minimize the need for face-to-face meetings" (COP Decision VIII.10). Such flexibility allows for online meetings and e-mail correspondences that significantly reduce the resource burden, in particular, for the Global South.

The four meeting format options in Table 2.3 provide the benefits and challenges that a global meeting may present. In addition to time zones, participants from the Global South have challenges in allocating resources (e.g., financial, personnel, etc.) to attend meetings in-person. Any organizer would need to carefully weigh the benefits and challenges to ensure equity, urgency of the meeting, and the likely effectiveness of the format of the meeting (Table 2.3).

Table 2.4 aims to match certain meeting types that may work well in an online, in-person, hybrid, and asynchronous discussion models. Chapter 1 of this book covers a rich history of the meetings that the UN hosted over two decades, until the adoption of the BBNJ Agreement. Innovative approach used during the COVID-19 pandemic (i.e., live stream discussion with asynchronous platform for discussions) kept the momentum of the negotiations, while keeping the participants and the organizers safe. Future meetings could apply the lessons learned from the intergovernmental conference meetings to determine the types of platforms that may best suit the meeting (Table 2.4).

2.3.6 Implications for the MGR Provisions

The MGR Framework and its provisions were thought to be the most complex and politically sensitive throughout the intergovernmental conference (Humphries et al., 2025). Therefore, effective implementation of the MGR Framework will benefit from the transparency and inclusivity factors discussed above: (a) timely and open access to relevant information, (b) inclusive meetings, (c) open communication among the subsidiary bodies to support the COP's informed decision-making, and (d) considerations on the type of meetings that may best suit the technical discussions on the MGR Framework.

As it is practiced in the CBD's Subsidiary Body on Scientific, Technical and Technological Advice, detailed technical discussions on MGRs could benefit from e-mail/platform for asynchronous discussions prior to and post live-streamed, in-person, or hybrid meetings. Such practice will allow technical information to be available so that Parties and observers can be prepared to engage in discussions after having the opportunity to digest information and also, to have documentation that provides next steps so that work can continue intersessionally.

Further, it will require careful consideration to decide what topics can be discussed asynchronously. It will be imperative that such decisions are based on objective information. Finally, for technical matters that can benefit from a wide range of scientific disciplines, choosing the forum that allows for active engagement from all regions of the world would be best. In that regard, considering the factors in the above referenced table would be useful (Table 2.4).

2.4 Conclusion

A multilateral environmental agreement's infrastructure can have a significant impact on how its provisions are implemented. The BBNJ

Table 2.3 A comparison of benefits and challenges of three types of meeting format

	Benefits	Challenges
Online	• Minimal costs to participate • Depending on the platform, more participants can engage rather than in-person meeting • Small carbon footprint	• Need stable IT infrastructure and IT equipment • Inability for all participants to meet at a reasonable hour due to time zones • Difficult to read body language • Potential cyber security issues • No impromptu side discussions
In-person	• No time zone issues • Face-to-face meetings allows for negotiators to read body language • Difficult issues can be resolved in person, possibility outside of the UN • Impromptu side discussions between meetings	• Travel costs could be high • The costs to host are high (rental of conference room, equipment, translation, etc.) • Limits on conference room space • Limits on conference services • Potential in situ security issues • Large carbon footprint
Hybrid	• Minimal costs to participate allowing wider participation • Depending on the platform, many more participants can engage rather than an in-person meeting • Moderate carbon footprint	• Challenges identified for in-person and online meetings • Creates equity issues, where those with less resources are likely to be able to participate online only and lose the benefits of in-person participation
E-mail correspondence or an IT platform for asynchronous discussions	• Minimal costs to participate • Very small carbon footprint • Enables discussions to happen at the participants' convenience	• Potential cyber security issues • The tone of the written communication may be misread • Correspondences and discussion space are likely to be limited to English, as opposed to all six UN official languages

Table 2.4 Examples of types of meeting that may match well with online, in-person, hybrid, and asynchronous meeting models

	Meeting types
Online/live-streamed (exchange of documents possible)	• Informal discussion (e.g., informal intersessional meeting) • Meeting on administrative matters (e.g., preparatory commission organizational meeting)
In-person	• Formal meetings (e.g., intergovernmental conference, preparatory commission meetings)
Hybrid	• Informal discussion (e.g., informal intersessional meeting) • Meeting on administrative matters (e.g., preparatory commission organizational meeting)
E-mail/platform for asynchronous discussions	• Exchange of documents/information prior to and post live-streamed, in-person, or hybrid meetings (e.g., informal intersessional meeting)

Agreement's infrastructure uses a mixture of lessons learned from previous UNCLOS implementation agreements, ideas borrowed from multilateral environmental laws and practices, and innovative approaches in a fit for purpose manner. That said, the several details that remain to be agreed by the COP after the BBNJ Agreement enters into force will ultimately determine how effectively and efficiently the infrastructure will enable the Agreement's objectives.

The concepts of transparency and inclusivity are the bedrock of the BBNJ Agreement's infrastructure. Not only is there a specific clause on transparency within the BBNJ Agreement's institutional arrangements, but the concept is also operationalized in the functions of the infrastructure, particularly through the ClHM. The Secretariat will also be critical in maintaining transparency throughout its work, with the mandate to timely and openly provide information on the implementation of the BBNJ Agreement. There are review processes that the subsidiary bodies, including the subject matter-based committees, will need to comply with while engaging with a wide range of stakeholders. The detailed modalities of reporting by the secretariat, the COP, and the Parties will be essential to ensure that relevant information is timely provided.

Also, in addition to the terms of references for the various subsidiary bodies, there will be a long list of matters that the COP will need to determine at its first meeting after the BBNJ Agreement enters into force. Further, even if the BBNJ Agreement does not specifically require the COP to decide on the terms of reference for the Finance Committee and the ABS Committee, their functions will play critical roles in terms of financing the implementation of the Agreement. The need for holistic and coherent thinking required for an effective implementation of the BBNJ Agreement will lead the Parties to consider the terms of reference for all the infrastructure of the Agreement. In doing so, an analysis that focuses on desirable characteristics, positive and negative synergies, and trade-offs would support the COP to make an informed decision about the design of each aspect of the BBNJ Agreement's infrastructure.

The work and collaboration leading up to the first COP meeting will dictate the level of success that the COP will enjoy. To that end, there are many promising initiatives that are occurring, globally. The PrepCom meetings, including the organizational meeting, can prioritize the work that needs the greatest focus: (a) for the signatories to become Parties so that the critical mass of 60 Parties can be reached as soon as possible and (b) effective implementation of the Agreement.

The capacity development work by the interim Secretariat has provided foundational information that is openly accessible through a dedicated website. Further, in addition to support by countries, NGO, and philanthropies, the GEF authorization to use up to $34 million will accelerate the existing funding gap (GEF/24/2023). To ensure that there is clarity as to how the funding can be obtained, GEF Secretariat has issued initial guidelines (GEF/C.66/07). It is imperative that the Parties, champion countries, regions, and other stakeholders continue to keep the momentum and build on the political will that made it possible for nearly 200 countries to agree on the historic global binding agreement to conserve and sustainably use the largest biodiversity of the Earth.

***

References

Agreement relating to the implementation of Part XI of the United Nations Convention on the Law of the Sea of 10 December 1982 (A/RES/48/263).

Agreement for the Implementation of the Provisions of the United Nations Convention on the Law of the Sea of 10 December 1982 Relating to the Conservation and Management of Straddling Fish Stocks and Highly Migratory Fish Stocks (A/CONF.164/37).

Agreement under the United Nations Convention on the Law of the Sea on the Conservation and Sustainable Use of Marine Biological Diversity of Areas beyond National Jurisdiction. (A/RES/77/321). Accessed 8 Aug 2023

Agreement under the United Nations Convention on the Law of the Sea on the conservation and sustainable use of marine biological diversity of areas beyond national jurisdiction, Intergovernmental conference on an international legally binding instrument under the United Nations Convention on the Law of the Sea on the conservation and sustainable use of marine biological diversity of areas beyond national jurisdiction. Further resumed fifth session. (A/CONF.232/2023/4*). Accessed 19 June 2023.

Agreement under the United Nations Convention on the Law of the Sea on the Conservation and Sustainable Use of Marine Biological Diversity of Areas beyond National Jurisdiction, 19 June 2023, 1833 U.N.T.S. 31363.

Agreement under the United Nations Convention on the Law of the Sea on the Conservation and Sustainable Use of Marine Biological Diversity of Areas beyond National Jurisdiction. (A/RES/78/272). Accessed 29 April 2024

Ardron, J. A., Lily, H., & Jaeckel, A. (2023). Public participation in the governance of deep-seabed mining in the area. In R. Rayfuse, & A. Jaeckel (Eds.), Chapter 16, research handbook on international marine environmental law, 2nd ed.

Ardron, J. (2018). Transparency in the operations of the international seabed authority: An initial assessment. *Marine Policy, 95*, 324–331. https://doi.org/10.1016/j.marpol.2016.06.027

Blasiak, R., & Yagi, N. (2016). Shaping an international agreement on marine biodiversity beyond areas of national jurisdiction: Lessons from high seas fisheries. *Marine Policy, 71*, 210–216. https://doi.org/10.1016/j.marpol.2016.06.004

Brown, A. E. L. (2025). The place of intellectual property under the BBNJ Agreement. In F. Humphries (Ed.), *Decoding governance of marine genetic resources under the BBNJ Agreement*. Springer.

Chasek, P. (2024). Summary of the organizational session of the preparatory commission for the entry into force of the BBNJ agreement: 24–26 June 2024. *The Earth Negotiations Bulletin, 12*, 1–6.

Collective statement from Edinburgh high seas treaty symposium (2023). Accessed 15 May 2024.

Consolidated modus operandi of the subsidiary body on scientific, technical and technological advice, Annex III, Operations of the Convention, COP Decision VIII.10.

Convention on the privileges and immunities of the united nations, adopted by the UN general assembly on 13 February 1946. (UNGA Res. 22A(I) of 13 February 1946)

Gaebel, C., et al. (2024). Institutionalising science and knowledge under the agreement for the conservation and sustainable use of marine biodiversity of areas beyond national jurisdiction (BBNJ): Stakeholder perspectives on a fit-for-purpose scientific and technical body. *Marine Policy, 161*, 105998. https://doi.org/10.1016/j.marpol.2023.105998

Global Environment Facility. (2024). Initial guidelines for enabling activities and ratification support projects for the agreement under the united nations convention on the law of the sea on the conservation and sustainable use of marine biological diversity of areas beyond national jurisdiction (BBNJ) (GEF/C.66/07)

Harden-Davies, H. (2024). First to finish, what comes next? Putting capacity building and transfer of marine technology under the BBNJ agreement into practice. *Ocean Sustainability, 3*(3), 39.

Humphries, F., Berry, T., & Muraki Gottlieb, H. (2025). Bridging divides: The evolution of marine genetic resources governance beyond national jurisdiction. In F. Humphries (Ed.), *Decoding governance of marine genetic resources under the BBNJ Agreement*. Springer

International Seabed Authority. (2000). The decision of the council of the authority concerning the rules of procedure of the legal and technical commission, 13 July 2000, (ISBA/6/C/9). ISBA 6 C 9 LTC rules.PDF (isa.org.jm) Accessed 25 October 2024

International Seabed Authority. (2017). Final report on the periodic review of the International Seabed authority pursuant to article 154 of the United Nations Convention on the Law of the Sea, 8 February 2017, (ISBA/23/A/3). https://www.isa.org.

jm/wp-content/uploads/2022/06/isba-23a-3_1.pdf. Accessed December 2023

Letter dated 14 May 2013 from the Co-Chairs of the Ad Hoc Working Group of the Whole addressed to the President of the General Assembly. Report on the work of the Ad Hoc Working Group of the Whole on the Regular Process for Global Reporting and Assessment of the State of the Marine Environment, including Socioeconomic Aspects (A/68/82).

Letter dated 30 June 2023 from the President of the intergovernmental conference on an international legally binding instrument under the United Nations Convention on the Law of the Sea on the conservation and sustainable use of marine biological diversity of areas beyond national jurisdiction to the President of the General Assembly (A/77/945).

Muraki Gottlieb, H., Kacheleriess, D., & Slobodian, L. (2025). Understanding the preamble, principles and objectives of the BBNJ Agreement with a focus on the fair and equitable sharing of benefits from marine genetic resources. In F. Humphries (Ed.), *Decoding governance of marine genetic resources under the BBNJ Agreement.* Springer.

Nagoya Protocol on Access to Genetic Resources and the Fair and Equitable Sharing of Benefits Arising from their Utilization to the Convention on Biological Diversity. Accessed 29 October 2010.

Note by Secretariat, *Matters to be addressed at the first meeting of the Conference of the Parties to the Agreement under the United Nations Convention on the Law of the Sea on the Conservation and Sustainable Use of Marine Biological Diversity of Areas beyond National Jurisdiction* (2024). (A/AC.296/2024/3).

Paris Agreement to the United Nations Framework Convention on Climate Change, Dec. 12, 2015, T.I.A.S. No. 16–1104.

Pena-Neira, S., & Coelho, L.F. (2025). Traditional knowledge associated with marine genetic resources in areas beyond national jurisdiction. In F. Humphries (Ed.), *Decoding governance of marine genetic resources under the BBNJ Agreement.* Springer.

Prepared by the President of the Conference, 20 June 2023 (A/CONF.210/2023/6). All the documents associated with the 2023 resumed Fish Stocks Agreement Review Conference. https://www.un.org/depts/los/convention_agreements/review_conf_fish_stocks.htm. Accessed 13 May 2024.

Preparing the GEF to serve as part of the financial mechanism of the internationally legally binding instrument under the United Nations Convention on the Law of the Sea on the Conservation and Sustainable use of marine biological diversity of areas beyond national jurisdiction (BBNJ) (Decision 14/2023)

Rabone, M., Horton, T., Humphries, F., Lyal, C., Muraki Gottlieb, H., Scholz, A. H., Vanagt, T., & Jaspars, M. (2025). BBNJ Agreement: Considerations for Scientists and Commercial End Users of MGR at Research, Development and Commercialization Stages. In F. Humphries (Ed.), *Decoding governance of marine genetic resources under the BBNJ Agreement.* Springer.

Reinisch, A. (2009) Convention on the privileges and immunities of the united nations, 1946 and convention on the privileges and immunities of the specialized agencies, 1947—introductory note, united nations audiovisual library of international law. Accessed 15 May 2024.

Report of the resumed Review Conference on the Agreement for the Implementation of the Provisions of the United Nations Convention on the Law of the Sea of 10 December 1982 relating to the Conservation and Management of Straddling Fish Stocks and Highly Migratory Fish Stocks.

Secretary-General's Bulletin. (2021). Organization of the office of legal affairs. (ST/SGB/2021/1).

The Convention on Biological Diversity of 5 June 1992 (1760 U.N.T.S. 69).

The decision of the council of the authority concerning the rules of procedure of the legal and technical commission (ISBA/6/C/9, 13 July 2000), ISBA/6/C/9.

The UN division for ocean affairs and law of the sea provides documents associated with the informal consultations of states parties to the agreement. https://www.un.org/oceancapacity/content/unfsa-informal-consultations-states-parties. Accessed 24 March 2024.

The UN division for ocean affairs and law of the sea provides documents associated with the fish stocks agreement review conference. https://www.un.org/depts/los/convention_agreements/review_conf_fish_stocks.htm. Accessed 24 March 2024.

United Nations Convention on the Law of the Sea, Dec. 10, 1982, 1833 U.N.T.S. 397.

United Nations Framework Convention on Climate Change, May 9, 1992, S. Treaty Doc No. 102-38, 1771 U.N.T.S. 107.

United Nations Charter.

United Nations Conference on Environment and Development (A/RES/44/228).

Hiroko Muraki Gottlieb is a licenced lawyer in the USA with a diversified career in global businesses, intergovernmental organizations, non-governmental organizations and academic institutions. She brings a unique blend of expertise in climate change and biodiversity conservation strategy, environmental law (domestic and international), policymaking, regulatory compliance and stakeholder engagement. She is the representative for the Ocean and led the International Council of Environmental Law to the BBNJ Agreement negotiations as the Head of Delegation and holds appointments with the Department of Organismic and Evolutionary Biology at Harvard University and Elisabeth Haub School of Law at Pace University.

Jeff A. Ardron has more than thirty years of experience in marine policy and conservation science. Having worked for governments, intergovernmental organizations, academia and non-governmental organizations, he is currently Africa Oceans Director for The Nature Conservancy and is based in Mombasa, Kenya. He holds a PhD in ocean sciences from University of Southampton, UK, and an MSc in environmental management from Royal Roads University, Canada.

Abbe Brown is a professor in Intellectual Property Law at the University of Aberdeen. Before returning to academia, she practised as an intellectual property and commercial litigator at leading firms in London, Melbourne and Edinburgh. Abbe has a strong interest in the ocean and in interdisciplinary research and is a member of the World Commission on Environmental Law and the Deep Ocean Stewardship Initiative.

Marine Genetic Resources Beyond National Jurisdiction: The Expansive Scope of the BBNJ Agreement

3

Fran Humphries

Abstract

This chapter interprets the scope and key definitions in Part II Marine Genetic Resource governance of the Biodiversity Beyond National Jurisdiction (BBNJ) Agreement. The purpose of this chapter is to: (a) aid practitioners and policy makers' understanding of the rationale underlying the treaty obligations and key gaps in interpretation; (b) outline how scope and definitions shape the treaty framework and institutional arrangements; and (c) provide practical considerations for how scope and definitions might be implemented in practice by Parties. Through a textual analysis of the treaty and its evolution during preparatory and intergovernmental committee meetings, it critically analyzes scope in three dimensions: jurisdiction (geography, sovereignty and relationships with other instruments, frameworks and bodies); subject matter (physical materials, sequence information, traditional knowledge, activities and exclusions); and temporal scope (including retrospectivity). The chapter analyzes the principles of 'without prejudice', 'non-appropriation', 'due diligence' and 'not undermining' in the treaty context. It concludes that the BBNJ Agreement has an expansive scope requiring clarification and calibration by treaty bodies for a common approach to implementation by Parties.

Keywords

BBNJ agreement · Benefit sharing notification · High Seas Biodiversity Treaty · Marine genetic resources · Digital sequence information · Geographical scope · Temporal scope · Subject matter scope · Traditional knowledge · Sovereignty · Collection · Utilization · Without prejudice principle · Non-appropriation principle · Not undermining principle · Due diligence principle

3.1 Introduction

The 2023 *Agreement under the United Nations Convention on the Law of the Sea on the Conservation and Sustainable use of Marine Biological Diversity of Areas Beyond National Jurisdiction* (BBNJ Agreement) is groundbreaking in its scope (UNGA, 2023). Part II of the BBNJ Agreement provides an international framework for marine genetic resources (MGR) of areas beyond national jurisdiction (ABNJ)

F. Humphries (✉)
Griffith Law School, Griffith University, Nathan, Queensland 4011, Australia
e-mail: fran.humphries@griffith.edu.au

F. Humphries (ed.), *Decoding Marine Genetic Resource Governance Under the BBNJ Agreement*, Sustainable Development Goals Series, https://doi.org/10.1007/978-3-031-72100-7_3

comprising the high seas—up to 60% of the world's ocean surface (Elferink et al., 2022; Rogers et al., 2021)—generally the deep seabed beyond the limits of national jurisdiction—called the Area. Clarifying the scope of subject matter and activities within clear geographical and temporal boundaries is crucial for Part II the BBNJ Agreement, which needs to fit into an existing puzzle of institutions, frameworks and bodies (IFBs).

The *United Nations Convention on the Law of the Sea* (UNCLOS)[1] recognizes legal and geopolitical boundaries between the deep seabed and the water column and between marine areas within national jurisdiction (AWNJ) and areas beyond national jurisdiction (ABNJ). In practice, the same MGR may be found both within and beyond national jurisdiction and in multiple national jurisdictions, creating complexity for aquatic genetic resource governance across jurisdictional areas (Humphries et al., 2024). Entities undertaking research and development (R&D) may use a combination of MGR from different origins in their work (Rogers et al., 2021), in multiple forms including the information from genetic resources such as digital sequence information on marine genetic resources (DSI) (Rabone et al., 2019) and draw from a range of knowledge including scientific and traditional knowledge (Mulalap et al., 2020). Negotiators of the BBNJ Agreement therefore needed to ensure that: (1) definitions associated with Part II MGR governance complement and do not conflict with those already established by existing treaty regimes; and (2) the scope of application of the BBNJ Agreement is clearly defined to avoid overlap and confusion for implementation.

The aim of this chapter is to interpret and analyze the scope and relevant definitions of Part II to highlight gaps in interpretation that may be shaped by treaty bodies such as the Conference of the Parties (CoP), Access and Benefit Sharing (ABS) Committee and the Scientific and Technical Body (STB) (see Chap. 2 of this collection Muraki Gottlieb et al., 2025a) or by State practice in coming years. It analyzes how relevant definitions relate to other articles in Part II, which are analyzed in greater detail in subsequent chapters in this edited collection. This chapter is a textual legal analysis based on the ordinary meaning of the terms in their context considering the BBNJ Agreement's object and purpose and any applicable rules of international law,[2] with an analysis supported by relevant United Nations documents and academic literature.

The general objective of the BBNJ Agreement is to 'ensure the conservation and sustainable use of marine biological diversity of [ABNJ] for the present and in the long term' through effective implementation of UNCLOS and further international cooperation and coordination (art 2). Part II has four specific objectives relating to: (1) the fair and equitable sharing of benefits arising from activities with respect to MGR and DSI on MGR of ABNJ for the conservation and sustainable use of marine biodiversity; (2) building capacity of Parties to carry out these activities; (3) generating knowledge, scientific understanding and technological innovation through the development and conduct of marine scientific research (MSR); and (4) the development and transfer of marine technology (art 9).

The four elements in Part II–V of the BBNJ Agreement were negotiated separately but as a package, so that one element could not be agreed without the other. In practice, their application to a given MGR activity may overlap at times. In some cases, MGR might be collected from ecologically or biologically significant marine areas such as the ecosystem of the Sargasso Sea (Roe et al., 2022). If collection activities occur in a high seas area that is subject to an area-based management tool (ABMT), Part III obligations may apply. Capacity Building and the Transfer of Marine

[1] Parts VII and IX of *United Nations Convention on the Law of the Sea*, opened for signature 10 December 1982, 1833 UNTS 397 (entered into force 16 November 1994) ('UNCLOS').

[2] *Vienna Convention on the Law of Treaties*, opened for signature 23 May 1969, 1155 UNTS 331 (entered into force 27 January 1980) art 31.

Technology (CBTMT) under Part V with a focus on equity are relevant to benefit sharing from the use of MGRs and DSI (see Harden-Davies et al., 2022). In many cases, the environmental impacts of sample collection are relatively minimal, but in some cases, collection activities may invoke the Environmental Impact Assessment (EIA) processes under Part IV depending on the species, area and activity in question. For example, most species associated with marine biotechnology have not been assessed for their vulnerability status by the International Union for the Conservation of Nature (IUCN) (Blasiak et al., 2023) and protected species or those from sensitive areas such as hydrothermal vents might attract EIA obligations under the BBNJ Agreement after its entry into force.

A framework for access and benefit sharing (ABS) requiring authorization and sharing the benefits from the use of MGRs within national jurisdiction is set out by the *Convention on Biological Diversity* (CBD)[3] and *Nagoya Protocol on Access to Genetic Resources and the Fair and Equitable Sharing of Benefits Arising from their Utilization to the Convention on Biological Diversity*[4] (Nagoya Protocol). The conservation and sustainable use of genetic resources, including DSI, is a central feature of the *Kunming–Montreal Global Biodiversity Framework* (GBF) adopted by the CoP to the CBD in 2022 (UNEP, 2022a). The same CoP decided to establish a multilateral mechanism for benefit sharing from the use of DSI (UNEP, 2022b). The CBD and Nagoya Protocol recognize States' sovereign rights to their biological resources, and in practice, national legislative, administrative and policy measures will suit their national interests to set out requirements for ABS on a bilateral basis [CBD art 15, Nagoya Protocol arts 5(2), 6(2), 6(7)]. As a result, scope and definitions for ABS regimes have varied widely under national laws (Humphries et al., 2021a). In contrast, the BBNJ Agreement's multilateral approach depends on having agreed scope and definitions, so that Parties can implement consistent legislative, administrative and policy measures for notifications of activities (see Humphries et al., 2025a), transparency (see Langlet and Vadrot, 2023) and benefit sharing (see Broggiato et al., 2025) requirements that shares a common language and interpretation. A common or consistent language between Parties[5] reduces loopholes and conflicts about the subject matter and activities governed under the benefit sharing scheme.

In this complex and multifaceted context, this chapter analyzes the scope of the BBNJ Agreement in three dimensions: jurisdictional scope (Sect. 3.2), subject matter scope (Sect. 3.3) and temporal scope (Sect. 3.4). Jurisdictional scope defines coverage based on location of resources and activities (geographical scope), relationships between Parties, relationships between Parties and the subject matter (sovereignty and sovereign rights) and relationship between the BBNJ Agreement and other IFBs. It answers questions like what is included in ABNJ and whether and how the BBNJ Agreement applies to activities in AWNJ that utilize or affect MGR or DSI of ABNJ. Subject matter scope refers to the components and activities covered by the BBNJ Agreement in general and those that are subject to notification, benefit sharing and monitoring requirements in particular. It addresses questions such as the definition and scope of MGR, DSI and traditional knowledge, regulated activities and the relationship with MSR and marine technology regimes under UNCLOS. Temporal scope addresses the extent to which the BBNJ Agreement applies

[3] *Convention on Biological Diversity*, opened for signature 5 June 1992, 1760 U.N.T.S. 79 (entered into force 29 December 1993).

[4] *Nagoya Protocol on Access to Genetic Resources and the Fair and Equitable Sharing of Benefits Arising from their Utilization* (Nagoya Protocol), opened for signature 29 October 2010, [2012] ATNIF 3 (entered into force 12 October 2014).

[5] A Party means a 'State or regional economic integration organization that has consented to be bound by this Agreement and for which this Agreement is in force' (art 1(11)). A regional economic integration organization is defined under article 1(12) and would include the European Union if it becomes a Party to the treaty.

retroactively, given the current existence of MGR of ABNJ and associated data and knowledge in national repositories and databases and use and commercialization of such resources at the time of negotiation and adoption of the BBNJ Agreement. These dimensions of scope are addressed through general provisions on scope and exceptions (arts 3 and 4), relationship with other agreements, legal rights and privileges (arts 5 and 6), specific provisions on application of Part II (arts 10 and 11), and definition of terms (art 1). The chapter concludes that some aspects of scope need more urgent clarification from BBNJ Agreement bodies than others if stakeholders wish to align their R&D practices and commercial operations with treaty intent (Sect. 3.5).

3.2 Jurisdictional Scope (Articles 3, 5, 6 and 11)

As with other chapters in this edited collection, this chapter uses the shorted term 'BBNJ Agreement'. Other short titles for the BBNJ Agreement include BBNJ Treaty, High Seas Biodiversity Treaty and High Seas Treaty. Mendenhall and Bateh (2024) argue that the use of the latter term biases interpretation by excluding the biodiversity focus of the agreement, misrepresents the scope by ignoring the seabed and elevates the freedom of the seas principle to the detriment of the common heritage of humankind principle. They argue that the use of 'High Seas Treaty' may shape implementation as the bias can affect States' subsequent practice regarding interpretation and application of the BBNJ Agreement as a whole. As an implementing agreement of UNCLOS, which must be interpreted and applied in a manner that does not undermine other relevant legal instruments and frameworks (art 5), negotiating States were careful to clarify the jurisdictional scope of the BBNJ Agreement. This included not only an article on the 'scope of application' (art 3), but also articles clarifying: sovereignty and sovereign rights ('without prejudice' principle under article 6); claims over the jurisdiction and resources (principle of 'non-appropriation' under article 11); relationships with other instruments, frameworks and bodies ('not undermining' principle under article 5) and rights and interests of coastal and other states ['due diligence' principle under article 11(3)]. This section interprets and analyzes these provisions, offering insights into key questions of interpretation and implementation that remain unresolved.

3.2.1 Geographical Scope

Article 3 Scope of Application: This Agreement applies to areas beyond national jurisdiction.
Article 1 Definition: '"Areas beyond national jurisdiction" means the high seas and the Area.'

The BBNJ Agreement applies to ABNJ, meaning the high seas and the Area [art 1(2)]. UNCLOS frames the high seas (Part VII) and the Area (Part XI) as those areas that are not within national jurisdiction, in other words, it is a negative definition. The high seas water column are 'all parts of the sea that are not included in the exclusive economic zone, in the territorial sea or in the internal waters of a State, or in the archipelagic waters of an archipelagic State' (UNCLOS art 86). The seabed and ocean floor and subsoil thereof beyond the limits of national jurisdiction (noting some seabed below the high seas may be within national jurisdiction in the case of the extended continental shelf regime) constitutes 'the Area'. The Area has special governance arrangements for mineral resources governed by the International Seabed Authority (ISA) (UNCLOS Part XI). Given that State jurisdiction may be recognized beyond the exclusive economic zone (EEZ) to the extent of the continental shelf (UNCLOS art 76) and that there are questions about jurisdictional boundaries in light of rising sea levels (Mendenhall et al., 2019; Rothwell & Stephens, 2023), ABNJ's negative definition is fluid, creating some ambiguity about whether MGRs in the jurisdictional boundaries fall within the CBD/Nagoya Protocol or BBNJ Agreement frameworks.

From IGC1 (September 2018), there was disagreement about whether the application of the instrument should be to both the Area and the high seas (UNGA, 2018a, p. 21). Some delegates suggested that scope should cover MGR of the Area only, while others regarded MGR to be sufficiently regulated under UNCLOS (UNGA, 2018a, p. 21). At IGC2 (March–April 2019) views differed about whether MGRs should be governed by a single regime or by different regimes for those of the high seas and those of the Area (UNGA, 2019a, p. 5), but by IGC3 (August 2019) there was convergence on defining geographical scope as ABNJ, including both areas (UNGA, 2019b, p. 5). There was ongoing discussion, however, about whether to refer to MGRs 'of', 'accessed in', 'originating from' or 'collected in' those areas or a combination of options (UNGA, 2019b, p. 5, see 17 May 2019 draft art 8 UNGA, 2019c). The significance of the location terms 'of', 'in' and 'from' related to whether the scope of the instrument and its benefit sharing mechanism was restricted to activities only occurring within ABNJ (accessed in) or also activities within national jurisdiction after the MGR were collected, such as utilization (see Sect. 3.5 below). The final text settled on the term 'of' for the phrase MGR 'of' ABNJ when referring to genetic material and information (DSI) within geographical scope and the term 'in' for traditional knowledge associated with MGRs 'in' ABNJ, when referring to access to traditional knowledge (art 13). This clarification, however, does not sufficiently settle questions about subject matter that fall within the geographical scope as the following analysis suggests.

BBNJ Agreement bodies will need to further clarify whether all references to MGR mean MGR actually collected from ABNJ or also those simply originating from ABNJ (where ABNJ is the known distribution). Most of Part II provisions use the shorthand phrase 'marine genetic resources and digital sequence information on marine genetic resources of areas beyond national jurisdiction' instead of: (a) MGR of ABNJ; and (b) DSI on MGR of ABNJ. While the placement of ABNJ in the phrase means that DSI is tethered to MGRs originating from ABNJ, arguably the placement of MGRs alone at the start of the phrase does not similarly tether MGRs to ABNJ. The definition of MGR does not mention (or connect MGR to) ABNJ, as MGR and ABNJ are defined separately (see Sect. 3.1 below). The phrase is likely to be interpreted as both MGRs of ABNJ and DSI on MGRs of ABNJ, but the shorthand term might be an interpretive loophole that a State may attempt to exploit if they seek to regulate benefits under the BBNJ Agreement from MGRs that are utilized in AWNJ but originally from both in ABNJ and AWNJ (art 1(8)), although any attempt to apply obligations to MGR of AWNJ is likely to be met with strong protest, noting the limited geographical scope in article 3 (ABNJ). The use of the connector term 'of' ABNJ does not assist with clarification because it does not mean that provisions only apply to MGRs directly 'in' or 'from' ABNJ. While this may be implied for the notification activities that use the phrase 'MGRs of ABNJ' for 'collection in situ' and 'utilization', it is less clear for the benefit sharing provisions, which use the shorthand phrase 'marine genetic resources and digital sequence information on marine genetic resources of areas beyond national jurisdiction' (see Humphries et al., 2025a).

Further, the phrase MGRs *of* ABNJ, rather than MGRs *from* ABNJ may cause complexity for species that migrate between jurisdictional areas. This is because 'of' suggests scope is determined by their known distribution rather than where the resources in question were collected. Article 12 seems to suggest a linear approach to R&D where collection activities will be within ABNJ and regulated 'utilization' activities will follow for the same resource within national jurisdiction (Humphries et al., 2025a). In practice, however, a research project may relate to MGRs collected from AWNJ and from ABNJ from the same cruise that was the subject of the pre-collection notification (Rabone et al., 2025), posing a challenge for identifying whether the particular MGR falls within scope of the BBNJ regime. Further, a Party that does not opt out of the retroactive effect of Part II (see Sect. 3.4 below) will have 'utilization' trigger without geographical evidence of a 'collection' notification and the associated BBNJ Standardized Batch Identifier (BBNJ Identifier), making it hard to demonstrate that the MGR was actually collected from ABNJ. The BBNJ Identifier is a stable and unique identifier that is issued by the Clearing-House Mechanism (CHM) and linked to the collection event in ABNJ and other databases that include the identifier (e.g., in the metadata of MGR from the collection that was entered into a repository) (Lawson et al., 2025). The idea is that by attaching an identifier to a batch (or bulk) collection from ABNJ rather than every sample, which was a proposal in earlier draft texts, then information about the location of MGR and DSI from the collection may be traced back to the time and location of the original bulk collection (Humphries et al., 2020; Oldham & Thambisetty, 2023). As argued in Sect. 3.5 below, BBNJ treaty bodies will need to clarify how geographical scope concerning MGRs 'of' ABNJ relates to the 'utilization' trigger. In other words, whether it is based on the known distribution of the MGR or requires evidence that the actual MGR incorporated in a genetic resource product or process was actually sourced from ABNJ. This will help to give Parties certainty about when their obligations are triggered and the subject matter to which their national laws will apply, as distinct to their national laws concerning MGRs from AWNJ.

The debate during IGCs about whether and how to include Intellectual Property provisions (see Brown, 2025), highlights the importance of identifying the origin or source of the MGR in R&D. In 2024, members of the World Intellectual Property Organization (WIPO) agreed on the *WIPO Treaty on Intellectual Property, Genetic Resources and Associated Traditional Knowledge* (WIPO Treaty) (WIPO, 2024a, b). The objectives under draft article 1 are to 'enhance the efficacy, transparency and quality of the patent system with regard to genetic resources and traditional knowledge associated with genetic resources' and to 'prevent patents from being granted erroneously for inventions that are not novel or inventive...' (WIPO, 2024a, b, art 1). WIPO Treaty article 3 requires that, 'where the claimed invention in a patent application is based on genetic resources, each Contracting Party shall require applicants to disclose: (a) the country of origin of the genetic resources, or, (b) in cases where ...[this] information ... is not known to the applicant, or where ...[there is no country of origin], the source of the genetic resources.' The Chair's explanatory notes on an earlier draft provide as an example for category (b) genetic resources 'in areas beyond national jurisdiction such as the high seas' (WIPO, 2023a, p. 10), signaling an intention to encompass MGR of ABNJ within scope of the WIPO framework. WIPO Treaty article 3 contains a similar disclosure provision for associated traditional knowledge in a patent application. The WIPO Treaty is likely to significantly impact the operation of the BBNJ Agreement traditional knowledge obligation under article 13 (see Pena-Neira and Coelho, 2025).

The WIPO Treaty, if adopted by Parties, is likely to significantly contribute to information sharing for the 'utilization' notification system and any future modalities for benefit sharing connected with utilization. In practice, it is likely that each Party to the WIPO Treaty will determine what the disclosure obligation means in the context of their national laws and whether it extends to patented products and processes that are based on MGR of ABNJ and DSI on MGR of ABNJ. There are already around 30 disclosure regimes in place under national laws, which vary significantly in terms of scope and relationship with ABS regimes and other instruments and the extent to which the invention must be materially/directly based on the genetic resource or traditional knowledge (WIPO, 2023a). As the disclosure requirement is likely to apply to MGR of ABNJ, BBNJ treaty bodies will need to closely monitor the scope of the WIPO Treaty and what it means for the BBNJ Agreement obligations (including the monitoring/transparency measures), processes and institutions, benefit sharing, interoperability between patent databases and the CHM.

The notification system and elements of the benefit sharing system is dependent upon ascertaining the origin of MGRs of ABNJ. The challenge is how to create triggers with legal definitions about scientific processes—how do you know that the claimed invention are based on the specific properties of the resource? Earlier drafts of the WIPO text proposed a distinction between those that are 'materially/directly' based on genetic resources and those that are not (WIPO, 2023b, draft art 3). A similar option is distinguishing between whether a genetic trait, etc., from a genetic resource has a functional effect for the genetic resource invention (e.g., patented product) or when the genetic material is merely present without a functional role in the final product (Humphries, 2015). The challenge for such an approach would be how to clearly identify the genes[6] involved, which is difficult for polygenic traits for selective breeding but possibly more straight forward for transgenic interventions (Humphries et al., 2024). There are problems with legally defining material/direct or functional/passive with enough precision to explain complex scientific processes. BBNJ treaty bodies may learn lessons from the WIPO forum about how to manage these thorny issues relating to how to ascertain origin and the connection between the MGR or associated traditional knowledge and the final product or output arising from its 'utilization'. This may take time, however, and BBNJ treaty bodies may need to forge ahead with procedures for ascertaining the extent to which genetic products/processes incorporate MGR before BBNJ Agreement obligations are triggered.

Finally, the BBNJ Agreement's scope of application does not clarify the extent to which its obligations are applied to the airspace above and the marine biodiversity within the airspace, including marine seabirds that traverse between the water column and the airspace above (IUCN, 2022). The UNCLOS freedom of the high seas reflected in the BBNJ Agreement includes the freedom of overflight [UNCLOS art 87(1)(b)]. This freedom applies to commercial, government or military aircraft and only seeks to identify the right to use airspace above the high seas—it does not regulate how that airspace is used (other than if piracy is involved) (Rothwell & Stephens, 2023, p. 166). The definition of MGR is not restricted to genetic resources from species in the water column and could be read broadly to include genetic material of seabirds. BBNJ treaty bodies would need to clarify whether Part II of the BBNJ Agreement or the CBD regime would apply to the 'collection' and 'utilization' of seabird MGR of ABNJ. While areas in outer space are covered by a separate outer space treaty regime (see Berry, 2023), the freedom of overflight does not distinguish between types of aircraft and could apply to spacecraft when flying over ABNJ within regulated airspace (Rothwell & Stephens, 2023). It is likely that 'marine' in the MGR definition refers to resources being from (or 'of') marine environments, rather than simply non-marine organisms that happen to be in ABNJ. This means that it is unlikely that any organisms on or in spacecraft could fall within scope of Part II of the BBNJ Agreement.[7] Similarly, it is unlikely to capture organisms that happen to be on board cruise vessels on the high seas such as organisms that did not originate from the marine environment (Humphries et al., 2021b).

[6] Or even conceptualize what a gene means in science and law—see Lawson (2022).

[7] Other parts of the BBNJ Agreement such as EIA and ABMT may apply for example, if space junk affects marine biodiversity of ABNJ.

3.2.2 Sovereignty and Rights and Interests of States

Article 5 Relationship between this Agreement and the Convention and relevant legal instruments and frameworks and relevant global regional, subregional and sectoral bodies: (1) This Agreement shall be interpreted and applied in the context of and in a manner consistent with the Convention. Nothing in this Agreement shall prejudice the rights, jurisdiction and duties of States under the Convention, including in respect of the exclusive economic zone and the continental shelf within and beyond 200 nautical miles.
Article 6 Without prejudice: This Agreement, including any decision or recommendation of the Conference of the Parties or any of its subsidiary bodies, and any acts, measures or activities undertaken on the basis thereof, shall be without prejudice to, and shall not be relied upon as a basis for asserting or denying any claims to, sovereignty, sovereign rights or jurisdiction, including in respect of any disputes relating thereto.
Article 11 Activities with respect to marine genetic resources of areas beyond national jurisdiction: …(3) Collection in situ of marine genetic resources of areas beyond national jurisdiction shall be carried out with due regard for the rights and legitimate interests of coastal States in areas within their national jurisdiction and with due regard for the interests of other States in areas beyond national jurisdiction, in accordance with the Convention. To this end, Parties shall endeavor to cooperate, as appropriate, including through specific modalities for the operation of the Clearing-House Mechanism determined under article 51, with a view to implementing this Agreement.

(4) No State shall claim or exercise sovereignty or sovereign rights over marine genetic resources of areas beyond national jurisdiction. No such claim or exercise of sovereignty or sovereign rights shall be recognized.

(5) Collection in situ of marine genetic resources of areas beyond national jurisdiction shall not constitute the legal basis for any claim to any part of the marine environment or its resources.

From the early stages of negotiations, there were discussions about whether to ensure a common approach for MGRs within and beyond national jurisdiction, 'taking into account an ecosystem approach and without prejudice to the rights and jurisdiction of coastal states' (UNGA, 2018a, p. 21). There were suggestions to 'focus on the place of access of the resources rather than the natural habitat of the resources, meaning that if access took place [in ABNJ], the instrument would apply, while other instruments, such as the [CBD and Nagoya Protocol] would apply if access took place in areas under national jurisdiction' (UNGA, 2018a, p. 21). As outlined in Sect. 2.1 above, the final text wording is less clear about how the geographical scope relates to the 'utilization' activity that is relevant for notification and benefit sharing. As this section outlines, the provisions concerning sovereignty, non-appropriation and State interests can further assist with clarifying the jurisdictional scope. The next section (Sect. 2.3) analyzing the 'not undermining' principle can further assist with understanding jurisdiction based on the relationship between the BBNJ Agreement and other relevant instruments, frameworks and bodies (IFBs).

The CBD (art 3, 15) and Nagoya Protocol (art 6) recognize a State's sovereign rights to the genetic resources within its national jurisdiction and that authority to determine access to them rests with the national governments subject to national legislation. In practice, while most ABS laws regulate native biological resources (for which they are the country of origin), some regulate exotic materials within their jurisdiction and some States regulate biological resources in publicly and (sometimes) privately owned ex situ facilities, even if they are not the country of origin of the resources (Humphries et al., 2021a). While the CBD framework does not apply to biological resources located in ABNJ, within scope are processes and activities 'regardless of where their effects occur, carried out under [a Party's] jurisdiction or control, within the area of its national jurisdiction or beyond the limits of national jurisdiction' (CBD art 4). This has the potential to cause confusion about the legal status of MGR of ABNJ that are accessed from ex situ facilities and utilized within national jurisdiction. The final BBNJ Agreement extends its reach to collection activities in ABNJ as well as 'utilization' and 'access' activities within national jurisdiction, requiring a series of articles (including 5, 6 and 11) to maintain a delicate balance between facilitating conservation, sustainable use and equity regarding marine biodiversity in ABNJ and respecting the sovereignty, sovereign rights and interests of States.

The provisions on sovereignty and sovereign rights aim to avoid claims (and not recognize them if they are claimed) whereas the principle of non-appropriation emphasizes that no single State has exclusive rights over the resources. There was ongoing debate at the IGCs about whether the BBNJ Agreement should state 'without prejudice' and 'non-appropriation' principles (UNGA, 2019b, p. 6). Rationales for these provisions include the need: (a) to maintain the *status quo* under UNCLOS with respect to the high seas and the Area regimes (UNCLOS Part VII and XI); and (b) to promote the fair and equitable access to MGR and the benefits that arise from their use by preventing powerful countries from monopolizing their exploitation and depriving other countries of their fair share of the benefits (de la Concepción, 2024).

Sovereignty and Sovereign Rights—'Without Prejudice'

There was early support for the inclusion of a 'without prejudice' clause in the instrument, drawing from article 142 UNCLOS and article 4 of the *Agreement for the Implementation of the Provisions of the United Nations Convention on the Law of the Sea of 10 December 1982 relating to the Conservation and Management of Straddling Fish Stocks and Highly Migratory Fish Stocks* (UNFSA)[8] (UNGA, 2018a). In this context, 'without prejudice' in articles 5(1) and 6 of the BBNJ agreement means that nothing in the implementing agreement shall prejudice the rights, jurisdiction and duties of States under UNCLOS and that the agreement must be interpreted and applied in the context of, and in a manner consistent with, UNCLOS (see e.g., UNFSA art 4).

Article 6, which applies to the whole BBNJ Agreement, emphasizes that the BBNJ Agreement, including decisions of the CoP and its subsidiary bodies shall be without prejudice to and not be relied upon as a basis for 'asserting or denying any claims to, sovereignty, sovereign rights or jurisdiction, including in respect of any disputes relating thereto.' Under UNCLOS, States exercise sovereignty over their internal waters and over their territorial waters (subject to the right of innocent passage) and have a bundle of distinctive coastal State jurisdiction and sovereign rights in the contiguous zone, EEZ and continental shelf (see Rothwell and Stephens, 2023). No State may validly subject any part of the high seas to its sovereignty and claim or exercise sovereignty or sovereign rights over any of the Area or its resources [UNCLOS arts 89 and 137(1)]. Article 6 seems to be motivated by concerns of States that the BBNJ Agreement should not be used as a platform to further any sovereignty claims for AWNJ. This is supported by article 60(9) and (10) of the dispute settlement clause, which provides that nothing in the treaty shall be interpreted as conferring jurisdiction upon a court or tribunal over any dispute that concerns 'the concurrent consideration of the legal status of an area as within national jurisdiction, nor over any dispute

[8] Article 142 of UNCLOS and article 4 of the Agreement for the Implementation of the Provisions of the United Nations Convention on the Law of the Sea of 10 December 1982 relating to the Conservation and Management of Straddling Fish Stocks and Highly Migratory Fish Stocks (United Nations Fish Stocks Agreement). See President's Report 20 September 2018, UNGA (2018a), p. 22.

concerning sovereignty or other rights over continental or insular land territory or a claim thereto of a Party.'

Regarding the BBNJ Agreement not being relied upon as the basis of *denying* claims, the effects of climate change, melting global icecaps and resulting sea level rises are affecting the amount of land above sea level for the purpose of calculating the baseline of boundaries (IPCC, 2021). Baselines for determining the extent of a State's EEZ are normally linked to the low-water line, some of which are becoming increasingly submerged over time (Rothwell and Stephens, 2023). UNCLOS does not sufficiently deal with these inevitably significant sea level rises when it comes to determining base lines, creating uncertainty and unfairness for affected coastal States if their sovereign rights are affected by any loss of maritime boundaries (Rothwell and Stephens, 2023). Implications for the treaty is that ABNJ would increase if sea level rise affects the delineation of EEZ boundaries. Pacific Island Forum leaders issued the 2021 *Declaration on Preserving Maritime Zones in the Face of Climate Change-Related Sea-Level Rise* (PIF Declaration) to present a package of State practice of preserving their rights and entitlements in their maritime zones despite sea level rise (Anggadi, 2022). It is unclear whether the language concerning 'denying' claims under article 6 was an attempt to preserve the status quo while State practice evolves on the maritime boundary sea level rise issue but treaty bodies would need to further clarify the effect of jurisdictional scope, which is tied to these issues. What is clear is that the BBNJ Agreement respects the 'sovereignty, territorial integrity and political independence of all States' (preamble).

The relationship between the BBNJ Agreement and the Antarctic Treaty System (ATS)[9] poses important and unique jurisdictional questions concerning sovereignty, sovereign rights and competence for governance of MGR of the Antarctic Treaty Area (ATA). Under the Antarctic Treaty,[10] all claims for sovereignty over Antarctica's waters, land and resources are on hold (art IV) and instead governance is dependent upon cooperation between Contracting Parties through the Antarctic Treaty Consultative Meeting (ATCM) (art IX(1)). Although the assertion and denial of claims have been contested in the Antarctic Treaty Area (Dodds, 2011), BBNJ Agreement article 6 preserves the status quo with respect to sovereignty and sovereign rights in the context of ABNJ. However, as Sect. 2.3 outlines below, there are important questions of exclusive competence for governance of MGR in the Southern Ocean that need to be addressed by both the ATCM and the BBNJ CoP before determining whether the BBNJ Agreement bears any influence over MGR governance in this unique jurisdictional area.

Rights and Interests of States—'Due Regard'

Respect for the rights and jurisdiction of coastal States over all areas under their national jurisdiction was recognized from the start of the negotiations (UNGA, 2018a, p. 21). The 'without prejudice' principle is important to maintain the status quo of rights and interests, but it will raise interesting implementation challenges that BBNJ treaty bodies will need to address as they further clarify jurisdictional scope. For example, under UNCLOS, coastal States have rights to sedentary species of the continental shelf (UNCLOS art 77), but as Mossop and Schofield (2020) argue, there are no provisions in the BBNJ Agreement that 'directly address the issue of coastal State rights to sedentary species on the continental shelf beyond 200 nautical miles in the context of biotechnology and bioprospecting' (p. 5). Species may be sedentary for only part of their life cycle and a coastal State may have sovereign rights over only some species in the one benthic ecosystem (e.g., those on an extended continental shelf in AWNJ but not those in the High Seas water column above), meaning the ecosystem may be regulated under two regimes (Mossop & Schofield, 2020).

There was ongoing debate at the IGCs about how to address the rights and legitimate interests

[9] https://www.ats.aq/index_e.html.

[10] *Antarctic Treaty,* opened for signature 1 December 1957, 402 UNTS 5778 (entered into force 23 June 1961).

of coastal States with respect to MGRs found in both ABNJ and AWNJ (UNGA, 2019b, p. 6). By IGC3, there seemed to be general convergence that the prior consent of coastal States concerned 'would not be required for activities that may result in the utilization of MGR found in both ABNJ and AWNJ' (UNGA, 2019b, p. 6). The concept of 'due regard' was included in the zero draft text as a solution (UNGA, 2019c, draft art 9(2)), but there were ongoing discussions about 'whether coastal States—whether concerned or adjacent—should be notified and consulted nevertheless' (UNGA, 2019b, p. 6). The components of 'due regard' are not settled, but it includes incorporating the interests, rights and duties of states in decision-making (Mendenhall et al., 2019).

The final text article 11(3) restricts these 'due regard' considerations to only collection in situ of MGR of ABNJ whereas previous drafts related to all activities [e.g., UNGA, 2019c, draft art 9(2)]. However, the text includes not only due regard for 'rights and legitimate interests of coastal states within *their* national jurisdiction', but also 'with due regard for the interests of other states *in* [ABNJ]'. Ambiguity remains around:

- Which coastal State rights and legitimate interests it applies to?
- Which State interests it applies to ('in' ABNJ) and why these are not restricted to 'legitimate' interests as it is for coastal States?
- Whether such interests extend to the information components (DSI) of the collected MGR of ABNJ?
- How it relates to other activities with respect to MGRs of ABNJ ('utilization' and access to MGR and DSI in repositories and databases)?
- Whether there are corresponding obligations for coastal States and other States concerning the rights and interests for which decision-making is to have 'due regard'?

BBNJ treaty bodies may need to consider these and other questions and clearly articulate what 'due regard' means concerning collection activities with respect to MGR of ABNJ.

Principle of 'Non-appropriation'

Application of the non-appropriation principle in relation to MGR of ABNJ was not settled until the last stages of negotiations. It remained in brackets in draft texts until the December 12, 2022, draft text that was presented to the resumed session of IGC5 in 2023, in which the term non-appropriation was removed entirely (UNGA, 2022a, draft art 9), only to re-appear in the final text under a separate provision under Part II. The final article 11(4) provides that:

> No State shall claim or exercise sovereignty or sovereign rights over marine genetic resources of areas beyond national jurisdiction. No such claim or exercise of sovereignty or sovereign rights shall be recognized.

The language of 'State' rather than 'Party' under the non-appropriation provision is significant. Part II obligations are imposed on a Party, which is a State or regional economic integration organization for which the Agreement is in force (art 1(11)(12)). A treaty can only bind a non-Party to the agreement with its consent, unless the rule in the treaty becomes binding to a non-Party as a customary rule of international law (*Vienna Convention on the Law of Treaties* arts 34, 38). Article 11(4) is consistent with article 241 UNCLOS where 'marine scientific research activities shall not constitute the legal basis for any claim to any part of the marine environment or its resources' and article 137(1) where no 'claim or exercise of sovereignty or sovereign rights nor such appropriation shall be recognized' in the area. This affirms the customary rule that the high seas and the Area (and the resources within) cannot be subject to appropriation on the basis of MSR conducted there, and protects the sovereign rights of States in the resources of their continental shelf and EEZ (Rothwell & Stephens, 2023). This indicates that negotiators intended the provision to state international law with respect to Parties and non-Parties to the treaty.

Article 11(5) was carefully crafted to confine the principle of non-appropriation to the activity of collection, meaning that no State can claim ownership over (or appropriate) MGR collected from ABNJ. The rationale for the final wording was not documented in the President's reports, but from a practical perspective, collection activities can be geographically tied to ABNJ, whereas 'utilization' generally occurs in AWNJ (until for example, automation technologies evolve to undertake 'utilization' of DSI directly on the high seas). Arguably it also recognizes a balance between MGR of ABNJ as part of the global commons, while still recognizing the freedom of the high seas, including freedom of MSR, where Parties and their juridical persons may freely 'utilize' MGRs and DSI on MGR of ABNJ, subject to the benefit sharing regime. In an effort to reach compromise between governing principles for Part II, the BBNJ Agreement does not clarify if MGR of ABNJ are the common heritage of humankind or if the freedom of the seas applies.[11] Article 11(6) makes it clear that activities with respect to MGR and DSI of ABNJ 'are in the interests of all States and for the benefit of all humanity, particularly for the benefit of advancing the scientific knowledge of humanity and promoting the conservation and sustainable use of marine biological diversity, taking into particular consideration the interests and needs of developing states.' The final wording suggests an intention to distance this from UNCLOS article 140, which earlier drafts of this provision more closely resembled.

Confining article 11(5) to the activity of collection arguably preserves the *status quo* concerning international intellectual property regimes, which was a controversial topic during the IGC negotiations (see Brown, 2025). Prior to the BBNJ Agreement, States had argued that UNCLOS provisions on MSR and technology transfer were ineffective in achieving equity because in promoting cooperation for the development and transfer of marine technology, they must have due regard for all legitimate interests (UNCLOS art 267, Humphries, 2017). Legitimate interests of holders and suppliers of marine technology included patents over biological resource inventions (Jørem and Tvedt, 2014), perpetuating the gulf between technology rich and technology poor States. To address the divide, States proposed intellectual property provisions including restrictions on patent applications and requiring the disclosure of origin in patent applications (UNGA, 2019c). For many States, omitting reference to intellectual property in the treaty text was not negotiable and after compromises on other provisions at the last IGC, the substantive provision on intellectual property was removed (Brown, 2025). In other words, legitimate interests of those utilizing MGR of ABNJ for the purpose of creating patented products or processes are preserved and are unaffected by the principle of non-appropriation.

Data Sovereignty and Self-Determination

The terminology of sovereignty[12] is arguably broad enough in the text to encompass not only national sovereignty but also data sovereignty, the meaning of which is not settled in international law, nor the literature (Couture & Toupin, 2019; Hummel et al., 2021). Some interpretations present data sovereignty as a crucial dimension of national sovereignty (Irion, 2012) and others present it as a crucial element of self-determination of Indigenous peoples (Kukutai & Taylor, 2016). This is complicated by data stored in the cloud which reveals little information about geographical location and does not have a legal tether to national boundaries (Hummel et al., 2021). Identifying origin of sequence information is complicated for benefit sharing regimes and could include the country of sequencing, country of upload, country of the database server or the country of

[11] Instead, the text reached compromise by applying both principles of the common heritage of humankind and the freedom of scientific marine research together with other freedoms of the high seas to the whole BBNJ Agreement [art 7(b) and (c)]. See Muraki Gottlieb et al. (2025b).

[12] An analysis of what 'sovereignty' means is beyond the scope of this chapter. See Berry (2023).

download (Humphries et al., 2021b). The BBNJ Agreement attempts to identify the legal tether through the use of the BBNJ Identifier and metadata requirements, but this does not resolve the question of data sovereignty when the sequence is from an organism that migrates between jurisdictional areas. When developing the modalities of the DSI on MGRs of ABNJ, the BBNJ treaty bodies need to be explicit about legal authority or control over data about DSI on MGRs of ABNJ and its relationship with the non-recognition of sovereignty and sovereign rights with respect to the marine environment of ABNJ and its resources, including MGR.

Questions of data sovereignty in Part II of the BBNJ Agreement include who has the right to control the use of data once MGR are dematerialized and DSI is entered into national databases?[13] What are the tensions between the objectives of Indigenous data sovereignty and the open data movement (see Oguamanam, 2020, Carroll et al., 2020 and Lawson et al., 2025)? As concepts of data sovereignty and strategies to manage it evolve, BBNJ Agreement bodies will need to provide clarity about how they apply to the MGR framework so that policy makers are not talking past each other when implementing national laws.

In summary, while the negotiated principles concerning sovereignty/sovereign rights, non-appropriation and interests of States seem settled in the BBNJ Agreement text, there will be ongoing complexities for how they are implemented in practice. Sea level rise and questions of 'ownership' and control over MGRs, DSI and traditional knowledge within different jurisdictional areas or geographical locations will continue to push the boundaries of the legal fictions underlying borders between maritime zones and borders between the tangible and intangible aspects of biological resources. BBNJ treaty bodies are likely to be clarifying the operation of the 'without prejudice', 'non-appropriation' and 'due regard' principles for many years to come during the treaty implementation phase.

[13] See Sect. 3.3.3 .

3.2.3 'Not Undermining' Other Instruments, Frameworks and Bodies

Article 5 Relationship between this Agreement and the Convention and relevant legal instruments and frameworks and relevant global regional, subregional and sectoral bodies:

(1) This Agreement shall be interpreted and applied in the context of and in a manner consistent with the Convention...
(2) This Agreement shall be interpreted and applied in a manner that does not undermine relevant legal instruments and frameworks and relevant global, regional, subregional and sectoral bodies and that promotes coherence and coordination with those instruments, frameworks and bodies.
(3) The legal status of non-parties to the Convention or any other related agreements with regard to those instruments is not affected by this Agreement.

Areas beyond national jurisdiction have a complex web of IFBs that affect the governance of living resources and marine biodiversity (see Langlet & Vadrot, 2023; Humphries et al., 2025b). Likewise, there are a range of IFBs that govern ABS of biological resources, DSI and traditional knowledge accessed or utilized within national jurisdiction (see Kachelriess et al., 2025). Consequences of regime complexity include on the one hand forum shopping, where actors attempt to take advantage of legal inconsistencies and undermine the regimes, but on the other hand, it can lead to an increase in overall resource availability for capacity building and compliance (Langlet & Vadrot, 2023). In this complex setting and to maintain the delicate balance between existing and new international frameworks, the 'not undermining principle' was crucial. Yet, in another case of 'constructive ambiguity', negotiators did not come to a consensus on its specific meaning (Mendenhall & Bateh, 2024).

Article 5 aims to clarify the relationship between the BBNJ Agreement, UNCLOS and other legal IFBs, but there will be ongoing discussions about how this is applied in practice. In IGC2, there 'seemed to be convergence toward interpreting and applying the Convention' and MGR Part II 'as a single instrument' and support for a general section applying to the whole agreement to reflect this (UNGA, 2019a, p. 5). There were, however, ongoing discussions about whether UNCLOS or the BBNJ Agreement would prevail in the event of any inconsistency (UNGA, 2019a, p. 5) and little evidence in the President's reports about the outcome. There is also little evidence about whether the BBNJ Agreement would prevail in the event of an inconsistency between Part II and other ABS and intellectual property regimes that may have overlapping jurisdiction. This includes the recent WIPO Treaty that ostensibly applies to the disclosure of origin or source of MGR and DSI on MGR of ABNJ or Traditional Knowledge on which a claimed invention is based (WIPO, 2024a art 3, see Sect. 2.1 above). It also applies to the GBF DSI multilateral mechanism, which is likely to include DSI on MGR of ABNJ in databases that are part of the mechanism (see Sect. 3.3 below and Kachelriess et al., 2025). Under article 15(5) of the BBNJ Agreement, the ABS Committee may consult and facilitate the exchange of information with these and other relevant IFBs on activities under its mandate, including benefit sharing, the use of DSI and lessons learned.

Scanlon (2018) suggests at least two different ways to understand the 'not undermining' principle with two different effects (p. 406–407). A first interpretation is to not undermine the authority of existing bodies by leaving their mandates untouched, so that subject matter under concurrent jurisdiction continue to operate under existing legal frameworks. A second interpretation arguably disempowers existing bodies by requiring the BBNJ Agreement to not undermine the *effectiveness* of IFBs, which could include improving the implementation or effectiveness of existing instruments. The removal of bracketed text 'does not undermine [the effectiveness of]' from earlier drafts of the 'not undermining' article[14] and the addition of the wording 'and that promotes coherence and coordination with those instruments, frameworks and bodies', suggests support for the former interpretation. Provisions for ABMTs in Part III have stronger language in support of the former interpretation with added wording 'respect the competence of' (e.g., art 22(2)), which is absent from Part II. This may suggest that the general provision under article 5(2) that applies to MGR but omits similar explicit language of 'respecting the competence of' IFBs suggests support for either (first or second) options of interpretation. Langlet and Vadrot (2023) argue that the formulation in the 'not undermine' may be too wide for Part II as indicated by fisheries exclusion (see Sect. 3.6 below) from scope of the BBNJ Agreement to not undermine the mandate of FAO or Regional Fisheries Management Organizations (RFMO), with the risk that many species of fish may be unregulated and unprotected (Langlet & Vadrot, 2023). As a provision that relates to public international law more broadly, it may not be within the power for the CoP to articulate precisely what the principle of 'not undermining' means with respect to Part II. However, the way it is worded in different parts of the BBNJ Agreement raises the question of whether the principle may be applied differently for the four elements of the treaty text, which could benefit from further guidance by the CoP.

The importance of clarifying the meaning of 'not undermining' can be illustrated by the complexities of governance of MGR of the Southern Ocean within the Antarctic Treaty Area (ATA). This hinges on: (1) questions of geographical and subject matter scope between the BBNJ

[14]For example, draft article in the text of June 1, 2022, draft article 4(3) (UNGA 2022b). By the time of resumed IGC5, the term 'the effectiveness of' had been removed from the text—see draft text December 12, 2022, article 4 (UNGA 2022a).

Agreement and ATS; and (2) the extent to which the ATCM has exercised its competency in governing MGR, DSI and Traditional Knowledge in this jurisdictional area. Legal complexities include contested claims of sovereignty and sovereign rights (see Sect. 2.2), which raise the question of whether the scope of the BBNJ Agreement would extend to all waters in the Antarctic Treaty Area or the Southern Ocean or only those areas that are not subject to ATA claims (see Scott, 2022).

Concerning the first question, the Antarctic Treaty applies to the area south of 60^0 South Latitude (art VI), including parts of the Southern Ocean, whereas other treaties in the ATS have jurisdictional competence based on jurisdictional area and subject matter (e.g., marine living resources and Antarctic seals). Scott (2022) outlines the range of IFBs in the ATS jurisdictional area, each with different competencies and geographical or subject matter scope. She argues that the application of the application of Part XI of UNCLOS concerning the regime for mineral resources in the seabed beyond the jurisdiction of States (the Area) to the Antarctic Treaty Area remains ambiguous (Scott, 2022). The Antarctic Treaty clarifies its relationship to the high seas regime: 'nothing in the [Antarctic Treaty] shall prejudice or in any way affect the rights or the exercise of the rights, of any State under international law with regard to the high seas within' the Antarctic Treaty Area (art VI). However, if the BBNJ Agreement did apply to all of the Southern Ocean, it is unclear whether it would apply to MGR both on the seabed and in the water column. Further, Sects. 2.1 and 3.2 highlight ambiguities concerning the scope of subject matter falling with the geographic scope of the BBNJ Agreement, including questions around whether birds and other marine life that spend much of their time outside the marine environment (and traverse jurisdictional areas) would fall within the definition of MGR. For birds, seals and other marine live in Antarctica, these ambiguities are compounded by the geographical uncertainty about the precise geographical areas where the BBNJ Agreement might apply.

Concerning the second question, since at least 2009, the ATCM has affirmed that the Antarctic Treaty System is the appropriate framework for managing bioprospecting (collection, use and arguably benefit sharing) in the Antarctic Treaty Area (ATCM, 2009, 2013). In other words, it continues to assert that the Antarctic Treaty System has the exclusive competence for managing MGRs within its jurisdictional area. Humphries (2018) argued that despite asserting competence for MGR governance, the absence of specific bioprospecting regulation meant that benefits from bioprospecting generally flowed either to a claimant State or to a State where the resources were located ex situ depending on the scope of their national ABS laws, rather than for the benefit of Antarctica. It is unlikely that an ACTM decision to not have a benefit sharing regime would be sufficient for exercising its mandate to be 'not undermined' by Part II because a lacuna in the benefit sharing regime for the high seas would be incompatible with the BBNJ Agreement under UNCLOS. Given the number of IFBs in the ATA and potentially different claims of competence (Haward, 2021), it may be some time before benefit sharing arrangements will be settled in the ATA. However, how the competence may be exercised in practice is likely to be debated for many years to come because of the complex jurisdictional, including sovereign rights, issues.

In summary, jurisdictional scope of Part II is a complex balancing of geographical, legal and political considerations, including the rights and interests of States and relationships with IFBs. This delicate balance is tied to the subject matter, geographical and temporal scope of the framework which, as the following sections argue, raise some unanswered questions. This is likely to pose challenges when the subject matter occurs in AWNJ and ABNJ in different time scales throughout the R&D pipeline. As with all treaties, there will be challenges with how to manage non-participation of States to the BBNJ Agreement and the issue that more than one international framework may apply to the same ABNJ subject matter.

3.3 Subject Matter Scope (Articles 1, 4, 7, 8, 9, 10 and 11)

The G77 and China group was heavily influential in shaping the scope of Part II toward 'fair and equitable' benefit sharing, the qualifier term of which was included in the title of Part II at its behest (de la Concepción, 2024). One of the most highly contested issues from the first to the last IGCs was whether the BBNJ Agreement, and benefit sharing more specifically, 'should apply to only MGR collected in in situ or also to those accessed ex situ and in silico and digital sequence data and/or information, as well as to derivatives' (UNGA, 2019b, p. 6). In later draft texts, the language of in silico was removed in favor of the narrower term DSI but disagreement about the subject matter scope, including the regulated materials and activities, continued until the very end of negotiations (de la Concepción, 2024). It is safe to say that getting the whole BBNJ Agreement over the line depended on the last-minute compromises around the subject matter scope and related processes of Part II for the purpose of fair and equitable benefit sharing.

This section outlines the evolution of the scope of subject matter and its significance for implementation of Part II. It analyzes the lack of definition of 'marine scientific research' and the inclusion of the definition of 'marine technology', and how this might impact the scope of capacity building and benefit sharing. The section interprets and analyzes definition of the physical materials (MGR), intangible aspects (DSI and traditional knowledge) and regulated activities with respect to MGR and DSI on MGR of ABNJ within the context of the substantive obligations. It concludes with observations about exclusions from scope and questions of subject matter scope that remain for the BBNJ treaty bodies to address in coming years.

3.3.1 Marine Scientific Research and Marine Technology

Article 9 Objectives. The objectives of this Part are:

(a) The fair and equitable sharing of benefits arising from activities with respect to marine genetic resources and digital sequence information on marine genetic resources of areas beyond national jurisdiction for the conservation and sustainable use of marine biological diversity of areas beyond national jurisdiction;
(b) The building and development of the capacity of Parties, particularly developing States Parties, in particular the least developed countries, landlocked developing countries, geographically disadvantaged States, small island developing States, coastal African States, archipelagic States and developing middle-income countries, to carry out activities with respect to marine genetic resources and digital sequence information on marine genetic resources of areas beyond national jurisdiction;
(c) The generation of knowledge, scientific understanding and technological innovation, including through the development and conduct of marine scientific research, as fundamental contributions to the implementation of this Agreement;
(d) The development and transfer of marine technology in accordance with this Agreement (emphasis added).

Article 7 General Principles and Approaches:

In order to achieve the objectives of this Agreement, Parties shall be guided by the following principles and approaches:

...(b) The principle of the common heritage of humankind which is set out in the Convention;

(c) The freedom of marine scientific research, together with the other freedoms of the high seas.

Article 8 International Cooperation:

...(3) Parties shall promote international cooperation in marine scientific research and in the development and transfer of marine technology consistent with the Convention in support of the objectives of this Agreement (emphasis added).

'Marine scientific research'—undefined.

Article 1(10)' "Marine technology" includes, inter alia, information and data, provided in a user-friendly format, on marine sciences and related marine operations and services; manuals, guidelines, criteria, standards and reference materials; sampling and methodology equipment; observation facilities and equipment for in situ and laboratory observations, analysis and experimentation; computer and computer software, including models and modelling techniques; related biotechnology; and expertise, knowledge, skills, technical, scientific and legal know-how and analytical methods related to the conservation and sustainable use of marine biological diversity.'

Part II aimed to address implementation gaps for MSR and marine technology regimes under UNCLOS with respect to activities leading to R&D and commercialization from marine biodiversity in ABNJ (Humphries et al., 2025). The United Nations noted that Part XIII (Marine Scientific Research), Part XIV (Development and Transfer of Marine Technology) and other relevant UNCLOS provisions did not adequately deal with the fair and equitable sharing of benefits arising from activities with respect to MGR, including bioprospecting and commercial product development, nor capacity building of technologically disadvantaged states (see UNGA, 2005). This gap is reflected in Part II objectives that focus on fair and equitable benefit sharing, capacity building, the generation of knowledge including through MSR and the development and transfer of marine technology (art 9). While provisions on MSR and marine technology were primarily debated under Parts V and cross-cutting elements of the BBNJ Agreement (e.g., Harden-Davies et al., 2022, 2024), the role of the UNCLOS MSR and marine technology regimes influenced the final Part II framework and will continue to influence its implementation.

3.3.1.1 Marine Scientific Research

Underlying the debate about how to manage MSR under the BBNJ Agreement were long running North–South tensions over exploitation of living and non-living ocean resources. As mainly industrialized States in the 1950s and 1960s expanded MSR effort to advance science, protect the environment, manage fisheries and locate oil and gas resources, other States were concerned that unrestricted MSR on the high seas (a much larger area pre-UNCLOS) would place them at a considerable disadvantage for resource exploitation and hinder their economic development if used to justify unreasonable restrictions on marine pollution (Rothwell & Stephens, 2023). The final UNCLOS framework significantly expanded the scope of coastal State jurisdiction over MSR (Rothwell & Stephens, 2023). It recognized that 'all States, irrespective of their geographical location, and competent international organizations have the right [in conformity with UNCLOS] to conduct marine scientific research in' the Area and the water column beyond the limits of the EEZ (UNCLOS arts 256 and 257) and the right to conduct MSR in other zones subject to the rights and duties of other States under UNCLOS (UNCLOS art

238). The MSR regime includes general principles of conduct, the promotion of international cooperation, State obligations to make available knowledge resulting from MSR and the promotion of knowledge sharing (Part XIII). The freedom of the high seas includes freedom of scientific research, subject to other rights and responsibilities under UNCLOS (UNCLOS art 87). Marine scientific research concerning the Area must be, among other things, for peaceful purposes, for the benefit of humankind as a whole and to promote international cooperation through programs benefiting technologically less developed States and disseminating the results of research when available (UNCLOS art 143).

Similar ideological lines were drawn for the BBNJ Agreement debate over regulating access to, use of, and sharing the benefits from MGRs (including DSI) of ABNJ for R&D and commercialization purposes, which are primarily exploited by technologically advanced countries. By IGC3, there continued to be different views on whether 'marine scientific research should be excluded from the material scope of application of the agreement' (UNGA, 2019b, p. 6). Conversely, there were discussions about whether the activities that fall within scope of Agreement 'should be limited to marine scientific research or also include other activities' (UNGA, 2019b, p. 6). Wrapped up in the discussions were diverging positions about whether the principle of the common heritage of humankind or the freedoms of the high seas, including marine scientific research would prevail for biodiversity beyond national jurisdiction (see de la Concepción, 2024; Muraki Gottleib et al., 2025b). In the end, both principles appear in the text without clarification about how they guide implementation of Part II, and a distinction was made between MSR as an objective of Part II rather than regulated subject matter per se. Instead 'activities with respect to' MGR and DSI is the subject matter,[15] which contributes to R&D and commercialization of biological resources beyond MSR.

Despite this distinction, the interaction between the MSR regime and the Part II 'activities' regime will require further clarification from BBNJ treaty bodies during implementation. Given that jurisdictional scope maintains the status quo for the MSR and technology transfer regimes in UNCLOS,[16] and that rights under the MSR regime apply to States irrespective of whether they are a Party to UNCLOS (Rothwell & Stephens, 2023), R&D on marine biodiversity may be subject to a range of regimes. In an era of scientific collaboration from institutions of multiple States for the one project investigating the properties of MGR, this may cause confusion about which obligations apply to the conduct of their research.

3.3.1.2 Development and Transfer of Marine Technology

The framework for the development and transfer of marine technology is primarily under Part V and Part VI of the BBNJ Agreement, with some overlaps in Part II. The objectives of Part V on 'Capacity Building and the Transfer of Marine Technology' (CBTMT) includes supporting developing States Parties through capacity building and the development and transfer of marine technology in achieving the objectives relating to other Parts, including Part II MGR. The objectives of Part II include 'the development and transfer of marine technology in accordance with this Agreement' (art 9(d)). The preamble also recognized that support for 'developing States Parties through capacity building and the development and transfer of marine technology are essential elements for the attainment of the objectives of the conservation and sustainable of marine biological diversity' of ABNJ (preambular para 6). Harden-Davies et al. (2024) argue that the BBNJ Agreement's long-term vision of the importance of fully realizing technology development and transfer for inclusive, equitable and effective cooperation and participation in [BBNJ Agreement] activities' (art 45), together with the inclusion of equity as a guiding principle and approach (art 5) reflects a two-way nature of partnerships and an advancement of the existing framework under UNCLOS (p. 2).

[15] See Sect. 3.3.5 .

[16] See Sect. 3.2 .

The definition in most of the draft texts of 'marine technology' had similarities with the International Ocean Commission's (IOC) definition (IOC, 2005). Harden-Davies et al. (2022) argued that the IOC's definition of marine technology (focusing on technology to undertake MSR) was too narrow for the BBNJ Agreement by excluding technologies required for monitoring, control and surveillance (p. 908). However, the final definition includes 'information and data, provided in a user-friendly format, on marine sciences and related marine operations and services...related biotechnology...' (art 1(10)). Capacity building is undefined, which commentators see as both a gap and an opportunity to innovate (e.g., Harden-Davies et al., 2022). This is because the BBNJ Agreement includes a mandate for the CoP, informed by the CBTMT Committee to respond to technological developments (Harden-Davies et al., 2022). However, the indicative list of the types of CBTMT includes the development of technical, scientific and research and development programs, including biotechnological research activities (Annex II(g)). While the text does not go so far as stating that marine technology includes genetic resource inventions, it indicates that the processes and know-how for biotechnology activities does fall within scope. The language in Part V generally relates to 'transfer of marine technology' unless it concerns the broader agreement focus (including Part II) on the '*development* and transfer of marine technology' (emphasis added). Subject to clarification from BBNJ treaty bodies, this indicates that Part II deals with capacity building for development of biotechnologies, whereas Part V primarily deals with the transfer of existing technologies.

3.3.2 Marine Genetic Resources and Samples

"'Marine genetic resources" means any material of marine plant, animal, microbial or other origin containing functional units of heredity of actual or potential value.'

'Sample'—undefined.

References to MGR and 'samples' are material subject matter for various substantive obligations under of the notification, monitoring and benefit sharing regimes. There is an abundance of UN documents and literature about interpreting the meaning of genetic resources for the purposes of its regulation under international ABS frameworks (e.g., Lawson, 2022; Muller, 2015; Tvedt and Schei, 2013). Key areas of debate include the extent to which derivatives and intangible aspects of genetic resources such as DSI fall within scope of the term (Lawson et al., 2025). Countries are developing a new international framework for DSI is subject to the new multilateral framework (see Sect. 3.3 below) and under the CBD/Nagoya Protocol frameworks, it is up to each Party to decide the extent to which derivatives fall within scope.[17] However, arguably the slight differences between the BBNJ Agreement and CBD the formulation of the MGR definition and the absence of a definition for 'sample' may open the debate about the material scope of Part II obligations during its implementation.

The President's Aid to Negotiations for IGC2 included MGR definition options distinguished by whether derivatives and DSI are included within scope of the term (UNGA, 2018b, p. 4). By IGC3, the draft text included separate definitions for genetic resources and genetic material, with the latter including bracketed text that 'it does not include material made from material, such as derivatives, or information describing material, such as genetic sequence data' [UNGA, 2019c, draft art 1(8), 1(9)]. At IGC3, there was discussion about consistency between the BBNJ Agreement and other ABS framework definitions and there was 'general convergence that geographical aspects should not be included in the terms "marine genetic resources" and "marine genetic materials"' (UNGA, 2019b, p. 8). Instead, the text defines ABNJ and denotes the geographical MGR that are within scope in the provisions.[18] By IGC4, any reference to derivatives or DSI in the definition of MGRs had

[17] See Sects. 3.3.3 and 3.3.5 .

[18] See Sect. 3.2.1 .

been removed from the separate definitions of marine genetic material and MGR.

The 'marine' element of the MGR definition is not further defined. However, it denotes a subset of genetic resources under a specialized instrument, similar to the Plant Treaty carving out certain plant genetic resources and the WHO forum carving out certain virus genetic resources (Kachelriess et al., 2025). It appears to mean resources that are marine in origin[19] rather than those simply located in ABNJ marine environment, such as a high seas passenger cruise with virus and other biological material on board (Humphries et al., 2021b). While MGR appears to be a subset of genetic resources, the final text definition of MGR uses a mixture of CBD definitions of 'genetic resources' and 'genetic material'. In the CBD these are:

- '"genetic resources" means genetic material of actual or potential value';
- '"genetic material" means any material of plant, animal, microbial or other origin containing functional units of heredity' (CBD art 2).

Other ABS frameworks,[20] including the WIPO Treaty for disclosure of origin of genetic resources (WIPO, 2024a art 3), include separate definitions for resources and materials. The reason for merging these two definitions in the BBNJ context is not clear in the President's reports but arguably will require further clarification by BBNJ treaty bodies as to its meaning and significance.

While ostensibly the final definition appears to have the elements of the CBD definition, the merging of the two definitions may result in diverging interpretations from the latter because it arguably shifts the object of value that is to be regulated. Genetic 'material' generally refers to the physical substances such as cells that contain genetic information. 'Resources' relate to the actual or potential value of the object. However, commentators argue that unlike other natural resources such as mineral resources, timber or fish for human consumption where the physical material is used *as* the resource, in the case of *genetic* resources, the genetic or biochemical information for the purpose of R&D is the resource being valued and used (Deplazes-Zemp, 2018; Lawson, 2022). This is because researchers in, for example, biodiscovery, breeding or biotechnology, search for traits or compounds that produce a desired effect (from the genetic information), and subsequent R&D usually no longer depends on supply of the original physical material (Deplazes-Zemp, 2018). In other words, it is the genetic information that is the object of actual or potential value.

One consequence of merging the CBD's two definitions is that the actual or potential value in the BBNJ Agreement directly relates to the physical materials, rather than the genetic information it contains. Not including DSI within the MGR definition demonstrates an intention of delegates to carve out DSI from the scope of MGR, so that they could more carefully identify which obligations relate to physical materials and which relate to both physical materials and DSI as subject matter. However, this separation of information and material within the definition of MGR itself is a legal fiction that does not necessarily correspond to the scientific reality of the value of genetic resources within the R&D pathway as outlined above. The MGR definition may be interpreted so narrowly that notification and certain benefit sharing outcomes might only be triggered in circumstances when the physical materials such as the cells or tissue are the value of the research, rather than their DNA, traits and

[19] See, however, Sect. 3.2.1 on geographical scope and whether birds fall within scope simply because of their proximity of the marine environment, even if they are also found on land.

[20] For example, the *International Treaty on Plant Genetic Resources for Food and Agriculture*, opened for signature November 3, 2001, 2400 UNTS 303 (entered into force 29 June 2004) (Plant Treaty). The definition under the WHO framework is more specific to H5N1 viruses: World Health Assembly, *Pandemic Influenza Preparedness: Sharing of Influenza Viruses and Access to Vaccines and Other Benefits*, Report by the Open-Ended Working Group of Member States on Pandemic Influenza Preparedness: Sharing of Influenza Viruses and Access to Vaccines and Other Benefits A64/8 Sixty-Fourth Assembly (2011).

other genetic information necessary for producing an effect for a genetic product. The effect of this may also be to undermine information and benefit sharing by limiting the pre-collection notification to only the collection of the physical material as the thing of value, rather than deployment of technologies that directly extract the genetic information from ABNJ.[21]

The BBNJ Agreement definition that confines scope to physical material containing functional units of heredity is unlikely to future-proof the treaty because it may not accommodate the unique life forms of the deep sea and advances in biotechnology. Heredity may be understood as the transmission of particular characteristics (or information) between generations, or the recurrence of phenotypes across generations (Mossio & Pontarotti, 2022). The 2005 report of the Secretary General to the UN Generally Assembly on preparatory issues for what was to become the BBNJ Agreement noted that every cell of every living organism contains 'functional units of heredity', so the term can include individual organisms as well as DNA and other molecules (such as RNA and proteins) extracted from a plant, animal or microbe (UNGA, 2005, para 6). However, not every cell has DNA (e.g., red blood cells), not every transmissible particle has a nucleus (e.g., prions) but these and other components may be the 'resource' of value for biodiscovery R&D purposes. Most of the life forms in extreme environments like deep sea hydrothermal vents have not been identified and may transmit cellular information in a different way (Hiyoshi et al., 2011). Would these living organisms that are of special interest and value for bioprospecting but without 'functional units of heredity' fall within scope? The BBNJ treaty bodies will need to clarify the meaning of 'functional units of heredity' in the context of the MGR definition and the object of value in the MGR definition.

It is interesting that the qualifier in the MGR definition that it does not include derivatives was removed at the same time that the definitions for genetic resources and genetic material were merged.[22] This may suggest that delegates intended to confine the scope of the MGR definition itself to not include for example chemical compounds or synthetic materials. Instead, the BBNJ Agreement took a similar approach to the Nagoya Protocol and included the undefined term 'derivatives' within the definition of 'utilization', so that derivatives as subject matter only relates to the 'utilization' notification, rather than the collection notifications and 'access' provisions outlined in Sect. 3.5 below (see also Rabone et al., 2025). The debate about the meaning of 'genetic material' and the extent to which it encompasses derivatives and synthetic materials lasted decades after the CBD entered into force (see e.g., Tvedt & Schei, 2013). Given the unusual organisms produced by extreme environments in the deep sea and the rapid pace of synthetic biology, there is likely to be ongoing work for BBNJ treaty bodies to clarify which obligations apply to derivatives and, for that matter, synthetic biology producing products based on or inspired by MGR of ABNJ. Understanding how derivatives will be managed under the framework will depend on its relationship to the activity of 'utilization' (Sect. 3.5).

Finally, the use of the term 'sample' instead of MGR and the absence of a definition for 'sample' might create ambiguity for the scope of the subject matter under the notification and benefit sharing obligations. The term 'sample' is referenced in six places of the text as the subject matter of substantive notification and benefit sharing obligations. Under the notification mechanism, Parties must ensure that 'samples of marine genetic resources' in repositories can be identified as originating from ABNJ and

[21] See Sect. 3.3.5 .

[22] In the November 18, 2019 (A/CONF.232/2020/3) text, the definitions are separate under article 1 and 'derivatives' are in bracketed text for article 8 concerning what the provisions of the agreement apply to. By the next draft—June 1, 2022, (A/CONF.232/2022/5), the definitions of materials and resources were merged into one definition and 'derivative' was moved to the definition of biotechnology.

Parties must ensure that the CHM has information about 'where the original sample that is the subject of utilization is held' [arts 12(6) and 12(8)(c)]. 'Access to samples and sample collections' is listed as a form of benefit sharing, which may be subject to reasonable costs associated with maintaining the 'relevant gene bank…in which the sample…is held' [arts 14(2)(a) and (4)(b)]. Only one instance connects the 'sample' to the MGR. The ordinary meaning of sample is 'a small part or quantity intended to show what the whole is like' (Oxford Dictionary). A sample may be interpreted broadly to include the water or sediment sample in which a variety of MGRs and other non-genetic materials are contained. Or it may mean a portion of the individual MGR, whether or not that portion contains functional units of heredity. Samples might denote something tangible or intangible. The inclusion of this undefined and ambiguous term in the substantive obligations may cause confusion for determining subject matter within scope and challenges for implementation because it may be far broader than the narrow interpretation of the physical genetic material with functional units of heredity under the MGR definition.

Ultimately, Parties are obliged to implement Part II obligations under national laws. State practice may offer substance to the terms of MGR and samples in the context of the BBNJ Agreement. However, given that MGR of areas within national jurisdiction vary according to national interest, it is important for BBNJ treaty bodies to give guidance to Parties and users about the MGRs and samples in scope. This includes clarity about:

- whether marine means those located in the marine environment or in the airspace above or those originating from the marine environment in ABNJ;
- whether MGR means those from species with known distribution in ABNJ or only those collected (so that action of collection determines whether it is within scope rather than the geographical scope of ABNJ per se);
- whether living organisms that are of special interest and value for bioprospecting but technically without 'functional units of heredity' fall within scope;
- whether the threshold for derivatives is different from MGR when it comes to the 'utilization' notification trigger; and
- whether a 'sample' means tangible or intangible subject matter and whether it is confined to physical genetic material with functional units of heredity.

Clarifying these questions is important for distinguishing those obligations that only apply to MGR, rather than those applying to MGR as well as DSI, which is outlined in the next section.

3.3.3 Digital Sequence Information

'Digital Sequence Information'—undefined.

The BBNJ Agreement is the first treaty text to specifically include DSI in the context of sharing of benefits associated with genetic resources. Rationale for its inclusion is that increasingly, genetic R&D can be conducted without the need for access to the physical materials and sequencing is a routine and necessary step for most biotechnology applications from which benefits would flow (see Rogers et al., 2021). The G77 and China group argued that to not include DSI as subject matter would be to create a loophole, rendering benefit sharing in effective (de la Concepción, 2024). The term DSI is used in a variety of ABS fora to denote a subset of information associated with genetic resources. These fora include the CoP to the CBD, Food and Agriculture Organization to the United Nations (FAO) Governing body for the *International Treaty on Plant Genetic Resources for Food and Agriculture*[23] (Plant

[23] *International Treaty on Plant Genetic Resources for Food and Agriculture,* opened for signature November 3, 2001, 2400 UNTS 303 (entered into force June 29, 2004).

Treaty) and the World Health Organization's World Health Assembly for the Pandemic Influenza Preparedness (PIP) Framework (Kachelriess et al., 2025; WHA, 2011). Each of these fora have grappled with loopholes in their ABS frameworks caused by the dematerializations of the genetic resource (see Lawson et al., 2025).

The definition and scope of DSI in the BBNJ Agreement will develop concurrently with other international ABS fora. In December 2022, the CoP to the CBD agreed as part of the GBF that DSI on genetic resources needs a distinctive solution for benefit sharing (UNEP, 2022b, para 3). It outlined principles and a process to further develop and operationalize a multilateral mechanism, including a global fund (paras 9 and 16). These principles included being 'efficient, feasible and practical', 'generate more benefits...than costs', 'not hinder research and innovation', 'be consistent with open access to data', 'be mutually supportive of other access and benefit sharing instruments' and 'take into account the rights of indigenous peoples and local communities, including with respect to the traditional knowledge associated with genetic resources that they hold' (para 9). Most of the practicalities of the mechanism are yet to be decided, including how it will manage DSI that falls within scope of other ABS mechanisms, including BBNJ (Scholz et al., 2023, 2024). However, the CoP to the CBD decision had a major influence on the compromises reached at the last IGC to include DSI within scope (de la Concepción, 2024), with most of the practicalities to be determined by the CoP to the BBNJ Agreement, including the definition and scope of DSI.

The CoP to the CBD did not define the scope of DSI under the DSI multilateral mechanism. However, the 2020 report to the Ad Hoc Technical Expert Group on Digital Sequence Information on Genetic Resources on the scope of digital sequence information is often cited as a guide for categorizing four main categories from most narrow to broadest:

- Group 1—Narrow: concerning DNA and RNA;
- Group 2—Intermediate: concerning (DNA and RNA) + proteins;
- Group 3—Intermediate: concerning (DNA, RNA and proteins) + metabolites;
- Group 4—Broad: concerning (DNA, RNA, protein, metabolites) + traditional knowledge; ecological interactions, etc. (UNEP, 2020, para 11).

The report noted that group 4 includes information with the weakest proximity to the underlying genetic resource and extends to behavioral data and downstream subsidiary information. There is no indication in the BBNJ Agreement text which of these or other grouping defines the scope of obligations under Part II. There is also no clarity in UN documents at the time of writing about whether DSI on MGRs means the same thing as DSI on genetic resources under the GBF proposed DSI multilateral mechanism. Understanding the scope and nature of DSI for the purposes of Part II obligations is one of the key uncertainties for implementation of the BBNJ Agreement, which will have a significant impact on how the notification, transparency and benefit sharing provisions will be implemented in practice (Broggiato et al., 2025; Humphries et al., 2025a; Langlet and Vadrot, 2023).

Related to the scope of DSI is understanding how information and knowledge associated with genetic resources that is not categorized as DSI will be managed. Understanding marine biodiversity requires rigorous metrics for the abundance, distribution and interactions between organisms, species and ecosystems and the linking of biodiversity, environmental, social and economic data (Muller-Karger et al., 2023). A network of environmental and genetic sequence databases already exists, but full utilization of these resources is limited by interoperability and access issues (Blasiak et al., 2023). However, some of this biodiversity information may be provided to the CHM as an outcome of utilization or as part of the other BBNJ framework elements of EIA, ABMTs or CBTMT.

A key gray area for the CoP to consider is how to balance open access to and sharing of DSI on MGR of ABNJ with the protection of traditional knowledge associated with those MGR and the protection of Indigenous rights and data sovereignty (see Sect. 2.2 above) that is increasingly acknowledged in a range of international fora, including the WIPO and human rights organizations (see Sect. 3.4 below). BBNJ treaty bodies will need to explain how DSI fits within scope of the article 13 traditional knowledge obligation, and if not, explain why not. As the following section argues, there is considerable uncertainty around the scope of the traditional knowledge obligation.

3.3.4 Traditional Knowledge

Article 13 Traditional Knowledge of Indigenous Peoples and Local Communities Associated with Marine Genetic Resources in Areas Beyond National Jurisdiction

Parties shall take legislative, administrative or policy measures, where relevant and as appropriate, with the aim of ensuring that traditional knowledge associated with marine genetic resources in areas beyond national jurisdiction that is held by Indigenous Peoples and local communities shall only be accessed with the free, prior and informed consent or approval and involvement of these Indigenous Peoples and local communities. Access to such traditional knowledge may be facilitated by the Clearing-House Mechanism. Access to and use of such traditional knowledge shall be on mutually agreed terms.

The BBNJ Agreement has a stand-alone article 13 on access to traditional knowledge, which follows the Nagoya Protocol system of prior informed consent and mutually agreed terms for access, but no overt provisions on benefit sharing (see Pena-Neira and Coelho, 2025). There was considerable debate about the scope and application of this provision during the IGCs that sheds light on the intent of the final provision. From IGC1, there were suggestions that the prior informed consent (PIC) 'of indigenous and local communities whose traditional knowledge was used to unlock the value of marine genetic resources should also be sought,' but no attempt to define the term 'traditional knowledge' in the text (UNGA, 2018a, p. 22). By IGC3, there was ongoing debate about whether a provision should relate to traditional knowledge generally or be confined to knowledge 'that is useful for unlocking the value'[24] of MGR. At IGC3, there was still debate about whether the scope of the traditional knowledge provision should extend to both access and benefit sharing (see UNEP, 2019b). The draft text discussed at IGC4 had a stand-alone provision on access (not benefit sharing) that introduced language 'associated with' MGRs, but with ongoing debate about whether it related to MGRs 'collected' or 'accessed' in ABNJ (draft art 10bis UNEP, 2019d). Between IGC 3 and IGC4, the language in the provision departed from the rest of Part II concerning MGR 'of' ABNJ by instead using MGR 'in' ABNJ, with no explanation in the President's report for the change in meaning. The key terms in article 13 remain undefined—'traditional knowledge associated with marine genetic resources in areas beyond national jurisdiction', 'Indigenous Peoples and local communities', 'free, prior and informed consent or approval and involvement' of these IPLCs and 'mutually agreed terms' and 'access to and use of'.

BBNJ treaty bodies will need to clarify the subject matter scope of the traditional knowledge obligation. Like the CBD and Nagoya Protocol, 'traditional knowledge' is undefined in the BBNJ Agreement and is likely to be determined under national laws or by the Indigenous Peoples and local community that holds it. The CBD's use of the term 'knowledge, innovations

[24] See bracketed text in draft text May 17, 2019, draft article 10(6) UNEP (2019c).

and practices' relates to biodiversity at the ecosystem, species and genetic levels (CBD art 8(j) and annex I), whereas the Nagoya Protocol more narrowly confines its use to 'traditional knowledge *associated with* genetic resources (emphasis added) (Nagoya Protocol art 7)'. The final BBNJ Agreement treaty text has been similarly confined to 'traditional knowledge... associated with' MGRs in ABNJ, but also adds a new spatial component (i.e., the connection with ABNJ). Negotiating countries settled on 'associated with' rather than 'useful for unlocking the value' of MGRs as in earlier drafts (A/CONF.232/2019/6 draft art 10(6)), arguably removing the need for proving a causal link between the use of the knowledge and the value-add in the product or process that is based on MGR in ABNJ.

The breadth of the requisite 'association' between traditional knowledge and MGRs is unclear in the text. Mulalap et al. (2020) argue that there are three main types of traditional knowledge of particular relevance to the BBNJ Agreement. These are: 'traditional knowledge based on the connectivity of species and marine processes (active and passive) between ABNJ and coastal waters; traditional knowledge emerging from environmental management best practices in coastal waters that can be models for similar measures in ABNJ; and traditional knowledge derived from traditional instrument-free navigation between coastal communities and across ABNJ' (Mulalap et al., 2020). The extent to which these categories fall within scope of article 13 is unclear, but the removal of the requirement for 'unlocking the value of MGRs' in earlier drafts suggest it is broader than only traditional knowledge about the genetic attributes of the MGR, but also includes traditional knowledge about *activities and observations* associated with MGR, or knowledge systems may even treat the MGRs themselves as traditional knowledge (Lawson et al., 2024b). Significantly, as DSI is not included in the MGR definition, the absence of DSI as subject matter in the traditional knowledge obligation signals the intent of drafters to confine the scope of the obligation to the physical materials or collection activities (Pena-Neira and Coelho, 2025). However, this may be contrary to other international agreements such as the *United Nations Declaration on the Rights of Indigenous Peoples* (UNGA, 2007), many national laws and IPLC protocols that treat Indigenous knowledge as a system without an artificial demarcation between the physical and intangible components of biological forms (Lawson et al., 2024b). This may create a significant loophole for misappropriation of knowledge if people use traditional knowledge as a research lead for further development of genetic properties using genomic data techniques instead of the physical materials.

BBNJ treaty bodies will need to clarify the geographical and temporal scope of the traditional knowledge obligation. Early drafts included the terms MGRs 'collected' or 'accessed' in ABNJ. While the rationale for using the term 'in' instead of 'of' MGR is not documented in the President's report, it suggests an intent to limit the scope of the obligation to knowledge associated with MGRs actually collected from ABNJ, rather than MGRs whose known distribution is in ABNJ. The only other instance of the preposition 'in' is for the activity 'collection in situ' of MGR in ABNJ (art 12). This suggests a restrictive geographical and temporal scope requiring a demonstrated link between the traditional knowledge and identifiable MGR located in ABNJ at the time of the collection activity before the obligation is triggered. Under this restrictive interpretation, the obligation would apply in circumstances where traditional knowledge is used to find or target MGR located within ABNJ for investigating certain characteristics of the MGR or for understanding MGRs, species and ecosystems located within ABNJ. It might not apply in circumstances where previously collected MGR of ABNJ are subject to 'utilization' (see Sect. 3.4 on temporal scope) because one reading of the text is that it is confined to the collection activity. The use of the term 'use of' traditional knowledge instead of the defined trigger 'utilization of' is a further indication that drafters intended to confine scope to the collection activity. The obligation provides that 'access to and use of such traditional knowledge shall be on mutually agreed terms', indicating that 'use' is qualified by 'access', so that you need 'access' before 'use'. This narrow interpretation, however, may lead to loopholes, inequity and misappropriation of traditional knowledge associated with MGR because the 'utilization' activity may be outside scope of national ABS laws under both the BBNJ Agreement and Nagoya Protocol frameworks. A broader interpretation is that the obligation is triggered by access to the knowledge and/or use of the knowledge associated with MGR whose known distribution is ABNJ, and not restricted by the collection activity, nor the temporal considerations for retrospectivity concerning 'utilization'. It is unclear how both the narrow and broad interpretations might work in practice, but Parties would benefit from guidance by the CoP in their interpretation (see Rabone et al., 2025).

Other key terms in article 13 are constructively ambiguous to ensure Parties can implement their obligations in accordance with their own circumstances. The definition of IPLCs and 'free, prior and informed consent' (FPIC) are undefined and may be determined under national law, and emerging international practices such as FPIC under UNDRIP. Recognizing that procedures for identifying the correct traditional knowledge holders and procedures for FPIC may vary considerably under national laws (Aime & Robinson, 2023), early drafts of the text suggested the CHM 'may act as an intermediary to facilitate access to such traditional knowledge' (article 10bis UNGA, 2019d). This was diluted in the final text as more of a passive role for the CHM open access platform to simply provide links to databases 'including those pertaining to relevant traditional knowledge of' IPLCs [art 51(3)(c)], rather than an active or intermediary role. The BBNJ treaty bodies will need to clarify how the BBNJ framework deals with traditional knowledge already in the public domain or where the traditional knowledge holder cannot be found. Like the Nagoya Protocol, 'mutually agreed terms' (MAT) is not defined. MAT generally means agreed terms and conditions as the basis of the consent for access to the knowledge, which may include benefit sharing conditions, but benefit sharing is not required under article 13 traditional knowledge obligation, nor the benefit sharing mechanism under article 14.

In practice, it will be up to Parties to interpret the scope and application of traditional knowledge under article 13 according to their national laws and circumstances. However, BBNJ treaty bodies may consider providing guidance, in partnership with IPLCs, on the intended geographical, temporal and subject matter scope of access to and use of traditional knowledge associated with MGR in ABNJ for a more consistent approach during implementation. The 'constructive ambiguity' poses a major challenge for Parties to take legislative, administrative and policy measures with respect to traditional knowledge associated with MGR in ABNJ and to close loopholes for compliance in non-Party States, given the absence of user compliance measures in the BBNJ Agreement that would complement the Nagoya Protocol approach to PIC and MAT.

3.3.5 Activities with Respect to MGR: Collection, Utilization and Access

'"Collection in situ" in relation to marine genetic resources, means the collection or sampling of marine genetic resources in areas beyond national jurisdiction'

'"Utilization of marine genetic resources" means to conduct research and development on the genetic and/or biochemical composition of marine genetic resources, including through the application of biotechnology, as defined in paragraph 3...'

'"Biotechnology" means any technological application that uses biological systems, living organisms, or derivatives thereof, to make or modify products or processes for specific use.'

"Access"—undefined.

"Derivatives"—undefined.

"Sampling"—undefined.

Part II of the BBNJ Agreement regulates 'activities with respect to' MGR and DSI on MGR of ABNJ (arts 10 and 11). Article 10 is analyzed in Sect. 3.4 below, and article 11 is analyzed extensively in Chap. 5 of this edited collection (Humphries et al., 2025a). To understanding the meaning of activities within the scope of the agreement, this section focuses on the definitions of the main activities that are regulated under Part II, namely 'collection in situ', 'utilization of marine genetic resources' and 'access' to MGR and DSI in repositories and databases.

To understand regulated activities in the final text, it is helpful to outline the activities that evolved or were removed during the years of negotiation. The President's Aid to Discussions for IGC1 raised questions about the way 'access' would be addressed, if at all (UNGA, 2018c, p. 4). If access were to be regulated, there were two models proposed—a licensing or permit-based model or a notification-based model and terms and conditions for regulated access including capacity building, transfer of marine technology, access to samples and contributions to benefit sharing (2018a, p. 22). By IGC2, there continued to be disagreement about whether to regulate access and different views about whether to address all activities or access for certain purposes only (UNGA, 2019a, p. 8). During IGC3, a contentious issue concerned the 'definition of "access", as views differed on whether this referred to the collection of marine genetic resources in situ or also to access ex situ and in silico' (UNGA, 2019b, p. 6). The final agreed model was the notification-based model with a range of terms and conditions attached to activities relating to 'collection in situ' and 'utilization of marine genetic resources' (art 12). While the subject matter activity of 'access ex situ' was removed as an access modality, language concerning 'access' to MGR and DSI in repositories and databases remained and was narrowed to only refer to specific reporting requirements of relevance to information sharing and future benefit sharing modalities (Humphries et al., 2025a). The term 'access' was also retained for the traditional knowledge obligation (see Sect. 3.4 above).

'Collection in situ' comprises two activities—collection and sampling—which are not defined and can assume their ordinary meaning. 'In situ' is not defined but the CBD defines 'in situ conditions' as 'where genetic resources exist within ecosystems and natural habitats...' (art 2), which may assist with interpretation in the BBNJ context. Collection could include taking the physical materials from the marine environment. The undefined term 'sampling' may relate to the activity of taking samples (which depends on the meaning of samples as outlined in Sect. 3.2 above) and other activities like identifying or verifying MGR from the batch collection. It may be a term broad enough to include environmental DNA (eDNA) technologies autonomous in situ monitoring, such as using eco-genomic sensors which allow in situ genetic analysis without retaining the physical samples (Hansen et al., 2020). Use of the term MGRs 'in' ABNJ instead of MGRs 'of' ABNJ combined with the link between 'collection' and 'in situ' indicates a temporal and geographical requirement, so that the collection or sampling that triggers obligations under the collection notification do not apply to collection of MGR of ABNJ from ex situ (or in situ) conditions within national jurisdiction. However, there is uncertainty about whether 'sampling' used in its ordinary scientific meaning includes gathering information about the MGRs without actually gathering physical materials. This interpretation could extend the collection notification to the collection of MGR *and* DSI on MGR of ABNJ, subject to confirmation from the CoP (Humphries et al., 2025a).

A major innovation of the BBNJ Agreement is its BBNJ Identifier that will be automatically generated by the CHM upon prior notification of the collection in situ (see Sect. 2.1). The term BBNJ Identifier is undefined. Chapters 5, 6, 7, 12 and 14 of this edited collection go into detail about the interpretation and potential operation of this innovation.[25] Its meaning,

[25] Broggiato et al. (2025), Humphries et al. (2025a), Langlet and Vadrot (2023), Lawson et al. (2024a, b, 2025), Rabone et al. (2025).

scope, form and procedures will need to be clarified by BBNJ treaty bodies at the time the treaty enters into force, if not before by the Preparatory Commission.

Regarding 'utilization of marine genetic resources', the final text uses language identical to the definitions of 'utilization' and 'biotechnology' in the Nagoya Protocol (Nagoya Protocol art 2). The June 2022 draft text had another option that extended 'utilization' to 'information on' MGRs (UNGA, 2022b, draft art 1). Delegates removed information components from the final definition and instead clarified under the substantive obligations (e.g., art 12(8)) the circumstances under which 'utilization' extends to DSI. The August 2022 draft text added the term 'and commercialization' at the end of the definition (UNGA, 2022c, draft art 1), which could have been interpreted as meaning only R&D commercial purposes fall within scope of utilization. Its removal in the December 2022 draft and final text signaled that the activity of 'utilization' would capture both commercial and non-commercial uses of MGRs in R&D. This is clear for example in article 12(8), which uses the inclusive term 'utilization, including commercialization' as a trigger for notification.

As in the Nagoya Protocol, the only mention in the BBNJ Agreement of derivatives falling within scope of subject matter is in the definition of 'biotechnology', which arguably is associated with the activity of 'utilization' and not the activity of collection. This means that unless the definition of MGR is interpreted broadly to include derivatives of MGR (see Sect. 3.2 above), the notification obligations triggered by the activity of collection as well as access provisions relating to MGR and DSI in repositories would arguably not apply to the derivatives of MGR. Given the consistency between the Nagoya Protocol and the BBNJ Agreement in the definitions for 'utilization' and 'biotechnology', the considerable body of knowledge analyzing the scope of these terms, including their application to synthetic biology, may be relevant for interpreting the treaty text.

Understanding how derivatives will be managed under the framework will depend on its relationship to the activity of 'utilization'. 'Utilization' means 'to conduct research and development on the genetic and/or biochemical composition of marine genetic resources, including through the application of biotechnology" (art 1(14)). This means the type of use that falls within scope of the notification trigger is some form of investigation into the properties of the MGR—the genetic or biochemical composition. The CoP would need to give guidance to users about the types of activities that may fall within this category; activities of genetic manipulation would clearly fall within scope, but so might other activities that investigate the genetic or biochemical composition, such as taxonomic or conservation research, although characterisation activities (e.g. identification of MGR from the batch collection) are more likely to fall within 'sampling' activities of the collection *in situ* activities and associated notifications. The CoP may need to review the application of the 'utilization' notification to ensure that it supports (and does not deter) conservation research for marine biodiversity of ABNJ through light touch reporting under article 12(8). MGR derivatives fall within scope of the utilization notification if the 'utilization' involves 'biotechnology' which means 'a technological application that uses biological systems, living organisms, or derivatives thereof, to make or modify products or processes for specific use' (art 1(3)). In other words, derivatives are not included in the MGR definition and are only brought into the notification system when 'biotechnology' is involved as a form of utilization. This suggests that there is a different threshold for derivatives than for MGR concerning the type of activity that falls within utilization—i.e., for the purpose of making or modifying products or processes for specific use, rather than simply investigating composition. Subject to confirmation from the CoP, the 'utilization' trigger for derivatives (as opposed to MGR or DSI) might not extend to pure research or mere investigation into genetic or biochemical composition but instead requires some form of human intervention or manipulation before it triggers the notification (see Rabone et al., 2025).

Neither the MGR definition, 'utilization' definition nor substantive treaty provisions provide guidance about the extent to which an MGR or derivative needs to be present in a MGR product for notification and benefit sharing obligations to apply to activities producing that product. For example, genetically engineered AquAdvantage Salmon was produced by micro-injecting a promoter from an ocean pout antifreeze protein gene and a protein-coding sequence from a Chinook salmon growth hormone into wild Atlantic salmon fertilized eggs (Fletcher et al., 1988). These inputs played a functional role in the final MGR product by expressing the traits for which they were selected. If, however, the promoter from the ocean pout collected from the High Seas (Northwestern Atlantic Ocean) was simply present in the final invention but not expressed or 'switched on' to perform its function, would this require notification of 'utilization' of MGR under article 12? While this is an issue not clarified under the CBD's approach to ABS, there may be some guidance from other areas of law such as intellectual property. For example, some early European court decisions indicated that a patented product may only attract protection when a patented trait is performing its function at the time of the alleged infringement (rather than simply being present in the patented product).[26] This is a moving area of law and policy at national and international levels. For consistency in approaches to implementing utilization notification requirements, it will be necessary for BBNJ treaty bodies to provide clarification or advice on a similar benchmark for determining the reach of the treaty obligations with respect to 'utilization' of MGRs and DSI.

Although the BBNJ Agreement has a unique system of notification and benefit sharing, it incorporates the CBD's concept of ABS through Part II provisions (as denoted by the name of the ABS Committee under art 15) but not a similar approach to 'access' (other than for article 13 on traditional knowledge, see Sect. 3.4 above). The term 'access' is used in article 12 notification obligation and article 14 benefit sharing obligation. Like the CBD and Nagoya Protocol, the term 'access' is undefined in the treaty text. Article 12 uses access in the context of:

- repository and database aggregate reports on access to MGRs and DSI for the ABS Committee, which will contribute to the benefit sharing system (see Broggiato et al., 2025); and
- Party information about modalities envisaged for access to MGRs and DSI being 'utilized' within its jurisdiction.

Article 14 uses access in the context of:

- depositing MGR and DSI that are the subject of 'utilization' in publicly assessable repositories (for subsequent 'access' by others);
- non-monetary benefits being shared in the form of access to samples, sample collections and DSI in accordance with international practice; and
- access to MGRs and DSI in repositories and databases being subject to reasonable conditions including reasonable costs associated with providing access to the MGR, data or information and opportunities for access on fair and most favorable terms to researchers and research institutions from developing States.

While 'access' is undefined, it is clearly distinct from the 'utilization' activity triggering notification obligations.

In summary, the activity of 'collection in situ' is bounded by geographical scope of being an activity conducted in ABNJ, although the extent to which it applies to DSI on MGR of ABNJ collected or sampled directly from ABNJ is uncertain. The use of the term 'access' generally refers to third party or subsequent access to MGRs or DSI held by repositories and databases

[26] Humphries (2015) at 213, citing *Monsanto Technology LLC* v *Cefetra BV* (C-428/08)[2010] E.C.R. 1-6765; [2012] 3 C.M.L.R. 7 at [50].

(which may be intermediaries or users of MGR or DSI). In contrast, the activities relevant for notification and benefit sharing are carried out by the original researchers/entities (for collection activities) and the original or downstream users of MGRs and DSI (for 'utilization' activities). However, unless the BBNJ treaty bodies clarify the terms 'access', 'sampling' and the extent to which an MGR or derivative needs to be present in a MGR product or process for 'utilization' to apply, there may be a similar patchwork approach to regulating MGR and DSI on MGR of ABNJ as that under national ABS laws (see Humphries et al., 2021a).

3.3.6 Exclusions from Scope

The application and wording of exclusions from scope of Part II was the subject of considerable debate throughout each of the IGCs. In accordance with articles 10(2) and (3), Part II obligations do not apply to:

- 'Fishing regulated under relevant international law and fishing-related activities'; or
- 'Fish or other living marine resources known to have been taken in fishing and fishing-related activities …[from ABNJ, except where they are] regulated as utilization under this Part'; or
- A Party's military activities including military activities by government vessels and aircraft engaged in non-commercial service.

One of the most contracted debates concerned the treaty's application to fish and fishing-related activities, within the broader debate around 'not undermining' other IFBs (see Sect. 2.3). The concern was that the MGR framework could undermine the status quo in relation to the extensive regulatory and governance of harvest activities and the trade of fish and other organisms as a commodity under UNCLOS and a range of other international agreements (IISD, 2019, see Marciniak, 2017).

'Fishing and fishing-related activities' are undefined in the text. Much of the discussion for this exclusion centered on harvest fisheries, particularly with respect to the mandate of Regional Fisheries Management Organisations (RFMOs) (IISD, 2019) but with very little, if any, reference to aquaculture activities on the high seas. As technologies become more accessible, open ocean aquaculture, including automated roaming sea cages in high seas areas (Wei et al., 2020) may become more common. Aquaculture may be the subject of R&D located within ABNJ into the genetic or biochemical composition of MGR through for example breeding or biotechnology for bio-solutions to climate change or the culture of organisms for pharmaceutical leads (Jones et al., 2022). The BBNJ Agreement in part addresses the regulatory vacuum for high seas aquaculture through its ABMT, EIA and CBTMT frameworks, but the ambiguity about whether the activity is excluded or included in Part II creates uncertainty concerning collection, utilization and benefit sharing from high seas aquaculture (Humphries, 2017). If it is excluded, 'collection in situ' and 'utilization of MGRs of ABNJ' might continue to be governed in accordance with the freedom of the high seas with the benefits flowing to the technologically rich countries. If it is included, BBNJ treaty bodies would need to clarify how the notification and monitoring systems would apply because there may be 'utilization' of MGRs within ABNJ without 'collection in situ' to which the BBNJ Identifier attaches. The activity does not fall within the lineal model of R&D biodiscovery envisaged by the drafters, which complicates treaty implementation (Rabone et al., 2025).

The second exclusion relates to 'fish or other living marine resources' which are also not defined. The text reflects the use of the term in UNCLOS of 'living resources', which is not defined but has a broader meaning than biological resources, genetic resources or genetic material with functional units of heredity (see Deplazes-Zemp, 2018). The wording clarifies that the exception only applies to collection activities and not for the purpose of 'utilization'. The reason for this is to avoid the loophole where organisms are originally collected for the purpose of harvest but then later used for R&D into genetic properties that falls within the notification mechanism under article 12(8).

Previous drafts excluded the 'use of biological resources as a commodity' [e.g., UNGA, 2022b, draft art 8(2)], which would have clarified that MGRs only fall within scope of Part II if they are being used for R&D for their genetic attributes (not as a bulk commodity for trade). Similar ambiguity in the CBD and Nagoya Protocol has led to a variety of approaches under national ABS laws that extend the reach of some laws to biotrade activities and not only R&D of their genetic attributes (Humphries et al., 2021a). Clarification from BBNJ treaty bodies about how the BBNJ Agreement manages biological resources used as a commodity (outside the fisheries context) will be relevant to clarification of the 'utilization' definition and clarification about whether the 'value' in the definition of MGRs relates to the physical material or the genetic information (attributes) that is useful for the R&D activity (see Sect. 3.2 above).

Finally, the exclusion for a Party's military activities including those undertaken by government vessels and aircraft engaged in non-commercial service was a sensitive issue during negotiations, due to its relevance to national security and for exercising sovereignty and sovereign rights. UNCLOS has similar exclusions or special treatment of military or government vessels (e.g., art 298), with ongoing cases about interpretation, including the extent to which a relevant activity is a military activity or a law enforcement activity (Shi, 2024). Article 10(3) read in conjunction with article 4 means that the MGR framework applies to government vessels but not to military vessels or government vessels being used for military activities, although Parties must ensure that 'these vessels or aircraft act in a manner consistent, so far as is reasonably and practicable, with this Agreement' (art 4). Part II obligations with respect to the 'utilization' of MGR and DSI 'shall apply to a Party's non-military activities (art 10(3)), which can only be carried out exclusively for peaceful purposes [art 11(7), but it is silent with respect to 'collection' activities]. There is little, if any, data about how much commercial or non-commercial MGR research is undertaken by government vessels, and there are several questions of interpretation requiring further clarity from the CoP for the military exemption, including:

- What are the factual and legal differences between government research activities, law enforcement activities and military activities?
- To what extent do Part II 'collection' requirements apply to a Party's non-military activities by government vessels?

Depending on the answer to these and other questions, it is likely that there will continue to be little transparency around the use of government military vessels and aircraft for research activities in ABNJ.

In summary, the exclusions from scope were the subject of highly sensitive and protracted negotiations to ensure that Part II did not restrict existing or future (harvest) fishing activities and government military activities. To reduce the scope of both exemptions, the text outlines qualifiers concerning 'utilization' of MGR and DSI on MGR of ABNJ. Rather than clarify how the exemptions might operate in practice, these qualifiers raise further questions about the nature and scope of activities to which the exclusions apply. Given their sensitive subject matter, it is likely that clarity from the CoP will be tied up in discussions for years to come.

3.4 Temporal Scope (Article 10)

Article 10: Application

Article 10(1): The provisions of this Agreement shall apply to activities with respect to marine genetic resources and digital sequence information on marine genetic resources of areas beyond national jurisdiction collected and generated after the entry into force of this Agreement for the respective Party. The application of the provisions of this Agreement shall extend to the utilization of marine genetic resources and digital sequence information on marine genetic resources of areas beyond national jurisdiction collected or generated before entry into force, unless a Party makes an exception in writing under article 70 when signing, ratifying, approving, accepting or acceding to this Agreement…

Early in the negotiating process, there was agreement about the importance of including language on the temporal scope of the BBNJ Agreement (UNGA, 2019b, p. 6).Previous drafts of the current article 10 did not clarify the temporal scope. Until IGC4, there were discussions about whether MGR (and associated DSI) collected before the entry into force of the BBNJ Agreement, but accessed afterward would fall within scope (UNGA, 2019b, p. 6). As a last-minute deal to get the treaty over the line, countries agreed that: (a) MGR and DSI on MGR that are collected from ABNJ will apply after the entry into force of the relevant Party (i.e., regulated under national law); but (b) those previously collected or generated prior to entry into force will fall within scope if they fall within the BBNJ Agreement's provisions on 'utilization', but that countries can opt out of creating measures or complying with this aspect of retrospectivity. Unlike (a), category (b) does not specify the relevant date as entry into force 'for the respective Party', indicating that the utilization provisions will apply to MGR/DSI collected/generated prior to the date the BBNJ Agreement enters into force (rather than the date the relevant Party regulating 'utilization' becomes a Party). The *Vienna Convention on the Law of Treaties* provides that unless a different intention appears from the treaty, 'its provisions do not bind a party in relation to any act or fact which took place or any situation which ceased to exist before the date of the entry into force of the treaty with respect to that party' (art 28). This means that even without relying on the opt-out clause, the BBNJ Agreement only creates obligations that apply after entry into force for that Party (prospectively). Retrospectivity or otherwise for the 'utilization' provisions will depend on how the Party implements their obligations under national law.

The question of whether new and continued uses of previously accessed genetic materials fall within scope of the CBD and Nagoya Protocol's framework has generated much uncertainty for Parties in their implementation of their obligations under national ABS laws (Greiber et al., 2012). One argument for retrospectivity is that the R&D pipeline for biological leads can be at least 20 years. Confining scope only to uses from the date the BBNJ Agreement came into force would mean that few benefits would flow into the mechanism in the next 20 or 30 years, undermining the effectiveness of Part II (de la Concepción, 2024). Also experience from other international ABS frameworks has shown that not including previously accessed materials within the scope of the measures that focus on 'utilization' can create loopholes and enforcement challenges.[27] Arguments against retroactivity include increased cost, resourcing and workloads for MGR collections for the species collected before the treaty enters into force. Rules around prospectivity or retroactivity are particularly relevant for DSI, which is usually open access and stored in large multinational databases, for which the origin of DSI on MGR of ABNJ is unlikely to be stated prior to the entry into force of the BBNJ Agreement (Scholz et al., 2024).

The agreed compromise to include an opt out for retrospectivity also creates a practical dilemma for the research community as there will be different regimes applying to MGR and DSI on MGR of ABNJ, creating confusion for reporting and benefit sharing requirements

[27] See e.g. CITES Resolution Conf. 13.6 (Rev.CoP18), preambular paragraph 2; CITES SC Doc. SC.41.12, paragraphs 38 and 39.

(Rabone et al., 2025). There is no guidance about how retrospectivity relates to MGR collected prior to UNCLOS, given the BBNJ Agreement is an implementing agreement of UNCLOS. In some cases, there may be multiple regimes and rules applicable to utilization of the same MGR of ABNJ in collections (or DSI on MGR of ABNJ in databases) in different countries, depending on the date that each of the Parties interpret and apply the retroactivity provision from a range of possible options. Obligations associated with 'utilization' of MGR and DSI may apply:

- after the BBNJ Agreement enters into force (120 days after 60 Party ratifications);
- after the BBNJ Agreement enters into force for that Party (i.e., the date of ratifying it);
- before the BBNJ Agreement enters into force but after UNCLOS entered into force;
- before the BBNJ Agreement enters into force but after UNCLOS entered into force for that Party (i.e., the date of ratifying it);
- before UNCLOS entered into force;
- before UNCLOS entered into force entered into force for that Party (i.e., the date of ratifying it); and
- on some other date that national law prescribes as the date, which might be different for MGR compared with DSI.

The election also means that depending on how a Party ratifies the BBNJ Agreement, the same country can have multiple regimes applicable to their MGR collections—for example, those collected after BBNJ Agreement, those collected between UNCLOS and the BBNJ Agreement and those collected prior to the BBNJ Agreement. As the BBNJ Identifier will only be generated for a collection after the BBNJ Agreement enters into force, BBNJ treaty bodies will need to clarify how information about MGR and DSI under these different temporal conditions will be collected and used by Parties and BBNJ treaty bodies.

In summary, temporal scope of Part II of the BBNJ Agreement will be the subject of much policy discussion for years to come. It will require careful consideration by policy makers about how to not only navigate the temporal scope but understand how the temporal aspects are affected by subject matter and jurisdictional scope. Clarification by the CoP on the scope of its temporal provisions will be crucial, because questions remain about the extent to which it applies to 'access' to MGR and DSI in repositories (which is a distinct subject matter category to 'utilization'—see Sect. 3.5) and traditional knowledge associated with MGR in ABNJ. For example, would retrospectivity apply if traditional knowledge is 'accessed' during the 'utilization' activity about MGR collected prior to entry into force of the BBNJ Agreement for a given Party (see Sect. 3.4 above)? Answers to these and other questions may have significant consequences for shaping how rights are protected and benefit sharing is achieved under the BBNJ Agreement.

3.5 Conclusion

In contrast to ABS measures for MGR, DSI and traditional knowledge under the CBD/Nagoya Protocol bilateral (sovereignty) approach that accommodates national circumstances, the BBNJ Agreement's multilateral approach depends on having a consistent interpretation of scope and definitions. A common or consistent language reduces loopholes and conflicts about the jurisdiction and subject matter, including activities, governed under the notification, transparency and benefit sharing schemes. The above analysis raises key legal loopholes for clarification by BBNJ treaty bodies, some of which may have an impact on substantial obligations, some of which may create minor points of confusion about how the framework may operate and some may be dismissed as mere semantics. Table 3.1 summarizes some of the key issues for clarification by BBNJ treaty bodies from the analysis in this chapter. On the other hand, the BBNJ treaty bodies and Parties may prefer to maintain the 'constructive ambiguity' of some of the provisions that can offer space for advancing the interests of Parties as they implement their

Table 3.1 Key issues for further clarification by BBNJ treaty bodies

Scope aspect	Key issues for further clarification by BBNJ treaty bodies
Jurisdictional scope—geographical	Is the trigger for MGR of ABNJ the known distribution of the MGR or do the obligations require evidence that product of R&D incorporated (or materially/directly based on) the MGR actually sourced from ABNJ (even if its known distribution is also AWNJ)?
Jurisdictional scope—political	How will the principles concerning sovereignty/sovereign rights and non-appropriation be implemented in practice, including questions around the jurisdictional effects of sea level rise and political issues around data sovereignty, including Indigenous data sovereignty?
Jurisdictional scope—legal	How will the concept of 'due regard' to the rights and legitimate interests of coastal States and the interests of other States in ABNJ work in practice, including clarification about its effect on 'utilization' and 'access' to MGRs and DSI in repositories and databases? What is the meaning of the 'not undermining principle' in the BBNJ agreement context—not undermining the mandate of instruments, frameworks and bodies (IFBs) or not undermining the *effectiveness* of IFBs, where BBNJ agreement might improve the effectiveness of the frameworks but undermine their mandate? How will Part II scope overlap with UNCLOS Marine Scientific Research (Part XIII) and development and transfer of marine technology (Part XIV) be managed?
Subject matter scope—MGRs and derivatives	Are there any consequences for merging the two CBD definitions of genetic resources and genetic materials for the MGR definition, thereby confining the value of the 'resource' to the physical material rather than the genetic information it contains for R&D purposes? How does the definition and scope of MGRs relate to derivatives, synthetic biology and organisms discovered in ABNJ extreme environments that may not contain functional units of heredity? What is the meaning of the term 'sample' as the subject matter of substantive notification and benefit sharing obligations—does it denote the ordinary scientific meaning which is broader than the narrow interpretation of MGR of physical genetic material with functional units of heredity?
Subject matter scope—DSI	What is the scope and nature of 'digital sequence information' and how will it be managed in practice under the notification, transparency and benefit sharing mechanisms? How will: (a) genetic information outside the subset of DSI; and (b) Traditional Knowledge associated with DSI on MGR in ABNJ be managed under Part II?
Subject matter scope—traditional knowledge	What is the scope and nature of 'traditional knowledge associated with marine genetic resources in areas beyond national jurisdiction'? What is the geographical and temporal scope and requisite 'association' link between the knowledge and the MGR 'in' ABNJ—how does it relate to previously collected MGR that are 'utilized' within national jurisdiction?
Subject matter scope—activities—collection in situ	What is the meaning of 'sampling' within the definition of 'collection in situ' and does the scope extend to in situ genetic analysis without the need for physical collection? What is the scope and operation of the BBNJ Standardized Batch Identifier that attaches to the pre-collection notification of collection in situ?
Subject matter scope—activities—'utilization' of MGRs	To what extent would the originally collected MGR from ABNJ or its derivative or its DSI need to be present in a value-added product that is the subject or outcome of 'utilization of MGR'? Does 'utilization' refer to all kinds of commercial and non-commercial uses of MGR or DSI or only those that generate new insights or knowledge?

(continued)

Table 3.1 (continued)

Scope aspect	Key issues for further clarification by BBNJ treaty bodies
Subject matter scope—activities—'access' to MGR or DSI in repositories or databases	What is the meaning of 'access' and does its scope only extend to third party or subsequent collection and/or utilization of MGR or DSI if it can demonstrate a clear link to the original collection of MGR from ABNJ (or is it sufficient that its known distribution is ABNJ)?
Subject matter scope—exclusions from scope	What is the scope of excluded 'fish', 'other living marine resources' and 'fishing-related activities' and does this extend to for example, all uses of MGR as a commodity and high seas aquaculture activities? What are the boundaries of the military activity exemption and how can loopholes be avoided for government vessels undertaking research?
Temporal scope	Does the opt-out provision for retroactivity of the 'utilization' activity obligations relate to the date the BBNJ Agreement enters into force (120 days after 60 Parties) or when it enters into force for a given Party and can the same country have multiple regimes applicable to their MGR collections, e.g., those collected after BBNJ Agreement, those collected between UNCLOS and the BBNJ Agreement and those collected prior to the BBNJ Agreement? How will information about MGR and DSI under these different temporal conditions be collected and used by BBNJ treaty bodies and Parties if the BBNJ Identifier is only generated for a collection event after the BBNJ Agreement enters into force?

obligations. Other broader international law concepts such as 'without prejudice', 'due regard', 'non-appropriation' and the 'not undermining' principle may be beyond the mandate of the CoP and require cooperation and coordination between IFBs and evolving State practice.

Some areas of ambiguity require more urgent attention than others. A major innovation that is also a vulnerability for effective implementation of the regime are the obligations that reach through to activities of MGR of ABNJ and DSI on MGR of ABNJ that occur within national jurisdiction, such as 'utilization' and 'access' to MGR and DSI in repositories and databases or traditional knowledge. A priority for clarification of jurisdictional and subject matter scope is how the provisions interact with IFBs with a mandate for access, utilization and benefit sharing of MGR (including their associated DSI and associated traditional knowledge) with a geographical distribution in AWNJ and ABNJ. Resolving these issues may be more complicated for DSI and traditional knowledge which are the subject of ongoing debate in other international fora. An associated priority for clarification is how the BBNJ Agreement will manage multiple temporal scopes for 'utilization' of MGR and DSI. Other ambiguities relating to collection and sampling activities within ABNJ (including the BBNJ Identifier) and exclusions from scope may be easier to clarify through BBNJ treaty body interpretations and guidelines for implementation.

Despite ambiguity of meaning and scope of some aspects of jurisdiction, subject matter and temporal scope, practitioners and policy makers can apply a commonsense approach to assist with implementation until further clarification is forthcoming. The intention of Part II of the treaty was to fill the governance gaps as outlined in its objectives—fair and equitable benefit sharing, building capacity (especially for developing countries) to carry out Part II activities, the generation of knowledge, scientific understanding and technological innovation and the development and transfer of marine technology. There is already a body of knowledge about key terms used in the BBNJ Agreement borrowed from other international ABS agreements. While policy makers iron out the detail for implementation, practitioners might consider aligning their collection, utilization and data management practices with the clearer aspects of the treaty framework, as outlined in this chapter and other chapters in this edited collection.

An efficient framework for ABNJ that effectively governs MGR, DSI and traditional knowledge associated with MGR requires a deep understanding of jurisdictional, subject matter and temporal scope to avoid overlaps with other IFBs with competence in areas both within and beyond national jurisdiction. Agreement on the scope of Part II MGR governance was a spectacular achievement of diplomacy and creative thinking, considering Part II was the main impasse throughout negotiations. The result was an ambitious and expansive application of scope—where jurisdiction includes activities in ABNJ and AWNJ, subject matter includes physical MGR samples and DSI and traditional knowledge used for a range of activities and temporal scope includes prospective and retroactive components. This expansive scope will assist Parties to achieve equity, conservation and sustainable use objectives for biodiversity beyond national jurisdiction.

References

Aime, E., & Robinson, D. (2023). Indigenous biocultural rights and the Blue Mountains: Local and international policy challenges. *Geographical Research, 61*(4), 413–428.

Anggadi, F. (2022). Establishment, notification, and maintenance: The package of state practice at the heart of the Pacific islands forum declaration on preserving maritime zones. *Ocean Development and International Law, 53*(1), 19–36.

ATCM. (2009). *Collection and ISE of Antarctic Biological Material* ATCM 32.

ATCM. (2013). *Biological prospecting in Antarctica* ATCM 36.

Berry, T. (2023). *Sovereignty and the limits of international law: Regulating areas beyond national jurisdiction*. Taylor & Francis.

Blasiak, R., Jouffray, J. B., Amon, D. J., Claudet, J., Dunshirn, P., Søgaard Jørgensen, P., et al. (2023). Making marine biotechnology work for people and nature. *Nature Ecology and Evolution, 7*(4), 482–485.

Broggiato, A., Dunshirn, P., Jaspars, M., & Pena-Neira, S. (2025). Monetary and non-monetary benefit sharing under the BBNJ Agreement. In F. Humphries (Ed.), *Decoding governance of marine genetic resources under the BBNJ Agreement*. Springer.

Brown, A. E. L. (2025). The place of intellectual property under the BBNJ Agreement. In F. Humphries (Ed.), *Decoding governance of marine genetic resources under the BBNJ Agreement*. Springer.

Carroll, S. R., Garba, I., Figueroa-Rodríguez, O. L., Holbrook, J., Lovett, R., Materechera, S., et al. (2020). The CARE principles for indigenous data governance. *Data Science Journal, 19*, 43–43.

Couture, S., & Toupin, S. (2019). What does the notion of "sovereignty" mean when referring to the digital? *New Media and Society, 21*(10), 2305–2322.

de la Concepción, R. T. (2024). Negotiating fair and equitable sharing of benefits in the BBNJ agreement: Role of the Group of 77 and China. *Marine Policy, 163*, 106085.

Deplazes-Zemp, A. (2018). 'Genetic resources', an analysis of a multifaceted concept. *Biological Conservation, 222*, 86–94.

Dodds, K. J. (2011). Sovereignty watch: Claimant states, resources, and territory in contemporary Antarctica. *Polar Record, 47*(3), 231–243.

Elferink, A. O., Lucia, V. D., & Nguyen, L. N. (2022). Areas beyond national jurisdiction: Looking at and beyond the BBNJ process. In R. Churchill & A. O. Elferink (Eds.), *International law and marine areas beyond national jurisdiction* (pp. 1–13). Brill Nijhoff.

Fletcher, G., Shears, M., King, M., Davies, P., & Hew, C. (1988). Evidence for antifreeze protein gene transfer in Atlantic salmon (Salmo salar). *Canadian Journal of Fisheries and Aquatic Sciences, 45*(2), 352–357.

Greiber, T., Moreno, S. P., Åhrén, M., Carrasco, J. N., Kamau, E. C., Medaglia, J. C., Oliva, M. J., Perron-Welch, F., Ali, N., & Williams, C. (2012). An explanatory guide to the Nagoya protocol on access and benefit-sharing. In *Environmental policy and law paper No. 83, international union for conservation of nature, Switzerland*.

Hansen, B. K., Jacobsen, M. W., Middelboe, A. L., Preston, C. M., Marin, R., Bekkevold, D., et al. (2020). Remote, autonomous real-time monitoring of environmental DNA from commercial fish. *Scientific Reports, 10*(1), 13272.

Harden-Davies, H., Amon, D. J., Chung, T. R., Gobin, J., Hanich, Q., Hassanali, K., et al. (2022). How can a new UN ocean treaty change the course of capacity building? *Aquatic Conservation: Marine and Freshwater Ecosystems, 32*(5), 907–912.

Harden-Davies, H., Lopes, V. F., Coelho, L. F., Nelson, G., Veiga, J. S., Talma, S., & Vierros, M. (2024). First to finish, what comes next? Putting capacity building and the transfer of marine technology under the BBNJ agreement into practice. *NPJ Ocean Sustainability, 3*(1), 3.

Haward, M. (2021). Biodiversity in areas beyond national jurisdiction (BBNJ): The commission for the conservation of antarctic marine living resources (CCAMLR) and the United Nations BBNJ agreement. *The Polar Journal, 11*(2), 303–316.

Hiyoshi, A., Miyahara, K., Kato, C., & Ohshima, Y. (2011). Does a DNA-less cellular organism exist on Earth? *Genes to Cells, 16*(12), 1146–1158.

Hummel, P., Braun, M., Tretter, M., & Dabrock, P. (2021). Data sovereignty: A review. *Big Data and Society, 8*(1), 2053951720982012. https://doi.org/10.1177/2053951720982012

Humphries, F. (2015). Shellfish patents krill experimentation: Defences for sharing patented aquatic genetic materials in aquaculture. *European Intellectual Property Review, 37*, 210.

Humphries, F. (2017). A stewardship approach to 'legitimate interests' in deep sea genetic resources for use in aquaculture. *University of New South Wales Law Journal, 40*(1), 27–56.

Humphries, F. (2018). Banking on a patent solution for sharing ex situ genetic resources from Antarctic waters. In K. Adhikari & C. Lawson (Eds.), *Biodiversity, genetic resources and intellectual property: Developments in access and benefit sharing* (pp. 59–94). Routledge.

Humphries, F., Gottlieb, H. M., Laird, S., Wynberg, R., Lawson, C., Rourke, M., Tvedt, M. W., Oliva, M. J., & Jaspars, M. (2020). A tiered approach to the marine genetic resource governance framework under the proposed UNCLOS agreement for biodiversity beyond national jurisdiction (BBNJ). *Marine Policy, 122*, 103910.

Humphries, F., Laird, S., Wynberg, R., Morrison, C., Lawson, C., & Kolisnikova, A. (2021a). *Survey of access and benefit-sharing country measures accommodating the distinctive features of genetic resources for food and agriculture and associated traditional knowledge (FAO)*. http://www.fao.org/documents/card/en/c/cb6525en.

Humphries, F., Rourke, M., Berry, T., Englezos, E., & Lawson, C. (2021b). COVID-19 tests the limits of biodiversity laws in a health crisis: Rethinking 'country of origin' for virus access and benefit-sharing. *Journal of Law and Medicine, 28*, 684–706.

Humphries, F., Lawson, C., John, A. H., Benzie, S., & Morrison, C. (2024). *Access and benefit sharing in global aquaculture: Genetic resources, digital sequence information and traditional knowledge*. Edward Elgar.

Humphries, F., Jaspars, M., Lavelle, J., & Kachelriess, D. (2025a). The novel notification information system for marine genetic resources under the BBNJ Agreement. In F. Humphries (Ed.), *Decoding governance of marine genetic resources under the BBNJ Agreement*. Springer.

Humphries, F., Berry, T., & Muraki Gottlieb, H. (2025b). Bridging divides: The evolution of marine genetic resources governance beyond national jurisdiction. In F. Humphries (Ed.), *Decoding governance of marine genetic resources under the BBNJ Agreement*. Springer.

IISD. (2019). Earth negotiations bulletin: BBNJ IGC-3 final.

IOC. (2005). 'Criteria and guidelines on the transfer of marine technology' intergovernmental oceanographic commission advisory body of experts on the law of the sea (IOC Information Document No 1203), UNESCO.

IPCC. (2021). Summary for policymakers. In Masson-Delmotte, V., Zhai, P., Pirani, A., Connors, S. L., Péan, C., Berger, S., Caud, N., Chen, Y., Goldfarb, L., Gomis, M. I., Huang, M., Leitzell, K., Lonnoy, E., Matthews, J. B. R., Maycock, T. K., Waterfeld, T., Yelekçi, O., Yu, R., & Zhou, B. (Eds.), *Climate change 2021: The physical science basis. Contribution of working group I to the sixth assessment report of the intergovernmental panel on climate change.* https://www.ipcc.ch/report/ar6/wg1/downloads/report/IPCC_AR6_WGI_SPM_Stand_Alone.pdf

Irion, K. (2012). Government cloud computing and national data sovereignty. *Policy and Internet, 4*(3–4), 40–71.

IUCN. (2022). *IUCN commentary on the further revised draft text of an agreement under the United Nations Convention on the Law of the Sea on the conservation and sustainable use of marine biological diversity of areas beyond national jurisdiction* (A/CONF.232/2022/5) 8 August 2022. Prepared by IUCN World Commission on Environmental Law—Ocean Specialist Group, IUCN World Commission on Protected Areas—High Seas Specialist Group and Ocean team, IUCN Centre for Conservation Action, International Union for Conservation of Nature. https://www.iucn.org/sites/default/files/2022-08/igc5-iucn-commentary-on-bbnj-further-revised-draft.pdf

Jones, A. R., Alleway, H. K., McAfee, D., Reis-Santos, P., Theuerkauf, S. J., & Jones, R. C. (2022). Climate-friendly seafood: The potential for emissions reduction and carbon capture in marine aquaculture. *BioScience, 72*(2), 123–143.

Jørem, A., & Tvedt, M. W. (2014). Bioprospecting in the high seas: Existing rights and obligations in view of a new legal regime for marine areas beyond national jurisdiction. *The International Journal of Marine and Coastal Law, 29*(2), 321–343.

Kachelriess, D., Dunshirn, P., Langlet, A., Brown, A. E. L., & Scholz, A.H. (2025). Marine genetic resources and digital sequence information under the BBNJ Agreement: Interlinkages with other access and benefit sharing frameworks. In F. Humphries (Ed.), *Decoding governance of marine genetic resources under the BBNJ Agreement.* Springer.

Kukutai, T., & Taylor, J. (2016). *Indigenous data sovereignty*. Australian National University Press.

Langlet, A., & Vadrot, A. B. (2023). Not 'undermining' who? Unpacking the emerging BBNJ regime complex. *Marine Policy, 147*, 105372.

Lawson, C. (2022). Regulating information in molecules: The convention on biological diversity and digital sequence information. *Law, Technology and Humans, 4*(1), 18–48.

Lawson, C., Humphries, F., Jaspars, M., & Rabone, M. (2025). Data management and the 'BBNJ standardized batch identifier' under the BBNJ Agreement. In F. Humphries (Ed.), *Decoding marine genetic resource governance under the BBNJ Agreement.* Springer.

Lawson, C., Humphries, F., & Rourke, M. (2024a). Challenging the existing order of knowledge sharing governance with digital sequence information on genetic resources'. *Journal of Intellectual Property Law and Practice, 19*(4), 337–357.

Lawson, C., Humphries, F., & Rourke, M. (2024b). Genetic resources as culture and heritage: Repatriation and benefit sharing. *Melbourne University International Law Journal, 24*(1), 27–53.

Marciniak, K. J. (2017). New implementing agreement under UNCLOS: A threat or an opportunity for fisheries governance? *Marine Policy, 84*, 320–326.

Mendenhall, E., & Bateh, F. (2024). *'High seas treaty' name is inaccurate and should centre biodiversity (commentary)*. Mongabay News. https://news.mongabay.com/2024/02/high-seas-treaty-name-is-inaccurate-and-should-center-biodiversity-commentary/. Accessed 19 Feb 2024

Mendenhall, E., De Santo, E., Nyman, E., & Tiller, R. (2019). A soft treaty, hard to reach: The second intergovernmental conference for biodiversity beyond national jurisdiction. *Marine Policy, 108*, 103664.

Mossio, M., & Pontarotti, G. (2022). Conserving functions across generations: Heredity in light of biological organization. *The British Journal for the Philosophy of Science, 12*, 31. https://doi.org/10.1093/bjps/axz031

Mossop, J., & Schofield, C. (2020). Adjacency and due regard: The role of coastal States in the BBNJ treaty. *Marine Policy, 122*, 103877.

Mulalap, C. Y., Frere, T., Huffer, E., Hviding, E., Paul, K., Smith, A., & Vierros, M. K. (2020). Traditional knowledge and the BBNJ instrument. *Marine Policy, 122*, 104103.

Muller, M. R. (2015). *Genetic resources as natural information: Implications for the convention on biological diversity and Nagoya protocol*. Routledge.

Muller-Karger, F. E., Canonico, G., Aguilar, C. B., Bax, N. J., Appeltans, W., Yarincik, K., Leopardas, V., Sousa-Pinto, I., Nakaoka, M., & Aikappu, A. (2023). Marine Life 2030: Building global knowledge of marine life for local action in the Ocean Decade. *ICES Journal of Marine Science, 80*(2), 355–357.

Muraki Gottlieb, H., Ardron, J. & Brown, A.E.L. (2025a). BBNJ Agreement: A new infrastructure to foster benefit sharing of marine genetic resources. In F. Humphries (Ed.), *Decoding governance of marine genetic resources under the BBNJ Agreement.* Springer.

Muraki Gottlieb, H. Kachelriess, D. & Slobodian, L. (2025b). Understanding the preamble, principles and objectives of the BBNJ Agreement: A focus on the fair and equitable sharing of benefits of marine genetic resources. In F. Humphries (Ed.), *Decoding governance of marine genetic resources under the BBNJ Agreement.* Springer.

Oguamanam, C. (2020). Indigenous peoples, data sovereignty, and self-determination: Current realities and imperatives. *The African Journal of Information and Communication, 26*, 1–20.

Oldham, P., & Thambisetty, S. (2023). *ONEST: The middle way for monetary benefit sharing in BBNJ.* https://zenodo.org/record/7573700#.Y9KmR-zP27B

Pena-Neira, S. & Coelho, L.F. (2025). Traditional knowledge associated with genetic resources in areas beyond national jurisdiction. In F. Humphries (Ed.), *Decoding governance of marine genetic resources under the BBNJ Agreement.* Springer.

Rabone, M., Harden-Davies, H., Collins, J. E., Zajderman, S., Appeltans, W., Droege, G., et al. (2019). Access to marine genetic resources (MGR): Raising awareness of best-practice through a new agreement for biodiversity beyond national jurisdiction (BBNJ). *Frontiers in Marine Science, 6*(520), 535.

Rabone, M., Horton, T., Humphries, F., Lyal, C., Muraki Gottlieb, H., Scholz, A. H., Vanagt, T., & Jaspars, M. (2025). BBNJ Agreement: Considerations for scientists and commercial end users of MGR at research, development and commercialisation stages. In F. Humphries (Ed.), *Decoding governance of marine genetic resources under the BBNJ Agreement.* Springer.

Roe, H. S., Freestone, D., & Sapsford, F. (2022). The sargasso sea high seas EBSA after ten years: Is it still relevant and how has it helped conservation efforts? *Frontiers in Marine Science, 9*, 821182.

Rogers, A. D., Baco, A., Escobar-Briones, E., Currie, D., Gjerde, K., Gobin, J., et al. (2021). Marine genetic resources in areas beyond national jurisdiction: Promoting marine scientific research and enabling equitable benefit sharing. *Frontiers in Marine Science, 8*, 600.

Rothwell, D. R., & Stephens, T. (2023). *The international law of the sea* (3rd ed.). Hart.

Scanlon, Z. (2018). The art of "not undermining": Possibilities within existing architecture to improve environmental protections in areas beyond national jurisdiction. *ICES Journal of Marine Science, 75*(1), 405–416.

Scholz, A. H., Humphries, F., Vanagt, T., & Jaspars, M. (2023). *A new dawn for global benefit-sharing: Capitalizing on the global biodiversity framework for marine genetic resources from areas beyond national jurisdiction*. IUCN Policy Brief. https://www.iucn.org/sites/default/files/2023-02/bbnj_icg5bis_policy_brief_global_benefit_sharing_1.pdf

Scholz, A. H., Nunez-Vega, G., Weissgold, L., & Wussmann, K. (2024). The future of access and benefit-sharing: What next after the adoption of the global biodiversity framework and decision on digital sequence information? *Diversity, 16*(1), 27. https://doi.org/10.3390/d16010027

Scott, K. N. (2022). Governing the ocean commons: Lessons from the Antarctic. In H. J. Diamond, H. Doremus, & H. C. Yang (Eds.), *Common currents: Examining how we manage the ocean commons* (pp. 67–100). Brill Nijhoff.

Shi, X. (2024). Distinction between law enforcement activity and military activity in article 298 (1)(b) of UNCLOS: Commentary on recent international decisions. *Marine Policy, 167*, 106287.

Tvedt, M. W., & Schei, P. J. (2013). The term 'genetic resources': Flexible and dynamic while providing legal certainty? 1. In S. Oberthur (Ed.), *Global governance of genetic resources* (pp. 18–32). Routledge.

UNEP. (2020). Digital sequence information on genetic resources: Concept, scope and current use, Ad hoc technical expert group on digital sequence information on genetic resources, CBD/DSI/AHTEG/2020/1/3.

UNEP. (2022a). Decision adopted by the conference of the parties to the convention on biological diversity, 15/4 Kunming-Montreal global biodiversity framework. In *Conference of the parties to the convention on biological diversity, CBD/COP/DEC/15/4.* Accessed 19 Dec 2022.

UNEP. (2022b). Decision adopted by the conference of the parties to the convention on biological diversity, 15/9 digital sequence information on genetic resources. In *Conference of the parties to the convention on biological diversity, CBD/COP/DEC/15/9.* Accessed 19 Dec 2022.

UNGA. (2005). *Oceans and the law of the sea: Report of the secretary-general*, A/60/63/Add.1.

UNGA. (2007). *United nations declaration on the rights of indigenous peoples*, A/RES/61/295.

UNGA. (2018a). Statement by the President of the conference at the closing of the first session. In *Intergovernmental conference on an international legally binding instrument under the United Nations Convention on the Law of the Sea on the conservation and sustainable use of marine biological diversity of areas beyond national jurisdiction, first session, A/CONF.232/2018/7.* Accessed 20 Sept 2018.

UNGA. (2018b). President's aid to negotiations. In *Intergovernmental conference on an international legally binding instrument under the United Nations Convention on the Law of the Sea on the conservation and sustainable use of marine biological diversity of areas beyond national jurisdiction*, second session, A/CONF.232/2019/1*. Accessed 3 Dec 2018.

UNGA. (2018c). President's aid to discussions. In *Intergovernmental conference on an international legally binding instrument under the United Nations Convention on the Law of the Sea on the conservation and sustainable use of marine biological diversity of areas beyond national jurisdiction*, first session, A/CONF.232/2018/3. Accessed 25 June 2018.

UNGA. (2019a). Statement by the President of the conference at the closing of the second session. In *Intergovernmental conference on an international legally binding instrument under the United Nations Convention on the Law of the Sea on the conservation and sustainable use of marine biological diversity of areas beyond national jurisdiction*, second session, A/CONF.232/2019/5. Accessed 18 April 2019.

UNGA. (2019b). Statement by the President of the conference at the closing of the third session. In *Intergovernmental conference on an international legally binding instrument under the United Nations Convention on the Law of the Sea on the conservation and sustainable use of marine biological diversity of areas beyond national jurisdiction*, third session, A/CONF.232/2019/10*. Accessed 13 Sept 2019.

UNGA. (2019c). Draft text of an agreement under the United Nations Convention on the Law of the Sea on the conservation and sustainable use of marine biological diversity of areas beyond national jurisdiction: Note by the President. In *Intergovernmental conference on an international legally binding instrument under the United Nations Convention on the Law of the Sea on the conservation and sustainable use of marine biological diversity of areas beyond national jurisdiction*, third session, A/CONF.232/2019/6. Accessed 17 May 2019.

UNGA. (2019d). Revised draft text of an agreement under the United Nations Convention on the Law of the Sea on the conservation and sustainable use of marine biological diversity of areas beyond national jurisdiction: Note by the President. In *Intergovernmental conference on an international legally binding instrument under the United Nations Convention on the Law of the Sea on the conservation and sustainable use of marine biological diversity of areas beyond national jurisdiction*, fourth session, A/CONF.232/2020/3. Accessed 18 Nov 2019.

UNGA. (2022a). Further refreshed draft text of an agreement under the United Nations Convention on the Law of the Sea on the conservation and sustainable use of marine biological diversity of areas beyond national jurisdiction: Note by the President. In *Intergovernmental conference on an international legally binding instrument under the United Nations Convention on the Law of the Sea on the conservation and sustainable use of marine biological diversity of areas beyond national jurisdiction*, resumed fifth session, A/CONF.232/2023/2. Accessed 12 Dec 2022.

UNGA. (2022b). Further revised draft text of an agreement under the United Nations Convention on the Law of the Sea on the conservation and sustainable use of marine biological diversity of areas beyond national jurisdiction: Note by the President. In *Intergovernmental conference on an international legally binding instrument under the United Nations Convention on the Law of the Sea on the conservation and sustainable use of marine biological diversity of areas beyond national jurisdiction*, fifth session, A/CONF.232/2022/5. Accessed 1 June 2022.

UNGA. (2022c). Further refreshed draft text of an agreement under the United Nations convention on the law of the sea on conservation and sustainable use of marine biological diversity of areas beyond national jurisdiction (track changes). In *Intergovernmental conference on an internationally legally binding instrument under the United Nations convention on the law of the sea on the conservation and sustainable use of marine biological diversity of areas beyond national jurisdiction*, Fifth Session, A/CONF.232/2022/CRP.13/Add.1. Accessed 26 Aug 2022.

UNGA. (2023). Agreement under the United Nations Convention on the Law of the Sea on the conservation and sustainable use of marine biological diversity of areas beyond national jurisdiction. In *Intergovernmental conference on an international legally binding instrument under the United Nations Convention on the Law of the Sea on the conservation and sustainable use of marine biological diversity of areas beyond national jurisdiction*, further resumed fifth session, A/CONF.232/2023/4*. Accessed 19 June 2023.

Wei, Y., Wei, Q., & An, D. (2020). Intelligent monitoring and control technologies of open sea cage culture: A review. *Computers and Electronics in Agriculture, 169*, 105119.

WHA. (2011). Pandemic influenza preparedness: Sharing of influenza viruses and access to vaccines and other benefits. In *Report by the open-ended working group of member states on pandemic influenza preparedness: Sharing of influenza viruses and access to vaccines and other benefits*. World Health Assembly, A64/8 Sixty-Fourth Assembly.

WIPO. (2023a). *Chair's text of a draft international legal instrument relating to intellectual property, genetic resources, traditional knowledge and folklore, forty-seventh session*. WIPO/GRTKF/IC/47/7. Accessed 1 May 2023.

WIPO. (2023b). *Draft international legal instrument relating to intellectual property, genetic resources and traditional knowledge associated with genetic resources*. In WIPO/GRTKF/IC/SS/GE/23/5 Prov. 2. Accessed 8 Sept 2023.

WIPO. (2024a). Diplomatic conference to conclude an international legal instrument relating to intellectual property, genetic resources and traditional knowledge associated with genetic resources. In *WIPO treaty on intellectual property, genetic resources and associated traditional knowledge*. GRATK/DC/7.

WIPO. (2024b). Special session of the intergovernmental committee on intellectual property and genetic resources, traditional knowledge and folklore. In *Draft report prepared by the secretariat*. WIPO/GRTKF/IC/SS/GE/23/5 Prov. 2. Accessed 4 March 2024.

Fran Humphries has specialized in marine and biodiversity law and policy for over two decades in government, academia and consultancies. She is an associate professor at Griffith Law School, Griffith University, Australia, with a background in fisheries management and marine law. She has led large international research consultancies on access and benefit sharing of biological resources for governments, UN organizations and other institutions. She was a delegate for the International Council of Environmental Law delegation for the Biodiversity Beyond National Jurisdiction (BBNJ) Agreement negotiations.

Understanding the Preamble, Objectives and Principles of the BBNJ Agreement: A Focus on the Fair and Equitable Sharing of Benefits of Marine Genetic Resources

4

Hiroko Muraki Gottlieb, Daniel Kachelriess, and Lydia Slobodian

Abstract

An effective implementation of the Part II of The Agreement under the United Nations Convention on the Law of the Sea on the Conservation and Sustainable Use of Marine Biological Diversity of Areas Beyond National Jurisdiction (BBNJ Agreement) will require understanding the relevance of the legal interpretation tools referenced in the treaty text, as well as those intentionally not included (e.g., Vienna Convention on Law of Treaties). The Agreement's Part II provides a new framework of information and benefit sharing of Marine Genetic Resources of areas beyond national jurisdiction and associated Digital Sequence Information and Traditional Knowledge. This chapter for the edited collection "Decoding Marine Genetic Resource Governance under the BBNJ Agreement" explores the three types of interpretation tools used in the Agreement: (1) preamble, (2) objectives (general and subject matter specific) and (3) principles. Each legal tool plays an important and distinct role in providing insights to the negotiators' intent with a focus on Part II of the Agreement. The chapter urges stakeholders to understand the overall objective of the BBNJ Agreement, compromises made during the negotiations, and moreover, the innovative approaches employed in reaching consensus. Remembering and honoring the negotiators' efforts to emphasize the "how" in operationalizing benefit sharing could avoid prolonged discussions that could delay effectively implementing the obligations to protect the largest ecosystem of the Earth.

H. M. Gottlieb (✉)
Department of Organismic and Evolutionary Biology, Harvard University, Cambridge, MA, USA
e-mail: hmurakigottlieb@fas.harvard.edu

Elisabeth Haub School of Law at Pace University, White Plains, NY, USA

D. Kachelriess
High Seas Alliance, Vienna, Austria

IUCN World Commission On Environmental Law, Vienna, Austria

L. Slobodian
Environmental Law and Policy Program, Georgetown University Law Center, Washington D.C, USA

F. Humphries (ed.), *Decoding Marine Genetic Resource Governance Under the BBNJ Agreement*, Sustainable Development Goals Series, https://doi.org/10.1007/978-3-031-72100-7_4

Keywords

BBNJ agreement · High seas · Biodiversity · Conservation · Marine genetic resources · Digital sequence information · Traditional knowledge · Vienna convention on law of treaties · UNCLOS · Ocean governance

4.1 Introduction

Adopted in June 2023 after over twenty years of studies and negotiations, the *Agreement under the United Nations Convention on the Law of the Sea on the Conservation and Sustainable Use of Marine Biological Diversity of Areas Beyond National Jurisdiction* (1833 U.N.T.S. 31363) (BBNJ Agreement) is the third implementing agreement under the United Nations Convention on Law of the Sea (UNCLOS). The BBNJ Agreement will create new infrastructure and governance measures aimed to better implement the relevant UNCLOS obligations to protect nearly two-thirds of the global ocean. (Muraki Gottlieb et al., 2025).

The BBNJ Agreement uses three types of interpretation tools: (1) preamble, (2) objectives (general and subject matter specific) and (3) principles. Each legal tool plays an important and distinct role in providing insights to the negotiators' intent behind the text of the Agreement. This chapter explores how each tool is used in the BBNJ Agreement, with a focus on the substantive element, marine genetic resources (MGRs), including the fair and equitable sharing of benefits (BBNJ Agreement, Part II).

4.1.1 Vienna Convention on Law of Treaties

In terms of treaty interpretation, the *Vienna Convention on Law of Treaties* (VCLT, 1155 U.N.T.S. 2331) is considered to codify many tenets of customary international law (Williams, 2010). William notes that "it now appears that States have elected not to ratify [the VCLT] due to the belief that [it]—or at least some of its provisions—is considered to reflect customary international law" and that "the [International Court of Justice] and other international judicial bodies have held that several of [its provisions] constitute customary international law") (Williams, 2010).

Sect. 4.3 of VCLT provides guidelines and structure under the title, "Interpretation of Treaties." The section has three articles:

- General rules of interpretation (VCLT, art. 31).
- Supplementary means of interpretation (VCLT, art. 32).
- Interpretation of treaties authenticated in two or more languages (VCLT, art. 33).

VCLT's article on general rules of interpretation provides the basic understanding that a treaty must be "interpreted in good faith in accordance with the ordinary meaning to be given to the terms of the treaty in their context and in the light of its object and purpose." (VCLT, art. 31.1). From such foundation, the VCLT dives into variables. Specifically, the VCLT states that interpreters must consider the context by reviewing the text, preamble and annexes and "any agreement relating to the treaty which was made between all the parties in connection with the conclusion of the treaty" and "any instrument which was made by one or more parties in connection with the conclusion of the treaty and accepted by the other parties as an instrument related to the treaty." (VCLT, art. 31.2(a) and 31.2(b)). For the BBNJ Agreement, since it contains all the cited materials (i.e., text, preamble and annexes), the provision provides insights into how the Agreement must be read as a cohesive whole.

In addition to the context, VCLT requires the interpreter to consider the following factors in the application of treaty requirements:

(a) any subsequent agreement between the parties regarding the interpretation of the treaty or the application of its provisions.
(b) any subsequent practice in the application of the treaty which establishes the

agreement of the parties regarding its interpretation; and c. any relevant rules of international law applicable in the relations between the parties." (VCLT, art., 31.3 (a)–(c)).

Further, VCLT states that "a special meaning shall be given to a term if it is established that the parties so intended." (VCLT, art., 31.4). The factors in the application of the treaty requirements, particularly the "special meeting" given to a terminology will leave the interpreters with many questions because the term "digital sequence information" (DSI) is not defined in the BBNJ Agreement and there are ongoing negotiations on the definition of the terminology in various international fora (Rabone et al., 2025).

VCLT also provides guidance on treaties that are authenticated in two or more languages (VCLT, art. 33). Such guidance is particularly useful for the BBNJ Agreement, because the text is in the UN's six official languages (i.e., Arabic, Chinese, English, French, Russian and Spanish). However, the BBNJ Agreement has a provision with regards to authenticated language in its final provisions (BBNJ Agreement, Part XII). It states, "[t]he Arabic, Chinese, English, French Russian and Spanish texts of this Agreement are equally authentic." (BBNJ Agreement, art. 76). Such article ensures that there is no possibility of divergence. Further, the authenticated language was adopted by consensus after the agreed upon period of "scrubbing" the text to ensure coherence (Humphries et al., 2025a). Combining the text and the action of the negotiators provide strong support of no divergence in interpretation of the text in all six languages.

VCLT also takes into consideration a situation when the text and the context discussed above "leaves the meaning ambiguous or obscure" or "leads to a result which is manifestly absurd or unreasonable." (VCLT, art. 32(a) and 32(b)). In such circumstances, the interpreter may consider supplementary information, "including the preparatory work of the treaty and the circumstances of its conclusion, in order to confirm the meaning" (VCLT, art. 32). Therefore, VCLT confirms the importance of avoiding ambiguity or obscurity in drafting treaty language. VCLT also ensures that unintended consequences of the interpretation can be avoided, for example, inability for the regulated entity to be able to comply. Such considerations will be important, given the challenges that have been identified in interpreting the provisions of MGRs and the fair and equitable sharing of benefits, particularly, on DSI (Rabone et al., 2025).

The negotiators of the BBNJ Agreement had different positions on including the VCLT in the treaty text. During the final intergovernmental conference, in the preamble of the *further refreshed draft text of an agreement under the United Nations Convention on the Law of the Sea on the Conservation and Sustainable Use of Marine Biological Diversity of Areas Beyond National Jurisdiction, issued on December 12, 2022*, (A/CONF.232/2023/2), the text stated:

> [Recalling, with respect to non-parties to the Convention, that Part III, Section 4, of the Vienna Convention on the Law of Treaties sets out the rules on treaties and third States,]

However, the adopted text of the BBNJ Agreement does not reference the VCLT in any part of the Agreement, and certainly not in relation to non-parties to UNCLOS:

> *Recalling* that the legal status of non-parties to the Convention or any other related agreements is governed by the rules of the law of treaties,

The revisions to the BBNJ Agreement's adopted text on non-parties in the preamble were due to some countries' unease about the specific reference to VCLT's provisions on non-parties. Part III, Sect. 4.4 has five articles about non-Parties or "third States": (1) General rule regarding third States, (2) Treaties providing for obligations for third States, (3) Treaties providing for rights for third States, (4) Revocation or modification of obligations or rights of third States, and Rules in a treaty becoming binding on third States through international custom (VCLT, 1155 U.N.T.S 2331, art. 34–38). The general rule specifically states that neither obligation nor right is provided to non-Parties (VCLT, art. 34, 35, 36, and 37). However, regarding third States or a group of States or to all States, Article 36 states, "Its assent shall be presumed so long as the contrary is not indicated, unless the treaty otherwise provides." Also, VCLT's Article 38 states that "[n]othing in Articles 34–37 precludes a rule set forth in a treaty from becoming binding upon a third State as a customary rule of international law, recognised as such."

In addition to the preamble, the BBNJ Agreement has other provisions on non-Parties. Specifically, in Article 5, under the heading "relationship between this Agreement and the Convention and relevant legal instruments and frameworks and relevant global, regional, subregional and sectoral bodies," it states, "[t]he legal status of non-parties to the Convention or any other related agreements with regard to those instruments is not affected by this Agreement." (BBNJ Agreement, art. 5.1). There are also specific provisions on the settlement of disputes that are for non-Parties (BBNJ Agreement, art. 60.5, 60.6, and 60.7). Finally, Part X's title is specifically on non-Parties to the Agreement, where it states, "Parties shall encourage non-parties to this Agreement to become Parties thereto and to adopt laws and regulations consistent with its provisions" (BBNJ Agreement, art. 62), which is identical to one sub-article on non-Parties in the *Agreement for the Implementation of the Provisions of the United Nations Convention on the Law of the Sea of 10 December 1982 Relating to the Conservation and Management of Straddling Fish Stocks and Highly Migratory Fish Stocks* (A/CONF.164/37). (Fish Stocks Agreement, art. 33.1). These provisions show the particularly important aspects of the rights and obligations of non-Parties that the negotiators were able to incorporate into the final text of the BBNJ Agreement.

The BBNJ Agreement's adopted preamble text on non-parties that refers not to the VCLT, but the "rules of the law of treaties," possibly takes out the interpretation out of the VCLT's realm and gives some flexibility to those countries that do not support the VCLT or specifically, the VCLT's provisions on non-parties discussed above. Such preference by countries is understandable because while the VCLT is in force, there has been no universal ratification or signing/accession to VCLT. Most importantly, while there are 169 Parties to UNCLOS, there were a few negotiating countries that are non-Parties to UNCLOS that aimed to make sure that the BBNJ Agreement does not alter their status in any way.

Dörr, O. and Schmalenbach, K. explain that VCLT is not the only guidance on treaty interpretation:

> There are more rules of treaty interpretation applied in international practice and diplomacy than codified in Arts 31-33 VCLT. The Convention's rules of interpretation are not exclusive in a way that they prevent the interpreter from applying other principles compatible with the general rule laid down in Art 31. It is thus in his or her discretion to have recourse to established customary interpretation rules or at least to the wealth of material on treaty interpretation, which preceded the Convention (Dörr & Schmalenbach, 2018).

Some of the negotiators' conscious effort to exclude reference to VCLT in the BBNJ Agreement was limited to non-Parties. Further, the terminology, "rules of the law of treaties," could include VCLT. Generally, VCLT continues to be a useful tool in interpreting a treaty because of its broad use and flexibility. For those who are trying to understand negotiator intent, the VCLT is "an attempt to designate the elements to be taken into account in that process, and to assess their relative weight in it, rather than to describe, let alone prescribe, the process of interpretation itself." (Dörr & Schmalenbach, 2018). Regarding Article 31 of VCLT, titled "general rule of interpretation," Dörr and Schmalenbach state:

> These principles are mostly drawn from international judicial and arbitral practice, as it had developed since the late nineteenth century, and they were adopted by the ILC as a pragmatic compromise avoiding to follow one particular doctrine or theory of treaty interpretation. Also, since it considered the interpretation of documents to be to some extent an art, not an exact science, the Commission disavowed the idea of proposing an elaborate code or canon of interpretation, but deliberately confined itself to some fundamental rules recourse to which is, moreover, discretionary rather than obligatory (Dörr & Schmalenbach, 2018).

For the above reasons, this chapter will reference the VCLT, even though the negotiators landed on specifically not incorporating the reference to the

VCLT in the BBNJ Agreement, instead, preferring the undefined terminology, ""rules of the law of treaties." (BBNJ Agreement, Preamble).

4.1.2 The Role of Preamble

The role of preamble in treaty interpretation is not yet settled. The VCLT states that the "context for the purpose of the interpretation of a treaty shall comprise, in addition to the text, including its *preamble* and annexes' related agreements between parties" (emphasis added) (VCLT, art 31(2)). Some countries have argued that "preambles serve primarily diplomatic purposes, such as by permitting concessions to negotiating parties without creating legal obligations." (Hulme, 2016). Others view preambles as an integral part of holistic textual analysis and "repository for statements of object and purpose." (Hulme, 2016). In fact, Hulme cautions that uncertainty associated with the role of preamble creates potential pitfalls for countries that may use the preamble for or against a particular position and recommends that negotiators focus on the potential implications of the recitations in the preamble of a treaty. He states, "the possibility of outside interpretation should motivate states to pay close attention to the preamble and what goes into it, as its potential to exert substantive legal power combined with the unruly nature of object-and-purpose analysis can have far-reaching consequences."(Hulme, 2016.).

For the BBNJ Agreement, the negotiators were careful in choosing the topics and wording of paragraphs in the preamble. Through multiple revisions and consultations from 2016 through 2023, (Humphries et al., 2025a), the negotiators came to a decision on seventeen paragraphs that not only include recitations from UNCLOS, but also, innovative concepts in the context of a treaty, such as the nexus between climate change and the ocean, biodiversity loss, rights of Indigenous Peoples and local communities, and utilization of DSI on MGRs of areas beyond national jurisdiction. The BBNJ Agreement's preamble reflects the negotiators' aspirations to tackle issues of today and the future, while respecting the requirements under UNCLOS and international law.

It should be noted that due to the uncertainty associated with the role of a preamble, it is difficult to say just how much power it will hold in interpreting the BBNJ Agreement. Such lack of clarity may have, in fact, helped the negotiators to include provisions where no consensus could be reached. For example, the concept of "act[ing] as stewards of the ocean in ABNJ on behalf of present and future generations [...]," which had been included as a principle up until the President's "Further revised draft text" in June 2022 (A/CONF.232/2022/5) is part of the preamble. There is a view, however, that the BBNJ preamble is not just window dressing. Instead, it is an integral component of the instrument that encapsulates its motivations, purpose and objectives. Lothian argues that the BBNJ preamble should be used as a "key to open the mind of the makers," as it encapsulates the spirit, traditions and sentiments that underpin the commitments made during the negotiation process." (Lothian, 2023).

Ultimately, the answer to the question of the power of the preamble as applied to the BBNJ Agreement will need to wait until the preambular paragraphs are used in disputes. Hulme provides various examples where the treaty preamble was used in disputes, such as in the WTO and the U.S Shrimp-Turtle Decision, and regarding investment treaty preambles (Hulme, 2016). Until then, the Parties to the BBNJ Agreement may use the preambular paragraphs in their positions without being bound to the interpretation during bilateral and multilateral discussions as well as during the COP.

4.1.3 The Role of Objectives

Objectives in treaties assist with interpretation of the substantive obligations. Often phrased as "object and purpose" in literature, a definitive definition of the terminology does not exist (Jonas & Saunders, 2010). Jonas, D. S., & Saunders, T. N. reached a conclusion that in their view, the word "object" could "be a unitary concept referring to the goals that the drafters of the treaty hoped to achieve." (Jonas & Saunders, 2010). In this chapter, such concept will be the basis of the term "objective" in the discussion of the BBNJ Agreement. The "purpose" portion is discussed in the general principles and approaches section below in Sect. 4.1.4.

The BBNJ Agreement has two types of objectives:

(a) overall objective of the BBNJ Agreement and b. subject matter objectives (e.g., MGRs and the fair and equitable sharing of benefits, EIAs). The overall objective is to ensure the conservation and sustainable use of marine biological diversity of ABNJ "for the present and in the long term." (BBNJ Agreement, art. 2).
(b) Further, there are four subject matter objectives, including objectives of fair and equitable benefit sharing arising from activities with respect to MGRs and the associated digital sequence information (BBNJ Agreement, Part II, III, IV, V).

How the two types of objectives work together are discussed further in this chapter.

The VCLT has two different but complementary perspectives on objectives. One view is on the use of objectives in interpreting treaty provisions, which is discussed under the section on VCLT above. In addition, the VCLT provides a view of protecting the objectives of a treaty with the following mandates:

1. prohibition on country reservations that are incompatible with a treaty's objectives (VCLT, art. 19(c)).
2. prohibition for treaty signatories from defeating a treaty's objectives prior to ratification (VCLT, art. 18).

"Reservation" means "a unilateral statement, however phrased or named, made by a State or an international organization when signing, ratifying, formally confirming, accepting, approving or acceding to a treaty or by a State when making a notification of succession to a treaty, whereby the State or organization purports to exclude or to modify the legal effect of certain provisions of the treaty in their application to that State or to that international organization." (A/66/10, para 75, Article 1.1.1). The International Law Commission (ILC) provides an interpretation of reservations by stating that it "is to be interpreted as including reservations which purpose to exclude or to modify the legal effect of certain provisions of a treaty, or of the treaty as a whole with respect to certain specific aspects, in their application to the state or to the international organization which formulates the reservation." (A/66/10, para 75, art. 1.1.2). In light of the definition of reservation and the interpretation provided by the ILC, it seems intuitive that reservations that would be incompatible with a treaty's objectives would be prohibited. That said, there could be disputes about how a particular reservation is incompatible with a treaty's objectives, which could lead to prolonged disputes.

In practice, while one may wonder why a treaty may allow reservations, they are helpful when there are numerous negotiators with strong divergent views on various aspects of a multilateral agreement. As long as the reservation is not incompatible with the agreement's objectives, carving out or modifying certain obligations could allow the negotiators to reach a consensus or to have more countries or international organizations to become parties. However, declaring a valid reservation is not a simple task. First, there are reviews by the UN Secretary General before a reservation is accepted. (A/66/10, para 75, art. 3.5.5). Second, a country or an international organization can file an objection (A/66/10, para 75, art. 3.5.6). That said, the reservation does not need to be permanent. A country may, unless prohibited by the terms of the agreement, withdraw the reservation or even modify it (A/66/10, para 75, art. 3.5.7 and 3.5.8). Such flexibility gives opportunities for the countries that have filed a reservation to reconsider their positions in light of changes in national circumstances or diplomatic reasons, for example.

UNCLOS also prohibits the use of reservations in an article titled "reservations and exceptions" by stating, "[n]o reservations or exceptions may be made to this Convention unless expressly permitted by other articles of this Convention" (UNCLOS, art. 309). The BBNJ Agreement has an identical requirement on reservations, stating that: "[n]o reservations or exceptions may be made to this Agreement unless expressly permitted by other articles of this Agreement." (BBNJ Agreement, art. 70). The fact that UNCLOS and the BBNJ Agreement have identical concepts on reservations and exceptions is not surprising, indeed deliberate. UNCLOS used and adopted a "package deal" where the entire agreement is considered as a whole, without any selective carve outs (Lee, 2006). The BBNJ Agreement is a legal document to implement UNCLOS, and the negotiators specifically agreed that unless all the substantive elements are agreed, there would be no agreement and the negotiations certainly proceeded until a consensus was reached (Humphries et al., 2025a).

Regarding BBNJ Agreement's Part II (MGRs and the fair and equitable sharing of benefits) the Agreement allows a country or a regional economic integration organization to make an exception to the provision on the retroactive application of the requirements associated with the utilization of MGRs of ABNJ and associated DSI collected or generated *before* the entry into force of the Agreement (BBNJ Agreement, art. 10.1). In other words, should a Party not make an exception in writing pursuant to Article 70 when signing, ratifying, approving, accepting or acceding to the BBNJ Agreement, the notification and benefit sharing requirements will apply to utilization of MGRs collected and DSI generated before entry into force. For example, samples collected in 1850 in ABNJ, but utilized after the BBNJ Agreement enters into force in 2025 would need to meet the benefit sharing requirements in Part II unless the exception is valid for the relevant Party. As another example, if DSI associated with an MGR of ABNJ was generated in the year 2000 and utilized after the BBNJ Agreement enters into force, the utilization notification and benefit sharing requirements could apply. (BBNJ Agreement, art.14). Such retroactive application presents various technical and logistical challenges that the COP will need to tackle once the BBNJ Agreement enters into force (Rabone et al., 2025).

A second exception that can be made by a Party relates to monetary benefit sharing in Part II of the BBNJ Agreement. While BBNJ Agreement's Article 14.8 uses the term "declaration" that is used in Article 71, it is more of an exception under Article 70 since the declaration would be made *after* the entry into force and the COP adopts the modalities associated with monetary benefit sharing. As discussed below, Article 71 specifically states that declarations must be made at the time of signing, ratifying, approving, accepting or acceding to the agreement. Also, the declaration in Article 14.8 is a lag in time of payment, rather than no payment. It states, "[a] Party may make a declaration at the time the Conference of the Parties adopts the modalities stating that those modalities shall not take effect for that Party for a period of up to four years, in order to allow time for necessary implementation." (BBNJ Agreement, art. 14.8). Further, the declarant needs to make payments according to the process that is in prior to the COP adopting the new payment processes (BBNJ Agreement, art. 14.8). This article shows that the negotiators chose a middle-path that would allow payments to flow and allow for the declarant to ensure that the new payment processes can be implemented at the domestic level.

Both UNCLOS and the BBNJ Agreement also allow declarations and statements to be made when signing, ratifying, approving, accepting or acceding to the agreement (UNCLOS, art. 310 and BBNJ Agreement art. 71). The declarations allowed in both agreements are nearly identical (BBNJ Agreement includes regional economic integration organizations in addition to countries) and have a very narrow scope. Declarations are made "with a view, inter alia, to the harmonization of its laws and regulations with the provisions" of the agreement and "do not purport to exclude or to modify the legal effect of the provisions" of the agreement (UNCLOS, art. 310 and BBNJ Agreement art. 71). As long as the declarations and statements meet the requirements, they can be phrased or named in any way (UNCLOS, art. 310 and BBNJ Agreement art. 71).

In the case of the BBNJ Agreement, as of June 11, 2024, there are 90 signatories and seven Parties. Of the signatories and Parties, three countries, Chile, Federated States of Micronesia and the United Kingdom of Great Britain and Northern Ireland (UK) have submitted declarations. The UK's declaration was made at the time of signature and encompass its views on the BBNJ Agreement and the Antarctic Treaty, as well as the rights of Indigenous Peoples and local communities (UN Treaty Collection, BBNJ Agreement status). Chile made its declaration at the time it deposited the ratified BBNJ Agreement with the UN Secretary General and covered various topics, ranging from its sovereign rights, jurisdiction, and powers of coastal States under UNCLOS, not undermining the legal regimes of the Antarctic Treaty and its related instruments in force as well as select regional fisheries management organizations to specific concepts on settlement of disputes (UN Treaty Collection, BBNJ Agreement). The Federated States of Micronesia also made its declaration at the time it deposited the instrument of ratification and clarified that its ratification of the BBNJ Agreement "does not in any way constitute a renunciation of any of its rights or entitlements under international law, particularly as reflected" in UNCLOS. The declaration also stated that "UNCLOS does not impose obligation to keep baselines and outer limits of maritime zones under review nor to update charts or lists of geographical coordinates... and proclaims that the maritime zones of the Federated States of Micronesia, as established an notified to the Secretary-General of the United Nations... and the rights and entitlements that flow from them, shall continue to apply, without reduction, notwithstanding any physical changes connected to climate change-related sea-level rise." (UN Treaty Collection, BBNJ Agreement).

4.1.4 The Role of General Principles and Approaches

Legal principles are also an important tool under international law for a variety of reasons. Principles act as the impetus for the creation of rules and contributing to the process of legal reasoning, assisting in the interpretation and application of rules to a given problem (Eggett, 2019). In contrast to legal rules which are either valid or invalid, where principles are in conflict, "the competing principles are balanced against one another. An overridden principle is not deemed to be invalid and may itself prevail in a different situation." (Eggett, 2019). Eggett argues that general principles function as a source of law for the international community (although according to Payne, C.R., the concept of a "international community" itself is a problematic legal fiction (Payne, 2022). Another way to look at general principles (in contrast to mere principles) is that they "have an inherent gap-filling function, most commonly deployed to identify basic procedural rules where treaties and custom fail to provide answers." (Eggett, 2019).

While a useful tool, one of the challenges with general principles is that there are various meanings to the terminology, and determining the legal status (i.e., the extent of its binding nature) of a general principle may not be easy (Winter et al., 2017). Also, Winter suggests, "[a] s the term 'general' has many different meanings (such as overarching, abstract, common) the 'general principles of international law' might be better characterized as being the 'fundamental (or essential) principles of international law.'" (Winter et al., 2017). Winter further states, "it should be noted that the stricter the content of a principle is the more difficult it becomes to acknowledge it to be binding, and vice versa, the less binding the principle is the stricter its content can become." (Winter et al., 2017).

In the BBNJ Agreement, in addition to "general principles," the title also includes "approaches" in the title of Article 7 and in Article 7(e), which states, "[t]he principle of precautionary principle or approach, as appropriate" without further explanation as to what could be considered "as appropriate" or the distinction between "precautionary principle" and "precautionary approach." Peel concludes that the debate over the difference or lack thereof between "principle" and "approach" "masks deeper political differences that have more to do with divergent views about the significance of potential health or environmental risks than the details of the available scientific evidence." (Peel, 2004). Rather than to engage in prolonged discussion as to which concept should provide guidance, Peel suggests:

> Precaution should not require decision-makers to achieve the impossible and reach the 'right' decision in advance, regardless of uncertainties. Rather, the best chance for the international community to prevent serious environmental degradation in the future lies in imposing particular procedural constraints on regulatory decision-making that are designed to ensure scientific uncertainty is factored into the process and that science itself is not extended beyond the limits of its utility and capacity to inform decisions on risk regulatory measures (Peel, 2004).

As discussed above, there are various legal analyses in literature on general principles and approaches that provide interesting views. To ensure clarity in this chapter, however, the foundational understanding of "general principles and approaches" will be that they refer "broadly to a treaty's goals and the character of the means employed to achieve them." (Jonas & Saunders, 2010). Jonas, D.S. & Saunders, T.N. suggested such concept for the "purpose" part of the "object and purpose" terminology, but the understanding that "general principles and approaches" focus on the ways to achieve the objectives appear to be the clearest way to think about "general principles and approaches" in the BBNJ Agreement, because the overall objective of is "to ensure the conservation and sustainable use of marine biological diversity of areas beyond national jurisdiction, for the present and in the long term, through effective implementation of the relevant provisions of the Convention and further international cooperation and coordination." (BBNJ Agreement, art. 2).

The BBNJ Agreement has 14 "general principles and approaches" that are intended to apply to the relevant sections of the Agreement, including the four substantive elements: MGRs and the fair and equitable sharing of benefits (Part II), Area-Based Management Tools, including Marine Protected Areas (Part III), Environmental Impact Assessments (Part IV) and Capacity Building and Marine Technology Transfer (Part V). For MGRs—and the conclusion of the negotiations as a whole—the most significant negotiating point was on reaching a middle ground between principles of common heritage of humankind and freedom of the high seas that are articulated in UNCLOS, which is discussed below in Sect. 4.2 of this chapter. The countries decided on an innovative approach in the BBNJ Agreement to break the deadlock, where the general principles on the two concepts depart slightly from the original wording in UNCLOS, but the spirit of the original principles is reflected in the operative sections of the MGR provisions. The challenges and how the countries resolved their divergence of views are explored below. This chapter will also provide analysis of other relevant "general principles and approaches" that may be helpful in interpreting the MGR provisions in the BBNJ Agreement.

While not part of the "general principles and approaches," there are two stand-alone articles and the concepts thereof that are repeated throughout the BBNJ Agreement. First is Article 8, which is on international cooperation. The second is Article 5, which is on the relationship between the BBNJ Agreement and UNCLOS and relevant legal instruments and frameworks and relevant global, regional, subregional and sectoral bodies. While not titled as such, Articles 5 and 8 of the BBNJ Agreement fit the concept of principles, which is "treaty's goals and the character of the means employed to achieve them." (Jonas & Saunders, 2010): as explored further below, the two articles work synergistically to advance the Agreement's objectives.

4.1.5 Conclusion

While there are various analyses in literature on how the preamble, principles and objectives are understood in a treaty, this chapter shows that the BBNJ Agreement's negotiators included general principles and objectives in UNCLOS and other multilateral environmental agreements, and introduced new concepts that are believed to further the overall objective of the BBNJ Agreement. Such approach could address governance gaps for the present and in the long term, with an eye toward effective conservation and sustainable use of biodiversity beyond national jurisdiction. The final text of the BBNJ Agreement shows modern and innovative drafting that can also contribute to the progressive development of law. The sections below provide further detailed exploration of the BBNJ Agreement's preamble, overall objective and subject matter objectives, and general principles and approaches. Also, a summary of some of the implementation considerations are discussed. This chapter concludes with some reflections on how the three interpretive tools will support future implementation of the BBNJ Agreement's provisions, with a focus on MGRs and the fair and equitable sharing of benefits (BBNJ Agreement, Part II).

4.2 The Preamble, General Principles and Approaches, and Objectives on MGRS and the Fair and Equitable Sharing of Benefits

4.2.1 The Evolution of the Structure of the Preamble, General Principles and Approaches, and Objectives

In addition to the general objective in Article 2, the BBNJ Agreement has "general principles and approaches" that apply to the whole Agreement. Also, the BBNJ Agreement has subject matter objectives on the four substantive elements: MGRs and the fair and equitable sharing of benefits (Part II), Area-Based Management Tools, including Marine Protected Areas (Part III), Environmental Impact Assessments (Part IV) and Capacity Building and Marine Technology Transfer (Part V). However, the first textual basis for the negotiations issued by the Intergovernmental Conference's President in December 2018 (A/CONF.232/2019/1*) (President's aid to negotiations), provided a structure of the interpretive tools that evolved over time (Table 4.1).

It is worth noting that the President's aid to negotiations issued after the first intergovernmental conference showed a preliminary thinking of the negotiators in terms of legal tools to support interpretation of the future agreement. The proposed concepts were based on the Preparatory Committee (PrepCom) discussions that took place over two years prior to the start of the IGC. A report adopted by consensus captured the conclusions of the PrepCom, but with a caveat that nothing in the document would prejudice the intergovernmental conference negotiations (A/CONF.232/2018/1).

Over the course of the inter governmental conference, the negotiators changed the structure of the interpretive tools by making separate buckets for the Preamble, General Principles and Approaches, and the subject matter objectives. Such restructuring helped the negotiators differentiate the concepts that would: (a) shows the aspirations of the negotiations (i.e., Preamble), (b) apply to the BBNJ Agreement in its entirety (i.e., General Principles and Approaches), and (c). support the understanding of the subject matter specific obligations in the operative text. Many of the principles included in

Table 4.1 Comparison of the initial structure of the general principles and approaches in the President's Aid to Negotiations to the final structure of the BBNJ Agreement

President's Aid to Negotiations (December 2018)	BBNJ Agreement
General principles and approaches as well as subject matter objectives were included in one section. In a footnote, the president states, "[s]uggestions were made that some of these principles and approaches would be included in a separate article and some in the preamble." (A/CONF.232/2019/1*, fn2)	General principles and approaches are in a stand-alone section (Article 7) and some of the proposed concepts originally proposed in the President's aid to negotiations were separated into subject matter objectives (i.e., Part II, III, IV, and V) and in the preamble

the President's aid to negotiations in December 2018 (A/CONF.232/2019/1*) are recognizable in BBNJ Agreement's Article 7, including the common heritage of [hu]mankind, the recognition of the special case of small island developing States (SIDS), equity and the equitable sharing of benefits, best available scientific information, traditional knowledge and the precautionary approach/principle. The sections below provide some of the concepts that were in the President's aid to negotiations and how the concepts were ultimately used in the BBNJ Agreement.

4.2.2 The Evolution of the Preamble in the BBNJ Agreement

The preamble in the BBNJ Agreement is not just a recitation of existing obligations, objectives or principles. To the contrary, while the preamble incorporated some recitations from UNCLOS, it also features new concepts that show the aspirations of the negotiators to achieve the overall objective of the Agreement: "to ensure the conservation and sustainable use of marine biological diversity of areas beyond national jurisdiction, for the present and in the long term, through effective implementation of the relevant provisions of the Convention and further international cooperation and coordination." (BBNJ Agreement, art. 2). As discussed in the introduction section of this chapter, just how much power the preambulatory provisions will hold in the interpretation of the BBNJ Agreement will be better understood when (or if) there is a dispute.

This section explores the specific preambular paragraphs that may be helpful in the interpretation of the BBNJ Agreement's Part II provisions: MGRs and the fair and equitable sharing of benefits. The paragraphs are grouped by subject matter and each section offers a comparison between the Draft text issued by the President of the intergovernmental conference in (May 2019) (A/CONF.323/2019/6) (Draft Text) and the BBNJ Agreement as well as how they may support the interpretation of relevant Part II. The topics were chosen based on the subject specific objectives for Part II, which is discussed in Sect. 4.2.2 of this chapter.

4.2.2.1 Conservation and Sustainable Use

On conservation and sustainable use of marine biological diversity of areas beyond national jurisdiction, the BBNJ Agreement's preamble includes three paragraphs referenced below:

> *Conscious* of the need for the comprehensive global regime under the Convention to better address the conservation and sustainable use of marine biological diversity of areas beyond national jurisdiction,
>
> *Recognizing* the importance of contributing to the realization of a just and equitable international economic order which takes into account the interests and needs of humankind as a whole and, in particular, the special interests and needs of developing States, whether coastal or landlocked,
>
> *Committed* to achieving sustainable development

"Conservation and sustainable use" are concepts that are repeatedly used throughout the BBNJ Agreement, including in its title and the general obligation (BBNJ Agreement, art. 2). The first paragraph listed above had considerable support from the negotiators since the concept was included the Draft Text in May 2019 and it was consistently incorporated in subsequent drafts and into the BBNJ Agreement, albeit with modifications. In the Draft Text, the paragraph read, "*Stressing* the need for the comprehensive global regime to better address the conservation and sustainable use of marine biological diversity of areas beyond national jurisdiction," which was stronger than the above referenced paragraph.

The second paragraph was not introduced until late in the negotiations. It incorporates the phrase "just and equitable international economic order" from the fifth paragraph of UNCLOS' Preamble, with the change "mankind" to "humankind." Such change reflects the countries' stance on the importance of gender equality, which is the topic of Sustainable Development Goal 5. (A/RES/70/1). It is also noteworthy that the President and the Secretariat of the Intergovernmental Conference

Table 4.2 Comparison of the preambular paragraphs between the Draft Text and the BBNJ Agreement

Draft Text (May 2019)	BBNJ Agreement
Preamble *Recalling* the relevant provisions of the United Nations Convention on the Law of the Sea, including the obligation to protect and preserve the marine environment *Stressing* the need to respect the balance of rights, obligations and interests set out in the Convention *Stressing* the need for the comprehensive global regime to better address the conservation and sustainable use of marine biological diversity of areas beyond national jurisdiction *Respecting* the sovereignty, territorial integrity and political independence of all States *Desiring* to promote sustainable development *Aspiring* to achieve universal participation	**Preambular paragraphs relevant for MGRs** Conscious of the need for the comprehensive global regime under the Convention to better address the conservation and sustainable use of marine biological diversity of areas beyond national jurisdiction Recognizing the importance of contributing to the realization of a just and equitable international economic order which takes into account the interests and needs of humankind as a whole and, in particular, the special interests and needs of developing States, whether coastal or landlocked Recognizing also that support for developing States Parties through capacity building and the development and transfer of marine technology are essential elements for the attainment of the objectives of the conservation and sustainable use of marine biological diversity of areas beyond national jurisdiction Recalling the United Nations Declaration on the Rights of Indigenous Peoples Affirming that nothing in this Agreement shall be construed as diminishing or extinguishing the existing rights of Indigenous Peoples, including as set out in the United Nations Declaration on the Rights of Indigenous Peoples, or of, as appropriate, local communities Acknowledging that the generation of, access to and utilization of digital sequence information on marine genetic resources of areas beyond national jurisdiction, together with the fair and equitable sharing of benefits arising from its utilization, contribute to research and innovation and to the general objective of this Agreement Committed to achieving sustainable development

Note: The underlined paragraphs in the right column show the concepts relevant for MGRs included in the Draft Text

were mostly women and there was a significant number of women negotiators.

The third paragraph shows strong stance by the negotiators on sustainable development. It is notable that the Draft Text had weaker language, "desiring to promote sustainable development." The two paragraphs discussed above provide more granular concepts on sustainable development, which is based on conservation and sustainable use, as well as the need to addressing equality (e.g., economic, gender, etc.).

Overall, the above three paragraphs show the negotiators' aspirations that benefit sharing associated with MGRs will contribute to conservation and sustainable use and ultimately toward sustainable development to achieve the goals of Agenda 2030 and beyond.

4.2.2.2 Capacity Building and the Transfer of Marine Technology

Capacity building and the transfer of marine technology is one of the substantive elements of the BBNJ Agreement (BBNJ Agreement, Part V). It is considered the most important element for developing countries. The preambular paragraph below on capacity building and the transfer of marine technology was not introduced until late in the negotiations, introduced in the Further refreshed draft text of an agreement under the United Nations Convention on the Law of the Sea on the conservation and sustainable use of marine biological diversity of areas beyond national jurisdiction, 12 December 2022 (A/CONF.232/2023/2). (Further Refreshed Draft Text).

Recognizing also that support for developing States Parties through capacity building and the development and transfer of marine technology are essential elements for the attainment of the objectives of the conservation and sustainable use of marine biological diversity of areas beyond national jurisdiction

Capacity building and the transfer of marine technology is a substantive element of the BBNJ Agreement and part of the "2011 Package" (A/RES/72/249) that was the basis of the countries' agreement to consider a future agreement (Humphries et al., 2025a). The above paragraph confirms the importance of capacity building and transfer of marine technology and an understanding that the focused recipients would be developing States Parties. Interestingly, there is no definition of the "developing States Parties," although in the operative texts, the list of what is meant by "developing States Parties" are provided. For example, in the objectives of MGRs in Sect. 4.2.4 of this chapter, there is a long list of States that need capacity building (BBNJ Agreement, art. 9(b)).

4.2.2.3 Indigenous Peoples and Local Communities

There are two paragraphs on Indigenous Peoples and local communities in the BBNJ Agreement:

Recalling the United Nations Declaration on the Rights of Indigenous Peoples,

Affirming that nothing in this Agreement shall be construed as diminishing or extinguishing the existing rights of Indigenous Peoples, including as set out in the United Nations Declaration on the Rights of Indigenous Peoples, or of, as appropriate, local communities,

The text on United Nations Declaration on the Rights of Indigenous Peoples (A/RES/61/295) was first introduced in the Further revised draft text of an agreement under the United Nations Convention on the Law of the Sea on the conservation and sustainable use of marine biological diversity of areas beyond national jurisdiction, 1 June 2022 (A/CONF.232/2022/5) (Further Revised Draft Text). More detailed discussion on Indigenous Peoples and local communities and the relevance of traditional knowledge for the BBNJ Agreement is in Chap. 8 (Pena-Neira & Coelho, 2025).

Relevant to Part II of the BBNJ Agreement, the language on Indigenous Peoples and local communities is also found in Article 7(k), under the General Principles and Approaches, as well as in the text on MGRs in Article 13. When signing the BBNJ Agreement, the United Kingdom made the following declaration:

The United Kingdom's long-standing and well-established position, set out in its annual explanation of position at the UN General Assembly on the rights of indigenous people, is that human rights are held exclusively by individuals. With the exception of the right of self-determination (Common Article 1 of the two International Human Rights Covenants), the United Kingdom does not recognise collective human rights in international law. The United Kingdom consider this important in ensuring that individuals within groups are not left vulnerable or unprotected by allowing the rights of the groups to supersede the human rights of the individual. The United Kingdom therefore understands any internationally-agreed reference to the rights of indigenous peoples or local communities, including those in the UN Declaration on the Rights of Indigenous Peoples and, in the Agreement signed today, to refer to those rights bestowed by governments at the national level. The United Kingdom further understands the term "local communities" to be used consistently with the way it is used in the Convention on Biological Diversity."

Of the 90 signatories, the UK is the only country that made a declaration upon signing the BBNJ Agreement and communicated it to the UN Secretary General (i.e., depositary). How such declaration will be used in the implementation of the BBNJ Agreement at the national level or if other countries will follow is yet to be determined.

4.2.2.4 Digital Sequence Information

The inclusion of a paragraph on digital sequence information (DSI) was agreed at the very end of the negotiations (Humphries et al., 2025a). The text on DSI was not part of any of the previous draft texts and only appeared in the BBNJ Agreement. The paragraph states:

> *Acknowledging* that the generation of, access to and utilization of digital sequence information on marine genetic resources of areas beyond national jurisdiction, together with the fair and equitable sharing of benefits arising from its utilization, contribute to research and innovation and to the general objective of this Agreement

The significance and the resulting challenges associated with introducing DSI into the BBNJ Agreement is discussed throughout this book, including in Chap. 14 (Rabone et al., 2025).

4.2.3 General Principles and Approaches

As Table 4.3 shows, there were no general principles or approaches relevant for MGRs in the Draft Text (see the left column). In the BBNJ Agreement (see the right column in Table 4.3), there are eight relevant general principles and approaches for interpreting the MGR provisions in Part II. This Section will focus on the following topics: Reconciling the disagreements on the common heritage of mankind and the freedom of the high seas (Sect. 4.2.3.1), equity and fair and equitable sharing of benefits (Sect. 4.2.3.2), information and knowledge (Sect. 4.2.3.3) and special interests, circumstances and rights (Sect. 4.2.3.4). It also briefly discusses the precautionary principle vs. approach (Sect. 4.2.3.4) and several principles that had been proposed, but that were eventually not included in general principles and approaches of the BBNJ Agreement (Sects. 4.2.3.5–4.2.3.8).

4.2.3.1 Reconciling Disagreements on the Common Heritage of Mankind and the Freedom of the High Seas/Right to Conduct Marine Scientific Research

The Question on the Applicable Principle on MGRs

MGRs of areas beyond national jurisdiction are found in some of the most extreme

Table 4.3 Comparison of the list of general principles and approaches of the Draft Text to the final text of the BBNJ Agreement, with a focus on MGRs

Draft Text (May 2019) General principles and approaches	**BBNJ Agreement** Select general principles and approaches that are relevant for MGRs (Article 7)
[(a) Apply an integrated approach [/principle]] (b) Apply an approach that builds ecosystem resilience to the adverse effects of climate change and ocean acidification and restores ecosystem integrity (c) Act so as not to transfer, directly or indirectly, damage or hazards from one area to another or transform one type of pollution into another (d) Endeavor to promote the internalization of environmental costs and the use of economic instruments, taking into account the approach that the polluter should [, in principle,] bear the cost of pollution, with due regard to the public interest and without distorting international trade and investment [(e) Ensure accountability] [(f) Be guided by the principle of non-regression] [(g) Take into consideration flexibility, pertinence and effectiveness]	(b) The principle of the common heritage of humankind which is set out in the Convention (c) The freedom of marine scientific research, together with other freedoms of the high seas (d) The principle of equity and the fair and equitable sharing of benefits (i) The use of the best available science and scientific information (j) The use of relevant traditional knowledge of Indigenous Peoples and local communities, where available (k) The respect, promotion and consideration of their respective obligations, as applicable, relating to the rights of Indigenous Peoples or of, as appropriate, local communities when taking action to address the conservation and sustainable use of marine biological diversity of areas beyond national jurisdiction (m) Full recognition of the special circumstances of small island developing States and of least developed countries (n) Acknowledgement of the special interests and needs of landlocked developing countries

Note: There was no text proposed relevant for MGRs in the Draft Text (see the left column) as opposed to the BBNJ Agreement that has eight general principles and approaches that are relevant for MGRs.

environments on the planet, making them potentially unique and valuable as well as difficult and expensive to access (Alcock, 2014; Clark et al., 2016). Such difficulty creates inequity, as many developing States lack the technical and financial capacity to access and benefit from such resources. For that reason, the BBNJ Agreement's Part II is an essential substantive element of the Agreement, especially for the developing countries.

During the negotiations, there was an intractable divide between the developed and developing countries on which principle to apply to benefit sharing associated with MGRs: the common heritage of mankind principle or the freedom of the high seas principle? (Humphries et al., 2025a). The answer to this question would, the countries thought, end the debate about whether benefit sharing would be mandatory and that benefit sharing would include monetary benefit sharing, in addition to non-monetary benefit sharing. Should the countries agree that the common heritage of mankind principle applies, then the mandatory benefit sharing (including monetary benefit sharing) scheme would be incorporated in the BBNJ Agreement, as in the same manner in the implementing agreement on deep seabed mining (A/RES/48/263). Developed countries strongly opposed benefit sharing to be mandatory and moreover, to include mandatory monetary benefit sharing. The disagreements stalled the discussions on MGRs during the Preparatory Committee meeting, until the Facilitator urged the negotiators to focus on the "how" and "park" the debate over the two principles (Humphries et al., 2025a).

In terms of the application of the common heritage of mankind principle, the developing countries argued that the two articles below apply to MGRs:

> The Area and its resources are the common heritage of mankind (UNCLOS, art. 136)

> For the purposes of this Part "resources" means all solid, liquid or gaseous mineral resources in situ in the Area at or beneath the seabed, including polymetallic nodules; (...) (UNCLOS, art. 133(a))

The principle of common heritage of mankind has its origins in the recognition of 'common' interests of humankind in various multilateral agreements, including the *Antarctic Treaty (1959)* and the *Outer Space Treaty (1967)* (Loan, 2004). In 1970, the UN General Assembly adopted the principle governing the seabed and seafloor that specifically includes the principle of common heritage of mankind (A/RES/2749/XXV), and subsequently in UNCLOS (Part XI, art. 136) in 1982. The most prominent application of the common heritage of mankind principle is under Part XI of UNCLOS requiring that activities in the Area (i.e., the seabed and ocean floor and subsoil thereof, beyond the limits of jurisdiction, UNCLOS art. 1(1)) "be carried out for the benefit of mankind as a whole, irrespective of the geographical location of the States, whether coastal or landlocked." (UNCLOS, art. 136). Common heritage of mankind therefore combines both intragenerational with intergenerational dimensions of equity (Tiladi, 2015).

Under standard rules of textual interpretation, the phrase "the Area and its resources are common heritage of mankind" in Article 136 means "the Area is common heritage of mankind" and "its resources are common heritage of mankind." The definition of 'resources' in Article 133(a) does not limit the first part of the article, which states that "the Area is common heritage of mankind". If the Area is common heritage of mankind, then it and everything in it, including MGR, would be subject to the rights and limitations: non-appropriation, peaceful use, freedom of access/research, sustainable use and equitable sharing of benefits. Such interpretation would also be more consistent with the UN General Assembly's previous use of the principle, which referred to resources in general, not constrained to the definition in Art 133(a) of UNCLOS (Allen, 2001; Milicay, 2015).

On the other hand, developed countries argued that the limitation of 'resources' to mineral resources combined with a lack of specific provisions for non-mineral resources showed clear intent to limit the scope of the common heritage of mankind principle in UNCLOS to mineral resources. Such argument is consistent with the scientific understanding of the deep seabed at the time of negotiations (from 1973 to 1982), because it was a decade out from molecular biology being integrated into marine scientific research. Also, there was a prevailing view at the time of minimal, if any, diversity on abyssal seabeds due to lack of light, despite seminal studies dating back to 1972 (Dayton & Hessler, 1972) to indicate otherwise. With regards to activities relating to MGRs specifically, the developed countries also argued that research on MGRs would fall under UNCLOS Article 238, the right to conduct marine scientific research (Milicay, 2015).

The traditional law of the sea was famously born in the competition between Hugo Grotius concept of *mare liberum* (1605), the freedom of the seas, and John Seldon's refutation of the concept in *mare clausum* (1935) and included freedom of navigation in the high seas and innocent passage in territorial waters (Nandan & Dalaker, 2021). Both concepts were included in the later codification of the law of the sea in UNCLOS. The corresponding text in UNCLOS Article 87 on "freedom of the high seas" clarifies that the high seas are open to all States, whether coastal or landlocked and spells out the following freedoms:

(a) Freedom of navigation.
(b) Freedom of overflight.
(c) Freedom to lay submarine cables and pipelines subject to Part VI.
(d) Freedom to construct artificial islands and other installations permitted under international law, subject to Part VI.
(e) Freedom of fishing subject to the conditions laid out in Sect. 4.2 ['conservation and management of the living resources of the high seas'].
(f) Freedom of scientific research, subject to Parts VI and XIII.

As the above list show, the freedom of the high seas principle has certain conditions.

Marine scientific research provisions in UNCLOS (Part XIII) also have certain conditions. While UNCLOS starts with the paragraph, "[a]ll States, regardless of their geographical location, and competent international organizations have the right to conduct marine scientific research" (UNCLOS, art. 238), there is a list of conditions attached to such right, which includes mandates to conduct marine scientific research exclusively for peaceful purposes, conducting research with appropriate scientific methods and means compatible with UNCLOS, research must not unjustifiably interfere with other legitimate uses of the sea, and requirements on protection and preservation of the marine environment (UNCLOS, art. 240 (a)-(d)).

The Answer on the Applicable Principle on MGRs

The freedom of the high seas as a principle was presented as an option by the President's aid to negotiations in 2018 (A/CONF.232/2019/1), but the principle was not included in any of the later draft texts until it was re-introduced as a counterbalance to the demand to include the common heritage of mankind principle. The lack of discussion for most of the intergovernmental conference on the principles was to deter the anticipated deadlock based on the past discussions (Humphries et al. 2025a). As it was done during the Preparatory Committee, the Facilitator led the discussions focusing on the mechanism of benefit sharing. Such leadership allowed the negotiators to create a benefit sharing mechanism that allowed both aspects of the principles—freedom of the high seas/right to conduct marine scientific research and the common heritage of mankind—to be operationalized in a fit-for-purpose way (Chap. 14 of this book provides detailed explanation of benefit sharing scenarios). The result of the negotiations led to a creation of a benefit sharing with mandatory non-monetary and monetary benefit sharing, but a system that also reflects the best scientific practices and protects the rights of the Indigenous Peoples and local communities on relevant traditional knowledge.

After the benefit sharing mechanism was agreed by the countries, the negotiators tackled the choice of the legal principles. From a drafting perspective, one might wonder, why would the BBNJ Agreement need principles if the relevant concepts were fully operationalized through the provisions of the Agreement? In other words, the negotiators could have chosen not to include the common heritage of mankind or the freedom of the high seas principles. The purpose of the principles (as discussed in Sect. 4.1.4 of this chapter), arguably, is to provide guidance on the means to achieve the objective. Including principles that are not needed for the Parties to implement the means to achieve the objective of the MGR provisions seem unnecessary and possibly risks interpretation issues in the future. Given both the length and ferocity of this debate, it will come as no surprise that States were equally divided during the BBNJ Agreement negotiations between using common heritage of mankind as a guiding principle for the MGR framework and those that supported no specific reliance or the use of freedom of the high seas principles (Tiladi, 2015).

However, the negotiators landed on the following principles:

> The principle of the common heritage of humankind which is set out in the Convention (BBNJ Agreement, art. 7(b))
>
> The freedom of marine scientific research, together with other freedoms of the high seas (BBNJ Agreement, art. 7(c)

The compromise reflected in the above referenced text was only struck in the last moments of the negotiations, highlighting the great importance that both sides (developed and developing countries) attached to it. The text is possibly open to multiple avenues of interpretation, potentially achieving the goal of some countries not to resolve but to preserve the controversy. The inclusion of language "which is set out in the Convention" after the common heritage of humankind could be interpreted as an attempt to limit the application of common heritage of mankind to that of UNCLOS, but as discussed below, may not have achieved that objective.

Arguably, the final wording may be interpreted broadly to extend common heritage of mankind principle to MGRs based on three grounds. First, as outlined above, UNCLOS itself can be interpreted to mean that common heritage of mankind applies to MGRs in the Area. A second ground is that nothing in UNCLOS would restrict application of common heritage of mankind to MGRs in the water column, because common heritage is not incompatible with freedom of the high seas. Moreover, freedom of the high seas is a limited freedom, conditioned by the provisions of UNCLOS as discussed above and by other rules of international law.

Also, the wording of Article 7(b) in the BBNJ Agreement may leave open the possibility for the common heritage of mankind to be applied to all MGRs (and possibly more). The final text shared after the all-night marathon session read, "the principle of the common heritage of humankind which is set out in the Convention." The word 'which' (as opposed to the word "as" for example) indicates a non-restrictive clause. In this case, while the principle is set out in UNCLOS, its scope is not necessarily limited to that set out in UNCLOS.

Yet another avenue of interpretation is to acknowledge that one of the central tenets of common heritage of mankind is benefit sharing—to ensure a more equitable allocation of benefits from the ocean (Morgera, 2016). The operationalization of benefit sharing for MGRs in Part II of the BBNJ Agreement is an implicit recognition that common heritage of mankind principle applies. However, delegations opposing such interpretation have stressed that their agreement to benefit sharing under Part II represented a change of policy, not legal interpretation.

As the above arguments show, discussions based on principles deters the focus away from the benefit sharing provisions that could begin to provide the desired benefits. Debating on the wording of the principles will only delay the implementation of the Part II of the BBNJ Agreement. A better option would be to focus on the "how", with a goal to implement the new requirements in an effective and an efficient manner. The operative text lays out the steps for the Parties to take action on benefit sharing. Therefore, the Parties could concentrate on addressing some of the technical challenges that will arise in the implementation phase (Rabone et al., 2025).

4.2.3.2 Equity and Fair and Equitable Sharing of Benefits

The question of equity underlay the arguments over common heritage of humankind and freedom of the high seas, as discussed above in Sect. 4.2.3.1. The principle of equity in international law has several dimensions. Equity can serve as a source of law, an aid to interpretation of treaties, a means of settling international disputes, or even a justification for derogation from law (Francioni, 2024; Sohn, 1988; Titi, 2021). At its heart, equity requires justice, including allocation of resources, costs, and benefits (Lapidoth, 1987; Titi, 2021).

Equity appears across international law, from the obligation of "equitable and reasonable" use of shared watercourses to the need to protect the climate system for present and future generations "on the basis of equity" in *UN Framework for Climate Change Convention* (UNFCCC art. 3.1). The principle of intergenerational equity recognizes the rights of each generation to use and enjoy the natural resources of the planet, and the corresponding duty to conserve these resources for the future (Brown Weiss, 1992). Numerous multilateral agreements, such as the UNFCCC (art. 3), the *Convention on Biological Diversity* (1760 U.N.T.S. 79) (preamble, art. 2), the *Convention concerning the Protection of the World Cultural and Natural Heritage* (1037 U.N.T.S. 151) (*World Heritage Convention)* (art. 4), and the *United Nations Economic Commission for Europe Convention on the Protection and Use of Transboundary Watercourses and International Lakes,*] (art. 2.5(c)) explicitly reference obligations to future generations. The principle also appears, inter alia, in the Declaration of the United Nations Conference on the Human Environment (*Stockholm Declaration)*(principles 1, 2), and the *Rio Declaration* (principle 3). The complementary principle of intragenerational equity addresses inequality among countries, communities and individuals in the present generation, and is recognized, inter alia, in the Sustainable Development Goal (SDG) 10: Reduce inequality within and among countries.

In the context of biodiversity and genetic resources, the *Convention on Biological Diversity* and its *Nagoya Protocol* set out the "fair and equitable sharing of benefits" as a guiding objective. Benefit sharing can be seen as both an aspect of the broader principle of equity and a mechanism for achieving inter- and intragenerational equity (Morgera, 2014). Both the Nagoya Protocol and other access and benefit sharing (ABS) frameworks, such as the *International Treaty on Plant Genetic Resources for Food and Agriculture, 400 U.N.T.S. 300* (Plant Treaty) do not define what is meant by 'fair and equitable' and what fairness and equity should be measured against. Such lack of understanding makes it difficult to assess the achievement or otherwise of benefit sharing objectives. The design of the ABS is one of the reasons why there are few publicly available examples of benefit sharing that achieve conservation objectives (Laird et al., 2020).

The unique geopolitical circumstances of areas beyond national jurisdiction where there is no sovereignty over resources and no 'provider' of MGRs as a party to an ABS transaction (such as under the Nagoya Protocol) means that the fairness and equity principle is evolving. There is precedent for this evolution. While the meaning of the terms fairness and equity are undefined under the Plant Treaty, the principle is directed toward a specific objective—for sustainable agriculture and food security (Plant Treaty, art.1).

In the context of the BBNJ Agreement, the principle of equity and the fair and equitable sharing of benefits in Article 7(d) combined with the specific objectives of the Part II of the Agreement of fair and equitable sharing of benefits (BBNJ Agreement, art. 9(a)) means that the principle of fairness and equity in the context of MGRs is directed toward the purpose of "conservation and sustainable use of marine biological diversity of areas beyond national jurisdiction," which overlaps with the overall objective of the BBNJ Agreement in Article 2.

However, the BBNJ Agreement's MGR provisions go even further. The MGR provisions are designed to build capacity of all States, especially developing States to carry out activities with respect to MGRs, which will include having access to information, knowledge and marine technology (BBNJ Agreement, art. 9(b)-(d)). This means that the principle may evolve in the BBNJ Agreement context as "fair and equitable sharing of benefits" for people and planet, which aligns with the Kunming-Montreal Global Biodiversity Framework's (CBD/COP/DEC/15/4) goal. By invoking the broader principle of equity, the provision potentially encompasses inter- and intragenerational equity and associated obligations of conservation and sustainable use.

4.2.3.3 Information and Knowledge

The term "science" in global environmental governance refers to authoritative knowledge (Forsyth, 2021) that can help understand problems and inform solutions for a given issue. Many international agreements as well as national legislation (Sullivan et al., 2006) require actors to use best available science or similar iterations of the same concept. UNCLOS, for example, requires States to base their high seas fisheries conservation and management measures on "the best scientific evidence available" (UNCLOS, art. 119(a)) and the *UN Fish Stocks Agreement* reiterates such mandate (UNFSA, art. 5(b)).

Despite—or because—of its widespread use, there is no agreed definition of what constitutes "best available science." Some multilateral agreements, such as the *Convention in International Trade of Endangered Species* have non-binding and non-exhaustive guidance on methodologies and sources of information (Resolution Conf.16.7). In the BBNJ Agreement, Article 7(i) states, "the use of best available science and scientific information" in the General Principles and Approaches, without a definition of the concept. The COP may consider including in the terms of reference or the scope of work of the STB, so that the subsidiary body can provide recommendation that may be helpful to gain clarity as to what is considered best available science and scientific information (Muraki Gottlieb et al., 2025).

The BBNJ Agreement's General Principles and Approaches also includes the following:

> The use of relevant traditional knowledge of Indigenous Peoples and local communities, where available (BBNJ Agreement, art.7(j)).

The concept of traditional knowledge is used throughout the BBNJ Agreement. A detailed exploration of the role of the traditional knowledge of Indigenous Peoples and local communities is in Chap. 8 of this book. Here, we note simply that the Article 7(j) quoted above reflects the effort by some of the negotiators to ensure that the BBNJ Agreement reflected the important role that traditional knowledge plays in conservation and sustainable develpment of ocean resources in areas beyond national jurisdiction. Further, in relation to the benefit sharing associated with MGRs, the BBNJ Agreement has a separate article and a regime that is different from the notification and benefit sharing process for scientific research. (BBNJ Agreement, art. 13). Such distinction was made to ensure that the benefit sharing associated with traditional knowledge would be compatible with the Nagoya Protocol so that the rights of Indigenous Peoples and local communities over their traditional knowledge would be protected and compensated.

4.2.3.4 Grouping of Developing Countries: Small Island Developing States, Least Developed Countries, and Landlocked Developing Countries

The BBNJ Agreement's General Principles and Approaches include articles on different categories of developing countries:

> Full recognition of the special circumstances of small island developing States and of least developed countries (BBNJ Agreement, art. 7(m))

> Acknowledgement of the special interests and needs of landlocked developing countries (BBNJ Agreement art. 7(n))

The concept of the "special circumstances of small island developing States" was first introduced at the UN Conference on Environment and Development in June 1992 in Rio de Janeiro (A/RES/44/228). The report from the conference states, "[s]mall island developing States, and islands supporting small communities are a special case both for environment and development. They are ecologically fragile and vulnerable. Their small size, limited resources, geographic dispersion and isolation from markets, place them at a disadvantage economically and prevent economies of scale." (A/Conf.151/26/Rev.1 (Vol.I) Chapter 17.123). Indeed, Coelho argues that "[t]he enhancement of the scientific and technological capabilities of Small Island Developing States (SIDS) is fundamental to addressing vulnerabilities arising from their special circumstances." (Coelho, 2024).

Least developed countries (LDCs) identification was created by the UN General Assembly's resolution in 1971 (A/RES/2768/XXVI). The UN's definition of LDCs is "countries that have low levels of income and face severe structural impediments to sustainable development." (UN DESA, 2024). The countries are considered the most disadvantaged among the developing countries. UN DESA states that there are 45 countries in this category and most are in Africa, but some are Pacific small island development States (SIDS) (UN DESA, 2024). The UN provides various support measures for the countries that qualify to be in the LDC category: international trade; development cooperation; and support for participation in international forums (UN DESA, 2024).

There are 32 landlocked developing countries (LLDCs) in Africa, Asia, Europe and South America (Vienna Programme of Action). There are approximately 440 million people who live in LLDCs and they face challenges due to their geographical location (Vienna Programme of Action). The Vienna Programme of Action, a decade long work ending in 2024 includes an objective to "promote unfettered, efficient and cost-effective access to and from the sea by all means of transport, on the basis of the freedom of transit, and other related measures, in accordance with applicable rules of international law."

The BBNJ Agreement's MGR regime's objectives do not provide how the MGR benefits will be allocated among the long list of developing countries:

> The building and development of the capacity of Parties, particularly developing States Parties, in particular the least developed countries, landlocked developing countries, geographically disadvantaged States, small island developing States, coastal African States, archipelagic States and developing middle-income countries, to carry out activities with respect to marine genetic resources and digital sequence information on marine genetic resources of areas beyond national jurisdiction (BBNJ Agreement, art. 9(b)).

The developing countries that fall into the grouping in Articles 7(m) and 7(n) are likely to voice their "full recognition" in the case of SIDS and LDCs and "acknowledgement" of LLDCs to ensure that they are prioritized.

4.2.3.5 Precautionary Principle or Precautionary Approaches?

During the BBNJ Agreement's negotiations, there were discussions about whether certain principles were more accurately 'approaches' under international law, which is why the title of Art. 7 includes the terms "principles and approaches." The distinction or lack thereof, is discussed above in Sect. 4.1.4 of this chapter. The lack of consensus on the use of principle or approaches are reflected in BBNJ Agreement's Art. 7(e), which mentions both the precautionary principle and precautionary approach without clarification. Such compromise allowed the negotiators to move toward concluding the discussions. Art. 7(e) states, "[t]he precautionary principle or precautionary approach, *as appropriate*," (emphasis added). The text suggests that there may be a distinction between the two and such lack of clarity could lead to disputes. That said, as discussed in the conclusion of this chapter, it would be beneficial for the Parties and the COP to leave out the political divisions aside and focus on what implementation activities can best achieve the overall objective of the BBNJ Agreement: conservation and sustainable use of biodiversity in areas beyond national jurisdiction. Such a stance is particularly important as the ocean faces unprecedented and accelerating degradation.

4.2.3.6 Transparency

The principle of "transparency" is not included in the General Principles and Approaches of the BBNJ Agreement (art.7) that applies to the entire Agreement, but it is included in various sections of the Agreement, including in MGR provisions. For example, transparency is an objective of the access and benefit sharing committee in Art. 15.1. Art. 16 on monitoring and transparency provide the means of *achieving* transparency. Further, the concept of transparency is an important aspect of the institutional arrangements in Part VI, where it has a specific article for the COP to "promote transparency in decision-making processes and other activities carried out" pursuant to the BBNJ Agreement (BBNJ Agreement, art. 48.1). The article on transparency has three other subparagraphs that focus on inclusive meetings and specifically mandates the COP to perform the following actions: "dissemination of information and the facilitation of, the participation of, and consultation with, relevant global, regional, subregional and sectoral bodies, Indigenous Peoples and local communities with relevant traditional knowledge, the scientific community, civil society and other relevant stakeholders as appropriate and in accordance with the provisions of this Agreement." (BBNJ Agreement, 48.3). The transparency mandates also require for the COP to provide timely access to all relevant information (BBNJ Agreement, 48.4). Finally, all meetings of the COP and its subsidiary bodies must be open to observers in line with the COP's rules of procedure unless otherwise decided by the COP (BBNJ Agreement, 48.2). Combined, the BBNJ Agreement has interpretive tools and operationalized mandates that provide a strong foundation to ensure transparency regarding MGRs and the fair and equitable sharing of benefits.

4.2.3.7 Common but Differentiated Responsibilities

Other principles listed in the President's Note were removed entirely over the course of the negotiations, which may also contribute to understanding the intent of the MGR provisions. For example, the principle of "common but differentiated responsibilities" (CBDR), which is not mentioned in UNCLOS, but was included as a principle ten years later in the UNFCCC and the 2015 Paris Agreement (Vanderheiden, 2021). Chen states, "[t]he CBDR principle recognizes that each country should take their own responsibilities because GHG emissions issue has been contributed by all countries, but these responsibilities might be differentiated in accordance with countries' social and economic conditions because not all countries contributed equally." (Chen, 2021, 2). Even though CBDR is not specifically mentioned in the BBNJ Agreement, in Part II (MGRs) and in other sections, such as in Part V (Capacity building and the transfer of marine technology), there are prominent distinctions of responsibilities and benefits between developed and developing countries where benefits always flow from those countries with resources to countries with less. In effect, the BBNJ Agreement operationalized the CBDR without specifically using the terminology to address equity concerns.

4.2.3.8 Adjacency

Another principle mentioned in the President's aid to negotiations on MGRs that was not included in the BBNJ Agreement is "adjacency." The concept is not in UNCLOS but appears in the UN Fish Stocks Agreement (art. 7.1(a)). Adjacency in the UN Fish Stocks Agreement is a concept that addresses the rights of relevant coastal States "with respect to straddling fish stocks, the relevant coastal States and the States whose nationals fish for such stocks in the adjacent high seas area shall seek, either directly or through the appropriate mechanisms for cooperation provided...to agree upon the measures necessary for the conservation of these stocks in the adjacent high seas area." (UNFSA, Art. 7.1(a)). Mossop and Schofield (2020) provide the following explanation on adjacency, which is, "coastal States should be given "greater influence over management of those ABNJ resources to which they lie adjacent."

Adjacency considerations are not explicitly included anywhere in Part II (MGRs), in contrast to Part III (Area Based Management Tools, including Marine Protected Areas) and IV (Environmental Impact Assessments) in the BBNJ Agreement, both of which give special consideration to adjacent coastal States. Inclusion of adjacency for Part II was considered during the intergovernmental conference (i.e., President's aid to negotiations, Part III art.1.2(n), Draft Text, Part II art. 11, Revised Draft Text, Part II, art. 11), but the negotiators ultimately opted to leave it out from Part II. Mossop and Schofield (2020) give some insights as to why inclusion of the adjacency concept would be problematic: (a) the requirement to provide "due regard" for coastal States in art.9(2) lacks clarity because there is no specific information as to what needs to be done or what exactly it means, (b) the option for prior notification and consultation or consent from the coastal State where activities might result in the utilization of MGRs found both within and beyond national jurisdiction in art.10(5) would be problematic because it may be difficult to know the overall distribution of a marine specie in question.

While BBNJ Agreement's MGRs requirements do not have an explicit provision on adjacency, it does have a more diluted text that will cause implementation and compliance challenges. Article 11.3 states:

> Collection in situ of marine genetic resources of areas beyond national jurisdiction shall be carried out with due regard for the rights and legitimate interests of coastal States in areas within their national jurisdiction and with due regard for the interests of other States in areas beyond national jurisdiction, in accordance with the Convention. To this end, Parties shall endeavour to cooperate, as appropriate, including through specific modalities for the operation of the Clearing-House Mechanism determined under article 51, with a view to implementing this Agreement.

The above text is more vague and expansive than the adjacency concept discussed above. How would research institutions or others that will collect samples in ABNJ know which coastal States and other States have "right and legitimate interests"? What does it mean to carry out the activities with "due regard"? Without more guidance, the lack of clarity will cause unintentional non-compliance by the regulated community. To effectively implement the requirement in 11.3, the BBNJ Agreement's Scientific and Technical Body could provide recommendations to the COP on some of the challenges and options for compliance that will foster scientific research and development (Muraki Gottlieb et al., 2025).

4.2.4 Objectives on MGRs

The BBNJ Agreement has a general objective (art.2) that applies to the Agreement as a whole and tospecific subject matter specific objectives. The latter would be read in light of the former, but the subject matter specific objectives would supersede the general objectives in the case of conflict or inconsistencies. This section will focus on two aspects on objectives: (a) complementarity between the overall objective of the BBNJ Agreement and the objectives specific to Part II (Table 4.4); and (b) consideration on peaceful purpose.

4.2.4.1 Complementarity of the BBNJ Agreement's General Objective with the Objectives Specific to Part II (MGRs)

As the list of the objectives for MGRs in the Draft Text shows (Table 4.4, left column), the first textual proposals did not include the concept of conservation. The list of objectives specifically focused on economic benefits. Since the overall objective in BBNJ Agreement's Article 2 include conservation (in addition to sustainable use), arguably, it could also apply to the MGR objectives. That said, the lack of

Table 4.4 Comparison of the list of objectives in the Draft Text to the BBNJ Agreement

Draft Text (May 2019) Objectives for MGRs and the fair and equitable sharing of benefits	**BBNJ Agreement** Objectives for MGRs and the fair and equitable sharing of benefits (Article 9)
[(a) Build the capacity of developing States Parties, in particular least developed countries, landlocked developing countries, geographically disadvantaged States, small island developing States, coastal African States and developing middle-income countries, to access and utilize marine genetic resources of areas beyond national jurisdiction] [(b) Promote the generation of knowledge and technological innovations, including by promoting and facilitating the development and conduct of marine A/CONF.232/2019/6 8/46 19-08171 scientific research in areas beyond national jurisdiction, in accordance with the Convention] [(c) Promote the [fair and equitable] sharing of benefits arising from the utilization of marine genetic resources of areas beyond national jurisdiction] [(d) Promote the development and transfer of marine technology [, subject to all legitimate interests, including, inter alia, the rights and duties of holders, suppliers and recipients of marine technology]] [(e) Contribute to the realization of a just and equitable international economic order]	(a) *The fair and equitable sharing of benefits arising from activities with respect to marine genetic resources and digital sequence information on marine genetic resources of areas beyond national jurisdiction for the conservation and sustainable use of marine biological diversity of areas beyond national jurisdiction* (b) *The building and development of the capacity of Parties, particularly developing States Parties, in particular the least developed countries, landlocked developing countries, geographically disadvantaged States, small island developing States, coastal African States, archipelagic States and developing middle-income countries, to carry out activities with respect to marine genetic resources and digital sequence information on marine genetic resources of areas beyond national jurisdiction* (c) *The generation of knowledge, scientific understanding and technological innovation, including through the development and conduct of marine scientific research, as fundamental contributions to the implementation of this Agreement* (d) *The development and transfer of marine technology in accordance with this Agreement*

Note: Unlike the general principles and approaches in Table 4.3, all the concepts introduced in the Draft Text (left column) were incorporated into the BBNJ Agreement with certain modifications

specific link to conservation in the Draft Text could have confirmed the economic focus on ABS regimes. Regarding ABS in other contexts, Humphries et al. states, "[t]oday, it is more common for governments to consider benefits for conservation to result indirectly from sharing of materials, scientific data, forms of capacity building and contractual agreements between parties. However, there is little or no published research providing evidence that the ABS transaction per se is an effective tool for conservation of genetic resources (as opposed to equitable use)." (Humphries et al., 2020). For that reason, the authors provided an option for a BBNJ Agreement's benefit sharing mechanism that had a specific link to conservation.

The BBNJ Agreement's MGR objectives specify that "the fair and equitable sharing of benefits arising from activities with respect to marine genetic resources and digital sequence information on marine genetic resources of areas beyond national jurisdiction for the conservation and sustainable use of marine biological diversity of areas beyond national jurisdiction" (BBNJ Agreement, art. 9(a)), which clearly complements the general objective in the Agreement(art. 2). The objective of conservation is also reflected in the BBNJ Agreement's paragraphs on the activities with respect to MGRs (art. 11.6) and on the fair and equitable sharing of monetary benefits (art . 14.5).

4.2.4.2 Peaceful Purposes

In terms of the concept, "peaceful purposes," it is not part of the objectives of Part II of the BBNJ Agreement (BBNJ Agreement, art.9). However, the concept appears in the Agreement's Article 11.7 under the title, "activities with respect to marine genetic resources of areas beyond national jurisdiction.":

> Activities with respect to marine genetic resources and digital sequence information on marine genetic resources of areas beyond national jurisdiction shall be carried out exclusively for peaceful purposes.

However, the BBNJ Agreement's Article 10.3 carves out military activities from the BBNJ Agreement by stating:

> [t]he obligations of this Part shall not apply to a Party's military activities, including military activities by government vessels and aircraft engaged in non-commercial services. The obligations in this Part with respect to the utilization of marine genetic resources and digital sequence information on marine genetic resources of areas beyond national jurisdiction shall apply to a Party's non-military activities.

Read together, the two articles may give significant leverage for a Party to conduct activities associated with MGRs that would be contrary to Article 11.7's goal of peaceful purposes. For further discussion of this concept, see Chap. 10 (Ardron et al., 2025).

4.3 Considerations for Implementation

4.3.1 Focus on the "How"

From an interpretation point of view, it was the operationalized aspects that drove the negotiators to draft the relevant language in the Preamble, General Principles and Approaches, and the specific objectives for Part II. The effect of such drafting resulted in the greater focus on paragraphs that operationalized benefit sharing through specific processes, and the interpretation tools were drafted either as a confirmation of the operative paragraphs (i.e., freedom of the high seas and the common heritage of humankind) or as a matter of compromise (i.e., FHH digital sequence information).

As for the digital sequence information provisions in the Preamble and Part II, objectives were a last-minute effort to conclude the negotiations and do not provide much guidance. Rather, interpretation of the obligations will need to be filled by the BBNJ Agreement's Scientific and Technical Body and reflecting the ongoing discussions in other international fora, such as at the Convention on Biological Diversity (Rabone et al., 2025). By design, the BBNJ Agreement provides flexibility and gives the Scientific and Technical Body an important role to ensure that rapidly evolving science and policy on digital sequence information can be future proofed.

4.3.2 Layering Various Articles on the Same Topic to Strengthen Implementation

For some topics, the BBNJ Agreement takes the approach of layering or repeating the concepts in the Preamble, General and General Principles and Approaches to provide guidance on the operationalization of the aspirations of the countries that adopted the text by consensus. The three examples below show that a single topic is referenced in three legal interpretation tools (i.e., Preamble, General Principles and Approaches, and MGR specific objectives). Moreover, Part II provides specific processes that operationalizes the Preamble, General Principles and Approaches and the MGR specific objectives.

Hulme states, in terms repeating concepts used in the preamble, "by repeating and expanding upon its statements of object and purpose in later provisions, they lifted from it the burden of doing heavy legal work. This approach of intentional redundancy is one that could be replicated by treaty drafters today to tame the object-and-purpose analysis mandated by the VCLT." (Hulme, 2016). Indeed, in the BBNJ Agreement, the drafters certainly used this approach, albeit the focus was always on the operative text in the MGR requirements (Table 4.5).

Table 4.5 "Layering" examples of text in the Preamble, General Principles and Approaches and MGR specific objectives that work synergistically with the operative paragraphs

Topic	Preamble	General Principles and Approaches	MGR specific objectives	MGR requirements
Indigenous Peoples and local communities	Para.7 and 8	Art.7(j) and 7(k)	9(c)	Art. 13
digital sequence information	Para 12	Art. 7(i)	9(a)	Art. 11, 12, 14, 15,16
Capacity building and the transfer of marine technology	Para 6	Art. 7(m), (n)	9(b)	Art. 14, Part V

4.3.3 Importance of International Cooperation and Coordination in Implementation: Not Undermining

In terms of implementation, an important aspect is reading Article 5 (relationship between the BBNJ Agreement and UNCLOS and relevant and legal instruments and frameworks and relevant global, regional, subregional and sectoral bodies) and Article 8 (international cooperation) together to meet the overall objective of the BBNJ Agreement, which is to foster conservation and sustainable use of BBNJ (art.2). International cooperation and coordination are included in the general objective of the BBNJ Agreement (art. 2) to ensure the conservation and sustainable use of BBNJ. While the concepts in Articles 5 and 8 are not part of the General Principle and Approaches, they fit the definition of principles discussed in Sect. 4.1.4 of this chapter which is, "treaty's goals and the character of the means employed to achieve them." (Jonas & Saunders, 2010).

The significance of reading the two articles together is that the term "not undermining" in Article 5 is not defined in the BBNJ Agreement. Langlet and Vadrot have conducted a detailed study of the number of relevant intergovernmental organizations that could be impacted by the BBNJ Agreement and reports that there are 52 existing international organizations (Langlet & Vadrot, 2023). Article 8 has two sub-articles on this point:

> Parties shall cooperate under this Agreement for the conservation and sustainable use of marine biological diversity of areas beyond national jurisdiction, including through strengthening and enhancing cooperation with and promoting cooperation among relevant legal instruments and frameworks and relevant global, regional, subregional and sectoral bodies in the achievement of the objectives of this Agreement.
>
> Parties shall endeavour to promote, as appropriate, the objectives of this Agreement when participating in decision-making under other relevant legal instruments, frameworks, or global, regional, subregional or sectoral bodies.

By interpreting the term "not undermine" in Article 5 with the concepts in Article 8, the implementation will also meet the general objective of the BBNJ Agreement in Article 2.

A more detailed exploration of the infrastructure of the BBNJ Agreement that provide insights on the institutional arrangements is available in Chap. 2 of this book (Muraki Gottlieb et al., 2025).

4.4 Conclusion

As explored in this chapter, the BBNJ Agreement has layered or repeated the concepts in the Preamble, General Principles and Approaches and MGR specific objectives to provide guidance on the operationalization of the benefit sharing of MGRs (BBNJ Agreement, Part II), which support the unified vision and aspirations of the nearly 200 countries that adopted the text by consensus. That said, regarding MGRs, the negotiators focused on the "how" first, then worked on the Preamble, General Principles and Approaches and the MGRs specific objectives. As explored in this chapter such negotiation method allowed the countries to break the political gridlock and to progress toward reaching a consensus.

The negotiating countries' agreement of the package approach, "all the four substantive elements would be negotiated together as a package" created a need for the negotiators to ensure that the most politically difficult topic would not lead to a lack of a global binding agreement to conserve and sustainably use marine biological diversity beyond national jurisdiction. (Humphries et al., 2025a). Over the two decades that the studies and negotiations took for the BBNJ Agreement to be adopted by consensus, the deterioration of ocean health had accelerated to an unprecedented rate and resulted in devastating changes. The Area-Based Management Tools, including Marine Protected Areas and Environmental Impact Assessment would provide significant conservation and protective measures for nearly 70% of the global ocean. Further, the capacity building and the transfer of marine technology could address the much-needed scope and the amount of support that could contribute to a healthier ocean. A robust implementation of the BBNJ Agreement will make significant contributions to achieving the ocean related conservation targets, including the goals of the UNFCCC, the Kunming-Montreal Biodiversity Framework, and the Agenda 2030 (UN Secretary General, 2023). The rights of the Indigenous Peoples and local communities and their contributions for the future of the ocean would also be addressed. A summary of the guidance on the practical aspects of implementation on the issues below and the chapters that explore further on specific topics can be found in in Chap. 1 of this book:

- Implementation of fair and equitable sharing of benefits.
- Best practices in use of best available science.
- Implementation associated with requirements on traditional knowledge of Indigenous Peoples and local communities.
- Implementation on ensuring the special circumstances and special interests and needs of groups of countries (i.e., SIDS, LDCs and LLDCs) (Humphries et al., 2025a).

Given the critical nature of the BBNJ Agreement, there was urgency for the countries to conclude the negotiations and agree on a robust BBNJ Agreement that could make lasting positive changes for the current and future generations (Humphries et al., 2025a).

Now with the adopted BBNJ Agreement with 91 signatories and seven Parties (as of June 26, 2024), it is imperative for the countries and the regional economic integration organizations to work on the next important steps. In the coming years towards implementation and during implementation, stakeholders need to understand the overall objective of the BBNJ Agreement and compromises made during the negotiations. Moreover, the innovative thinking in reaching a consensus and building in flexibility to future-proof the BBNJ Agreement can provide inspiration for future negotiation of agreements. In implementing the BBNJ Agreement, remembering and honoring the negotiators' efforts to emphasize the "how"—operationalizing benefit sharing of MGRs—in Part II of the Agreement, could avoid prolonged discussions that could delay effectively implementing the new obligations. Rehashing the disagreements over the Preamble, General Principles and Approaches and MGRs specific objectives that stalled the negotiations would cost precious time on the task at hand: the conservation and sustainable use of the largest ecosystem of the Earth. Rather, focusing on resolving some of the challenging technical implementation questions will lead to providing the benefit sharing that the negotiators, especially the developing countries, desired from the historic BBNJ Agreement.

***

References

Agreement for the Implementation of the Provisions of the United Nations Convention on the Law of the Sea of 10 December 1982 Relating to the Conservation and Management of Straddling Fish Stocks and Highly Migratory Fish Stocks (A/CONF.164/37).

Agreement relating to the implementation of Part XI of the United Nations Convention on the Law of the Sea of 10 December 1982 (A/RES/48/263).

Agreement under the United Nations Convention on the Law of the Sea on the Conservation and Sustainable Use of Marine Biological Diversity of Areas beyond National Jurisdiction, 19 June 2023, 1833 U.N.T.S. 31363.

Alcock, A. (2014). From seabed to World Wide Web: an overview of marine zoological sampling, data processing and potential production of digital marine Faunas. Chapter 17. In M. F. Watson, C. H. C. Lyal & C. A. Pendry (Eds.), *Descriptive taxonomy. The foundation of biodiversity research* (pp. 214–225). Cambridge University Press. https://doi.org/10.1017/CBO9781139028004.022

Allen, C. (2001). Protecting the oceanic gardens of Eden: International law issues in deep-sea vent resource conservation and management.

Ardron, J., Kachelriess, D., Lyal, C. H. C., Nwapi, C., Rabone, M., & Sirakaya, A., & Swaddling, A. (2025). Considerations concerning state ratification of the BBNJ Agreement. In F. Humphries (Ed.), *Decoding governance of marine genetic resources under the BBNJ Agreement.* Springer.

Broggiato, A., Dunshirn, P., Jaspars, M. & Pena-Neira, S. (2025). Monetary and non-monetary benefit sharing under the BBNJ Agreement. In F. Humphries (Ed.), *Decoding governance of marine genetic resources under the BBNJ Agreement*. Springer.

Brown, A. E. L. (2025). The place of intellectual property under the BBNJ Agreement. In F. Humphries (Ed.), *Decoding governance of marine genetic resources under the BBNJ Agreement.* Springer.

Chen, Y. (2021). Reconciling common but differentiated responsibilities principle and no more favourable treatment principle in regulating greenhouse gas emissions from international shipping. Marine Policy *123*, 104317. https://doi.org/10.1016/j.marpol.2020.104317

Clark, M. R., Consalvey, M., & Rowden, A. A. (2016). *Biological sampling in the deep sea.* John Wiley & Sons.

Coelho, L. F. (2024). The practice of the Caribbean SIDS on the consent regime for marine scientific research under UNCLOS: Trends, gaps, and recommendations. *Ocean Development and International Law*, 1–29. https://doi.org/10.1080/00908320.2024.2332304

Convention on Biological Diversity, 5 June 1992, 1760 U.N.T.S. 79.

Convention concerning the Protection of the World Cultural and Natural Heritage, 17 December 1975, 1037 U.N.T.S. 151.

Convention on the Protection and Use of Transboundary Watercourses and International Lakes, 17 March 1992, 1936 U.N.T.S. 269.

Dayton, P. K., & Hessler, R. R. (1972). Role of biological disturbance in maintaining diversity in the deep sea. *Deep Sea Research and Oceanographic Abstracts, 19*(3), 199–208. https://doi.org/10.1016/0011-7471(72)90031-9

Decision adopted by the Conference of the Parties to the Convention on Biological Diversity, Kunming-Montreal Global Biodiversity Framework, 19 December 2022, CBD/COP/DEC/15/4.

Declaration of the United Nations Conference on the Human Environment, 16 June 1972.

Declaration of Principles Governing the Sea-Bed and the Ocean Floor, and the Subsoil Thereof beyond the Limits of National Jurisdiction (A/RES/2749/XXV).

Dörr, O., & Schmalenbach, K. (Eds.) (2018). *Vienna convention on the law of treaties* (p. 560). ILC and Commission in the quote refer to the International Law Commission. https://doi.org/10.1007/978-3-662-55160-8_34

Draft text of an agreement under the United Nations Convention on the Law of the Sea on the conservation and sustainable use of marine biological diversity of areas beyond national jurisdiction, 17 May 2019 (A/CONF.323/2019/6).

Eggett, C. (2019). The role of principles and general principles in the 'constitutional processes' of international law. *Netherlands International Law Review, 66*, 197–217.

Forsyth, T. (2021). J.-F. Morin, & A. Osrini (Ed.), *Essential concepts of global environmental governance* (2nd ed). Earthscan from Routledge.

Francioni, F. (2024). Equity in International Law. Oxford Public International Law. https://opil.ouplaw.com/display/10.1093/law:epil/9780199231690/law-9780199231690-e1399. Accessed February 13, 2024.

Further revised draft text of an agreement under the United Nations Convention on the Law of the Sea on the conservation and sustainable use of marine biological diversity of areas beyond national jurisdiction, 1 June 2022 (A/CONF.232/2022/5).

Further refreshed draft text of an agreement under the United Nations Convention on the Law of the Sea on the conservation and sustainable use of marine biological diversity of areas beyond national jurisdiction, 12 December 2022 (A/CONF/.232/2).

Guide to Practice on Reservation to Treaties. (2011). Adopted by the International Law Commission at its sixty-third session, in 2011, and submitted to the General Assembly as a part of the Commission's report covering the work of that session (A/66/10, para 75). The report appears in Yearbook of the International Law Commission, 2011, vol. II, Part Two.

Handbook on the Least Developed Country Category: Inclusion, graduation and special support measures, UN Department of Economic and Social Affairs, May 2024.

Hulme, M. H. (2016). Preambles in treaty interpretation. *University of Pennsylvania Law Review, 164*(5), 1281–1343, 1286.

Humphries, F. (2025). Marine genetic resources beyond national jurisdiction: the expansive scope of the BBNJ Agreement. In F. Humphries (Ed.), *Decoding governance of marine genetic resources under the BBNJ Agreement.* Springer.

Humphries, F., Berry, T., & Muraki Gottlieb, H. (2025a). Bridging divides: The evolution of marine genetic resources governance beyond national jurisdiction. In F. Humphries (Ed.), *Decoding governance of marine genetic resources under the BBNJ Agreement.* Springer.

Humphries, F., Jaspars, M., Lavelle, J., & Kachelriess, D. (2025b). The novel notification information system for marine genetic resources under the BBNJ Agreement. In F. Humphries (Ed.), *Decoding governance of marine genetic resources under the BBNJ Agreement.* Springer.

Humphries, F., Muraki Gottlieb, H., Laird, S., Wynberg, R., Lawson, C., Rourke, M., Tvedt, M. W., Oliva, M. J., & Jaspars, M. (2020). A tiered approach to the marine genetic resource governance framework under the proposed UNCLOS agreement for biodiversity beyond national jurisdiction (BBNJ). *Marine Policy, 122*, 103910.

International Treaty on Plant Genetic Resources for Food and Agriculture, 3 November 2001, 400 U.N.T.S. 303

Jonas, D. S., & Saunders, T. N. (2010). The object and purpose of a treaty: Three interpretive methods. *Vand. J. Transnat'l l., 43*(565), 571.

Kachelriess, D., Dunshirn, P., Langlet, A., Brown, A. E. L. & Scholz, A.H. (2025). Marine genetic resources and digital sequence information under the BBNJ Agreement: Interlinkages with other access and benefit sharing frameworks. In F. Humphries (Ed.), *Decoding governance of marine genetic resources under the BBNJ Agreement*. Springer.

Lavelle, J. & Wynberg, R. (2025) Benefit sharing under the BBNJ Agreement in Practice. In F. Humphries (Ed.), *Decoding governance of marine genetic resources under the BBNJ Agreement.* Springer.

Laird, S., Wynberg, R., Rourke, M., Humphries, F., Muller, M. R., & Lawson, C. (2020). Rethink the expansion of access and benefit sharing. *Science, 367*(6483), 1200–1202. https://doi.org/10.1126/science.aba9609

Langlet, A., Dunshirn, P., Jaspars, M., Humphries, F & Kachelriess, D. (2025). Monitoring and transparency aspects of MGR-utilization under the BBNJ Agreement. In F. Humphries (Ed.), *Decoding governance of marine genetic resources under the BBNJ Agreement.* Springer.

Langlet, A., & Vadrot, A. B. M. (2023). Not 'undermining' who? Unpacking the emerging BBNJ regime complex. *Marine Policy, 147*, 105372. https://doi.org/10.1016/j.marpol.2022.105372

Lapidoth, R. (1987). Equity in international law. In *SSRN Scholarly Paper*. Rochester, NY. https://papers.ssrn.com/abstract=2672190

Lee, M. (2006). The interrelation between the law of the sea convention and customary international law. *San Diego Int'l L.J., 7*, 405. Available at: https://digital.sandiego.edu/ilj/vol7/iss2/7

Loan, J. (2004). The common heritage of mankind in Antarctica: An analysis in light of the threats posed by climate change. *New Zealand Yearbook of International Law, 1*, 149.

Lothian, S. (2023). The BBNJ preamble: More than just window dressing. *Marine Policy, 153*, 105642. https://doi.org/10.1016/j.marpol.2023.105642

Milicay, F. (2015). The common heritage of Mankind: 21st century challenges of a revolutionary concept. In L. del Castillo (Ed.), *Law of the sea from Grotius to the international tribunal for the law of the sea.*

Morgera, E. (2014). Conceptualizing benefit-sharing as the pursuit of equity in addressing global environmental challenges. In: *SSRN Scholarly Paper.* Rochester, NY. https://doi.org/10.2139/ssrn.2524003

Morgera, E. (2016). The need for an international legal concept of fair and equitable benefit-sharing. *European Journal of International Law, 27*(2), 353–383.

Mossop, J., & Schofield (2020) Adjacency and due regard: The role of coastal States in the BBNJ treaty. *Marine Policy, 122*, 103877.

Muraki Gottlieb, H., Ardron, J., & Brown, A. E. L. (2025). BBNJ Agreement: A New Infrastructure to Foster Benefit Sharing of Marine Genetic Resources. In F. Humphries (Ed.), *Decoding governance of marine genetic resources under the BBNJ Agreement.* Springer.

Nagoya Protocol on Access to Genetic Resources and the Fair and Equitable Sharing of Benefits Arising from their Utilization to the Convention on Biological Diversity, 29 October 2010.

Nandan, S. N., & Dalaker, K. E. (2021). Reflections on the making of the modern law of the Sea National University of Singapore Press, 2020. xxi + 289 pp.

Paris Agreement to the United Nations Framework Convention on Climate Change, 12 December 2015, 3156 U.N.T.S. 54113.

Payne, C. R. (2022). Responsibility to the international community for marine biodiversity beyond national jurisdiction. *Cambridge International Law Journal, 11*(1), 24–50.

Peel, J. (2004). Precaution—A matter of principle, approach or process? *MelbJlIntLaw 19*; (2004) *Melbourne Journal of International Law, 5*(2), 483

Pena-Neira, S., & Coelho, L.F. (2025). Traditional knowledge in genetic resources in areas beyond national jurisdiction. In F. Humphries (Ed.),

Decoding governance of marine genetic resources under the BBNJ Agreement. Springer.

President's aid to negotiations, 2018. A/CONF.232/2019/1*.

Rabone, M., Horton, T., Humphries, F., Lyal, C., Muraki Gottlieb, H., Scholz, A. H., Vanagt, T., & Jaspars, M. (2025). BBNJ Agreement: Considerations for scientists and commercial end users of MGR at research, development and commercialisation stages. In F. Humphries (Ed.), *Decoding governance of marine genetic resources under the BBNJ Agreement.* Springer.

Report of the United Nations Conference on Environment and Development, Annex I, Rio Declaration on Environment and Development, 12 August 1992, A/CONF.151/26 (Vol/I).

Sohn, L. (1988). Equity in international law. *Proceedings of the Annual Meeting (American Society of International Law), 82*, 277–291.

Sullivan, P. J. Acheson, J. M. Angermeier, P. L. Faast, T. Flemma, J. Jones, C. M., Knudsen, E. E. Minello, T. J., Sector Wunderlich, D. H., & R. Zanetell, B. A. (2006). Defining and implementing best available science for fisheries and environmental science, policy, and management.

The Nagoya Protocol on access to genetic resources and the fair and equitable sharing of benefits arising from the utilization to the Convention on Biological Diversity, 29 October 2010, 3008 U.N.T.S. 3

Titi, C. (2021). The function of equity in international law. Oxford University Press. https://doi.org/10.1093/oso/9780198868002.001.0001

Tladi, D. (2015). The common heritage of mankind and the proposed treaty on biodiversity in areas beyond national jurisdiction: The choice between pragmatism and sustainability. *Yearbook of International Environmental Law, 25*, 113–132.

Transforming our world: The 2030 Agenda for Sustainable Development, 25 September 2015 (A/RES/70/1).

United Nations Conference on Environment and Development, 22 December 1989 (A/RES/44/228).

United Nations Convention on the Law of the Sea, 10 December 1982, 1833 U.N.T.S. 397.

United Nations Convention on the Law of the Non-Navigational Uses of International Watercourses, 21 May 1997, 1155 U.N.T.S. 331

United Nations Declaration on the Rights of Indigenous Peoples, 2 October 2007 (A/RES/61/295).

United Nations Vienna Convention on the Law of Treaties, 23 May, 1969, 1155 U.N.T.S. 331.

United Nations Framework Convention on Climate Change, 9 May 1992, 1771 U.N.T.S. 107.

United Nations Secretary General, Statement at the Intergovernmental Conference on an International Legally Binding Instrument under the United Nations Convention on the Law of the Sea on the Conservation and Sustainable Use of Marine Biological Diversity Beyond National Jurisdiction, 19 June 2023.

United Nations Treaty Collection provides a comprehensive list of countries' status of signatures, accessions, and ratifications of the BBNJ Agreement at https://treaties.un.org/Pages/ViewDetails.aspx?src=TREATY&mtdsg_no=XXI-10&chapter=21&clang=_en#EndDec. Last visited May 23, 2024.

United Nations Treaty Collection provides a comprehensive list of countries' status of signatures, accessions, and ratifications of VCLT at https://treaties.un.org/Pages/showDetails.aspx?objid=080000028003902f. Last visited April 3, 2024.

United Nations Treaty Collection provides a comprehensive list of countries' status of signatures, accessions, and ratifications of UNCLOS as well as declarations are at https://treaties.un.org/Pages/ViewDetailsIII.aspx?src=TREATY&mtdsg_no=XXI-6&chapter=21&Temp=mtdsg3&clang=_en. Last visited May 21, 2024.

Vanderheiden, S. (2021). Common but differentiated responsibilities. In J.-F. Morin, & A. Osini (Eds.), *Essential concepts of global environmental governance*. Earthscan from Routledge.

Vienna Convention on the Law of Treaties, 23 May 1969, 1155 U.N.T.S 2331.

Vienna Programme of Action for Landlocked Developing Countries for the Decade 2014–2024.

Weiss, E. B. (1992). In Fairness to future generations and sustainable development. *American University International Law Review, 8*, 1, 19–26.

Williams, S. (2010). A. Orakhelashvili, & S. Williams (Eds.), *Introduction to 40 years of the Vienna convention on the law of treaties.*

Winter, G. M., & Salomon, T. M. (eds.), (2017). *Handbook on marine environment protection, international principles of marine environmental protection*, Ch.30. https://doi.org/10.1007/978-3-319-60156-4_30

Hiroko Muraki Gottlieb is a licenced lawyer in the USA with a diversified career in global businesses, intergovernmental organizations, non-governmental organizations and academic institutions. She brings a unique blend of expertise in climate change and biodiversity conservation strategy, environmental law (domestic and international), policymaking, regulatory compliance and stakeholder engagement. She is the representative for the Ocean and led the International Council of Environmental Law to the BBNJ Agreement negotiations as the Head of Delegation and holds appointments with the Department of Organismic and Evolutionary Biology at Harvard University and Elisabeth Haub School of Law at Pace University.

Daniel Kachelriess is an expert on oceans, fisheries, wildlife law and policy and followed the negotiations of the BBNJ Agreement as part of the High Seas Alliance and as a member of the IUCN World Commission on

Environmental Law. He continues to advise the High Seas Alliance and other organizations on aspects of the BBNJ Agreement, including on Marine Genetic Resources, including the fair and equitable sharing of their benefits. His previous roles include Executive Director of Sea Shepherd Legal, a non-profit law firm, and the Marine Species Officer of the CITES Secretariat.

Lydia Slobodian is the director of the Environmental Law and Policy Program at Georgetown Law, where she teaches International Environmental Law. Prior to Georgetown, she was a senior legal officer at IUCN and led the IUCN delegation to the negotiations on Marine Biodiversity in Areas Beyond National Jurisdiction.

5 The Novel Notification Information System for Marine Genetic Resources Under the BBNJ Agreement

Fran Humphries, Marcel Jaspars, Jessica Lavelle, and Daniel Kachelriess

Abstract

The new treaty on the conservation and sustainable use of marine biodiversity in areas beyond national jurisdiction promotes equity through the fair and equitable sharing of benefits from activities with respect to marine genetic resources (MGR) and digital sequence information (DSI). To obtain information about these activities, the treaty provides a framework for a 'light touch' notification system, leaving many questions about how it might work in practice. Using textual legal analysis, this chapter draws from the negotiation history to offer a detailed interpretation of articles 11 and 12 of the treaty and the key features of the notification system: collection and utilization notifications and procedures for access to MGR and DSI in repositories and databases. It offers practical considerations about how its innovative features such as the BBNJ Standardized Batch Identifier may work in practice and summarises some priority areas for clarification if the notification system is to be effective for years to come as technologies change.

Keywords

BBNJ agreement · Biodiversity · Conservation · Sustainable use · Benefit sharing · Marine genetic resources · Digital sequence information · Notification · Collection · Utilization

F. Humphries (✉)
Griffith Law School, Griffith University, Queensland, Australia
e-mail: fran.humphries@griffith.edu.au

M. Jaspars
Marine Biodiscovery Centre, Department of Chemistry, University of Aberdeen, Aberdeen, UK

J. Lavelle
Department of Environmental and Geographical Science, University of Cape Town, Rondebosch 7701, South Africa

D. Kachelriess
IUCN World Commission on Environmental Law, High Seas Alliance, Vienna, Austria

5.1 Introduction

The new notification system for marine genetic resources (MGRs) and digital sequence information (DSI) on MGRs of areas beyond national jurisdiction (ABNJ) is a significant achievement for a jurisdictional area that does not recognise sovereignty or sovereign rights. Articles 11 and 12 of the *Agreement under the United Nations Convention on the Law of the Sea on the Conservation and Sustainable use of Marine Biological Diversity of Areas Beyond National*

F. Humphries (ed.), *Decoding Marine Genetic Resource Governance Under the BBNJ Agreement*, Sustainable Development Goals Series, https://doi.org/10.1007/978-3-031-72100-7_5

Jurisdiction (BBNJ Agreement) (UNGA, 2023a) set out the requirements for notifications concerning MGRs by users to the Clearing House Mechanism (CHM). The notification system plays a vital role in linking MGR collection and utilisation activities within the scope of Part II MGR governance of the BBNJ Agreement, with information sharing, benefit sharing, capacity building and the transfer of marine technology. While it is a component of the BBNJ Agreement's monitoring and transparency system under article 16, the notification system's innovations, including the BBNJ Standardized Batch Identifier (BBNJ Identifier), aim to link Part II with other governance systems and databases of genetic resources and DSI). This chapter aims to: analyse the evolution and interpretation of articles 11 and 12; identify key innovations and gaps in policy that need clarification by the BBNJ Agreement bodies; and offer considerations for practical implementation of the obligations.

This chapter demonstrates how the notification system arose out of a compromise during treaty negotiations for a relatively light touch system that avoided a resource-intensive access permit regime, while still facilitating information and benefit sharing. Historically, only a few high-income countries have benefitted from being able to access MGRs more readily in ABNJ, while low and middle-income countries have been challenged by capacity constraints including the high cost of deep-sea vessels and sampling technology (Blasiak et al., 2020; Stokstad, 2018). Such differentials in access have contributed to widening disparities in ocean knowledge and commercialisation opportunities. Further, while best scientific practice encourages the depositing of samples in repositories and DSI in databases, these are not always findable or publicly accessible, hindering access and utilisation of MGRs and DSI from ABNJ (Garrity et al., 2009; Slobodian et al., 2015). The BBNJ Agreement seeks to reduce these disparities through the notification system both by facilitating the opportunity for low and middle-income countries to express their interest in upcoming research cruises and facilitating access to MGR samples and DSI from ABNJ by all through transparent record-keeping, including their location and modalities for access.

The access and benefit sharing (ABS) concept under various international fora influenced investigation into a mechanism for ABNJ under the *United Nations Convention on the Law of the Sea* (UNCLOS[1]) that could promote equity and fairness from the use of biological resources (Laird et al., 2020). The *Convention on Biological Diversity*[2] (CBD) and its implementing agreement the *Nagoya Protocol on Access to Genetic Resources and the Fair and Equitable Sharing of Benefits Arising from their Utilization*[3] (Nagoya Protocol) set out international frameworks for a bilateral approach to ABS under national law or policy. This framework recognises the sovereignty of a State Party over its genetic resources (CBD art 15(1)) and its authority to regulate access to genetic resources for activities within the scope of the framework within its jurisdiction (CBD art 15, Nagoya Protocol arts 6 and 7). If a Party has access measures, it generally requires the prior informed consent (PIC) of the provider of the genetic resource or traditional knowledge associated with the genetic resource (CBD art 15, Nagoya Protocol arts 6 and 7). It is up to the Party to determine under its national legislation who has rights and obligations as the 'provider'. Monetary and/or non-monetary benefits arising from the utilization of the genetic resources or associated traditional knowledge must be shared in a fair and equitable way with the resource/knowledge provider (CBD art 15, Nagoya Protocol art 5), which is usually done

[1] *United Nations Convention on the Law of the Sea*, opened for signature 10 December 1982, 1833 UNTS 397 (entered into force 16 November 1994).

[2] *Convention on Biological Diversity*, opened for signature 5 June 1992, 1760 UNTS 79 (entered into force 29 December 1993) ('*CBD*').

[3] *Nagoya Protocol on Access to Genetic Resources and the Fair and Equitable Sharing of Benefits Arising from their Utilization,* opened for signature 29 October 2010, [2012] ATNIF 3 (entered into force 12 October 2014) ('*Nagoya Protocol*').

through contracts between the user and provider or through a benefit sharing fund (see, e.g. Humphries et al., 2021a, 2021b, 2021c). Global monitoring and information sharing under this bilateral system is primarily achieved through a system of Internationally Recognised Certificates of Compliance (proof of PIC), designated checkpoints (to collect or receive information about provenance, utilization and regulatory compliance in specific cases), checkpoint communiques (reports to the CHM) and State Party reports to the CHM about regulations and activities within their jurisdiction (Nagoya Protocol arts 17 and 18).

Other ABS frameworks take a multilateral approach, where the sharing of benefits is not dependent on authorisation for access and negotiation of sharing benefits with the provider on a case-by-case basis. The United Nations Food and Agriculture Organization (FAO) *International Treaty on Plant Genetic Resources for Food and Agriculture* (Plant Treaty)[5] and the World Health Organization' Pandemic Influenza Preparedness (PIP) framework[4] both have a contractual approach for benefit sharing but break the nexus between benefit sharing and access (Lawson et al., 2019). For example, Contracting Parties to the Plant Treaty agree to take measures to provide access to plant genetic resources in the multilateral system for the purpose of utilization and conservation for research, breeding and training for food and agriculture using a standard material transfer agreement (SMTA) including benefit sharing provisions on information exchange, access to and transfer of technology, capacity building and sharing of monetary benefits including through a fund (Plant Treaty arts 12 and 13). In other words, unlike the CBD model, users are not required to negotiate benefit sharing terms on a case-by-case basis in return for access. The PIP Framework also uses a SMTA (contractual) model at the core of its multilateral system (PIP Framework 5.4). Both have information sharing mechanisms (Lawson et al., 2019) and they apply to genetic resources from areas within national jurisdiction.

Whereas benefit sharing was especially tied to MGR governance as one of the four pillars of the BBNJ Agreement from the preparatory committee (UNGA, 2018a), regulating 'access' was a contentious issue from the start of the negotiations. The President's Aid to Negotiations included three options to regulate access: 'that access be governed by the provisions of [UNCLOS], that access be undertaken in accordance with the instrument, with provision made for access modalities, or not addressing access in the instrument' (UNGA, 2019a p. 8, 2018b Sect. 3.2.1). As a diplomatic solution, many of the provisions associated with an access framework were incorporated into a general provision concerning 'activities' with respect to MGRs of ABNJ (art 11), whereas a separate provision focused on access modalities phrased as 'notification on activities' (art 12). This avoided contentious language of 'access', while still enabling a framework for access activities in substance, albeit in the form of a notification procedure (with conditions) and not in the form of PIC or an authorisation or permit procedure.

The purpose of this chapter is to interpret and analyse articles 11 (regulated activities) and article 12 (notification mechanism) to: (a) aid stakeholder understanding of the rationale behind obligations; and (b) analyse how the final notification mechanism might be implemented in practice. Section 5.2 outlines the negotiation history of articles 11 and 12 and an overview of the final access modalities. The following sections analyse the articles through a textual analysis with support from academic sources and other treaty documents and offer insights into how gaps in interpretation might be addressed during implementation of the BBNJ Agreement. Section 5.3 analyses article 11 (regulating activities and the remaining sections analyse the notification mechanism—Sect. 5.4 (pre-collection notification), Sect. 5.5 (post-collection notification), Sect. 5.6

[4]World Health Assembly Resolution 64, Pandemic Influenza Preparedness: Sharing of Influenza Viruses and Access to Vaccines and Other Benefits, 2011 (Sixty-fourth World Health Assembly, WHA64/5, 2011).

[5]International Treaty on Plant Genetic Resources for Food and Agriculture, opened for signature, 3 November 2001, 2400 UNTS 303 (entered into force 26 June 2004).

(utilization notification) and Sect. 5.7 (access to MGR and DSI in repositories or databases). The chapter concludes that there are many innovative components of the BBNJ Agreement framework for collecting and utilizing MGRs and DSI on MGR of ABNJ and summarises priority areas for clarification if the system is to be effective once the treaty enters into force.

5.2 Evolution of the Notification System

The Facilitator of Part II noted in her report for ICG1 that countries disagreed about whether to regulate 'access' but there was convergence that marine scientific research should not be hampered regardless of which system was eventually decided (UNGA, 2018c p. 22). Underlying the debate were differences in state positions about the principles that should govern regulation—the common heritage of humankind or freedom of the high seas, including freedom of marine research, the analysis of which is beyond the scope of this chapter (see Muraki Gottlieb et al., 2025). A particular impasse for negotiations was whether to include within the agreement's scope regulation of ex situ (e.g. MGRs in repositories) or in silico (MGRs in digital form) and if so, how. Other proposals included having different levels of access for vulnerable or ecologically significant areas and different procedures for regulating access for scientific and commercial purposes (UNGA, 2018c p. 22), which would have added considerable complexity and did not end up in the final text. The unique geopolitical nature of ABNJ where there are no sovereign rights over the oceans and its resources (UNCLOS arts 89 & 137(1)) meant that more creative options were required to govern collection and use of MGRs and DSI.

5.2.1 Negotiation History of Articles 11 and 12

Discussion at IGC1 to IGC4 concerning the President's Aid to Discussions (UNGA, 2018d) and early drafts of the text proposed different access procedures depending on where the MGRs are sourced or originate:

- access options for in situ (collection) of MGR:
  - prior notification to a treaty body; or
  - a permit with terms and conditions set under the treaty; or
  - a licence with terms and conditions set under the treaty.
- ex situ access to MGRs 'shall be free and open';
- 'access to in silico information and data shall be facilitated'; and
- access to traditional knowledge associated with MGRs of ABNJ is accessed with the PIC or approval and involvement of Indigenous Peoples and Local Communities and mutually agreed terms (UNGA, 2019b Sect. 3.2.1, 2019b draft art 10, 2019c draft art 10).

The two models discussed for collection in situ were a licensing or permit-based model 'which might borrow elements from the sponsoring State system for the Area' and a notification-based model 'which would require notification of sampling or collection activities [in ABNJ] to a designated entity under the instrument before or after the activities' (2018c, p. 22). Questions arose during negotiations about who the authorising body would be and where the financial and other resources would come from for maintaining a permit or licensing system. In the end, permit-based models such as those administered by the ISA in the deep seabed (the Area)[6] or by the *Convention on International Trade in Endangered Species of Wild Fauna and Flora* (CITES) for specimens of species listed on its Appendices taken in ABNJ[7] were disregarded.

[6] See the *1994 Agreement relating to the Implementation of Part XI of the United Nations Convention on the Law of the Sea of 10 December 1982*, opened for signature 28 July 1994, 1836 U.N.T.S. 42 (entered into force 28 July 1996).

[7] See *Convention on International Trade in Endangered Species of Wild Fauna and Flora*, opened for signature 3 March 1973, 993 U.N.T.S., 243 (entered into force 1 July 1975) see art. I (c), (e), art III. (5) and art IV 6–7.

By the end of IGC 4, only the notification option remained, based on a policy decision that a 'light touch' would be the most efficient and cost-effective approach to 'access' in situ, rather than an authorisation model (see Mendenhall et al., 2022).

Countries disagreed about whether to regulate ex situ MGRs and the proposed access modalities if it was to be regulated (UNGA, 2019d). The proposed modality for 'ex situ access' from the President's Aid to Discussions was 'free and open' (UNGA, 2018d), which remained in the draft texts until the end of IGC 4 when the modality was removed for the 1 June 2022 text (UNGA, 2022a). None of the definitions in the draft texts clarified whether ex situ meant 'outside ABNJ' or MGRs in repositories or how ex situ MGRs would be distinguished from in situ MGR, particularly because as soon as they are collected and landed on board a vessel in ABNJ, they may be deemed to be within the jurisdiction of the vessel flag state and therefore ex situ—for which access is 'free and open' under this early proposal (Humphries et al., 2020). There was little clarity about what 'free and open' would mean in practice for benefit sharing and about the relationship with domestic ABS laws and repositories that already control trans-jurisdictional ex situ MGRs located in national jurisdiction (Humphries et al., 2020). By ICG 5 these questions were side-stepped when the approach changed from regulating access to requiring information to be sent to the CHM regarding access to ex situ MGR and the modalities envisaged for access ex situ for utilization within a State Party's jurisdiction (UNGA, 2022b draft art 9). The thorny legal definitional issue for 'ex situ' MGRs was avoided by removing reference to the term in the final text but the approach remained the same, albeit expanded to DSI—where modalities for access to MGRs and DSI arising from utilization within national jurisdiction are to be notified to the CHM (art 12(8)(d)). The implications for this are explored in Sects. 5.6 and 5.7 below.

Whether to include and have access provisions for DSI was one of the most contentious issues during negotiations (de la Concepción, 2024). The President's Aid to Discussions included the term 'digital sequence data' (DSI) as an option for the scope of the BBNJ Agreement (UNGA, 2018d Sect. 3). The draft texts for ICG2-4 focused on the term 'in silico', requiring that access to in silico information and data 'be facilitated' (UNGA, 2019b draft art 10). There was little clarity about what DSI meant and what 'facilitated' would mean in practice. Negotiating countries were divided between those who wanted to maintain the status quo and leave DSI outside the scope for reasons of practicality, risks of interference with open access to data and scientific or economic freedom, and those who wanted regulation for reasons of fairness, equity and capacity building. During BBNJ Agreement negotiations there were concurrent discussions in other international ABS fora about whether to govern (and if so, how) the access and use of DSI and the fair and equitable sharing of benefits from its use (Kachelriess et al., 2025).

The timing of the Kunming Montreal Global Biodiversity Framework (GBF) and its decision to create a multilateral mechanism for benefit sharing from DSI (UNEP, 2022) as well as the strong position by the G77 and China group[8] broke the deadlock at the BBNJ Agreement negotiations (de la Concepción, 2024). Consequently, states reached a compromise, which included DSI as subject matter in many provisions of Part II. Cognizant that the CBD had not yet decided on the modalities of its benefit sharing framework (UNGA, 2023b) and that regulating DSI in a manner incompatible across different international ABS fora would impair scientific research and increase regulatory burden and undermine countries' agreement to cooperate, negotiators included both general (e.g. art 8 duty to cooperate and art 5 relationship with other frameworks) and specific provisions (art 14(9) future modalities for benefit sharing) to avoid inconsistency. The latter provision allows for future modalities for

[8] Group of 77 is the largest intergovernmental organization of the 'Global South' in the United Nations to promote their collective interests—https://www.g77.org/doc/.

benefit sharing for DSI decided under the GBF to be taken into account by the CoP when deciding on future modalities for benefit sharing under the BBNJ Agreement. The final wording in the treaty text does not include DSI as subject matter of the pre-collection notification but requires Parties to ensure that the repository where DSI is or will be deposited is included in the post-collection notification. There are also information requirements for DSI under arts 12(6), (7) and (8) and proposals for automation in the notification and identifier systems as a light touch practical approach to governance (see Sect. 4.4 below).

Throughout negotiations, draft treaty texts proposed an approach to accessing traditional knowledge that incorporated the Nagoya Protocol concept of PIC and mutually agreed terms, with slight tweaks in the wording and placement of the obligation and with the addition of 'free' (free prior and informed consent) (UNGA, 2018b Sect. 3.2.1). By IGC4, the provision had moved to its own stand-alone provision and by the final text, there was agreement that access to such traditional knowledge may be facilitated by the Clearing-House Mechanism (UNGA, 2019c draft art 10bis). Chapter 8 of this edited collection goes into detail about the BBNJ Agreement framework for ABS of traditional knowledge and the considerations for implementation of this provision (Pena-Neira & Coelho, 2025). The notification mechanism examined in the present chapter will refer to the access system for traditional knowledge where relevant.

Early draft texts also proposed an option for requiring the prior consent or notification and consultation of coastal States for activities that may result in the utilization of MGRs found in areas both within and beyond national jurisdiction (UNGA, 2019b draft art 10(5), 2019c draft art 10(5)). There was considerable discussion at IGC 1–3 on this issue with some coastal States arguing that they have a special role to play in ABNJ governance that are adjacent or ecologically linked to areas within their national jurisdiction (Dunn et al., 2017; UNGA, 2019a p. 11), while other States argued that there is no accepted principle under UNCLOS that coastal States have priority over other States in ABNJ (see Mossop & Schofield, 2020). By the end of IGC3, 'there seemed to be general convergence that the prior consent' of the concerned coastal state would not be required but discussions continued about whether coastal, concerned or adjacent states 'should be notified and consulted nevertheless' (UNGA, 2019d p. 6; 2019c draft art 10). In IGC5 the issue was resolved by solely relying on the UNCLOS concept of 'due regard', requiring that collection of in situ MGRs 'should be carried out with due regard for the rights and legitimate interests of coastal states in areas within their national jurisdiction' with the procedures under which Parties 'shall endeavour to cooperate including through specific modalities for the operation of the Clearing-House Mechanism determined under article 51' during BBNJ Agreement implementation (art 11(3)).

5.2.2 Overview of the Agreed Notification Mechanism

The final modalities for (a) collection, (b) utilization and (c) access to MGRs and where practicable, DSI in repositories and databases fall within a notification and information sharing system. Notification obligations fall on Parties to the BBNJ Agreement and not directly on private individuals or entities, which is the norm for international treaties. Parties are required to implement national laws, policies or administrative measures to ensure information is notified to the CHM (art 12(1)). The treaty text does not specify whether only the Party may notify the CHM or whether there will be procedures for natural or juridical persons (e.g. scientists, repositories, corporations) to notify the CHM directly. This means that while ultimate responsibility for notification rests with a government, each country's laws may vary in the national procedures for fulfilling the obligations such as who has a duty to notify and how. Obligations concerning (c) are spread throughout the notification and benefit sharing provisions of the treaty (articles 12 and 14), requiring information

Table 5.1 Overview of modalities including timeframes for notification under the notification system

Notification category	Procedure	Article reference
Pre-collection notification of MGRs of ABNJ	Specified information must be notified to the CHM 'six months or as early as possible prior to the collection in situ of'[9] of MGRs of ABNJ;	12(2)
	where there is a material change' to information provided to the CHM prior to the planned collection, updated information must be notified to the CHM 'within a reasonable time and no later than the start of collection in situ, where practicable.'	12(4)
	CHM must automatically generate a BBNJ Identifier for the pre-collection notification	12(3)
Post-collection notification of MGRs of ABNJ	Specified information and the BBNJ Identifier must be notified 'as soon as it becomes available, but no later than one year from the collection in situ of' MGR of ABNJ	12(5)
Notification as aggregate reporting for access to MGR samples and DSI in repositories and databases	Parties must ensure that samples of MGR and DSI that are in repositories or databases under their jurisdiction can be identified as originating from ABNJ (no timeframe specified)	12(6)
	Parties must ensure that repositories and, to the extent practicable, databases under their jurisdiction prepare on a biennial basis an aggregate report on access to MGR and DSI linked to their BBNJ Identifier, and make the report available to the ABS Committee	12(7)
Utilization notification of MGRs and DSI (if practical) of ABNJ	Specified information must be notified to the CHM 'as soon as information becomes available' about 'utilization,[10] including commercialization' of MGRs of ABNJ and where practicable DSI on MGRs of ABNJ by natural or juridical persons in a Party's jurisdiction. Information includes the results of utilization, details of the post-collection notification where available, where the original sample is held, the modalities for third-party access to MGRs and DSI and associated data management plan and information on product sales and further development if available	12(8)

to be shared with the Access and Benefit Sharing (ABS) Committee set up under article 15 (see Sect. 5.7 below).

While the notification mechanism includes within its scope MGRs and, where practicable, DSI, it does not directly include traditional knowledge associated with MGRs in ABNJ, which is governed under article 13 (Pena-Neira & Coelho, 2025). However, as the information requirements are minimum standards of information that must be forwarded to the CHM, there is nothing preventing the inclusion of information about proposed and current use of Traditional Knowledge with respect to a notifiable activity in article 12.

Article 12 is divided into substantive obligations and procedural requirements for notifying the BBNJ Agreement's Clearing-House Mechanism about specified activities concerning specified subject matter (Table 5.1). Substantive obligations set out standards that must be met

[9] 'Collection in situ' means 'the collection or sampling of' MGRs of ABNJ art 1(4).

[10] Utilization means to 'conduct research and development on the genetic and/or biochemical composition of marine genetic resources, including through the application of biotechnology [any technological application that uses biological systems, living organisms, or derivatives thereof, to make or modify products or processes for specific use]' art 1(3) & (4).

through the conduct of treaty Parties whereas procedural obligations generally relate to how Parties would go about fulfilling their duties (Brunnée, 2017). There is not yet a body of knowledge about which obligations are substantive and which are procedural in the BBNJ Agreement, which have different consequences for breach (*Vienna Convention on the Law of Treaties* art 60). The boundaries between the two are often fluid and operate symbiotically (Brunnée, 2017). As there is not yet state practice in relation to BBNJ Agreement obligations, Table 5.1 simply refers to the relevant category of notification and procedure without classifying them as substantive and procedural obligations.

The BBNJ Agreement imposes a duty on Parties to take the necessary legislative, administrative or policy measures to ensure that information is notified to the Clearing-House Mechanism (CHM) for the activities outlined in Table 5.1 during the specified timeframes.

The four categories of notifications under the system can only be properly understood in the context of the whole BBNJ Agreement, including the functions of the treaty bodies (Chap. 2 of this collection Muraki Gottlieb et al., 2025), the benefit sharing system (Chaps. 6 and 13 of this collection—Broggiato et al., 2025; Lavelle & Wynberg, 2025) and the monitoring and transparency system (Chap. 7 of this collection—Langlet et al., 2025). Figure 5.1 demonstrates how the notification categories and procedures work together in the notification system. In a simple scenario where collection is intended as part of a research cruise to ABNJ, there is a lineal progression through the notification steps—from pre-collection to post-collection and consequent utilization. As Chap. 14 of this book demonstrates (Rabone et al., 2025), the reality of R&D pipelines can be varied, with scenarios ranging from use of MGRs of ABNJ or DSI collected prior to the BBNJ Agreement coming into force, to the use of DSI on MGR of ABNJ without any associated collection, to the whole R&D process being carried out autonomously and with the use of artificial intelligence.

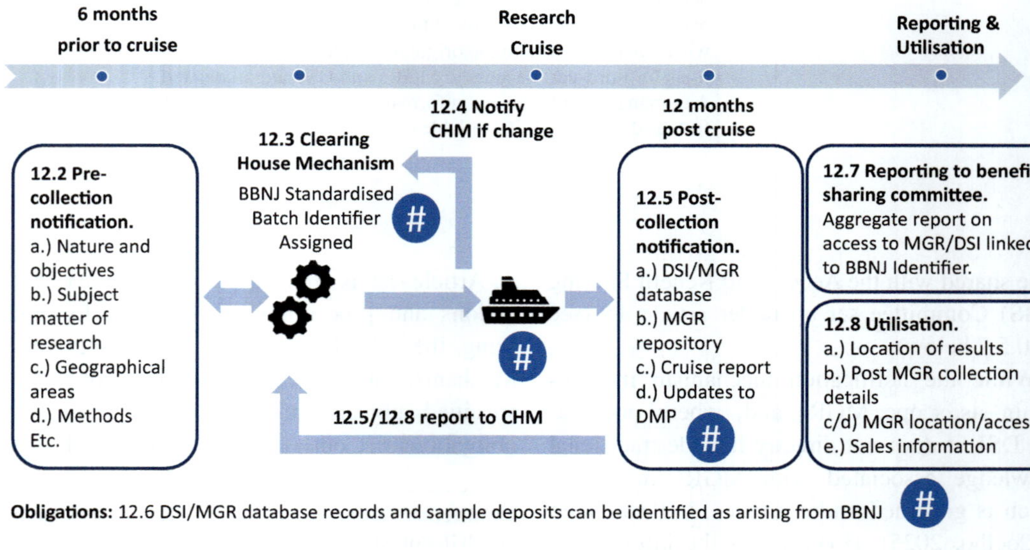

Fig. 5.1 Key elements of the notification provision

5.3 Regulating Activities (Article 11)

Article 11: Activities with Respect to Marine Genetic Resources of Areas Beyond National Jurisdiction
(1) Activities with respect to marine genetic resources and digital sequence information on marine genetic resources of areas beyond national jurisdiction may be carried out by all Parties, irrespective of their geographical location, and by natural or juridical persons under the jurisdiction of the Parties. Such activities shall be carried out in accordance with this Agreement....

Article 11 is based on UNCLOS language of the right to conduct marine scientific research (UNCLOS art 238).[11] Article 11 ensures that the activities with respect to MGRs of ABNJ and DSI on MGRs of ABNJ must be in accordance with the whole BBNJ Agreement and not only in accordance with Part II obligations. This ensures, for example, that activities are interpreted in accordance with Part I general provisions (including objectives), Part VI institutional arrangements (including the roles of the treaty bodies) and Part VII financial resources and mechanism and other administrative provisions (Parts VIII-XII). It may also ensure that notification and benefit sharing mechanisms under Part II are interpreted in light of the other substantive elements of the BBNJ Agreement (Part III Area Based Management Tools, Part IV Environmental Impact Assessment, and Part V Capacity Building and the Transfer of Marine Technology) and vice versa. The following textual interpretation of article 11 focuses on the subject matter, geographical scope, responsibilities and duties of cooperation and due regard for legitimate interests as well as the effect and purpose of the activities under the BBNJ Agreement.

5.3.1 Subject Matter and Activities

The subject matter scope of Part II (the regulated matter) is closely tied to the activities being conducted with the matter. While the BBNJ Agreement applies to MGR of ABNJ and DSI on MGR of ABNJ as the regulated matter (art 2), Part II designs a new governance framework for activities with this subject matter that are conducted by a Party's nationals both within ABNJ (collection activities) and in areas within national jurisdiction (AWNJ) ('utilization' and 'access' activities). This is consistent with the CBD framework that covers the components of biological diversity (e.g. MGR) *in* AWNJ but also processes and activities within a Party's control within the area of its national jurisdiction or beyond the limits of national jurisdiction (CBD art 4).

Earlier drafts of article 11(1) designated specific activities to which the MGR framework applies (e.g. 'collection in situ', 'utilization') of MGRs 'originating from' ABNJ (UNGA, 2022a draft art 8(a)). The final article 11 only refers to activities 'with respect to' MGR and DSI on MGR of ABNJ without defining what it means by activities. On the one hand, this might be interpreted narrowly to mean the activities specified in the notification, monitoring/transparency and benefit sharing provisions of 'collection in situ' and 'utilization' as defined under article 1. It may also extend to third-party 'access' to MGRs and DSI which is undefined.[12] Third-party access refers to a transfer of possession of the MGR or DSI to an entity other than the original collector of the MGR from which the physical materials or DSI came. For example, the MGR may have been collected by researcher A, deposited in a repository within national jurisdiction and subsequently transferred to commercial researcher B for their use. Here, researcher

[11] 'All States, irrespective of their geographical location, and competent international organizations have the right to conduct marine scientific research subject to the rights and duties of other States as provided for in this Convention' (UNCLOS art 238).

[12] Phrase mentioned in arts 12(7), 12(8) and 14(2).

B is taken to have had third-party access to the MGR.

On the other hand, 'activities' may be interpreted to broadly encompass any activity in ABNJ that relates to MGR of ABNJ (e.g. deep-sea aquaculture). The qualifier of article 11 is that the activities 'shall be carried out in accordance with this Agreement', suggesting it is not limited to activities carried out in accordance with Part II. This could include consideration of other Parts of the treaty, including Area Based Management Tools (ABMT) (Part III), Environmental Impact Assessments (EIA) (Part IV) and Capacity Building and Transfer of Marine Technology (CBTMT) (Part V) when regulating subject matter under Part II. Support for this interpretation is the activities that are further elaborated in subsequent paragraphs of article 11—state cooperation, advancing scientific knowledge, promoting conservation and sustainable use of marine biological diversity and carrying out activities exclusively for peaceful purposes. This may stretch the interpretation of article 11 beyond negotiator's intention, but Parties and resource users would benefit from clarity by the CoP on the interpretation of the breadth of activities that are regulated under Part II and the relationship with other parts of the treaty.

The absence of express reference to Traditional Knowledge in article 11 could indicate that Part II does not make a new governance framework for Traditional Knowledge but preserves the *status quo* with the existing framework under the Nagoya Protocol and national laws that regulate the use of traditional knowledge through PIC and mutually agreed terms (Pena-Neira & Coelho, 2025). On the other hand, the connection between Traditional Knowledge and MGRs by the use of the term 'associated with' MGRs in ABNJ in article 13, may suggest that article 11 requires that all provisions under Part II apply to activities carried out with respect to Traditional Knowledge associated with MGRs in ABNJ, although this was not alluded to during negotiations. The relationship between articles 11 and 13 needs further clarification by the treaty bodies and committees during implementation.

5.3.2 Responsibilities Under the BBNJ Agreement

Part II applies to activities carried out by Parties within national jurisdiction and beyond national jurisdiction and by 'natural or juridical persons under the jurisdiction of' States. The latter phrasing assumes that humans will be undertaking the collection and utilization, which potentially limits the application of the framework by not recognising advances in artificial intelligence for collection, research, development and commercialisation. The rapid technological developments in fully autonomous and uncrewed collection and research methods in ABNJ (see Sloyan et al., 2019) demonstrate the need to ensure that this framework relates to activities carried out by natural and juridical persons as well as entities driven by artificial intelligence (AI).

The implications for future-proofing the BBNJ Agreement framework in the advent of AI and autonomous vessels were well known at the time of drafting the final text. The existence of remotely operated and autonomous underwater vehicles as an important part of the research equipment in ABNJ was clear to policy makers well before the General Assembly charged the Ad Hoc Informal Working Group with the task of identifying policy gaps in 2004 (UNGA, 2005, paras 60 and 63–65). In recent years there have been rapid technological developments in fully autonomous collection and research methods in ABNJ (Sloyan et al., 2019). For example, the Woods Hole Oceanographic Institution is combining autonomous submersibles, 3D imaging cameras, autonomous sensors and DNA sampling techniques to study the deep seabed, including pioneering work in environmental DNA (eDNA) and carbon isotope analysis, for real-time information about deep-sea species without the need for human collection.[14] Using these techniques, DNA does not need to be collected from the organism itself—only from the

[14] https://twilightzone.whoi.edu/work-impact/technology/.

aquatic environment. The eDNA needs comparison with a known library or database of genetic sequences or barcodes.[15] Synthetic biology techniques using DSI from databases do not need access to the physical materials to produce new biological products, and this field is also expanding in the areas of machine learning and automation (Carbonell et al., 2019). The pre- and post-collection notifications are not limited to activities undertaken by humans so that the BBNJ Agreement is future-proofed for technological developments in autonomous sampling in ABNJ research. The omission of DSI in the trigger for pre- and post-collection notifications, however, may create the need for further guidance by the CoP when sequencing technology becomes portable enough to be deployed in ABNJ, which may not be far away.[13] The BBNJ Agreement is likewise not future proofed for automated research activities under the 'utilization notification' requirement (see Sect. 5.6).

It might be argued that developers, operators or controllers of autonomous vessels are state-based and would be captured by the Parties with responsibilities under the BBNJ Agreement. However, the legal status and legal responsibility of and for AI is unsettled in many countries (e.g. Mesevic & Skamo, 2023) and the extent to which the treaty's territorial approach to assigning responsibility can accommodate technical advances will be the subject of much debate. Court cases in several countries under different areas of law are testing the idea about whether AI has legal personality (i.e. is a juridical person). For example, recent intellectual property cases in Australia and several other countries have found (for different reasons) that AI cannot be an 'inventor' for the purposes of a patent claim (see, e.g. Afshar, 2022; Merritt, 2023). Chapter 14 of this edited collection demonstrates the loopholes in the notification system that would be created if AI activities were not captured by Part II (Rabone et al., 2025).

5.3.3 Cooperation Between Parties

> Article 11(2): 'Parties shall promote cooperation in all activities with respect to... [MGRs and DSI on MGRs of ABNJ]'.

State cooperation is one of the fundamental principles of UNCLOS (e.g. arts 118 and 242). Articulation of this principle both under article 11 and in the stand-alone article on 'international cooperation' under article 8 fills a gap with respect to cooperation in the context of the BBNJ framework (UNGA, 2005). When the United Nations General Assembly (UNGA) started to explore the idea of a treaty in 2005, it recognised that a limited number of institutions worldwide own or operate vehicles that are able to engage in deep-sea research and advocated international cooperation in sharing logistics in scientific exploration, strengthening the 'autonomous marine scientific research capability of developing states' and the flow of scientific data and information resulting from marine scientific research (UNGA pp. 25 and 51).[16] With technological advances in all activities within the scope of Parts II ('collection', 'utilization' and 'access'), part of the philosophy behind the benefit sharing concept is to cooperate to create conditions for the use of the MGRs, DSI or results of research by others (Tvedt, 2020).

5.3.4 Due Regard for State Interests

> Article 11(3) Collection in situ of marine genetic resources of areas beyond national

[13] See https://nanoporetech.com/products/sequence/minion; https://emea.illumina.com/company/news-center/press-releases/2022/bc68d667-9740-4990-890f-0ae6ff584665.html.

[15] Ocean Exploration Trust Nautilus project 'There's something in the water' https://nautiluslive.org/sites/default/files/documents/2020-04/eDNA%20There%27s%20Something%20In%20the%20Water_%20Combined.pdf.

[16] See UNCLOS article 244 concerning the publication and dissemination of information and knowledge.

jurisdiction shall be carried out with due regard for the rights and legitimate interests of coastal States in areas within their national jurisdiction and with due regard for the interests of other States in areas beyond national jurisdiction, in accordance with the Convention. To this end, Parties shall endeavour to cooperate, as appropriate, including through specific modalities for the operation of the Clearing-House Mechanism determined under article 51, with a view to implementing this Agreement.

During negotiations, there was considerable debate about whether coastal States have a special role or greater rights in parts of ABNJ that are adjacent to their Exclusive Economic Zone (Mossop & Schofield, 2020). Earlier draft texts had bracketed proposals that States Parties must ensure that activities,

> that may result in the utilization of marine genetic resources found in areas both within and beyond national jurisdiction are subject to the prior [consent] [,] [notification and consultation] of the coastal States [and any other relevant State] concerned, with a view to avoiding infringement of the rights and legitimate interests of [that] [those] State[s].] (2019b draft art 10(5)).

Reasoning for this greater role for coastal States included the interconnectivity between high seas and coastal waters (and the MGR found in both jurisdictional areas) resulting in a particular interest in how areas adjacent to their waters are managed (Mossop & Schofield, 2020). An alternative view was that the law of the sea does not contain a principle of priority for adjacent coastal States over other States in protecting biodiversity in the high seas (Mossop & Schofield, 2020). By ICG5 a solution to break the deadlock was proposed—focusing on the UNCLOS concept of 'due regard'. While the term under UNCLOS lacks an authoritative definition, it may be interpreted as comprising two components: (1) consideration of the interests, rights and duties of other states and (2) incorporating these into decision making (Mendenhall et al., 2019). However, questions remain about how the concept of due regard would work in practice in the context of MGRs of ABNJ, which rights and interests are to be regarded and which 'other States' interests trigger due regard. Parties must endeavour to cooperate on this issue 'including through specific modalities for the operation of the Clearing-House Mechanism' (art 11(3)).

The Clearing-House Mechanism under article 51 will primarily be an open-access platform with modalities for its operation to be determined by the CoP (Muraki Gottlieb et al., 2025). So, in effect, the modalities about how the pre- and post-collection notifications will be carried out with due regard for the rights and legitimate interests of coastal States (and interests of other States) are likely to be determined by the CoP. Significantly, article 11(3) only concerns collection in situ, so arguably the principle of due regard within the scope of Part II is not triggered by the 'utilization' notification—only collection. The provision appears to be confined to collection of the physical MGRs, which may spark debate about whether the DSI of the MGRs is also relevant for due regard, given the rise of in situ sequencing technologies (see Sect. 5.4 above). As due regard in this context is arguably confined to coastal States immediately adjacent to the relevant collection in question, the provision is likely to spark many years of discussion about which State's interests are regarded and what this may mean for their interests.

5.3.5 Sovereignty and Sovereign Rights

Article 11(4) No State shall claim or exercise sovereignty or sovereign rights over marine genetic resources of areas beyond national jurisdiction. No such claim or exercise of sovereignty or sovereign rights shall be recognized.

Article 11 (5) Collection in situ of marine genetic resources of areas beyond national jurisdiction shall not constitute the legal basis for any claim to any part of the marine environment or its resources.

Article 11 should be read in conjunction with article 6, which applies to all parts of the BBNJ Agreement:

> This Agreement, including any decision or recommendation of the Conference of the Parties or any of its subsidiary bodies, and any acts, measures or activities undertaken on the basis thereof, shall be without prejudice to, and shall not be relied upon as a basis for asserting or denying any claims to, sovereignty, sovereign rights or jurisdiction, including in respect of any disputes relating thereto (article 6).

These provisions maintain the *status quo* under UNCLOS with respect to sovereignty, sovereign rights and legal claims in ABNJ. UNCLOS recognises State sovereignty over their international and territorial waters and sovereign rights to exploit natural resources within that State's jurisdiction (exclusive economic zone and continental shelf) in accordance with their duty to protect and preserve the marine environment (art 192) but 'no State may validly purpose to subject any part of the high seas to its sovereignty' (art 89, See Berry, 2023). Regarding the Area,

> No State shall claim or exercise sovereignty or sovereign rights over any part of the Area or its resources, nor shall any State or natural or juridical person appropriate any part thereof. No such claim or exercise of sovereignty or sovereign rights nor such appropriation shall be recognised (art 137).

Regarding parts of the Southern Ocean and its resources that are both within the Antarctic Treaty Area and ABNJ, sovereign rights are on hold (Antarctic Treaty art IV). Humphries (2025) outlines some of the considerations for geographical scope of the BBNJ Agreement in relation to the Antarctic Treaty Area.

In the context of regulated activities, article 11(4) attempts to clarify the sovereignty status for all States and not only Parties to the BBNJ Agreement. It means that States cannot legitimately claim rights to benefit sharing as a 'provider' under their national access and benefit sharing (ABS) laws over MGRs of ABNJ that are located or 'utilized' within their jurisdiction. This is because under the CBD, genetic resources within the scope of ABS 'are only those that are provided by Contracting Parties that are countries of origin or such resources or by the Parties that have acquired the genetic resources in accordance with [the CBD]' (CBD art 15). The CBD does not have jurisdiction over the components of biodiversity (i.e. MGR) in ABNJ (see Sect. 5.3.1 above)[17] and article 11 removes beyond doubt that those located within national jurisdiction retain their independent status from a State's jurisdictional control or legal authority. The only State power recognised is the power to make the necessary legislative, administrative or policy measures for implementing the framework under Part II.

Part II, including article 11, is silent about sovereignty or self-determination of Indigenous Peoples and Local Communities (IPLCs) with respect to their traditional knowledge and the MGRs that are associated with it. Article 13 provides for limited legal protection of rights of IPLCs for PIC and MAT with respect to their Traditional Knowledge associated with MGR in ABNJ but does not touch on self-determination (see Pena-Neira & Coelho, 2025). The history of international law is a struggle between often diverging norms of sovereignty, non-intervention, territorial integrity and self-determination, often from western or European perspectives (see MacFarlane & Sabanadze, 2013). The treaty bodies may need to reconcile these norms and consider all perspectives when establishing modalities for benefit sharing.

5.3.6 Benefits and Purpose of Activities

> Article 11(6) Activities with respect to marine genetic resources and digital sequence information on marine genetic resources of areas beyond national jurisdiction are in the interests of all States and for the benefit of all humanity, particularly for the benefit of advancing the scientific knowledge of humanity and promoting

[17] See Sect. 3.1 above.

the conservation and sustainable use of marine biological diversity, taking into particular consideration the interests and needs of developing States.

Article 11 (7) Activities with respect to marine genetic resources and digital sequence information on marine genetic resources of areas beyond national jurisdiction shall be carried out exclusively for peaceful purposes.

Article 11(6) may be interpreted to mean that the activities of 'collection', 'utilization' and third-party 'access' must be conducted in a way that does not put the interest of one State above another. While it is activities that must benefit all humanity, this provision does not go so far as to say that MGRs of ABNJ are the common heritage of humankind (see Muraki Gottlieb et al., 2025). The benefits are an inclusive list that mentions human and biodiversity benefits but with a human-centric focus. The language relating to advancing the scientific knowledge *of* humanity (instead of *for* humanity) is interesting but presumably includes knowledge from the outcomes of bioprospecting and biotechnology (e.g. drugs). Promoting the conservation and sustainable use of marine biological diversity is linked to the 'benefit of all humanity' but could be read broadly for the intrinsic value of nature. Presumably, this also includes advancing scientific knowledge at the ecosystems, species and genetic levels.[18] This article in conjunction with article 9(c) arguably implies the advancement of scientific knowledge from the regulated activities in Part II should serve implementation of the whole BBNJ Agreement, such as forming baseline data for EIA and ABMT from MGR research, with particular consideration to the interests and needs of developing States.

Article 11(6) alludes to but does not explicitly include the principle of common but differentiated responsibilities (CBDR) by taking into particular consideration the interests and needs of low and middle-income States. This principle of international environmental law, formalized at the 1992 United Nations Conference on Environment and Development (UNCED) (UNGA, 1992), acknowledges that responsibility among States for conservation of biodiversity is unequally distributed due to their differing contributions to global environmental degradation and varying abilities to address such degradation. In the context of the BBNJ Agreement, alluding to CBDR suggests that high-income States should acknowledge the responsibility they bear in view of the greater pressures they place on the ocean environment and of the resources and technologies they hold. Therefore, the activities of these Parties related to MGRs and DSI should support the advancement of scientific knowledge for conservation and sustainable use of marine biodiversity and the capacity of lower-income Parties to contribute to conservation and sustainable use, although not explicitly required to do so.

Article 11(6) is consistent with the international law principle of peaceful uses of seas and oceans. For example, under UNCLOS, the 'high seas shall be reserved for peaceful purposes' (UNCLOS art 88), the Area is 'open to use exclusively for peaceful purposes by all States' (UNCLOS art 141)[21] and marine scientific research in ABNJ must be conducted exclusively for peaceful purposes (UNCLOS arts 143 and 240). Antarctica (including the high seas out from the land mass[19]) 'shall be used for peaceful purposes only.'[20]

[18] See, for example, the CBD article 2 definition of sustainable use of the components of biodiversity at the ecosystems, species and genetic levels.

[19] Noting the 'constructive ambiguity of the Antarctic Treaty' with sovereign rights on hold but with some states that? claim territory asserting the high seas are beyond the limits of their maritime zones; see Titterton, H., & Haward, M. (2022). The Kerguelen Plateau: interactions between the Law of the Sea and the Antarctic Treaty. *Marine policy*, *138*, 104,993; Homan, A. (2006). Maritime Zones in Antarctica. *Austl. & NZ Mar. LJ*, *20*, 69.

[20] Antarctic Treaty art I. Noting 'nothing in the present Treaty shall prejudice or in any way affect the rights, or the exercise of the rights, of any State under international law with regard to the high seas within that area.' Antarctic Treaty art VI.

[21] And see generally the peaceful uses of the seas under art 301.

In summary, the creative sidestep of the thorny issue of regulating access to MGR and DSI of ABNJ resulted in an 'activities' provision that shines some light on the scope of regulated activities (subject matter and geographical scope) and reiterates cooperation between Parties and the due diligence principle of UNCLOS. However, the treaty bodies will need to clarify several gaps in interpretation including how article 11 relates to other Parts of the BBNJ Agreement and the Traditional Knowledge provision in article 13. It will need to future-proof the BBNJ Agreement by considering procedures in light of advancements in uncrewed autonomous vessels and AI R&D. Finally, as the analysis in Sects. 5.5 and 5.6 below demonstrates, there are gaps in interpreting the kinds of activities that fall within the notification system in article 12.

5.4 Pre-collection Notification Mechanism

> Article 12(2): 'The following information shall be notified to the Clearing House Mechanism six months or as early as possible prior to the collection in situ of marine genetic resources of areas beyond national jurisdiction...'.

The rationale for pre-collection notification is to collate information about the planned collection activity so that it can be shared through the CHM, to enhance transparency and for the benefits to start accruing. Most of the information requested is already available in a 'cruise plan' and follows good scientific practice, with the notification making this information publicly available (Broggiato et al., 2018; Rabone et al., 2025). This then allows others to assess whether there is potential for coordination (e.g. avoiding multiple overlapping cruises in the same region) and cooperation (e.g. for additional samples to be collected for research) and capacity building as part of benefit sharing (e.g. accepting cruise participants from developing States). This fulfils multiple aims such as gathering more comprehensive information about marine biodiversity of ABNJ and other requirements of the BBNJ Agreement including benefit sharing, capacity building and the transfer of marine technology. As Sect. 5.3.6 above argues, information from the notification mechanism may also contribute baseline data for other elements of the BBNJ Agreement—EIAs and ABMTs.

The pre-collection notification obligation is on Parties to notify the relevant information to the CHM. It is up to each Party to determine who within their jurisdiction has responsibility for the notification, which may be a government department or possibly private organisations or individuals, depending on how they implement their obligations under national law or policy. The notification only relates to the activity of 'collection in situ' of MGRs of ABNJ, which means the activities of 'collection or sampling' (art 2), neither of which are defined in the treaty.

The following information (a) to (j) is required under the treaty text as minimum requirements but there may be more information, guidelines, procedures or protocols developed by the CoP. Most of the listed requirements are consistent with the UNCLOS art 248 duty to provide information to the coastal State in the context of marine scientific research in the exclusive economic zone and on the continental shelf, which may aid in consistency of practice within and outside areas of national jurisdiction (Humphries 2025). Much of the information constitutes information sharing, technology transfer and capacity building, which are non-monetary benefits that meet the objectives of Part II and were important non-negotiables for the G77 and China group during negotiations (de la Concepción, 2024).

5.4.1 Trigger and Responsibility for the Pre-collection Notification

> (a) 'The nature and objectives under which the collection is carried out, including, as appropriate, any programme(s) of which it forms part';
> (b) 'The subject matter of the research or, if known, the marine genetic resources to be targeted or collected, and the purposes for which such resources will be collected';...

A key issue for the pre-collection notification obligation is determining the trigger for each notification and who is responsible for providing the information in article 12 for the notification. Unlike the Nagoya Protocol's concept of ABS, there is no pre-requisite for R&D with respect to the genetic attributes of the MGRs for the collection notification requirement to be triggered. The Nagoya Protocol provides an access framework for 'access to genetic resources for their utilization' (Nagoya Protocol art 6), which means R&D on the genetic and/or biochemical composition of genetic resources (Nagoya Protocol art 2). In other words, the access modality (e.g. prior informed consent) is triggered by R&D into the genetic material attributes of the biological resource and not, for example, a fish fillet for consumption.

In contrast to the Nagoya Protocol, the BBNJ Agreement has separate 'activity' (article 11) modalities for collection and for utilization, the latter of which contains a nearly identical definition to the Nagoya Protocol and is linked to R&D activities. The BBNJ Agreement definition of 'collection in situ' means 'collection or sampling' without further explanation of these terms and without connecting them to activities of R&D into their genetic attributes (Humphries, 2025). This indicates that the trigger for the pre-collection (and associated post-collection) notification is simply the collection activity, which may also encompass collection for purposes other than investigation into their genetic material potential (i.e. bioprospecting) such as basic research on geographic distribution which might be relevant for ABMT identification. Article 11(7) only refers to 'peaceful purposes' and does not confine the access modalities to only activities for bioprospecting purposes (see Sect. 5.3 above). While there may be cases where researchers are targeting specific MGRs for collection in ABNJ, most cruises (expeditions) carry out basic (fundamental) science and relatively few have dedicated bioprospecting projects, although months or years down the track there may be a change of sample use to bioprospecting (see Rogers et al., 2021). Collections using unmanned and autonomous vessels are increasingly being used in ABNJ, and while these trips often carry out basic science (see Yamahara et al., 2019), they are also likely to carry out targeted collections (see Rogers et al., 2021). Due to the broad trigger for collection notifications, even basic science without bioprospecting intent is likely to require notification, subject to clarification from the treaty bodies. The pre-collection notification applies regardless of whether the purpose is commercial or non-commercial research.

It is unclear from the text whether a notification is required for: (a) every cruise to ABNJ that may have multiple projects on board intending collection activities; (b) individual legs or projects within a cruise; or (c) every individual collection activity within a project or cruise. Collection of MGR in ABNJ is often carried out by 'cruises' headed by a cruise leader with spaces available for multiple project teams working on a range of collection and non-collection activities (see Rogers et al., 2021). Clarification about who is responsible for supplying the information for notification is important because options (b) and (c) would add considerable complexity to the system (e.g. attaching multiple BBNJ Identifiers to multiple collections from the same cruise) and duplication, overlaps or inconsistency in information for collections under multiple projects over different cruises and different time scales. Early draft texts simply attached information requirements to the activity of collection (UNGA, 2019c draft art 10) but later drafts connected the information requirements for notification to the 'nature and objectives of the project' (UNGA, 2022a, draft art 10). In the final text, however, the term 'project' was removed from this phrase but remained in other information descriptors for the requirements, such as 'the name(s) and sponsoring institution(s) and the person in charge of the project' (UNGA, 2022a draft art 10).

There is no explanation in the President's reports about why the term 'project' was partially removed in the final text, but it may indicate an intention to require one collection notification per cruise, rather than per project on the cruise. On the other hand, the usage of BBNJ 'batch' identifier, rather than 'cruise' identifier can be interpreted as intention to leave this particular decision to the BBNJ CoP for further guidance. The technical paper that introduced the batch identifier approach to IGC5 refers to 'systems that group or aggregate individual sequence accessions together by overall project (BioProject) and individual samples (BioSamples)' (Oldham & Thambisetty, 2023). An earlier idea for this bulk collection identifier approach that the International Council of Environmental Law delegation presented to IGC3 was to attach a notification to the cruise (Humphries et al., 2020).

Some cruises to ABNJ have multiple projects for MGR R&D on board but the general practice is for the cruise leader to fill out pre-cruise reports required by funders and other entities involved in the cruise (see Rogers et al., 2021). Requiring notification per cruise would essentially require information that is already compiled as part of this cruise report and be a 'lighter touch' for regulation than every project (Rabone et al., 2025). On the other hand, the cruise leader may not have access to the information required for the post-collection notification (see below) and create inefficiencies and information gaps in the monitoring system, particularly in scenarios where collection takes place in areas both within and outside of national jurisdiction during one cruise.

5.4.2 Information Requirements for the Pre-collection Notification

(c) 'The geographical areas in which the collection is to be undertaken';
(d) 'A summary of the method and means to be used for collection, including the name, tonnage, type and class of vessels, scientific equipment and/or study methods employed';
(e) 'Information concerning any other contributions to proposed major programmes';
(f) 'The expected date of first appearance and final departure of the research vessels, or deployment of the equipment and its removal, as appropriate';
(g) 'The name(s) of the sponsoring institution(s) and the person in charge of the project';
(h) 'Opportunities for scientists of all States, in particular scientists from developing States, to be involved in or associated with the project';
(i) 'The extent to which it is considered that States that may need and request technical assistance, in particular developing States, should be able to participate or to be represented in the project'...

Article 12(4) Where there is a material change to the information provided to the Clearing House Mechanism prior to the planned collection, updated information shall be notified to the Clearing-House Mechanism within a reasonable period of time and no later than the start of collection in situ, when practicable.

The notification must include the geographical areas in which the collection is to be undertaken (art 12(2)(c)). The requirement to specify the area of the proposed collection site is general in nature and unlike the post-collection requirement, it is not necessary to specify the longitude, latitude and depth of collection, as at this stage they would not be known. It is not limited to specifying geographical areas within ABNJ and States may (but are not required to) offer information about AWNJ from which they are collecting on the same trip. The wording of this requirement is also sufficiently general to take into account changes in cruise plans that happen during a research cruise due to a change in weather conditions or equipment failure.

Information from article 12(2)(d) to (g) is the type of information you would expect to see in a cruise plan for an ABNJ expedition and constitutes research best practice.[22] For article 12(2)(d), the size of the vessel would indicate its capacity for long deployments in remote areas of the ocean. Scientific equipment and study methods deployed refer to whether the cruise is for instance carrying out a transect (a line of observation stations sometimes (re)visited over a timeframe of years) or deep core samples amongst other methods. The equipment deployed depends on the study methods and may include methods such as plankton nets, trawls, box cores, multicores, piston cores or water sampling. Article 12(2)(e) requires information on contribution to major programmes referring to the fact that some cruise deployments are part of larger/longer projects such as the UK's Porcupine Abyssal Plain Sustained Observatory[23] that has been collecting data since 1985 to assess long term changes in the ocean. Such cross-referencing is helpful as it allows researchers to identify related data from other collection activities. Article 12(2)(f) requires information on cruise departure and return which is important for forward planning for those who may wish to join the cruise or benefit from its activities by obtaining samples. For longer term research, buoys and seafloor installation may be deployed and knowing when this is occurring is essential for other researchers to know so that they may request access to materials or data. The sponsoring institution and person in charge of the project under article 12(2)(g) is typically the national oceanography institution or funder but in the case of non-governmental organisations it will be the foundation that manages the vessel (e.g. Schmidt Ocean Institute and its vessel R/V Falkor) and person in charge will be the cruise leader or Principal Scientist on board the vessel who is in charge of coordinating all the projects being carried out during the cruise.

Requirements under article 12(2)(h) and (i) are forms of capacity building and technology transfer, which have been a feature of the notification provision since the zero draft (UNGA, 2019b draft art 10). Examples of, and approaches to, capacity building are analysed in detail under Chaps. 6 and 13 of this book (Broggiato et al., 2025; Lavelle & Wynberg, 2025). They are also the subject of stand-alone capacity building and transfer of marine technology (CBTMTT) obligations under Part V of the BBNJ Agreement, demonstrating that the treaty parts be interpreted alone and as a whole.

Throughout negotiations there were diverging views about whether CBTMT measures should be voluntary or mandatory (Harden-Davies & Snelgrove, 2020). Article 12(2) includes mandatory information about making opportunities available for scientists from other countries, including 'developing' States to be involved or associated with the project, and they need to be read in conjunction with article 14(2)(f) and (g) (Broggiato et al., 2025). 'Involvement' could include participation in expeditions, lab-based work, data analysis work and student exchanges (Rabone et al, 2019). These opportunities are already in practice as voluntary capacity building for many expeditions[24] but the BBNJ Agreement aims to ensure information about

[22] See, for example, the report by ALAYSE Anne-Marie (1987) HYDRONAUT cruise, RV Le Nadir, https://doi.org/10.17600/87004911.

[23] https://projects.noc.ac.uk/pap/.

[24] Examples include https://nektonmission.org/initiatives/knowledge-exchange/programmes and https://www.geomar.de/en/centre/research-in-cape-verde.

these opportunities is shared with the CHM, to implement some of the non-monetary benefit sharing provisions of article 14(2) (Broggiato et al., 2025).

While article 12(h) includes information about general opportunities for scientists (i.e. individuals) on offer, article 12(i) requires notifiers to assess the extent to which States (particularly developing States) needing technical assistance (rather than individuals) should be able to participate or be represented in the project. In other words, notifiers must report on opportunities for individuals as well as demonstrate that participation or representation in the project at the State level is feasible in practice. The treaty bodies will need to clarify what kind of evidence is required for this assessment, the consequences if it is considered the State is unable to participate or be represented in the project and whether this relates to opportunities for scientists to participate in the project, or more broadly. Until there is clarification, notifiers could at least state in the notification the technical assistance they are offering scientists to participate in the project (if any) and any further assistance they might need from the capacity building fund and mechanisms under the BBNJ Agreement. Article 12(h) could also provide the opportunity to facilitate mutually beneficial connections between industry-led sampling expeditions and those working in biodiversity conservation towards more collaborative global consortia for biodiscovery and conservation of ABNJ.

While the intended purpose of the BBNJ Agreement is to promote the conservation and sustainable use of marine biodiversity, no requirements are made to provide information in this regard in the pre-collection notification. However, the pre-collection notification could be a useful mechanism to encourage the provision of information about how the proposed collection and subsequent use will contribute to biodiversity conservation. Similarly, the post-collection notification (and utilization notification) could share the ways in which the collection was used for conservation and further opportunities for this purpose.

The rationale for requiring updates to the information in the pre-collection notification is because cruise activities often change in the lead up to the point of departure (see Rogers et al., 2021). This requirement to update is similar to the practice under the Nagoya Protocol's ABS clearing house mechanism,[25] although with much less detail for information requirements than the BBNJ Agreement text. Interpretation issues yet to be resolved include what would constitute a 'material change' to the information, a 'reasonable period of time' and the consequences for not updating the information.

5.4.3 Data Governance Responsibilities

> (j) 'A data management plan prepared according to open and responsible data governance, taking into account current international practice.'

Good data management is key for knowledge discovery, innovation and reuse of data and knowledge for ongoing benefits but is left to the discretion of the data or repository owner (Wilkinson et al., 2016). Wilkinson et al. 2016 proposed four foundational principles (FAIR) that have subsequently been adopted into the treaty text in relation to data management. Under article 11(2)(e), non-monetary benefits shall be shared in the form of among other things 'open access to findable, accessible, interoperable and reusable (FAIR) scientific data in accordance with current international practice and open and responsible data governance.' Wilkinson et al. explain that to be:

---

[25] https://absch.cbd.int/en/kb/tags/monitoring/The-flow-of-information-through-the-ABS-Clearing-House-to-support-monitoring-the-utilization-of-genetic-resources/5be4876871ac250001aadc45 (find specific reference to updating information).

- Findable, data are assigned a globally unique and persistent identifiable, described with rich metadata and registered or indexed in a searchable resource;
- Accessible, data are retrievable by their identifier using a standardized communications protocol which his open, free and universally implementable;
- Interoperable, data use 'formal, accessible, shared and broadly applicable language for knowledge representation'; and
- Reusable, data are described with accurate and relevant attributes and are associated with detailed provenance (Wilkinson et al., 2016).

The data management plan under the notification provision article 12 was a last-minute inclusion by the G77 and China group to further entrench the requirements for good data practices based on FAIR principles that will contribute to information and knowledge sharing (de la Concepción, 2024). There are two sets of obligations—the first is for a plan as part of the pre-collection notification and the second relates to modalities envisaged for access to MGRs and DSI that are being used under the 'utilization' notification trigger (arts 12(2) and 12(8)(d)). The latter needs to be read in conjunction with article 14(3) obligation to deposit MGR and DSI on MGR of ABNJ, together with their BBNJ Identifier in publicly accessible repositories and databases. This indicates that it is the role of the repositories/databases to develop a model for data management plans associated with access to MGR and DSI from the repository/database. In practice, there may be a role for governments to develop model data management plans (see Lawson et al., 2025).

Data management plans are standard requirements for many research funders and relate to how data is managed, including how to make it FAIR. Data management plans may include:

- when and where data need to be deposited;
- metadata standards (schemes developed by a community to enable the best possible description of a resource for their needs);
- a description of allocation of resources to data management; and
- aspects of data security, ethics, and confidentiality (Lawson et al., 2025).

A data management plan can vary by subject matter, activity and intended use (research, development and/or commercialisation of the MGR or DSI.[26]

The CARE principles for Indigenous data governance were also raised by non-government organisations at the negotiations but did not appear in the treaty text. CARE stands for:

- Collective benefit—data ecosystems designed and function to enable Indigenous peoples to derive benefits;
- Authority to control—enables Indigenous peoples to determine how data represents Indigenous Peoples, territories, resources, knowledge and geographical indicators;
- Responsibility—meaningful evidence of efforts to use data to support Indigenous people's self-determination and collective benefit;
- Ethics—Indigenous peoples' rights and wellbeing should be the primary concern at all stages of the data life cycle (Carroll et al., 2022).

The extent to which the CARE principles will be adopted into modalities for data governance under the BBNJ Agreement is yet to be determined by the treaty bodies.

5.4.4 BBNJ Batch Identifier Procedural Requirements

> Article 12(3) Upon notification referred to in paragraph 2 above, the Clearing-House Mechanism shall automatically generate a "BBNJ" standardized batch identifier.

[26] For examples of what data management plans typically include see—https://www.ukri.org/councils/bbsrc/guidance-for-applicants/what-to-include-in-your-application/data-management-plan/; https://embassy.science/wiki/Theme:67a453dc-7fa0-4f58-a869-748fea8dec7f; https://www.dcc.ac.uk/; https://ec.europa.eu/research/participants/data/ref/h2020/gm/reporting/h2020-tpl-oa-data-mgt-plan_en.docx.

The BBNJ Identifier was an innovation to preserve linkages between data concerning biodiversity including MGRs and DSI. It was a last-minute compromise between negotiation groups including G77 and China group that wanted clear chain of provenance for the purposes of connecting benefit sharing with notification, monitoring and compliance and like-minded countries that advocated against a track and trace monitoring system (de la Concepción, 2024). Under the BBNJ Agreement, the CHM will assign the BBNJ identifier to the pre-collection notification. It represents a bulk (batch) collection—a tag that identifies that material was collected in ABNJ during a particular collection event rather than individual biological resources collected, which would be impractical given that one collection could contain millions of organisms (Humphries et al., 2020). The idea is that as the BBNJ Identifier will be unique, persistent, authoritative and interoperable with other databases, it will be capable of linking relevant MGRs and DSI back to the original collection activity (Humphries et al., 2020; Oldham & Thambisetty, 2023). Many large databases already require the location of collection in the metadata[27] and can incorporate the BBNJ Identifier as a record in the metadata.

The BBNJ Identifier aims to be the link between the different categories of notification. Requirements to include the original BBNJ Identifier in the reporting information for post-collection and utilization notifications will ensure only one link to the pre-collection activity. BBNJ Agreement bodies have not yet clarified what form the BBNJ Identifier will take or the procedures for issuing one. Some of the practicalities will depend on clarification about whether the pre-collection notification applies per cruise, per leg or per project (see above). The scope and procedures for the BBNJ Identifier (when clarified) need to be clearly explained to stakeholders because the notification requirement for access to MGR samples and DSI in repositories and databases (Sect. 5.7 below) depends on an unbroken BBNJ Identifier link with the post-collection notification.

Given the role of the CHM to assign the BBNJ Identifier, it is worth noting that while the notification obligations will presumably apply from the time the treaty enters into force, the specific modalities for the operation of the CHM are to be determined by the COP at a future meeting. This mismatch in timing could likely cause challenges for Parties in the implementation of their obligations and Parties may wish to consider suitable interim arrangements either at the national level or at the international level, e.g. in the context of a Preparatory Commission. A precedent for such an approach would be the pilot phase of the access and benefit sharing clearing house of the Nagoya Protocol (see, e.g. UNEP, 2011) that ran from 2011 to the entry into force of the Nagoya Protocol on 12 October 2014.

To sum up, while the pre-collection notification mechanism contains several innovations, including digital identifiers and aims for a light touch to regulation, there are many unresolved questions for implementation that the treaty bodies and Parties will need to address. These include:

- Whether the pre-collection notification attaches to the cruise, leg of a cruise or the project?
- Who is responsible for the pre-collection notification in the case of a cruise? Assuming the pre-collection notification attaches to the cruise rather than the project, is it the Party that is the flag state of the cruise? The Party from which the cruise was launched (if it is different from the flag state)? The Party in whose jurisdiction the cruise expects to return and land the MGRs? And what about a muti-state research consortium which is the more common arrangement for MGR exploration?
- Who is responsible for the pre-collection notification in the case of autonomous unmanned vessels?

[27] E.g. https://www.insdc.org/news/insdc-spatiotemporal-metadata-minimum-standards-update-03-03-2023/.

- Whether a cruise would have one pre-collection notification for the whole trip that may include both collections of MGRs from within national jurisdiction as well as beyond national jurisdiction? As the pre-collection notification can only deal with the ABNJ collection, how would the data for collections from within national jurisdiction on the same trip be handled?
- If the cruise leader is responsible for the pre-collection notification, how would they keep track of the location of MGRs and DSI from the multiple projects on board for the original collection for the purpose of the post-collection notification? If the BBNJ Identifier is triggered by the pre-collection notification at the cruise level, how would it link to samples and data that belong to different projects within the cruise? Would there need to be another mechanism or tag for different projects leads to update the cruise leader with information for the post-collection notification?
- How would the notification work in practice with current cruise identifiers and reporting practices?
- What kind of evidence is required for the notifier's assessment that States needing technical assistance should be able to participate or be represented in the project? Is there an active duty for the notifier to assist States with technical and financial assistance to participate or is it simply an information requirement for the CHM to be aware that further assistance may be needed?

Some practical suggestions about how these gaps might be addressed were included in the analysis above, but it may be many years before all of the kinks are ironed out of this new mechanism.

5.5 Post-collection Notification Mechanism

Article 12(5) 'Parties shall ensure that the following information, along with the "BBNJ" standardized batch identifier, is notified to the Clearing-House Mechanism as soon as it becomes available, but no later than one year from the collection in situ of [MGRs of ABNJ]'.

(a) 'The repository or database where digital sequence information on marine genetic resources is or will be deposited';
(b) 'Where all marine genetic resources collected in situ are or will be deposited or held';
(c) 'A report detailing the geographical area from which marine genetic resources were collected, including information on the latitude, longitude and depth of collection, and, to the extent available, the findings from the activity undertaken.';
(d) 'Any necessary updates to the data management plan…'

As with pre-collection notification, the post-collection notification only applies to the activities of 'collection in situ' of MGRs of ABNJ, which means the activities of 'collection or sampling', neither of which are defined. The rationale for this provision is to close the loop on collection data with confirmation about the actual location of the batch collection in ABNJ, the actual locations of MGRs that were identified or sequenced from the batch collection and the repositories and databases where MGR and their associated DSI is deposited or held after they return to shore, and any updates to the data management plan.

One of the key issues that delayed agreement on the treaty text concerned how to manage 'access' to DSI. The compromise reached in the final notification provisions was to require researchers to document where they upload DSI of the genetic material they are analysing along with the BBNJ Identifier. It does not require that all genetic materials must be sequenced, nor require databases to issue an identifier to link the DSI with ABNJ as in previous draft iterations (e.g. UNGA, 2019b draft art 13(3)) but the obligation to ensure that this happens rests with the

Parties (article 12.6), and INSDC declared during the last IGC its readiness to do so.[28] One of the purposes of the BBNJ Identifier is to connect the original collection to this data. As long as the BBNJ Identifier is provided in the DSI record within the metadata, it would be redundant for databases to actively issue a unique identifier for the origin of the sequence. This requirement would be expected to be included in the data management plan of article 12(2)(j).

Requirements for access to MGR of ABNJ samples from ex situ facilities were the subject of much discussion during negotiations. Several commentators argued at negotiations and in publications about the high cost, technical and legal difficulties faced by researchers and repositories for requirements to deposit samples of MGRs in third party (intermediary) repositories, which would be responsible for curating and providing free access to the samples (Humphries et al., 2020; Rabone et al., 2019). The compromise wording in article 12 recognised that an important information sharing outcome is being able to locate the initial place where MGRs are held, similar to the World Health Organization's Influenza Virus Traceability Mechanism.[29] Maintaining records of the location of the original MGRs is scientific best practice and is an important step for an end-user traceability approach to have a link between subsequent use and the original collection (Humphries et al., 2021a, 2021b, 2021c). Some samples from the original bulk collection may not be identified as having value for research as an MGR and DSI may not be generated until well after the 12-month post-notification period. Article 12(5)(a) and (b) require Parties to ensure that the CHM has information about where MGR identified from the original collection and DSI are or will be deposited or held. In practice, this may relate to the remaining MGR from the original collection after the first analysis, which for some uses may lead to depletion or destruction of the original samples as part of the use, but there may still be information about where the DSI from those samples has been deposited. It does not require Parties to continue to report newly identified MGR or generated DSI after the 12-month period, which may suggest a reporting loophole, although the BBNJ Identifier link with the original collection will persist and further information about the location of MGR and DSI might be captured under the notification provisions relating to 'utilization' and 'access' to MGR and DSI in repositories/databases under articles 12(7), 12(8) and 14(3) and 14(4) (see Sects. 5.6 and 5.7 below).

The treaty requires that States must ensure that more specific geographical information with coordinates stays with the post-collection record, which is usual scientific best practice (Rabone et al., 2019; Rogers et al., 2021). The report must contain other information including the findings of the research, to the extent that they are available up to one year after collection. The wording recognised the scientific reality that many samples are not analysed at the time of collection but can take months or years to collate (Rabone et al., 2019). The softer language 'to the extent available' means that this is a best practice goal rather than a strictly enforceable requirement.

Parties are required to update any data management plan. This plan may include information such as the location of the DSI, but there are no other requirements to update records after a year post-collection. Subject to clarification by the treaty bodies, it appears that the data management plan envisaged under the utilization notification requirement is separate from the collection data management plan (Lawson et al., 2025). There are other unanswered questions with respect to the requirement under article 12(5)(d) including:

- Article 12(8)(d) read in conjunction with article 14(3) indicates it is the role of the repositories/databases to develop a model for data management plans associated with access to MGR and DSI from the repository/database, or is it

[28] See INSDC declaration to require mandatory country (source) and collection date metadata for new sequence records with out a valid exemption—https://www.insdc.org/news/insdc-spatiotemporal-metadata-minimum-standards-update-03-03-2023/.

[29] https://extranet.who.int/ivtm2/ See Humphries et al. (2021a, 2021b, 2021c).

the role of the Parties where MGRs/DSI are being accessed for 'utilization' to determine the modalities of data management plans (e.g. through a model plan) or each person conducting 'utilization' of the relevant MGR or DSI to create their own data management plan?
- To what extent will CARE principles for Indigenous data governance be adopted into modalities for data management plans?

While the post-collection notification requirement did not appear until later versions of the draft text, it is an important means of verifying information about the origin and location of the MGRs and DSI as a form of information sharing. For it to work effectively, there would need to be clarity about who is responsible to provide this information—would it be the cruise leader or project leader? Pursuant to article 11 and article 12(1), discussed above put the obligation on states to ensure that natural or juridical persons comply with the notification provisions, this decision seems to fall to Parties at the national level. Future guidance by the BBNJ CoP or its subsidiary bodies may be helpful to reduce the risk of complexity and inefficiency in the notification system resulting from differing state practices.

5.6 Utilization Notification Mechanism

Article 12(8) Where [MGRs of ABNJ] and where practicable, the [DSI] on such resources are subject to utilization, including commercialization, by natural or juridical persons under their jurisdiction, Parties shall ensure that the following information, including the "BBNJ" standardized batch identifier, if available, be notified to the Clearing-House Mechanism as soon as such information becomes available:

(a) Where the results of the utilization, such as publications, patents granted, if available and to the extent possible, and products developed, can be found;
(b) Where available, details of the post-collection notification to the Clearing-House Mechanism related to the marine genetic resources that were the subject of utilization';
(c) Where the original sample that is the subject of utilization is held;
(d) The modalities envisaged for access to marine genetic resources and digital sequence information on marine genetic resources being utilized, and a data management plan for the same;
(e) Once marketed, information, if available, on sales of relevant products and any further development.

The utilization mechanism differs from the collection notifications in the types of activities and subject matter that trigger information requirements for the CHM, and the information required. Part of the rationale for the utilization notification was that the information contained (in particular parts of (a) and (e)) could in future serve as a basis for establishing a tiered system for monetary benefit sharing, including a flat fee payment model, sharing of profits from commercialisation and capacity building—in other words, different tiers for which Party contributes based on the information collected through the notifications (Thambisetty et al., 2023). While Parties decided to adopt a simpler, 'decoupled' scheme as the initial modality for monetary benefit sharing, the idea of the tiered approach was retained in a list of examples of what future modalities for benefit sharing the BBNJ CoP could consider (art 14(7)) and the retention of the related information in the utilization notification could be interpreted as a potential future enabler to keep the option open. This provision was the subject of considerable debate at negotiations and underwent many iterations, including more onerous proposals for pre-collection notification and consultation of coastal States at ICG3 (UNGA, 2022a draft art 10(6)). The final notification category aims for a lighter touch by requiring information from the outcomes of R&D as well as commercialisation, with the important caveat of 'if available and to the extent possible'. This includes the results of research such as publications and patents, the location of the original sample, information on sales and a link to the post-collection notification. It is unclear why it requires data on the post-collection notification instead of the BBNJ Identifier that links both the pre- and post-collection events.

While 'utilization of marine genetic resources' is defined in the general provisions of the treaty (art 1(14)), the trigger for the 'utilization' event in the text is more open to interpretation, which has the potential to cause confusion for implementation without guidance from the CoP. On the one hand, there was a deliberate attempt during negotiations to limit the utilization notification to information about the results of utilization (R&D and commercialisation) rather than the notification of every use of MGR and DSI on ABNJ (see Sect. 5.2.1 above). On the other hand, Parties might argue the notification is triggered by the act of utilization itself, rather than the results. However, this latter interpretation might lead to an unworkable notification mechanism. There are considerable practical problems for requiring every utilization of MGRs and DSI, even those not leading to results as defined in (a), to be the subject of notification because the same user may be required to report every time they use the materials/data for existing uses (e.g., sending it for analysis to another lab), which may create a disproportionate burden on researchers. If the notification mechanism requires information for every use, the computing power required for the CHM to catalogue, review, and analyse raw data and information would be significant and may be deemed impractical. Previous drafts included a requirement to supply the information within three years of the utilization (UNGA, 2022a draft art 11 option II), but in a more practical approach for compliance and enforcement (considering outcomes from research may take decades), the final draft requires it 'as soon as such information becomes available'.

The text wording clarifies that the 'utilization' notification obligation is on Parties (rather than private business or databases) but it is unclear which Party would have the obligation. This is because the geographical link between the activity and the Party is not clear in the wording. One interpretation is that it is the Party where the 'utilization' occurs that needs to ensure that the relevant information is notified to the CHM, but it is unclear whether this is the location of the funder, project lead or specific 'utilization' activity, which may be in multiple countries (Rabone et al., 2025). There are practical challenges with this approach, including over-reporting (e.g. of patents and publications) for the same 'utilization' event (Rabone et al., 2025), particularly as it seems to require the reporting of the result of utilization, even if it is in other countries. Another interpretation is that each Party must notify information about the *results* of utilization in their jurisdictions (e.g. patents, publications) even if the utilization occurred elsewhere. There would be practical challenges with obtaining other information under this approach such as where the original sample subject to utilization is held. Depending on the final infrastructure of the CHM, the notification mechanism for utilization may be assisted through links to records of final products and processes, which may include intellectual property databases,[30] records within the checkpoint system of the Nagoya Protocol if the products/processes include MGR from a range of origins and records under other relevant national systems such as export, biosafety, and pharmaceutical frameworks (Lawson et al., 2025).

The terms and conditions for the 'utilization' notification need to be read in conjunction with the benefit sharing requirements under

[30] See Langlet et al. (2025) about the relationship between the BBNJ Agreement and the new international treaty requiring Parties to disclose the origin of genetic resources and traditional knowledge on which their patented inventions are based: *WIPO Treaty on Intellectual Property, Genetic Resources and Associated Traditional Knowledge* (WIPO 2024).

article 14, including the availability of samples for further 'utilization'. For example, the utilization notification provision requires information about the modalities envisaged for access to MGRs and DSI being utilized and a data management plan. Article 14(4) provides that access to MGR and DSI in repositories and databases 'may be subject to reasonable conditions' listed in the provision, including reasonable costs associated with maintaining the biorepository. Early drafts of the notification mechanism proposed mandatory deposit of samples and data in 'open source' platforms (UNGA, 2019b draft art 10(2)). There are clear advantages for increasing the sharing of samples and minimising repeated collection of the same sample (Broggatio et al., 2018). However, use of the MGR by the original collector may deplete the availability of samples and it may not be possible to take identical or multiple samples at the time of collection as duplicates (Humphries et al., 2020). There would be significant costs to curators for the maintenance and storage of physical specimens without significant infrastructure and resources from the BBNJ Agreement framework (Rabone et al., 2019). As a compromise, the sample/data deposit requirement was removed from the notification mechanism as a condition of access and included in the benefit sharing mechanism, where only those MGR and DSI that are subject to 'utilization' are to be deposited in publicly accessible repositories and databases (art 14(3)), and not all MGR collected from ABNJ. Unlike other requirements for 'utilization' under article 12, this requirement under article 14 retained the deadline of 'no later than three years from the start of such utilization.' This means that national laws would need to define what the 'start of such utilization' means for article 14, unlike the temporal trigger for article 12, which is 'as soon as information becomes available' for the utilization notification.

The obligations for pre-and post-collection notification are not restricted to activities 'by natural or juridical persons' unlike the utilization notification obligation. This suggests that while Parties have ultimate responsibility for notification, they might require information about collection and sampling by human or artificial intelligence means. However, as outlined in Sect. 5.3 above, the wording of article 11 of the BBNJ Agreement appears to restrict Part II obligations to Parties and their natural or juridical persons carrying out the activities in Part II. In many countries, the legal status of AI is either unclear or subject to decisions that AI does not have legal personality and cannot be considered a 'juridical person' (Mesevic & Skamo, 2023; Schwartz & Rogers, 2021). The concern that this legal distinction could unnecessarily limit the scope of activities for notification was expressed during negotiations (IUCN, 2022), but for an undocumented reason, negotiators removed this limiting factor for collection activities, yet not for 'utilization' activities. The CoP would need to clarify the extent to which collection and utilization using fully autonomous and AI would trigger the notification requirement.

In summary, the framework for the utilization notification mechanism was highly contested throughout the IGCs. On the one hand, it posed concerns for those States who wanted to avoid the BBNJ Agreement reaching through to activities within national jurisdiction and those that wanted to minimise the cost of infrastructure. On the other hand, it was a non-negotiable for States seeking capacity building, technology transfer and a fair and equitable share of any monetary benefits arising from MGR activities after collection. The final result was a relatively light touch framework with information requirements about the results of research but with significant gaps in how the system might work in practice.

5.7 Access to MGR and DSI in Repositories or Databases

Article 12(6) Parties shall ensure that samples of 'MGRs and DSI on MGRs of ABNJ] that are in repositories or databases under their jurisdiction can be identified as originating from [ABNJ], in accordance with current international practice and to the extent practicable.

Article 12(7) 'Parties shall ensure that repositories, to the extent practicable, and databases under their jurisdiction prepare, on a biennial basis, an aggregate report on access to [MGRs and DSI] linked to their "BBNJ" standardized batch identifier, and make the report available to the access and benefit sharing committee established under article 15'.

An important activity omitted from the final text is 'access ex situ', which was included in previous drafts of the text (e.g. UNGA, 2019b) to encompass samples and associated data and information used within national jurisdiction. This caused confusion about the time and location when the 'collection in situ' activity ceases and 'access ex situ' commences and overlapped with 'utilization' activities (Humphries et al., 2020). It would also have created a huge burden for physical collection, and a potentially impossible obligation to implement for databases. Its removal signalled an intention for only three key temporal points for notification for activities within the scope of Part II: (1) prior to collection/sampling in ABNJ, including updates; (2) post-collection; and (3) utilization of the MGRs and DSI within national jurisdiction. However, both the notification and benefit sharing mechanisms include obligations that require reporting of subsequent access to MGR and DSI in repositories or databases.[31] The term 'access' is undefined but the reporting provisions under article 12 clearly relate to reporting of 'access' for the purpose of aggregate data, rather than reporting every single 'access' activity.

Whereas article 12(5) requires Parties to provide information as to where MGRs and DSI of ABNJ are deposited or held within their jurisdiction, article 12(6) seeks to ensure that they are findable and accessible in accordance with FAIR principles.[32] The easy identification of MGRs and DSI of ABNJ is a vital precursor to tracking any later utilisation and fulfilling the non-monetary benefit sharing obligations set out in article 14(2) and 14(3) including publicly accessible samples, sample collections and DSI. The phrase 'in accordance with current international best practice' indicates that scientific practice may change in response to new technologies, in which case best practices may change.

As with article 12(5), the language in article 12(7) was a compromise between the financial/technical/legal constraints of requiring repositories to notify the CHM of every movement of MGRs and DSI for the purpose of traceability (track and trace) as envisaged in past drafts, with the need for maintaining a digital link between ABNJ and the metadata of samples and information (Langlet et al., 2025). In essence, once the CHM has been established, the assignment of the BBNJ Identifier through the notification system will provide the means of identification of MGR and DSI in repositories and databases. However, it is the obligation of Parties to ensure that natural or juridical persons depositing MGRs and DSI of ABNJ in repositories and databases within their jurisdiction include the BBNJ Identifier in the relevant metadata. Given the likely mismatch in timing between the treaty coming into effect and the CHM being operable, this article likely also seeks to ensure that until BBNJ Identifiers are assigned, MGRs and DSI of ABNJ are still findable once held ex situ. It also offers another way of finding the location of MGR and DSI that is not reported within the 12-month period of the post-collection notification.

As part of the benefit sharing system, article 14(3) obliges Parties to ensure MGRs and DSI together with their BBNJ Identifiers are deposited in publicly accessible repositories and databases no later than three years from the start of their utilisation. Given the responsibility of Parties to take the necessary administrative,

[31] The meaning of the terms 'collection in situ',' utilization', and 'access' are further analysed in Humphries (2025).

[32] FAIR means findable, accessible, interoperable and reusable. See Lawson et al. (2025).

legislative and policy steps to ensure information is notified to the CHM for all of article 12, including article 12(6) requiring Parties to ensure the origin of MGR and DSI can be identified and article 12(7) (aggregate reporting), it stands to reason that Parties could continue to facilitate the reporting of any access activities of those repositories and databases, even information beyond the repository/database aggregate reports. However, what is uncertain is what 'access' includes and when this notification requirement is triggered as access is not defined (see Humphries, 2025, Rabone et al., 2025). It is, however, differentiated from 'utilisation' for which obligations are set out in article 12(8). Guidance by the CoP is necessary to clarify the meaning to avoid confusion and overly bureaucratic reporting or under reporting. Key to the successful use of this access and utilisation information and any possible role it may have for the benefit sharing system will depend on the capability of the ABS Committee to interpret information about access to MGR and DSI in repositories and databases.

5.8 Conclusion

One of the overarching tensions in the process leading up to the adoption of the BBNJ Agreement was the question of the sharing of benefits from MGR from ABNJ. An important subset of that question, which greatly increased in importance after 'in principle agreement' to include monetary benefit sharing in the obligations, was how to set up mechanisms to facilitate and monitor the implementation of such benefit sharing (Langlet et al., 2023) and how to balance these against the wish of delegations to not create undue burdens on marine scientific research. The striking of this balance both at 'in principle' and operational levels became one of the final sticking points in the BBNJ negotiations. The notification provisions serve both as a direct means to deliver non-monetary benefits and as a means to collect the information to monitor implementation and inform future decision making on benefit sharing, including monetary benefit sharing, so it should be no surprise that these provisions were central to resolving this tension. The notification system, as reflected in article 12 in the final treaty text, with its innovations, but also carefully constructed ambiguities should therefore be understood as the outcome of these intense, final efforts to reach a balance between facilitating fair and equitable benefit sharing while minimizing, to the degree possible, the burden on marine scientific research.

While the text seems to have struck the right initial balance in the eyes of negotiating States, as reflected by the adoption by consensus of the treaty text, it is important to note that the tension between enabling benefit sharing while avoiding undue burden on marine scientific research continues to persist in many of the remaining questions and constructive ambiguities on how operationalize to the notification system in practice. It will be important in future discussions at the BBNJ CoP and its subsidiary bodies, including the ABS Committee, to continue to seek the balance that enabled the adoption of the BBNJ Agreement. While the above analysis outlined detailed questions and considerations for implementation of the pre-collection, post-collection and utilization notification triggers, as well as obligations for access to MGR and DSI in repositories and databases, some of the key outstanding issues for clarification on the notification modalities include:

- the level at which the notification system and associated BBNJ Identifier should be implemented, e.g. on the cruise, leg, or project level and who is responsible for the notification (the flag State, cruise leader, project leader, etc.);
- the practical implementation of the BBNJ Identifier;
- the triggers that require updating the pre-collection notification;

- the utilization notification, including:
  - whether the utilization notification trigger applies to every use, new third-party transfer uses or only to information about the results of research, the latter of which seems to be the plain reading of the text;
  - the relationship between articles 12 and 14 including how to define the 'start of such utilization' when triggering the requirement to deposit samples that are the subject of 'utilization'; and
  - clarification about how the data management plan in the utilization notification relates to data management required in other stages; and
- which kinds of activities beyond R&D activities of the genetic attributes of MGRs, if any, trigger the notification system.

Other areas for further guidance include future proofing the notification provisions by clarifying how they apply to various activities of autonomous vehicles and/or AI and providing considerations to Parties on how they may want to designate the responsibilities for making the notifications for actors under their jurisdiction. Managing data sovereignty (see Hummel et al., 2021) is an important emerging issue and the treaty bodies could clarify how the framework will manage data sovereignty and sovereignty or self-determination of IPLCs with respect to their traditional knowledge and the MGRs and DSI that are associated with it.

The BBNJ CoP or its subsidiary bodies may also consider further how the notification system could support identifying and embedding synergies with other parts of the agreement by encouraging submitters of information to:

- provide information about how the utilization of MGR and DSI relate to article 13 traditional knowledge obligation;
- provide information on the impact on in situ biodiversity and associated habitats and ecosystems, taking into account the outcomes of EIAs and Strategic Environmental Assessments already carried out under Part IV of the BBNJ Agreement;
- contribute to the conservation and sustainable use of biodiversity and ABMTs;
- contribute to capacity building and technology transfer under Part V of the BBNJ Agreement; and
- ensure equitable partnerships in marine scientific research and fair access to marine genetic resources.

There are many innovative components of the BBNJ Agreement framework for notifying the collection, utilization and aggregate access of MGRs and DSI of ABNJ. While the notification obligations will presumably apply from the time the treaty enters into force, many of the specific modalities for the operation of the CHM and the BBNJ Identifier are to be determined by the CoP at future meetings. This mismatch in timing is likely to cause challenges for Parties in the implementation of their obligations and Parties may wish to consider suitable interim arrangements either at the national level or at the international level, for example, through the Preparatory Commission work. A precedent for this approach is the Nagoya Protocol's pilot phase of the ABS clearing house. There are important lessons to be learned from other ABS fora about how to effectively implement and monitor information systems to ensure the notification system facilitates information and benefit sharing, while supporting scientific and commercial innovation. As MGR and marine biodiversity may flow through multiple jurisdictional areas during their life cycles, it is important for all notification and information systems to communicate seamlessly with each other.

References

Afshar, M. S. (2022). Artificial intelligence and inventorship-does the patent inventor have to be human? *Hastings Science and Technical LJ, 13*, 55.

Berry, T. (2023). *Sovereignty and the limits of international law: Regulating areas beyond national jurisdiction*. Routledge.

Blasiak, R., Wynberg, R., Grorud-Colvert, K., Thambisetty, S., et al. (2020). *The ocean genome: Conservation and the fair, equitable and sustainable use of marine genetic resources*. World Resources Institute.

Broggiato, A., Vanagt, T., Lallier, L. E., Jaspars, M., Burton, G., & Muyldermans, D. (2018). Mare geneticum: Balancing governance of marine genetic resources in international waters. *The International Journal of Marine and Coastal Law, 33*(1), 3–33. https://doi.org/10.1163/15718085-13310030

Broggiato, A., Dunshirn, P., Jaspars, M., & Pena-Neira, S. (2025). Monetary and non-monetary benefit sharing under the BBNJ Agreement. In F. Humphries (Ed.), *Decoding governance of marine genetic resources under the BBNJ Agreement.* Springer.

Brunnée, J. (2017). International environmental law and community interests: Procedural aspects. In E. Benvenisti, & G. Nolte (Eds.), *Community obligations in international law community interests across international law* (Oxford, 2018), 151–175 at 154. https://doi.org/10.1093/oso/9780198825210.001.0001

Carbonell, P., Radivojevic, T., & Garcia Martin, H. (2019). Opportunities at the intersection of synthetic biology, machine learning, and automation. *ACS Synthetic Biology, 8*(7), 1474–1477.

Carroll, S. R., Garba, I., Plevel, R., Small-Rodriguez, D., Hiratsuka, V. Y., Hudson, M., & Garrison, N. A. (2022). Using indigenous standards to implement the CARE principles: Setting expectations through tribal research codes. *Frontiers in Genetics, 13*(823309), 4.

de la Concepción, R. T. (2024). Negotiating fair and equitable sharing of benefits in the BBNJ agreement: Role of the Group of 77 and China. *Marine Policy, 163*, 106085.

Dunn, D. C., Crespo, G. O., Vierros, M., Freestone, D., Rosenthal, E., Roady, S., & Sloat, M. R. (2017) 'Adjacency: How legal precedent, ecological connectivity, and traditional knowledge inform our understanding of proximity' Nereus Scientific and Technical Briefs on ABNJ Series.

Garrity, G. M., Thompson, L. M., Ussery, D. W., Paskin, N., Baker, D., Desmeth, P., Ong, P. S., et al. (2009). Studies on monitoring and tracking genetic resources: An executive summary. *Standards in Genomic Sciences, 1*, 78–86.

Harden-Davies, H., & Snelgrove, P. (2020). Science collaboration for capacity building: Advancing technology transfer through a treaty for biodiversity beyond national jurisdiction. *Frontiers in Marine Science, 7*(40), 1.

Homan, A. (2006). Maritime zones in Antarctica. *Australian and NZ Marine LJ, 20*, 69.

Hummel, P., Braun, M., Tretter, M., & Dabrock, P. (2021). Data sovereignty: A review. *Big Data & Society, 8*(1), 2053951720982012, 1

Humphries, F. (2025). Marine genetic resources beyond national jurisdiction: the expansive scope of the BBNJ Agreement. In F. Humphries (Ed.), *Decoding governance of marine genetic resources under the BBNJ Agreement.* Springer

Humphries, F., Gottlieb, H. M., Laird, S., Wynberg, R., Lawson, C., Rourke, M., Tvedt, M. W., Oliva, M. J., & Jaspars, M. (2020). A tiered approach to the marine genetic resource governance framework under the proposed UNCLOS agreement for biodiversity beyond national jurisdiction (BBNJ). *Marine Policy, 122*, 103910.

Humphries, F., Rourke, M., Berry, T., Englezos, E., & Lawson, C (2021a). COVID-19 tests the limits of biodiversity laws in a health crisis: Rethinking "country of origin" for virus access and benefit-sharing. *Journal of Law and Medicine, 28*, 684–706.

Humphries, F., Rabone, M., & Jaspars, M. (2021b). Traceability approaches for marine genetic resources under the proposed ocean (BBNJ) treaty. *Frontiers in Marine Science, 8*, 661313.

Humphries, F., Laird, S., Wynberg, R., Morrison, C., Lawson, C., & Kolisnikova, A. (2021c) *Survey of access and benefit-sharing country measures accommodating the distinctive features of genetic resources for food and agriculture and associated traditional knowledge*, background study paper 70 (FAO).

IUCN, *IUCN commentary on the further revised draft text of an agreement under the United Nations Convention on the Law of the Sea on the conservation and sustainable use of marine biological diversity of areas beyond national jurisdiction* (A/CONF.232/2022/5) 8 August 2022. Prepared by IUCN World Commission on Environmental Law—Ocean Specialist Group, IUCN World Commission on Protected Areas—High Seas Specialist Group & Ocean team, IUCN Centre for Conservation Action, International Union for Conservation of Nature https://www.iucn.org/sites/default/files/2022-08/igc5-iucn-commentary-on-bbnj-further-revised-draft.pdf

Kachelriess, D., Dunshirn, P., Langlet, A., Brown, A. E. L., & Scholz, A.H. (2025). Benefit sharing provisions of the BBNJ Agreement for MGRs and DSI—interlinkages with other ABS frameworks. In F. Humphries, (Ed.) *Decoding governance of marine genetic resources under the BBNJ Agreement.* Springer.

Laird, S., Wynberg, R., Rourke, M., Humphries, F., Muller, M. R., & Lawson, C. (2020). Rethink the expansion of access and benefit sharing. *Science, 367*(6483), 1200–1202.

Langlet, A., Dunshirn, P., Kachelriess, D., & Currie, D. (2023). High Seas Allliance Non-paper: Marine Genetic Resources, including questions on the sharing of benefits—a brief on broad options for benefit sharing. https://www.highseasalliance.org/wp-content/uploads/2023/02/MBS-Options-paper-1.pdf?_gl=1*3tphdh*_ga*MjA5Mzc2NjUyMS4xNzAyNDIwODA5*_ga_QWVZ3ZNDMC*MTcxMDMzMDEzMy43OS4xLjE3MTAzMzAyNTEuNDcuMC4w

Langlet, A., Dunshirn, P., Jaspars, M., Humphries, F., & Kachelriess, D. (2025). Monitoring and transparency aspects of MGR-utilization under the BBNJ Agreement. In F. Humphries (Ed.) *Decoding governance of marine genetic resources under the BBNJ Agreement.* Springer.

Lavelle, J., & Wynberg, R. (2025). Benefit sharing under the BBNJ Agreement in practice. In F. Humphries (Ed.), *Decoding governance of marine genetic resources under the BBNJ Agreement.* Springer.

Lawson, C., Humphries, F., & Rourke, M. (2019). The future of information under the CBD, Nagoya protocol, plant treaty and PIP framework. *Journal of World Intellectual Property, 22*, 103.

Lawson, C., Humphries, F., Jaspars, M., & Rabone, M. (2025). Data management and the 'BBNJ standardized batch identifier' under the BBNJ Agreement. In F. Humphries (Ed.), *Decoding governance of marine genetic resources under the BBNJ Agreement.* Springer.

MacFarlane, N., & Sabanadze, N. (2013). Sovereignty and self-determination: Where are we? *International Journal, 68*(4), 609–627.

Mendenhall, E., De Santo, E., Nyman, E., & Tiller, R. (2019). A soft treaty, hard to reach: The second inter-governmental conference for biodiversity beyond national jurisdiction. *Marine Policy, 108*(103664), 4.

Mendenhall, E., De Santo, E., Jankila, M., Nyman, E., & Tiller, R. (2022). Direction, not detail: Progress towards consensus at the fourth intergovernmental conference on biodiversity beyond national jurisdiction. *Marine Policy, 146*, 105309.

Merritt, C. G. (2023). A compulsory solution to the machine problem: Recognizing artificial intelligence as inventors in patent law. *Vanderbilt Journal of Entertainment and Technology Law, 25*, 211.

Mesevic, I. R., & Skamo, A. (2023). The human inventor and AI between hall 9000 and DABUS. *South East European Law Journal (SEE Law Journal), 11*, 81–111.

Mossop, J., & Schofield, C. (2020). Adjacency and due regard: The role of coastal States in the BBNJ treaty. *Marine Policy, 122*, 103877.

Muraki Gottlieb, H., Ardron, J., & Brown, A. E. L. (2025). BBNJ Agreement: A new infrastructure to foster benefit sharing of marine genetic resources. In F. Humphries (Ed.), *Decoding governance of marine genetic resources under the BBNJ Agreement.* Springer.

Oldham, P., & Thambisetty, S. (2023) *ONEST: The middle way for monetary benefit sharing in BBNJ negotiations* (pp. 1–2). Available at https://zenodo.org/record/7573700#.Y9KmR-zP27B

Pena-Neira, S. & Coelho, L.F. (2025). Traditional Knowledge associated with marine genetic resources in areas beyond national jurisdiction. In F. Humphries (Ed.), *Decoding governance of marine genetic resources under the BBNJ Agreement.* Springer.

Rabone, M., Harden-Davies, H., Collins, J. E., Zajderman, S., Appeltans, W., Droege, G., Horton, T., et al. (2019). Access to marine genetic resources (MGR): Raising awareness of best-practice through a new agreement for biodiversity beyond national jurisdiction (BBNJ). *Frontiers in Marine Science, 6*(520), 535.

Rabone, M., Horton, T., Humphries, F., Lyal, C., Muraki Gottlieb, H., Scholz, A. H., Vanagt, T., & Jaspars, M. (2025). BBNJ Agreement: Considerations for scientists and commercial end users of MGR at research, development and commercialisation stages. In F. Humphries (Ed.), *Decoding governance of marine genetic resources under the BBNJ Agreement.* Springer.

Rogers, A. D., Baco, A., Escobar-Briones, E., Currie, D., Gjerde, K., Gobin, J., Harden-Davies, H., et al. (2021). Marine genetic resources in areas beyond national jurisdiction: Promoting marine scientific research and enabling equitable benefit sharing. *Frontiers in Marine Science, 8*, 600.

Schwartz, D. L., & Rogers, M. (2021). "Inventorless" inventions? The constitutional conundrum of AI-produced inventions. *Harvard Journal of Law and Technology, 35*, 531.

Slobodian, L., Kinna, R., Kambu, A, Ognibene, L. (2015). Bioprospecting in the global commons: Legal issues brief. nairobi: United nations environment programme. Division of Environmental law and conventions. Environmental law and governance branch.

Sloyan, B. M., Wanninkhof, R., Kramp, M., Johnson, G. C., Talley, L.D., Tanhua, T., McDonagh, E., et al. (2019) The global ocean ship-based hydrographic investigations program (GO-SHIP): A platform for integrated multidisciplinary ocean science. *Frontiers in Marine Science, 6*, 445.

Stokstad, E. (2018). Norwegian billionaire funds deluxe deep ocean research ship. *Science*. https://www.sciencemag.org/news/2018/11/norwegian-billionaire-funds-deluxe-deep-ocean-research-ship

Thambisetty, S., Oldham, P., & Claudio, C. (2023). The expert briefing document: A developing country perspective on the making of The BBNJ treaty (September 21, 2023). In *LSE Legal Studies Working Paper No. 30/2023.*

Titterton, H., & Haward, M. (2022). The Kerguelen Plateau: Interactions between the law of the sea and the antarctic treaty. *Marine Policy, 138*, 104993.

Tvedt, M. W. (2020). Marine genetic resources: A practical legal approach to stimulate research, conservation and benefit sharing. In *The Law of the Seabed* (pp. 238–254). Brill Nijhoff, 252.

UNEP. (2011). Report of the expert meeting on the modalities of operation of the access and benefit-sharing clearing house. UNEP/CBD/ABS/EM-CH/1/4. Expert meeting on the modalities of operation of the access and benefit-sharing clearing house. April 21, 2011.

UNEP. (2022). Decision adopted by the conference of the parties to the convention on biological diversity, 15/9 digital sequence information on genetic resources. In *Conference of the Parties to the Convention on Biological Diversity.* CBD/COP/DEC/15/9. December 19, 2022.

UNGA. (1992). Report of the United Nations conference on environment and development. In *Annex 1 Rio Declaration on Environment and Development.* United Nations General Assembly, 12 August 1992 A/CONF.151/26 (Vol. I).

UNGA. (2005). *Oceans and the law of the sea: Report of the secretary-general.* A/60/63/Add.1. July 15, 2005.

UNGA. (2018a). *Resolution adopted by the General Assembly on 24 December 2017: Internationally legally binding instrument under the United Nations Convention on the Law of the Sea on the conservation and sustainable use of marine biological diversity of areas beyond national jurisdiction.* A/RES/72/249. January 19, 2018.

UNGA. (2018b). *President's aid to negotiations,* Intergovernmental conference on an international legally binding instrument under the United Nations Convention on the Law of the Sea on the conservation and sustainable use of marine biological diversity of areas beyond national jurisdiction. Second session. A/CONF.232/2019/1*. 3 December 2018.

UNGA. (2018c). *Statement by the President of the conference at the closing of the first session,* Intergovernmental conference on an international legally binding instrument under the United Nations Convention on the Law of the Sea on the conservation and sustainable use of marine biological diversity of areas beyond national jurisdiction. First session. A/CONF.232/2018/7. 20 September 2018.

UNGA. (2018d). *President's aid to discussions,* Intergovernmental conference on an international legally binding instrument under the United Nations Convention on the Law of the Sea on the conservation and sustainable use of marine biological diversity of areas beyond national jurisdiction. First session. A/CONF.232/2018/3. 25 June 2018.

UNGA. (2019a). *Statement by the President of the conference at the closing of the second session,* Intergovernmental conference on an international legally binding instrument under the United Nations Convention on the Law of the Sea on the conservation and sustainable use of marine biological diversity of areas beyond national jurisdiction. Second session. A/CONF.232/2019/5. 18 April 2019

UNGA. (2019b). *Draft text of an agreement under the United Nations Convention on the Law of the Sea on the conservation and sustainable use of marine biological diversity of areas beyond national jurisdiction: Note by the President,* Intergovernmental conference on an international legally binding instrument under the United Nations Convention on the Law of the Sea on the conservation and sustainable use of marine biological diversity of areas beyond national jurisdiction. Third session. A/CONF.232/2019/6. 17 May 2019

UNGA. (2019c). *Revised draft text of an agreement under the United Nations Convention on the Law of the Sea on the conservation and sustainable use of marine biological diversity of areas beyond national jurisdiction: Note by the President,* Intergovernmental conference on an international legally binding instrument under the United Nations Convention on the Law of the Sea on the conservation and sustainable use of marine biological diversity of areas beyond national jurisdiction. Fourth session. A/CONF.232/2020/3. 18 November 2019.

UNGA. (2019d). *Statement by the President of the conference at the closing of the third session,* Intergovernmental conference on an international legally binding instrument under the United Nations Convention on the Law of the Sea on the conservation and sustainable use of marine biological diversity of areas beyond national jurisdiction. Third session. A/CONF.232/2019/10*. 13 September 2019.

UNGA. (2022a). *Further revised draft text of an agreement under the United Nations Convention on the Law of the Sea on the conservation and sustainable use of marine biological diversity of areas beyond national jurisdiction: Note by the President,* Intergovernmental conference on an international legally binding instrument under the United Nations Convention on the Law of the Sea on the conservation and sustainable use of marine biological diversity of areas beyond national jurisdiction. Fifth session. A/CONF.232/2022/5. 1 June 2022.

UNGA. (2022b). *Further refreshed draft text of an agreement under the United Nations Convention on the Law of the Sea on the conservation and sustainable use of marine biological diversity of areas beyond national jurisdiction: Note by the President,* Intergovernmental conference on an international legally binding instrument under the United Nations Convention on the Law of the Sea on the conservation and sustainable use of marine biological diversity of areas beyond national jurisdiction. Resumed fifth session. A/CONF.232/2023/2. 12 December 2022.

UNGA. (2023a). *Agreement under the United Nations Convention on the Law of the Sea on the conservation and sustainable use of marine biological diversity of areas beyond national jurisdiction,* Intergovernmental conference on an international legally binding instrument under the United Nations Convention on the Law of the Sea on the conservation and sustainable use of marine biological diversity of areas beyond national jurisdiction. Further resumed fifth session. A/CONF.232/2023/4. 19 June 2023.

UNGA. (2023b) Res. 78/155 Resolution adopted by the General Assembly on 19 December 2023: Implementation of the Convention on Biological Diversity and its contribution to sustainable development. A/RES/78/155. 21 December 2023.

Wilkinson, M. D., Dumontier, M., Aalbersberg, I. J., Appleton, G., Axton, M., Baak, A., Mons, B., et al. (2016). The FAIR guiding principles for scientific data management and stewardship. *Scientific Data, 3*(1), 1–9, 1.

WIPO. (2024). Diplomatic Conference to Conclude an International Legal Instrument Relating to Intellectual Property, Genetic Resources and Traditional Knowledge Associated with Genetic

Resources, *WIPO Treaty on Intellectual Property, Genetic Resources and Associated Traditional Knowledge*. GRATK/DC/7.

Yamahara, K. M., Preston, C. M., Birch, J., Walz, K., Marin, R., III., Jensen, S., Scholin, C., et al. (2019). In situ autonomous acquisition and preservation of marine environmental DNA using an autonomous underwater vehicle. *Frontiers in Marine Science, 6*, 373.

Fran Humphries has specialized in marine and biodiversity law and policy for over two decades in government, academia and consultancies. She is an associate professor at Griffith Law School, Griffith University, Australia, with a background in fisheries management and marine law. She has led large international research consultancies on access and benefit sharing of biological resources for governments, UN organizations and other institutions. She was on the International Council of Environmental Law delegation for the Biodiversity Beyond National Jurisdiction (BBNJ) Agreement negotiations.

Marcel Jaspars is a professor of Organic Chemistry at the University of Aberdeen where he leads the Marine Biodiscovery Centre which focusses on marine resources for novel pharmaceuticals, and to investigate fundamental questions in marine chemical ecology and biosynthesis. Marcel has been active at national and international levels to develop the science, its applications/industrial uptake and associated policy involved in marine biodiscovery and biotechnology. He provides scientific advice to the UK, EU and UN for global policy processes on ocean conservation and digital sequence information via reports, papers and taking part in discussion meetings.

Jessica Lavelle holds a PhD in environmental governance and is a research associate of the Bio-economy Research Chair at the University of Cape Town.

Daniel Kachelriess is an expert on oceans, fisheries, wildlife law and policy and followed the negotiations of the BBNJ Agreement as part of the High Seas Alliance and as a member of the IUCN World Commission on Environmental Law. He continues to advise the High Seas Alliance and other organizations on aspects of the BBNJ Agreement, including on Marine Genetic Resources, including the fair and equitable sharing of their benefits. His previous roles include Executive Director of Sea Shepherd Legal, a non-profit law firm, and the Marine Species Officer of the CITES Secretariat.

[illegible] Artificial Intelligence …

[illegible]

Fran Hunoldt [illegible]

Marcus Jaspers [illegible]

Monetary and Non-monetary Benefit Sharing Under the BBNJ Agreement

6

Arianna Broggiato, Paul Dunshirn, Marcel Jaspars, and Sergio Pena-Neira

Abstract

This chapter describes and interprets the system for sharing the benefits arising from activities with respect to marine genetic resources (MGR) and digital sequence information (DSI) on MGR of areas beyond national jurisdiction (ABNJ) that is being set up by the *Agreement under the United Nations Convention on the Law of the Sea on the Conservation and Sustainable use of Marine Biological Diversity of Areas Beyond National Jurisdiction*. The chapter provides an overview of discussions and rationales behind the negotiations of this system. It highlights the crucial compromises made to find a creative solution for a de-coupled monetary benefit sharing in the first phase of implementation, and to include digital sequence information in the scope of application and in the operational provisions. It also shows the contribution that the Agreement will make to the harmonisation and further improvement of best scientific practices in disclosing certain data in open access and in involving developing countries' scientists in the research pipelines. Moreover, the chapter shows how important novelties have been integrated and made legally binding: the deposit of available samples in publicly accessible repositories, thus amplifying access to MGR of ABNJ; and the use of a batch identifier to ensure that samples are linked to materials stored in collections and to the relevant data stored in databases, and that these three elements are retrievable with the use of one single batch identifier. The chapter concludes that several open questions remain, related to modalities for monetary benefit sharing to be decided by the Conference of the Parties, and to the requirement that the monetary benefit-sharing solution related to DSI is mutually supportive of and adaptable to other access and benefit-sharing instruments.

Arianna Broggiato: The opinions expressed are personal, remain the responsibility of the author in her individual capacity and do not necessarily represent the views of the European Union or its Member States.

A. Broggiato (✉)
European Commission, Brussels, Belgium
e-mail: arianna.broggiato@ec.europa.eu

P. Dunshirn
Department of Political Science, University of Vienna, Vienna, Austria

M. Jaspars
Marine Biodiscovery Centre, Department of Chemistry, University of Aberdeen, Aberdeen, UK

S. Pena-Neira
Universidad Mayor, Santiago, Chile

F. Humphries (ed.), *Decoding Marine Genetic Resource Governance Under the BBNJ Agreement*, Sustainable Development Goals Series, https://doi.org/10.1007/978-3-031-72100-7_6

Keywords

BBNJ agreement · Benefit sharing · Equity · Marine genetic resources · Digital sequence information · Monetary benefits · Non-monetary benefits · Capacity building

6.1 Introduction

The purpose of this chapter is to describe and interpret the system for sharing the benefits arising from activities with respect to marine genetic resources (MGR) and digital sequence information (DSI) on MGR of areas beyond national jurisdiction (ABNJ) that is being set up by the *Agreement under the United Nations Convention on the Law of the Sea on the Conservation and Sustainable use of Marine Biological Diversity of Areas Beyond National Jurisdiction* (BBNJ Agreement)[1] under article 14. The chapter provides an overview of discussions and rationales behind the formulation of this article, which aims to ensure that benefits, including monetary ones, are shared fairly and equitably without impairing research and innovation and without creating excessive administrative costs.

The chapter also aims to show that the BBNJ Agreement is an opportunity for better harmonisation of best scientific practices at the international level.

Fairness and Equity

Historically, the demand for benefit sharing related to genetic resources focused on the concepts of fair and equitable sharing.

"The fair and equitable sharing of benefits arising from activities with respect to marine genetic resources and digital sequence information on marine genetic resources of areas beyond national jurisdiction for the conservation and sustainable use of marine biological diversity of areas beyond national jurisdiction" is the first objective of Part II of the Agreement focusing on "Marine Genetic Resources, including the fair and equitable sharing of benefits" (article 9).

The concepts of fairness and equity have their origins in the philosophical works of Aristotle and have persisted through the Middle Ages into the contemporary world, as notably highlighted by various authors, with John Rawls (Rawls, 1971; Perelman, 1963) being of paramount significance. Equitable and fair resource sharing has been a foundational principle of the United Nations' framework for natural resources management since 1958 (Pena-Neira, 2017; UNGA Resolution 1314 (XIII)). Initially, it pertained to the allocation of rights between States and entities engaged in natural resource exploration within a state's territorial jurisdiction. The principle has been central to the evolution of genetic resource governance under the *Convention on Biological Diversity (CBD)* and the *Nagoya Protocol on Access to Genetic Resources and the Fair and Equitable Sharing of Benefits Arising from their Utilization to the Convention on Biological (Nagoya Protocol)*, which regulates access and benefit sharing related to genetic resources under national jurisdiction. Today, equitable and fair resource sharing extends to encompass marine genetic resources sourced from ABNJ, as outlined in the BBNJ Agreement. How gains and benefits are distributed has evolved significantly. In the context of the BBNJ Agreement, on one hand, the origins of these gains lie beyond the sovereign control of individual states, while on the other hand, they are intrinsically linked to a specific objective: the conservation of these resources. From a strict international law perspective, this presents a complex challenge. It raises inquiries about the procedural mechanisms for acquiring gains from these resources. Moreover, if these resources are transported to territories under the sovereignty of individual states,

[1] Intergovernmental Conference on an International Legally Binding Instrument under the United Nations Convention on the Law of the Sea on the Conservation and Sustainable Use of Marine Biological Diversity of Areas Beyond National Jurisdiction, Agreement under the United Nations Convention on the Law of the Sea on the Conservation and Sustainable Use of Marine Biological Diversity of Areas Beyond National Jurisdiction (2023) A/CONF.232/2023/4 (BBNJ Agreement).

the means by which these gains can be realized becomes a matter of scrutiny and deliberation.

The cornerstone for achieving equity and fairness in the distribution of benefits derived from MGR of ABNJ lies in establishing a robust framework. Nevertheless, it is imperative to keep in mind the other objective of the Agreement, which is the conservation of marine biodiversity, which includes MGR. In today's context, the principle of "equity" and the imperative of achieving "the fair and equitable sharing of benefits," as articulated in article 7, subsection "d," underscore the pervasive nature of these principles throughout the entirety of the Agreement. Furthermore, the concept of the fair and equitable sharing of benefits" permeates the Agreement as a whole, with the subject of gains and benefits featuring prominently within its provisions.

Two Opposing Views

The BBNJ Agreement is an implementing agreement to the United Nations Convention on the Law of the Sea (UNCLOS). UNCLOS provides the legal framework for all activities in the oceans and seas. However, it does not explicitly mention MGR in any part, nor specifically refer to benefit sharing (see Box 6.1).

Box 6.1

Benefit-Sharing in the Context of the Convention on the Law of the Sea

UNCLOS does not specifically provide for a benefit sharing regime with regard to MGR in ABNJ. However, Part XIII related to marine scientific research is relevant for this aspect, as it provides for some obligations to share non-monetary benefits arising out of scientific research, as illustrated below. With regards to benefit-sharing, the general provisions on marine scientific research (MSR) under UNCLOS require States and competent international organizations to:

- Promote international cooperation in MSR for peaceful purposes (Article 242.1 of the UNCLOS).
- Cooperate to create favourable conditions for the conduct of MSR (Article 243 of the UNCLOS).
- Make available by publication and dissemination through appropriate channels (a) information on proposed major programs and their objectives, and (b) knowledge resulting from MSR (Article 244.1 of the UNCLOS).
- Actively promote the flow of scientific data and information and the transfer of knowledge resulting from MSR, especially to developing states, as well as the strengthening of the autonomous MSR capabilities of developing states through programs to provide adequate education and training of their technical and scientific personnel (Article 244.2 of the UNCLOS).

The UNCLOS provisions on MSR apply both in areas within and beyond national jurisdiction, and also both to the high seas and the Area. These benefit sharing obligations are useful to partially address the uneven research means referred to above (Broggiato et al., 2018), and they were the basis of the non-monetary benefits that were negotiated.

The issues of MGR and benefit sharing were central to the political process that led to the start of the BBNJ negotiations. There were two opposing views (Marciniak, 2020; Muraki Gottlieb et al., 2025a): developing countries applying the principle of the common heritage of mankind to MGR of ABNJ, therefore claiming the sharing of the benefits, including monetary ones; and developed countries applying the principle of the freedom of the high seas and the freedom of research and opposing, in the beginning, the sharing of benefits.

Research in ABNJ is expensive, and it was clear during the political process that led to the negotiations that there are only a handful of States and funding agencies that have the means to undertake that research (Oldham et al., 2014). So far research in ABNJ has been undertaken only at the level of States' funding agencies with public funding. However, the advancement of technologies might allow private companies to start sampling in ABNJ in the future, and to undertake genetic sequencing directly in situ without the need to sample. Not only are there disparities between the two groups of countries (developing and developed ones) in financial and technological capabilities to access these resources, but also to use them for research and development. This is why the issue of fairness and whether or how benefits should be shared quickly became important points of negotiation. The underlying discussion over the two principles, common heritage of mankind versus freedom of the high seas, turned out to be too arduous to be solved, so the negotiations shifted to a practical discussion over what benefits could be shared and how.

In terms of what benefits could be shared, the discussions focused first on non-monetary benefits, such as support and involvement of developing countries in the research pipelines, sharing of information and data, capacity building, and transfer of marine technologies; and then, at a later stage, on monetary benefits. There was a strong opposition towards accepting the sharing of monetary benefits by developed countries. Besides the underlying difference in favoured legal principles, developed countries' argument against monetary benefit sharing focused on the fact that to date there is little evidence of commercialisation of products based on MGR of ABNJ from which to share monetary benefits (see Box 6.2). Developed countries argued that a monetary benefit-sharing system based on track and trace (Langlet et al., 2025), as it is usually done in the traditional Access and Benefits Sharing (ABS) context, only makes sense if benefits outweigh the administrative costs to establish the system, which in their views was doubtful given the little commercialisation. Developed countries also wanted to keep bureaucratic and financial burdens on scientific research and subsequent utilization at a reasonable level, to not hamper research and development. Developing countries argued that fair and equitable benefit sharing of MGR of ABNJ requires both non-monetary and monetary benefit sharing. Additionally, they saw traceability of activities relating to the utilisation of MGR of ABNJ as a pre-condition for such a system to future-proof the Agreement for the eventual upscaling of commercialisation. While various forms of track and trace or traceability have been proposed by different developing countries (Langlet & Dunshirn, 2023), they all insisted that evidence of scientific or corporate use needs to be collected and that the sharing of benefits should be tied to traced activities.

Box 6.2

Commercial Value of Marine Genetic Resources of Areas Beyond National Jurisdiction

The lack of data and precise assessments of the commercial value of MGR of ABNJ notably led the Ad Hoc Open-ended Informal Working Group to study issues relating to the conservation and sustainable use of marine biological diversity beyond areas of national jurisdiction to call, in 2006, in 2008, in 2012 and 2013, for more research to assess the actual or potential total economic value of MGR from ABNJ, to assist decision-making by providing indications of the expected economic and societal benefits of such resources, as well as providing supporting arguments for the possible regulatory measures to adopt. The Working Group recommended to conduct studies on:

(1) the nature and level of interests in marine biological diversity beyond areas of national jurisdiction, in particular commercial interest in genetic resources from the deep sea;

(2) the socio-economic value of marine biological diversity beyond areas of national jurisdiction (UN Doc A/61/65, Annex 2).

No such study has ever been formally carried out as part of the UN processes considering the issue. However, during the years of the Working Group, official UN documents have made references to a huge industry interest in the commercialisation of MGR from ABNJ, to numerous products already on the market (without references): these figures related to the profits of marine biotechnology (related to shallow waters to date) (Leary, 2018). Conflating the biotechnology potential of marine biodiversity in ABNJ with the actual experience of other existing sectors of the biotechnology industry, and with the market of marine biotechnology on shallow waters MGR, is in part what has fuelled the expectations of a new deep-sea gold rush (Leary, 2018).

The first attempt to investigate in that direction was a detailed study commissioned by the United Kingdom's Department for Environment, Food and Rural Affairs, and published in 2014 (Oldham et al., 2014). It investigated the nature and extent of scientific and commercial interest in MGR: the conclusions confirmed the emerging growing interest in MGR, including from deep-water locations, but could not precisely indicate any market estimate for products derived from MGR of ABNJ. The main findings revealed that most scientific research and commercial research and development, including research involving deep-water locations, focuses on marine organisms from habitats inside national jurisdictions. The available evidence suggested that marine scientific research in ABNJ concentrated around a limited number of sites relative to the scale of the deep-sea. Analysis of patent data revealed increasing reference to deep-sea marine organisms, but additionally that these organisms also frequently occur inside the exclusive economic zone (EEZ) and in terrestrial aquatic environments. This made it difficult to determine with precision whether a sample originated from ABNJ or within national jurisdictions. The study concluded that debates on the economic value of marine genetic resources of areas beyond national jurisdiction are focused on potential economic value rather than actual economic value, and that the interest in MGR from ABNJ was emergent (Oldham et al., 2014). An update of this study is currently being funded by the European Commission through the European Maritime, Fisheries and Aquaculture Fund (EMFAF)

The divergence of views softened in March 2023, when compromise was found to include monetary benefit sharing (see Box 6.3). However, disagreement over the general principles guiding the MGR Part and the provisions on benefit sharing more specifically was maintained until the very last hours of IGC5. Negotiators eventually agreed on the inclusion of the common heritage of humankind[2] principle alongside the freedom to conduct scientific research, alongside other freedoms, and, in more practical terms, on a gradually evolving system to ensure the fair and equitable sharing of benefits derived from MGR of ABNJ and associated DSI.

Box 6.3 illustrates the sticking and controversial points related to MGR of ABNJ on which delegations having two opposing views managed to compromise during the negotiations.

[2] Mankind was replaced by humankind to be gender inclusive.

Box 6.3

Summary of Significant Compromises Made During the Negotiations:

- Agreement on the "fair and equitable manner" of sharing the benefits (unbracketed at IGC4).
- Agreement on the use of the terminology "digital sequence information" with the agreement not to define it (at IGC5 final round).
- Agreement to include DSI within the scope of Part II and in most of the operational provisions of Part II (in the last hours of the final IGC session).
- Introduction of the concept of the "BBNJ standardized batch identifier" in article 12 (in the last hours of the final IGC session).
- Shift from only non-monetary BS to monetary BS (at the final IGC5 session).
- Agreement on the first phase of monetary benefit sharing through assessed contributions by developed Parties at IGC5 final round (article 14.6) (at IGC5 final round).
- Agreement on the second phase of monetary benefit sharing through the COP deciding on the modalities for the sharing of monetary benefits from the utilization of marine genetic resources and digital sequence information on marine genetic resources of areas beyond national jurisdiction at IGC5 final round (article 14.7 and para 8, 9 and 10) (in the last hours of the final IGC session).
- Agreement on the role of the ABS Committee as an advisory body on subjects matter related to benefit sharing (at IGC5 final round).

6.2 Overview of Key Aspects of the Benefit-Sharing Regime

Article 14 of the BBNJ Agreement is a comprehensive article illustrating almost all the obligations Parties must fulfill in terms of benefit sharing: one benefit-sharing obligation is included in article 12 on notification (Humphries et al., 2025). Article 14 includes the historic compromise that monetary and non-monetary benefits from the utilization of MGR of ABNJ and related DSI, including commercialization, shall be shared fairly and equitably. This compromise was reached at the very last moment in the negotiations, and it could unlock the deal in the whole Agreement.

Non-monetary benefit sharing can include access to samples, sample collection, and DSI; access to information on upcoming research projects; collaborative research opportunities; capacity building and transfer of marine technologies. Monetary benefits are financial resources destined to the Special Fund that will use them, on a project basis, for capacity building, support for conservation and sustainable use, and for the implementation of the Agreement by developing countries.

The monetary benefit-sharing system in the BBNJ Agreement is organised in two phases illustrated as part of the visualisation below (Visualisation 1). In the visualisation the line connecting biodiscovery with commercialisation is striped to indicate that this is not a linear flow in reality: in only rare cases biodiscovery translates into commercialisation of a product (Fig. 6.1).

In the first part of article 14, the main non-monetary benefits to be shared are listed (para 2). The most innovative obligation is contained in paragraph 3 and deals with the deposit of marine genetic resources of ABNJ and associated DSI in publicly accessible repositories and databases, together with their "BBNJ" standardized batch identifiers (BBNJ Identifiers) (Lawson et al., 2025) with a precise time

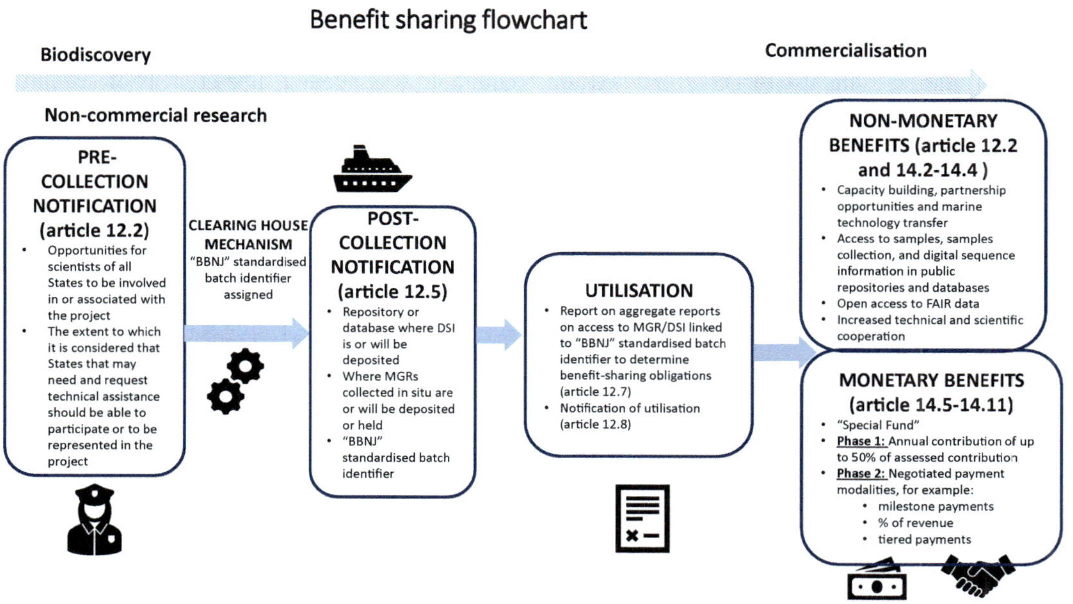

Fig. 6.1 The interplay between traceability and benefit sharing under the BBNJ Agreement

deadline (i.e. no later than three years from the start of utilization). Sharing DSI, here intended as genetic sequence data, is already standard practice in big international research consortia where these data are generally deposited in the International Nucleotide Sequence Database Collaboration (INSDC)[3] databases as they are generated or at the latest before publication of the results in an academic journal (Rohden et al., 2020). However, the obligation to share physical MGR samples is novel, as their existence is often not well documented, and sharing is mostly done on a case-by-case basis (Collins et al., 2021; Rabone et al., 2019; Rogers et al., 2021). Similarly, novel is the BBNJ Identifier, which, once implemented, should allow for a certain level of traceability of MGR of ABNJ in the future and for transparency of activities (Lawson et al., 2025).

The rest of the article illustrates the 'two-phase approach' for monetary benefit sharing. In the first phase, developed Parties will contribute to the Special Fund established under article 52 with an annual contribution (para 6). This contribution is capped at 50% of a Party's annual contribution to the BBNJ budget. After regular reviews and assessments of the monetary benefits by the Conference of the Parties (COP) (para 10), in a second phase, the COP shall decide on additional or alternative benefit-sharing modalities (para 7), which in the case of DSI should be mutually supportive of and adaptable to other access and benefit-sharing instruments (para 9). The article provides a non-exhaustive list of options to be considered during this stage.

The first phase reflects the fact that there is little evidence of commercial products derived from MGR of ABNJ from which monetary benefits could be shared (see Box 6.2). But it also recognizes that monetary and non-monetary benefit sharing are needed to level the playing field when it comes to MGR access, resources, and technological skills between different regions. The second phase will mirror the developments in the current state of the art of commercialisation thanks to the review of the monetary benefits from utilisation of MGR and DSI of ABNJ

[3] International Nucleotide Sequence Database Collaboration (insdc.org).

of para 10 (see below). The modalities which will eventually be adopted during this second phase in relation to the use of DSI should ensure consistency/compatibility with other international frameworks on access and benefit sharing, according to para 9 (see below).

6.3 In-Depth Interpretation of Article 14

Article 14

Fair and Equitable Sharing of Benefits

1. The benefits arising from activities with respect to marine genetic resources and digital sequence information on marine genetic resources of areas beyond national jurisdiction shall be shared in a fair and equitable manner in accordance with this Part and contribute to the conservation and sustainable use of marine biological diversity of areas beyond national jurisdiction.

The chapeau of article 14 introduces the general obligation to share the benefits arising from MGRs and DSI on MGR of ABNJ. The undefined terminology "activities with respect to MGR of ABNJ" (initially used in article 11—ex article 9) was proposed in the benefit-sharing article by the Chair during IGC5 (August 2022) as a possible compromise to resolve the divergence between using "collection", seen as too limited for the benefit-sharing provision, and the concept of "utilization" which was not acceptable for all the delegations. There was an attempt to define "activities with respect to MGR of ABNJ" during the last session (March 2023), but it was impossible to find an agreement on the long list of proposed activities. It is safe to argue that "activities" include both collection of MGR and utilization of MGR (Humphries, 2025).

The connotation of 'fair and equitable manner' has for long debated, as the concept comes from the CBD where fairness and equity are concepts to be implemented through the bilateral relationship between a user and a provider. As no provider country exists for MGR of ABNJ, a different multilateral conception of fair and equitable was needed. The content of article 14 is a first step in such a direction, even though many details still need to be further developed in upcoming COP meetings.

The link between the benefits and the conservation and sustainable use was also debated, as there was a divergence of positions on this issue. During the negotiations, discussions on MGR have focused on the commercial values of the resources. It was only at IGC4 that some observers, including scientists, started stressing that MGR certainly has economic potential, but that they are first of all the biological foundations of marine biodiversity: enhancing knowledge about them is crucial in terms of establishing great environmental baselines and adopting effective conservation measures. According to these considerations, MGR constitutes the fundamental scientific basis for policy measures to be adopted within the BBNJ framework, way beyond Part II of the Agreement.

2. Non-monetary benefits shall be shared in accordance with this agreement in the form of, inter alia:

(a) *Access to samples and sample collections in accordance with current international practice;*
(b) *Access to digital sequence information in accordance with current international practice;*
(c) *Open access to findable, accessible, interoperable and reusable (FAIR) scientific data in accordance with current international practice and open and responsible data governance;*
(d) *Information contained in the notifications, along with "BBNJ" standardized batch identifiers, provided in accordance with article 12, in publicly searchable and accessible forms;*
(e) *Transfer of marine technology in line with relevant modalities provided under Part V of this Agreement;*

(f) *Capacity building, including financing research programmes, and partnership opportunities, particularly directly relevant and substantial ones, for scientists and researchers in research projects, as well as dedicated initiatives, in particular for developing States, taking into account the special circumstances of small island developing States and of least developed countries;*
(g) *Increased technical and scientific cooperation, in particular with scientists from and scientific institutions in developing States;*
(h) *Other forms of benefits as determined by the Conference of the Parties, taking into account recommendations of the access and benefit-sharing committee established under article 15.*

Paragraph 2 introduces a non-exhaustive list of non-monetary benefits to be shared. There were debates on whether the sharing of these benefits should be compulsory, and on temporal aspects related to their implementation. The debates were not solved, and the literal interpretation of the text does not help. It can be safely argued that letters (a) to (d) are to be shared subsequently to every collection, as the compulsory nature of benefit sharing was not questioned in these cases. The debate over the compulsory nature focused more on the practical impossibility of a State to provide for everything that is listed in letters (e) to (f) and to do so at the same time (temporal aspect mentioned above). So, it can be argued that a Party has some discretion on what transfer of marine technology, what capacity building, and what cooperation to provide, as these are generic terms where choices need to be made. The different non-monetary benefits are described in detail below.

The benefits included in letters (a) to (c) are, to some extent, already being shared according to the best scientific practices, as illustrated below. The implementation of the BBNJ Agreement will then be a good opportunity to further improve this sharing of samples and data; and to harmonise at the international level, according to the best standards, these best practices and make them compulsory. In this sense, the BBNJ Agreement will benefit the scientific community worldwide.

(a) Access to samples and sample collections in accordance with current international practice;

Parties to the BBNJ Agreement have to provide, as a benefit to be shared, access to samples and to samples collections (usually stored in collections, museums, and other repositories), in accordance with current international practices. Considering the disparities in financial resources and technologies to access MGR of ABNJ illustrated above, access to collected samples is a way to provide access to MGR to countries without the same resources and technologies to collect them. Letter (a) has to be read in combination with para 3 of article 14, which obliges Parties to ensure that MGR are deposited in publicly accessible repositories (see below). Providing access to samples and data—letters (b) and (c) below—through publicly available repositories and databases ensures that scientists can use them and start their own research and development on their own basis, beyond the scientists initially involved in the research project.

"Samples" and "sample collections" are not defined by the BBNJ Agreement: to this end, scientific practices could help. It is safe to expect that this will also be a topic for clarification by the Scientific and Technical Body of the BBNJ Agreement.

Access to MGR in repositories and collections follows standard practice in taxonomic collections where materials are shared to identify species or carry out systematic studies. What is more recent is the ability to request materials from such repositories for other scientific and applied research such as biotechnological work to discover useful functions, products, and processes. This means that samples may be destroyed and 'used up' in the process and care needs to be taken to ensure that samples are not entirely used up and that research carried out on MGR is complementary—that is to prevent two sets of researchers carrying out exactly the same work on the same sample (Collins et al., 2021; Rabone et al., 2019). Additional considerations are that sample collections should be discoverable and their databases available for access via mechanisms such as the Global Biodiversity Information Facility (GBIF).[4] This may require additional resources to be dedicated to collections and their curation to ensure that this is possible, both in developed countries and low and middle-income countries. The second consideration is the additional burden on curators of collections and repositories to provide MGR to researchers, and reasonable costs for handling and sending materials should be permitted to account for this. Reporting of the use of MGR in collections and repositories as required by the Agreement (article 12.7—Humphries et al., 2025) entails additional work, and funds must be made available to allow for this to ensure that curators are not overburdened by this requirement.

The reference to current international practice is used in Part II of the BBNJ Agreement with the intent to endorse bottom-up approaches of best practices that are used or emerging in the scientific communities (Rabone et al., 2019; The interridge code of conduct on responsible behaviour at hydrothermal vents[5]). The goal of referring to current international practice was twofold: on one hand, to build on already existing best practices in use by scientists, and avoid reinventing the wheel and interfering with bottom-up approaches that are driven by the scientific communities; and, on the other hand, to future-proof the BBNJ Agreement and provide the necessary flexibility in implementation as technologies and practices change. It can also be argued that the BBNJ Agreement encourages the scientific community to develop and further harmonise international practices: in previous drafts of the monitoring and traceability article (16 ex 13) international practices were recognised by the COP as international standards and therefore given authority. However, this text was not included in the final Agreement text, because the article on monitoring changed its focus.

(b) Access to digital sequence information in accordance with current international practice;

According to current international practice, access to DSI (at the moment of writing the chapter and only for practical purposes of tentatively explaining the obligation contained in article 14.2b), the current international practice here described understands DSI as genetic sequence data, without prejudging different decisions that might be taken within the BBNJ Agreement) is provided through the deposition of DSI into one of the INSDC[6] databases,[7] preferably when the sequence data is generated (Amann et al., 2019) but before publication of the research arising from this DSI at the latest. This is standard practice (Collins et al., 2021; Rabone et al., 2019; Rogers et al., 2021) in the field. The novelty will be to record the BBNJ Identifier in the relevant field when the DSI is deposited and the INSDC accession number is assigned. Due to many forms and types of DSI and the importance of open access to scientific workflows, it is generally not straightforward to implement new data governance measures into existing database infrastructures. The BBNJ tag may aid such an implementation as it will not be technically complicated for databases to integrate this tag in its current efforts to increase reporting of origin information.[8]

[4] https://www.gbif.org/.

[5] http://194.254.225.67/de/node/16908.

[6] https://www.insdc.org/.

[7] National Center for Biotechnology Information (NCBI), European Bioinformatics Institute (EBI), DNA Data Bank of Japan (DDBJ).

[8] As announced by the INSCCD on 3 March/03/2023 https://www.insdc.org/news/insdc-spatiotemporal-metadata-minimum-standards-update-03-03-2023/.

(c) Open access to findable, accessible, interoperable and reusable (fair) scientific data in accordance with current international practice and open and responsible data governance;

Open access to scientific data is not only a prerequisite for science to advance but also a benefit that is being shared across countries. Maintaining scientific databases is an expensive activity in terms of financial resources but also in terms of human resources to ensure the quality of the data hosted. Most, if not all, of the funding agencies that are financing research in marine ABNJ are imposing open access to the data and the publications originating from that research (Broggiato et al., 2018).

Besides DSI, other scientific data should also be deposited in open-access databases that adhere to the FAIR principles (Findable, Accessible, Interoperable and Reusable). Which data types this includes still needs to be decided in the framework of the implementation of the Agreement. For instance, there are standard databases for transcriptome and proteome data, but for small molecule data, there are still many different databases for molecules of different origin. The situation for instrumental data acquired on proteins and small molecules is also complex (Wilkinson et al., 2016) and there are many different databases where data may be deposited. It is hoped that over time a clear set of databases will develop similar to the INSDC, and that international practices will develop for other sets of data. As required by the Agreement, much of the detail on where such scientific data is deposited will be recorded in the data management plan and its subsequent updates (Lawson et al., 2025).

(d) Information contained in the notifications, along with "BBNJ" standardized batch identifiers, provided in accordance with article 12, in publicly searchable and accessible forms;

Another important benefit to be shared relates to the information contained in the pre-collection and post-collection notifications (Humphries et al., 2025). Through the sharing of this information, benefit sharing can start accruing from the moment of planning a collection in ABNJ, for example by providing partnership opportunities and opportunities for technical and scientific cooperation to scientists, in particular from developing countries. This letter (d) should be read in conjunction with letters (f) and (g) of para 2.

Information in the notifications will normally be recorded as part of the cruise plan and final cruise report in case of cruises, and otherwise in specific pre-collection and post-collection notifications in case of different types of collections that new technologies might allow in the future. Currently, many national bodies that carry out marine scientific research deposit this information online, but to varying degrees and in varying levels of detail. Article 12 lists all the elements that have to be included in the notification: the pre-collection notification builds on article 248 UNCLOS illustrating the information to be shared with the coastal State by States and competent international organizations which intend to undertake marine scientific research in the exclusive economic zone or on the continental shelf of that coastal State. This information will be shared, in the BBNJ context, with the Clearing House Mechanism and should be linked to the BBNJ Identifier so that it becomes discoverable based on this identifier. Support will be needed for many developed, low, and middle-income countries to achieve this and to work to a common standard. Moreover, the CHM will have to properly establish a mechanism to allocate these partnerships and cooperation opportunities, with guidance from the COP.

Letter (d) stresses the link between the notification system and benefit sharing: article 12.8 on the post-collection notification refers to some important information to be deposited in the Clearing House Mechanism. This information constitutes already shared benefits themselves: knowing where to find available samples and data allows access to them and further use, as well as being able to retrieve the results of the utilisation. Moreover, benefit sharing starts accruing already at the level of the pre-collection notification [Article 12.2 (h) and (i)], where opportunities are given to scientists of all States, in particular to scientists from developing States, to be involved in or associated with the research project. This implies that spaces will have to be made available on the research cruises: so far, this is usually undertaken based on personal contacts (Rogers et al., 2021). Upon implementation, this needs to be harmonised and properly advertised at the international level to ensure transparency. Various stakeholders have strongly advocated for early involvement of developing countries' scientists already at the time of planning a cruise, which is important in terms of equity and fairness, but also as a fundamental capacity-building element to pass on important know-how on what kind of equipment to choose for certain research activities, as suggested by scientists.

(e) Transfer of marine technology in line with relevant modalities provided under Part V of this Agreement;
Capacity building, including financing research programmes, and partnership opportunities, particularly directly relevant and substantial ones, for scientists and researchers in research projects, as well as dedicated initiatives, in particular for developing States, taking into account the special circumstances of small island developing States and of least developed countries;
Increased technical and scientific cooperation, in particular with scientists from and scientific institutions in developing States;

Letters (e), (f) and (g) introduce non-monetary benefit sharing related to transfer of marine technologies, capacity building, and collaborative research: all of those are orientated in particular towards developing countries. The aim is to level the playing field to ensure that developing countries' scientists can profit from the non-monetary benefit sharing listed in letters (a)–(d). Access to samples and data needs to be paired with transfer of technologies (which is to be done according to Part V of the BBNJ Agreement), the necessary training, and other types of capacity building to be able to use the samples and the data. Bioinformatic skills are still lacking in marine scientists in general, and in particular in developing countries. Involvement in the research process at all different steps within efficient international collaborative opportunities is also crucial to strengthen the autonomous marine scientific research capabilities of developing States as required by article 244 UNCLOS.

Current capacity-building initiatives are ad hoc and piecemeal: lots of capacity-building projects are provided globally by various agencies (Harriet Harden-Davies et al., 2022), but there is no central repository of what agencies are carrying out work and where this is being done[9]: the Clearing House Mechanism of the BBNJ will provide coordination on this. The first attempt to look at how capacity building is organised in deep sea/marine research in ABNJ was done in 2014 (Oldham et al., 2014). The study showed that there was a need for more infrastructure and access to financial resources to enable training and research exchanges between developed and developing countries; and that an agreed long-term strategy for capacity building and technology transfer was required to overcome short-termism in support of deep sea research. An update of this study is currently being funded by the European Commission through the European Maritime, Fisheries and Aquaculture Fund (EMFAF)[10]: the outcomes of the study are expected in the second half of 2024.

[9] Coordination—Capacity Development//UNSD.

[10] Study on 'Marine Genetic Resources' Market Value and State of the Art of Commercialisation of Related Products in the Context of the BBNJ Negotiations. Reference number: CINEA/2022/OP/0017, available at 608281-2023—Result—TED (europa.eu).

In the views of the authors of this chapter, an initial task of the Access and Benefit Sharing Committee (Muraki Gottlieb et al., 2025b) should be to consider existing capacity-building efforts and provide a comprehensive overview of the possibilities and the gaps. This could then be used to develop research directions and involve relevant stakeholders, both funders and participants from low and middle-income countries in shaping a long-term strategy, with the aim of moving away from the donor/recipient language towards one of partnership.

Partnerships could be developed in areas critical to the BBNJ Agreement objectives such as using information on MGR to develop conservation measures relevant to the other parts of the Agreement. Needs assessments can be very valuable to establish baselines of research capacities and assist in the development of equitable partnerships that will be sustainable over the long term to boost research capabilities.[11] Training researchers in partner labs and on research vessels is essential in this regard. It is important to recognise that, when such researchers return to their home institutions, support must continue, and relevant facilities must be available for them to continue their work in their home institutions. Such facilities should be relevant to the location chosen and service and maintenance requirements of equipment should be considered, as very high-tech equipment often cannot be kept running in the longer term without this support (Ramutsindela & Mickler (eds.), 2020).

3. Parties shall take the necessary legislative, administrative or policy measures to ensure that marine genetic resources and digital sequence information on marine genetic resources of areas beyond national jurisdiction, together with their "BBNJ" standardized batch identifiers, subject to utilization by natural or juridical persons under their jurisdiction are deposited in publicly accessible repositories and databases, maintained either nationally or internationally, no later than three years from the start of such utilization, or as soon as they become available, taking into account current international practice.

Through the implementation of para 3 of article 14 Parties will achieve one of the main and most requested types of benefit sharing, which is to make samples available worldwide, therefore effectively ensuring access to MGR of ABNJ. As stressed above, this will ensure that samples and data will be further used by scientists for research and development, beyond the group of scientists initially involved in the research project. This is not standard practice at the moment: remaining or available samples are very often stored in each scientists' laboratories, and their existence is often not documented nor shared in the community. Sharing of samples that remain available is mostly done on a case-by-case basis (Rogers et al., 2021). Importantly, this obligation will be implemented through national measures which will also target private entities if they start sampling MGR in ABNJ.

The same applies to DSI derived from MGR samples of ABNJ, even though sharing such data in open-access databases, such as via INSDC, is already more common in research practice. Nevertheless, the time limit imposed by the article will improve the practice. Deposit of DSI can happen immediately if that is the practice in the research field. However, most scientists are depositing DSI in INSDC at the moment of publication of their research results in journals: the accession number obtained when depositing DSI must be included in the manuscript for publication, according to the main publishers' requirements.[12] Scientific publishers

[11] Science in Small Island Developing States—AOSIS.

[12] See for example, Nature's instructions regarding data: https://www.nature.com/sdata/publish/submission-guidelines#:~:text=For%20repositories%20using%20accessions%20(e.g.,further%20guidance%20after%20peer%2Dreview and https://www.nature.com/sdata/policies/data-policies, https://www.nature.com/sdata/policies/repositories#nuc.

have therefore power to foster and further advance scientific practices, now with support by existing international law.

The wording "subject to utilization by natural or juridical persons" refers to a scientist/user or a juridical person (company or other entity) who uses the samples and/or the data thereon derived. During the negotiations, the obligation to share samples was intended to apply to collectors to ensure that, once they do the research with the samples, they deposit the remaining available samples, if any are remaining given how research can be done (see above). In a previous draft text of the Agreement (A/CONF.232/2023/2 from 12 December 2022, and A/CONF.232/2023/CRP.1/Add.1 from 25 February 2023), article 11.3 contained the word "available", which was then deleted because redundant: if no samples are available, the obligation does not apply. In relation to DSI, the intention during the negotiations was to transform the scientific practice of sharing those data into a legal obligation (see above). The BBNJ scientific and technical body together with the ABS Committee will likely further investigate and eventually provide recommendations on how to better implement these obligations.

The assignment of a standardized BBNJ Identifier (Lawson et al., 2025) is a novelty in itself and will, once implemented and practiced, establish more transparency about MGR and DSI origins. As such, it could also function as an important lever for any estimations of monetary benefit sharing at aggregate level, which may be introduced during the second phase of the two-phase monetary benefit-sharing approach. Keeping the samples and the data linked to the BBNJ Identifier will be crucial to ensure transparency of information and for users to be aware that they will have to share benefits.

4. Access to marine genetic resources and digital sequence information on marine genetic resources of areas beyond national jurisdiction in the repositories and databases under a Party's jurisdiction may be subject to reasonable conditions, as follows:

(a) The need to preserve the physical integrity of marine genetic resources;

(b) The reasonable costs associated with maintaining the relevant gene bank, biorepository or database in which the sample, data or information is held;

(c) The reasonable costs associated with providing access to the marine genetic resource, data or information;

(d) Other reasonable conditions in line with the objectives of this Agreement; and opportunities for such access on fair and most favourable terms, including on concessional and preferential terms, may be provided to researchers and research institutions from developing States.

Paragraph 4 introduces elements from the practice of collections who are usually charging a symbolic fee to access the samples they have been curating [letter (a) and (b)] (Rabone et al., 2019). These symbolic fees are usually not applicable when the requests come from developing countries. This para introduces the possibility for the same practice to be applied by certain databases: it is currently not the case for INSDC, which is free.

Letter (d) is an additional clause to make sure that developing countries' users have access on fair and most favourable terms.

5. Monetary benefits from the utilization of marine genetic resources and digital sequence information on marine genetic resources of areas beyond national jurisdiction, including commercialization, shall be shared fairly and equitably, through the financial mechanism established under article 52, for the conservation and sustainable use of marine biological diversity of areas beyond national jurisdiction.

Paragraph 5 introduced the obligation to share the monetary benefits from the utilisation of MGR and DSI. Ensuring that monetary benefits are shared with a clear objective to foster conservation and sustainable use of marine biological diversity of ABNJ was an important compromise to get all delegations on board of the deal.

The reference to the financial mechanism established under article 52 clarifies that monetary benefits are shared through a Multilateral Fund (the Special Fund) which is regulated by the BBNJ Agreement itself, whose purposes are set in article 52.6. Monetary benefit sharing will be channelled mainly towards capacity building for developing countries to implement the Agreement and thus to conserve and sustainably use marine biological diversity. Monetary benefits are therefore not redistributed to Parties.

The Special Fund will also be financed by additional voluntary contributions by Parties and private entities wishing to do so.

The reference to commercialisation is an anticipation of what is elaborated in paragraph 7 and ensures that the modalities the COP might decide on can also be linked to the commercialisation of MGR of ABNJ, if and when this happens at a larger scale.

6. After the entry into force of this Agreement, developed Parties shall make annual contributions to the special fund referred to in article 52. A Party's rate of contribution shall be 50 per cent of that Party's assessed contribution to the budget adopted by the Conference of the Parties under article 47, paragraph 6 (e). Such payment shall continue until a decision is taken by the Conference of the Parties under paragraph 7 below.

7. The Conference of the Parties shall decide on the modalities for the sharing of monetary benefits from the utilization of marine genetic resources and digital sequence information on marine genetic resources of areas beyond national jurisdiction, taking into account the recommendations of the access and benefit-sharing committee established under article 15. If all efforts to reach consensus have been exhausted, a decision shall be adopted by a three-fourths majority of the Parties present and voting. The payments shall be made through the special fund established under article 52. The modalities may include the following:

(a) Milestone payments;

(b) Payments or contributions related to the commercialization of products, including payment of a percentage of the revenue from sales of products;

(c) A tiered fee, paid on a periodic basis, based on a diversified set of indicators measuring the aggregate level of activities by a party;

(d) Other forms as decided by the conference of the parties, taking into account recommendations of the access and benefit-sharing committee.

8. A Party may make a declaration at the time the Conference of the Parties adopts the modalities stating that those modalities shall not take effect for that Party for a period of up to four years, in order to allow time for necessary implementation. A Party that makes such a declaration shall continue to make the payment set out in paragraph 6 above until the new modalities take effect.

9. In deciding on the modalities for the sharing of monetary benefits from the use of digital sequence information on marine genetic resources of areas beyond national jurisdiction under paragraph 7 above, the Conference of the Parties shall take into account the recommendations of the access and benefit-sharing committee, recognizing that such modalities should be mutually supportive of and adaptable to other access and benefit-sharing instruments.

10. The Conference of the Parties, taking into account recommendations of the access and benefit-sharing committee established under article 15, shall review and assess, on a biennial basis, the monetary benefits from the utilization of marine genetic resources and digital sequence information on marine genetic resources of areas beyond national jurisdiction. The first review shall take place no later than five years after the entry into force of this Agreement. The review shall include consideration of the annual contributions referred to in paragraph 6 above.

Paragraphs 6 and 7 establish the two-phase monetary benefit-sharing system, which was introduced because Parties principally agreed on monetary benefit sharing towards the end of negotiations, yet with little agreement on the precise modalities beyond the de-coupled modality of the first phase.

In the first phase, once the budget for the implementation of the BBNJ Agreement is adopted by the COP, developed Parties will start paying an annual monetary contribution into the Special Fund: the contribution will be equal to 50% of their annual assessed contribution to the budget of the Agreement. This allows for a constant, predictable and reliable flow of financial resources coming into the Special Fund to be used for capacity building and conservation measures. This innovative way to provide for monetary benefit sharing is 'de-coupled' from any specific access or use of any specific MGR and therefore does not need a burdensome track and trace system. It is conceived to be monetary benefit sharing related to MGR of ABNJ, to be paid only by developed Parties, independently of whether they are sampling or utilising MGR of ABNJ.

In the second phase, different modalities for the sharing of monetary benefits will be negotiated in future COP meetings, which may tie monetary benefit sharing more closely to indicators of actual utilization of marine genetic resources. As listed in para 7, this may be connected, for instance, to milestone payments (7a); revenues related to the sale of products resulting from the use of MGR from ABNJ or related DSI (7b); or estimations of aggregate activities and their value for particular Parties (7c).

In the framework of the traditional ABS context, milestone payments are usually paid when a certain step occurs, for example, the deposit of a patent request, the granting of a patent, or the moment a product enters the market (Lavelle & Wynberg, 2025). The option of milestone payments is therefore based on checkpoints to monitor the steps where these payments are required. As it stands, the BBNJ Agreement does not prescribe such checkpoints.

The option of payments or contributions related to the commercialization of products implies that users that are commercialising a product that is the result of utilisation of MGR of ABNJ or is produced through the use of DSI on MGR of ABNJ, would pay a percentage of the revenues related to the sale of the products to the Special Fund. The reason would be that those entities that obtain financial benefits from the use of the MGR/DSI should share portions of those financial benefits to the Fund. If this option were to be considered by Parties, once commercialisation of MGR of ABNJ upscales, the following questions would need to be resolved:

1. How to determine the weight of the MGR and/or the DSI derived thereof in the final product? In other words, to what extent have the MGR and/or DSI contributed to the development of the product, and how crucial are they for the functionality of the product? For DSI this issue is even more difficult considering the integrative "mixing" of DSI in the process of many bioinformatic- and synthetic-biology-related R&D (see Box 6.4).
2. How to deal with the issue of stacking, which is the situation when a product is developed through the use of several genetic resources, including MGR, and/or DSI sourced from different origins (including areas within national jurisdiction)?
3. Are additional track and trace obligations needed to ensure the implementation of this option, especially for products that do not require registration and/or market approval?

4. How would State Parties implement this provision? taking into account that:
   a. Companies typically do not report revenue at product level;
   b. Sales would have to be broken down per jurisdiction, information that currently does not fall under any financial reporting obligations;
   c. Party versus non-Party application would have to be clarified, such as: would revenue in non-Party countries be excluded, or would products developed in a non-Party be exempted, also for sales within a Party's jurisdiction?

The option to estimate monetary benefits based on aggregate use estimators is a product of long-standing disagreement over the extent to which traceability is feasible and needed to fulfill the aims of the Agreement. Aggregate use might provide a compromise that would neither require heavy track and trace of individual MGR samples or data nor a complete decoupling of benefits from the frequency or intensity of actual use. Even though aggregate use indicators may not necessarily measure direct financial benefits arising from MGR use (e.g. profits derived from a commercial product), they could still serve as broad indicators of research and possible development.

Decisions by the COP on these modalities are to be adopted by consensus, but if consensus is not reached, decisions may be taken by a three-fourths majority (article 14.7). As this is a non-exhaustive list of modalities the COP, which will be advised by the Access and Benefit Sharing Committee, is free to decide which modality to adopt beyond them. Confirming the assessed contributions of phase one can also be a possibility.

In relation to the modalities for the sharing of monetary benefits from the use of DSI, however, the "modalities should be mutually supportive of and adaptable to other access and benefit-sharing instruments" (para 9). The request for mutual supportiveness comes from the fact that in December 2022, the CBD established a Multilateral Fund for the sharing of benefits arising from the use of DSI. This Fund is not linked to the CBD context only and is potentially open to other international frameworks dealing with DSI associated with genetic resources, such as the FAO Treaty on Plant Genetic Resources, the PIP framework (referred to as 'genetic sequence data' under the latter), or the BBNJ Agreement (Kachelriess et al., 2025). The ABS committee (Muraki Gottlieb et al., 2025b) will play a role in ensuring this consistency as, according to article 14.9, it will recommend modalities for the sharing of monetary benefits from the use of DSI to the COP, and, according to article 15.5, it can facilitate exchange with relevant legal instruments and frameworks and relevant bodies on activities under its mandate, including benefit sharing, the use of DSI on marine genetic resources, and lessons learnt.

In the views of the authors of this chapter, it would be important to ensure coherence of approaches among all these legal frameworks in dealing with benefit sharing related to DSI, both for a simple practical reason that DSI data from all different research sectors and geographical origins is stored in one network of databases (INSDC) and for the need to ensure that benefit-sharing measures are implementable by the users without excessive burdens. To implement different benefit-sharing measures related to DSI depending on DSI origin would require implementing different rules for different DSI. This may not be practical, as research is often done using DSI from various origins (see Box 6.4).

Box 6.4 Recommendations from the DSI Scientific Network

The DSI Scientific Network (*DSI Scientific Network—Giving the scientific community a voice on Digital Sequence Information*) is advocating for the benefit sharing rules for the different fora attempting to regulate DSI (CBD/NP, FAO/ITPGRFA, UNCLOS/BBNJ, WHO/PIP and CA+) to be harmonised (Halewood, et al., 2023; Scholz et al., 2022, 2023; A Harmonized System for Benefit-Sharing from DSI—DSI Scientific Network—Policy Brief, 2024).

According to this view, a sectoral approach to the use of DSI poses a number of potential practical challenges, since it does not reflect the way DSI is increasingly being made available and used in cross disciplinary, cross sectoral, research and development. A fragmented approach would slow down and hamper research, from which benefits are accruing, therefore it would likely reduce benefit sharing.

The DSI Scientific Network also argues that, if harmonisation does not happen, different subsets of DSI would be treated differently depending on their origin. In the worst case, databases could become fragmented reducing the value of the unified dataset available in the INSDC. The value of the data is not in a single sequence, but the ability to compare a query sequence to the whole database to enable for example the understanding of its evolutionary origin and its potential function. Products based on DSI are often based on sequences from different origins and are modified to increase the desired property. Tracing back the relative contribution that each sequence made to the final product would be impossible as stated in the GBF's DSI decision of 18 December 2022 which "Recognizes that tracking and tracing of all digital sequence information on genetic resources is not practical" (CBD/COP/15/L.30 https://www.cbd.int/doc/c/c181/12cf/d29ef8c3f-6bd4ec701699d9d/cop-15-l-30-en.pdf). Having different rules for the different silos of DSI would negatively impact the open access and interoperability of databases, the compliance of the users and the legal certainty.

For these reasons, the DSI Scientific Network sees a multilateral approach that recognises the integrity of the INSDC dataset as essential. The network suggests that "a harmonized multilateral system for DSI access and benefit sharing must be simple in order to ensure compliance and implementation, provide legal certainty to users, offer transparency on the benefits generated and cost less to enforce than the value of the benefits." (A Harmonized System for Benefit-Sharing from DSI—DSI Scientific Network—Policy Brief, 2024).

Para 8 introduces an interim period of 4 years in which a Party can continue paying the monetary contributions through the assessed contribution of the first phase, once the COP decides on the new modality, in order to take the time to implement the new modality.

Para 10 introduces the review and assessment of the system of monetary benefit sharing on a biennial basis, the first of which should take place no later than five years after the entry into force of this Agreement. This review is partially based on the initial review clause proposed by the EU and its Member States at IGC4 in March 2022.

Despite the long-standing polarised views on the issue of monetary benefit sharing, it became apparent during the negotiations that no deal would have been possible without a compromise on sharing them. The first attempt in this direction was made by the European Union and its Member States at IGC4 in March 2022 who proposed that, upon reviewing and assessing the extent of commercialization of products based on the utilization of marine genetic resources of areas beyond national jurisdiction, the Conference of the Parties could explore alternatives to identify the most appropriate processes for relevant financial contributions, if tangible and substantial monetary benefits arise therefrom. The review clause was included as one option in the further revised draft text in May 2022 (and then kept in brackets). Towards the end of IGC5 in August 2022, the like-minded countries agreed on a common approach to monetary benefit sharing in a non-paper, which proposed a de-coupled solution: a payment to the Special Fund by States Parties capped to a percentage (to be determined) of every Party's BBNJ budget contribution. The proposal also included the review clause giving the COP the duty to review the extent of commercialisation and decide on alternative modalities for monetary benefit sharing. The times were not ripe enough for a compromise and IGC5 was resumed in February 2023 (Marciniak et al., Oxford University Press forthcoming 2025). The concept of the review and assessment of monetary benefits from the utilisation of MGR and DSI is based on the review of commercialisation, even though the terminology shifted from commercialisation to monetary benefits, implying a much wider review including the consideration of the periodical reports received by the ABS committee according to article 16.2.

6.4 Considerations for Implementation

Implementation

11. Parties shall take the necessary legislative, administrative or policy measures, as appropriate, with the aim of ensuring that benefits arising from activities with respect to marine genetic resources and digital sequence information on marine genetic resources of areas beyond national jurisdiction by natural or juridical persons under their jurisdiction are shared in accordance with this Agreement.

State Parties will be required to introduce the necessary legislative, administrative, or policy measures for databases, repositories, funding agencies and users to comply with the benefit-sharing provisions. Some of the benefit-sharing obligations will need to be translated into national legislation to be implemented: notably, the obligation to deposit MGR and DSI, together with their BBNJ standardised Identifiers, in publicly accessible repositories and databases (para 3 article 14); the obligation to provide the non-monetary benefits listed in para 2 and the obligation for developed Parties to make the annual contribution as monetary benefit sharing (para 6).

The obligation to deposit MGR and DSI in publicly accessible repositories and databases will target scientists and public and private entities that will utilize MGR and DSI, as well as funding agencies and bodies who are financing this kind of research. Funding agencies will have a crucial role in implementing certain provisions of the BBNJ Agreement by translating certain obligations into criteria to get research grants, ensuring the implementation of the benefit sharing.

Additionally, Parties may introduce their own legislation to mobilize financial resources from entities within their jurisdiction to fulfill their monetary benefit-sharing commitments under this Agreement.

6.5 Conclusions

The conclusion of the BBNJ Agreement is a milestone for the governance and protection of the ocean and an important success of multilateralism during a difficult geopolitical situation. The compromise achieved in the benefit sharing part of the BBNJ Agreement is what made the package deal and thus the conclusion of the whole Agreement possible. This compromise is the result of the strong political will to have the Agreement in place, as well as of the dedication and preparation of all the delegations involved through the long political process till the end of the negotiations. Two elements were the most crucial for the deal in the last hours: the settlement on monetary benefit sharing and the inclusion of DSI within the scope and within the benefit-sharing part. The BBNJ Agreement is innovative in both these two aspects: it is the first agreement where a contribution is paid by a Party as monetary benefit sharing in a de-coupled way (not linked to the use of the GR); and it is the first internationally binding agreement including DSI within its scope and in the operational provisions, still without defining it.

In relation to DSI, the BBNJ Agreement maintains open access to valuable scientific information as it is in scientific practice, while also taking a step towards establishing equitable benefit sharing systems for DSI. This is a complicated endeavour that the CBD, the PIP framework and the FAO are equally trying to tackle in current discussions (Aubry et al., 2022).

Many open questions exist about the implementation of benefit sharing. They relate to the modalities of the second phase of the monetary benefit sharing system. These include questions about the extent to which these modalities will be tied to the use of MGR or DSI and the scope of traceability measures required to accomplish such ties. The BBNJ Identifier will likely play an important role, but it leaves open the question of how exactly this will be translated into questions of monetary value. It will also be important to clarify the role of the Clearing House Mechanism in terms of information sharing to make both the opportunities for benefit sharing and the shared benefits themselves visible and transparent.

Bibliographys

A Harmonized System for Benefit-Sharing from DSI—DSI Scientific Network—Policy Brief, 2024 Policy-Brief-Harmonization-of-DSI-BS-systems.pdf (dsiscientificnetwork.org).

Agreement under the United Nations Convention on the Law of the Sea on the Conservation and Sustainable Use of Marine Biological Diversity of Areas beyond National Jurisdiction, New York, 19 of June 2023, C.N.203.2023.TREATIES-XXI.10 of 20 July 2023.

Alliance of Small Islands States, Science in Small Island Developing States, April 23, 2021, Oceans Reports Science in Small Island Developing States—AOSIS.

Aubry, S., Frison, C., Medaglia, J. C., Frison, E., Jaspars, M., Rabone, M., Sirakaya, A., Saxena, D., & van Zimmeren, E. (2022). Bringing access and benefit sharing into the digital age. *Plants, People, Planet, 4*(1), 5–12. https://doi.org/10.1002/ppp3.10186

Broggiato, S. A.-H., Chiarolla, C., & Greiber, T. (2014). Fair and equitable sharing of benefits from the utilization of marine genetic resources in areas beyond national jurisdiction. *Marine Policy, 49*(43). https://doi.org/10.1016/j.marpol.2014.02.012i

Broggiato, A., Vanagt, T., Lallier, L. E., Jaspars, M., Burton, G., & Muyldermans, D. (2018). Mare geneticum: Balancing governance of marine genetic resources in international waters. *International Journal of Marine and Coastal Law, 33*(1), 3–33. https://doi.org/10.1163/15718085-13310030

CBD, COP-15 Decision 15/9, Digital sequence information on genetic resources (2022).

Collins, J. E., Rabone, M., Vanagt, T., Amon, D. J., Gobin, J., & Huys, I. (2021). Strengthening the global network for sharing of marine biological collections: recommendations for a new agreement for biodiversity beyond national jurisdiction. *ICES Journal of Marine Science*. https://doi.org/10.1093/icesjms/fsaa227. *ICES Journal of Marine Science, 78*(1), January-February 2021, 305–314. https://doi.org/10.1093/icesjms/fsaa227

Convention on Biological Diversity, opened for signature 5 June 1992, 1760 UNTS 79 (entered into force 29 December 1993).

Harden-Davies, H., Amon, D. J., Vierros, M., Bax, N. J., Hanich, Q., Hills, J. M., Guilhon, M., McQuaid, K. A., Mohammed, E., Pouponneau, A., Seto, K. L., Sink, K., Talma, S., & Woodall, L. (2022) Capacity development in the Ocean Decade and beyond: Key questions about meanings, motivations, pathways, and measurements. *Earth System Governance, 12*, 100138. ISSN 2589-8116, https://doi.org/10.1016/j.esg.2022.100138

Hartman Scholz, A., Humphries, F., Vanagt, T., & Jaspars, M. (2023) *A new dawn for global benefit-sharing: capitalizing on the global biodiversity framework for marine genetic resources from areas beyond national jurisdiction*. https://www.iucn.org/sites/default/files/2023-02/bbnj_icg5bis_policy_brief_global_benefit_sharing_1.pdf

Humphries, F. (2025). Marine genetic resources beyond national jurisdiction: The expansive scope of the BBNJ Agreement. In F. Humphries (Ed.), *Decoding governance of marine genetic resources under the BBNJ Agreement*. Springer.

Humphries, F., Jaspars, M., Lavelle, J. & Kachelriess, D. (2025). The novel notification information system for marine genetic resources under the BBNJ Agreement. In F. Humphries (Ed.), *Decoding governance of marine genetic resources under the BBNJ Agreement*. Springer.

Kachelriess, D., Dunshirn, P., Langlet, A., Brown, A. E. L., & Scholz, A.H. (2025). Marine genetic resources and digital sequence information under the BBNJ Agreement: Interlinkages with other access and benefit sharing frameworks. In F. Humphries (Ed.), *Decoding governance of marine genetic resources under the BBNJ Agreement*. Springer.

Langlet, A., & Dunshirn, P. (2023). Traceability options for marine genetic resource from areas beyond national jurisdiction. https://highseasalliance.org/resources/paper-traceability-options-for-marine-genetic-resource-from-areas-beyond-national-jurisdiction/

Langlet, A., Dunshirn, P., Jaspars, M., Humphries, F., & Kachelriess, D. (2025). Monitoring and transparency aspects of MGR-utilization under the BBNJ Agreement. In F. Humphries (Ed.), *Decoding marine genetic resource governance under the BBNJ Agreement*. Springer

Lavelle, J. & Wynberg, R. (2025). Benefit sharing under the BBNJ Agreement in Practice. In F. Humphries (Ed.), *Decoding marine genetic resource governance under the BBNJ Agreement*. Springer.

Lawson, C., Humphries, F., Jaspars, M., & Rabone, M. (2025). Data management and the 'BBNJ Standardized Batch Identifier' under the BBNJ Agreement. In F. Humphries (Ed.), *Decoding governance of marine genetic resources under the BBNJ Agreement*. Springer.

Leary, D. (2018). Marine genetic resources in areas beyond national jurisdiction: Do we need to regulate them in a new agreement? *Maritime Safe Law Journal, 5*, 22–47.

Marciniak, K. J. (2020). The legal status of marine genetic resources in the context of BBNJ negotiations: Diverse legal regimes and related problems. In T. Heidar (Ed.), *New knowledge and changing circumstances in the law of the sea* (pp. 40–64). Brill.

Marciniak, K., Broggiato, A., & Gobin, J. (Forthcoming 2025) *Marine genetic resources, Oxford University Press, BBNJ commentary*.

Mass. Perelman, C. (1963). Justice et raison, Bruxelles, Presses Universitaires du Bruxelles.

Muraki Gottlieb, H., Kachelriess, D. & Slobodian, L. (2025a). Understanding the preamble, principles and objectives of the BBNJ Agreement: A focus on the fair and equitable sharing of benefits of marine genetic resources. In F. Humphries (Ed.), *Decoding governance of marine genetic resources under the BBNJ Agreement*. Springer.

Muraki Gottlieb, H., Ardron, J. & Brown, A. E. L. (2025b). BBNJ Agreement: A new infrastructure to foster benefit sharing of marine genetic resources. In F. Humphries (Ed.), *Decoding governance of marine genetic resources under the BBNJ Agreement*. Springer.

Halewood, M., Bagley, M. A., Wyss, M., & Scholz, A. H. (2023). New benefit-sharing principles for digital sequence information. *Science Policy Forum, 382*(6670).

Oldham, P., Hall, S., Barnes, C., Oldham, C., Cutter, A. M., Burns, N., & Kindness, L. (2014). *Valuing the deep: Marine genetic resources in areas beyond national jurisdiction. DEFRA Contract MB0128*. London: DEFRA. https://doi.org/10.13140/2.1.2612.5605

Pena-Neira, S. (2017). *Recursos geneticos de animales, plantas y microorganismos y su regulacion internacional, Olejnik, Buenos Aires* (p. 178) ss.

Rabone, M., Harden-Davies, H., Collins, J. E., Zajderman, S., Appeltans, W., Droege, G., Brandt, A., Pardo-Lopez, L., Dahlgren, T. G., Glover, A. G., & Horton, T. (2019). Access to Marine Genetic Resources (MGR): Raising Awareness of Best-Practice Through a New Agreement for Biodiversity Beyond National Jurisdiction (BBNJ). *Frontiers in Marine Science, 6*, 520.

Ramutsindela, M., & Mickler, D. (Eds.) (2020). *Africa and the sustainable development goals, sustainable development goals series, 2020* https://doi.org/10.1007/978-3-030-14857-7_24 *257*.

Rawls, J. (1971). *A theory of justice*. Harvard University Press.

Rudolf, I. A., et al. (2019). Toward unrestricted use of public genomic data. *Science, 363*, 350–352.https://doi.org/10.1126/science.aaw1280, https://www.science.org/doi/10.1126/science.aaw1280

Rogers, A. D., Baco, A., Escobar-Briones, E., Currie, D., Gjerde, K., Gobin, J., Jaspars, M., Levin, L., Linse, K., Rabone, M., Ramirez-Llodra, E., Sellanes, J., Shank, T. M., Sink, K., Snelgrove, P. V. R., Taylor, M. L., Wagner, D., & Harden-Davies, H. (2021). Marine genetic resources in areas beyond national jurisdiction: Promoting marine scientific research and enabling equitable benefit sharing. *Frontiers in Marine Science, 8*, 667274.

Rohden, R. F., Huang, S., Dröge, G., Scholz, A. H., et al. (2020). Combined study on digital sequence information in public and private databases and traceability CBD/DSI/AHTEG/2020/1/4 31 January 2020. https://www.cbd.int/doc/c/1f8f/d793/57cb114ca40cb6468f479584/dsi-ahteg-2020-01-04-en.pdf

Scholz, A. H., Freitag, J., Lyal, C. H. C., et al. (2022). Multilateral benefit-sharing from digital sequence information will support both science and biodiversity conservation. *Nature Communications, 13*, 1086. https://doi.org/10.1038/s41467-022-28594-0

Study on 'Marine Genetic Resources' Market Value and State of the Art of Commercialisation of Related Products in the Context of the BBNJ Negotiations. Reference number: CINEA/2022/OP/0017, available at 608281-2023—Result—TED (europa.eu)]: the outcomes of the study are expected in the second half of 2023.

The interridge code of conduct on responsible behaviour at hydrothermal vents. http://194.254.225.67/de/node/16908

UNGA Resolution 1314 (XIII) Recommendations concerning international respect for the right of peoples and nations to self-determination, 12 December, 1958.

Wilkinson, M. D., Dumontier, M., Aalbersberg, I. J., Appleton, G., Axton, M., Baak, A., et al. (2016). The FAIR Guiding Principles for scientific data management and stewardship. *Scientific Data*, Article number: 160018.

Arianna Broggiato holds a PhD in international law and LLM in Environmental Law. She has specialized on the issue of access and benefit sharing to genetic resources, with a focus on marine genetic resources of areas beyond national jurisdiction, and has been working in this field in the last 15 years, focussing on bridging the gaps between involved stakeholders, and raising awareness about the different technical aspects. She joined the European Commission (DG MARE) in 2018 to join the EU team negotiating the international legally binding instrument under the United Nations Convention on the

Law of the Sea on the conservation and sustainable use of marine biological diversity of areas beyond national jurisdiction (BBNJ).

Paul Dunshirn is a PhD researcher at the ERC-funded research project Twin Politics, University of Vienna. His research explores global patent and science landscapes around marine genetic resources and digital sequence information (DSI), aiming to sketch pathways for equitable policymaking. Paul has served as an independent observer and advisor to various UN processes, such as the recently concluded negotiations on the Biodiversity Beyond National Jurisdiction (BBNJ) treaty and the efforts to regulate DSI under the Convention of Biological Diversity (CBD).

Marcel Jaspars is a professor of Organic Chemistry at the University of Aberdeen where he leads the Marine Biodiscovery Centre which focusses on marine resources for novel pharmaceuticals, and to investigate fundamental questions in marine chemical ecology and biosynthesis. Marcel has been active at national and international levels to develop the science, its applications/industrial uptake and associated policy involved in marine biodiscovery and biotechnology. He provides scientific advice to the UK, EU and UN for global policy processes on ocean conservation and digital sequence information via reports, papers and taking part in discussion meetings.

Sergio Pena-Neira (Ph.D. in Law, UNIA, Spain, 2014 and Postdocs in King´s College London, YTLCPPL, and University of Cambridge, LRCFIL) is an associate professor at Universidad Mayor, Chile. He has published widely articles, chapters and books on genetic resources, traditional knowledge and climate change.

7 Monitoring and Transparency Aspects of MGR-Utilization Under the BBNJ Agreement

Arne Langlet, Paul Dunshirn, Marcel Jaspars, Fran Humphries, and Daniel Kachelriess

Abstract

This chapter examines monitoring and transparency aspects in the governance of marine genetic resources (MGR) as outlined in Article 16 of the Agreement under the United Nations Convention on the Law of the Sea on the Conservation and Sustainable Use of Marine Biological Diversity of Areas Beyond National Jurisdiction (BBNJ). It focuses on the development, negotiation history, and operational dynamics of a monitoring and transparency system which combines notification and reporting processes with digital identifiers to streamline the tracing of MGR utilization. By comparing these new measures with existing international frameworks, the chapter assesses the potential gaps and future policy developments necessary to ensure the effectiveness of the monitoring and transparency mechanism in promoting equitable benefit-sharing and responsible use of MGR.

Keywords

BBNJ agreement · Marine genetic resources · Transparency · Monitoring · Biodiversity governance · Digital sequence information · Equitable benefit-sharing · Article 16 · Batch identifier

7.1 Introduction

An integral part of the Agreement under the *United Nations Convention on the Law of the Sea for the Conservation and Sustainable Use of Marine Biodiversity of Areas Beyond National Jurisdiction* (BBNJ Agreement) (UNGA, 2023a, 2023b) is the governance and regulation of marine genetic resources of areas beyond national jurisdiction (MGR). The objectives of MGR governance under the Agreement are to ensure fair and equitable benefit sharing, capacity building, knowledge generation and the development and transfer of marine technology (Art 9). However, one of the challenges is how to 'balance fair and equitable benefit-sharing of MGR [...] with as

A. Langlet (✉) · P. Dunshirn
Department of Political Science, University of Vienna, Vienna, Austria
e-mail: arne.langlet@univie.ac.at

M. Jaspars
Marine Biodiscovery Centre, Department of Chemistry, University of Aberdeen, Aberdeen, UK

F. Humphries
Griffith Law School, Griffith University, Nathan, Queensland, Australia

D. Kachelriess
IUCN World Commission on Environmental Law, High Seas Alliance (Advisor), Vienna, Austria

F. Humphries (ed.), *Decoding Marine Genetic Resource Governance Under the BBNJ Agreement*, Sustainable Development Goals Series, https://doi.org/10.1007/978-3-031-72100-7_7

few burdens on marine scientific research as possible' (Kachelriess, 2023 p. 9).[1] The monitoring and transparency system of the Agreement, as formulated in Article 16, intrinsically speaks to this matter. Art. 16 entitled 'Monitoring and transparency' lays out a system that aims to ensure fair and equitable benefit-sharing without posing unnecessary burdens on researchers. The article prescribes a framework for interactions between State Parties and bodies of the BBNJ Agreement, laying out the structure of a lean ('light touch') system aiming to ensure monitoring and transparency of activities with respect to MGR and digital sequence information (DSI) of MGR from ABNJ while being robust enough to facilitate fair and equitable benefit-sharing under Article 14 (Broggiato et al., 2025). It does not, however, explicitly provide obligations concerning monitoring and transparency of traditional knowledge associated with MGR of ABNJ, which is regulated under Article 13 (Pena-Neira & Coelho, 2025). The monitoring system works in conjunction with Article 12 on the 'Notification on activities [...]' which lays out details of the notification requirements referred to in Article 16 (1) (Humphries et al., 2025).

The text of art. 16 emerged out of complex negotiations and deliberations which touched upon the question in how far MGR represent a 'global common' from areas beyond national jurisdiction (Scovazzi et al., 2008). While there was early consensus that the BBNJ Agreement should incorporate some form of benefit-sharing, significant uncertainties persisted until late in the negotiations regarding the traceability system necessary for implementation. The necessity to strike a balance between the freedom of scientific discovery and the need to gather relevant data to ensure fair and equitable benefit-sharing was a key aspect underscored by numerous countries during the negotiations (Vadrot et al., 2022). Throughout the negotiations, diverse traceability options were deliberated: from a transparency system which largely codified current forms of transparency in scientific best practices to an extensive track and trace-oriented monitoring system.[2] The final text introduces a system combining elements of both—transparency-oriented tools including notification and reporting, and innovative monitoring infrastructure in the form of a digital identifier system.

This chapter explores the individual provisions constituting art. 16, their trajectory throughout BBNJ negotiation rounds and discusses how a system implementing these provisions could be envisioned while highlighting potential implications and open questions. While obligations are on Parties to the Agreement, the chapter conceptualizes monitoring and transparency aspects as potentially encompassing a range of measures imposed on different stakeholders under the jurisdiction of those Parties. These include notification obligations for researchers sampling MGR, as well as potential users of MGR and related data, reporting by Parties, and the establishment of a special Access and Benefit-Sharing Committee (ABS Committee). Such measures aim to promote what is generally accepted as best practices in research, development and commercialization for the documentation and sharing of scientific information. The chapter interprets and analyses the elements under art. 16 of the BBNJ Agreement related to monitoring and transparency aspects of utilization of MGR. Section 7.2 provides a summary of the historical evolution of the monitoring and transparency article during the BBNJ negotiations. Section 7.3 introduces an overview of the monitoring and transparency system, demonstrating its general logic and flow. It explores the different kinds of actors and activities addressed by these provisions and how various BBNJ implementing bodies would be involved. Sections 7.4 and 7.5 interpret and analyse the article's three elements in detail, namely information sharing procedures, Party reporting and BBNJ Agreement body reports. The conclusion draws together the analyses and highlights open

[1] See Chap. 6 of this collection for a discussion of the benefit-sharing mechanism (Broggiato et al., 2025).

[2] For an overview and discussion of traceability options see Langlet and Dunshirn (2023).

questions and uncertainties to be addressed to ensure smooth implementation of the monitoring and transparency system underpinning Part II of the BBNJ Agreement.

7.2 Evolution of BBNJ Monitoring and Traceability System

The need to regulate the use of genetic resources and ensuring the sharing of benefit arising from them 'whilst at the same time promoting scientific and commercial use of the genetic resources' (Laird et al., 2004 p. 148) is not new. Prior to the BBNJ negotiations, other international agreements had already developed different traceability systems for biological resources. The *Convention on Biological Diversity* (CBD), *Nagoya Protocol on Access to Genetic Resources and the Fair and Equitable Sharing of Benefits Arising from their Utilization to the Convention on Biological Diversity* (Nagoya Protocol), the Food and Agriculture Organization's (FAO) *International Treaty on Plant Genetic Resources for Food and Agriculture* (ITPGRFA) and the World Health Organization's *Pandemic Influenza Preparedness* (PIP) *Framework* each address issues of monitoring and transparency, but they do so in different ways due to their distinct scope and objectives (Kachelriess et al., 2025). Parties to the CBD agreed at the Conference of the Parties (CoP) in 2022 to work towards the establishment of a multilateral benefit-sharing mechanism for digital sequence information (DSI) on genetic resources (Scholz et al., 2024; UNEP, 2022). The monitoring and compliance procedures of the DSI multilateral framework are yet to be determined, but Parties will likely agree on some form of traceability system, including information on geographical origin and other relevant metadata in public databases (see Sect. 7.4.3.1). According to the CoP to the CBD, track and trace systems are not practical for all DSI (UNEP, 2022 para 5), which was echoed by negotiators in the BBNJ negotiations (Langlet et al., 2024). Experiences with other existing or emerging ABS traceability systems may have had some influence on the BBNJ negotiations but the unique jurisdictional and geo-political conditions in ABNJ resulted in a bespoke monitoring and transparency model under the Agreement's framework.

Art. 16 of the Agreement was substantially changed during the five intergovernmental conferences (IGC) between 2018 and 2023. Whereas the President's Aid to Discussions from the first Intergovernmental Conference (IGC), outlined the need for monitoring arrangements, it merely raised questions about how to monitor the utilization of MGR and who would conduct this monitoring (UNGA, 2018). The next iterations of the Treaty draft texts at ICG 2–4 added detail to the monitoring framework such as provisions for assigning identifiers to MGR, requiring notifications for access, and submitting periodic status reports and the role of a Clearing-House Mechanism, a Scientific and Technical Body (STB), or an obligatory prior electronic notification system managed by the secretariat (UNGA, 2019, 2022a). However, agreement on the final article 16 was dependent on the framework for collection, utilization and sharing the benefits from MGR and DSI of ABNJ. As there was relatively little agreement on the substantial parts of Part II, there was accordingly little movement in some of its key components such as the monitoring and transparency provisions during IGC 1–4.

Going into IGC 5, the President's draft text presented two distinct choices for the former Article 13 on the traceability system. Option 1 centred on a voluntary transparency system without monetary benefits, while Option 2 offered a comprehensive monitoring system that traces MGR from collection to commercialization to facilitate monetary benefit-sharing (UNGA, 2022b). These options represented the two diverging interests in Part II, namely those of negotiating groups advocating for a comprehensive system of 'track and trace' from the point of collection to the commercial product outcome and those of groups advocating for a light touch 'transparency' approach based on existing practices for information sharing.

However, during IGC 5 several new ideas and approaches were circulated to break the deadlock and converge the options towards common interests. Some negotiating groups indicated a willingness to move away from their preferred monitoring model outlined in Option 2, while other negotiating groups expressed openness to some form of monetary benefit-sharing which would entail a monitoring mechanism (Tiller et al., 2023). As a result, ideas attempting to find a middle way picked up discussions leading to the idea of so-called de-coupled benefit-sharing and aggregate ways to assess the use and commercialization of MGR. Such a system aims to document the results of research related to Marine Genetic Resources (MGR), including publications, patents, and products (PPP). This information could be aggregated so that information on the products, patents and publications stemming from MGR could be used to approximate benefit-sharing obligations, i.e. to focus on outcomes and circumvent the need to track the whole research process.[3] A comparable system, focussing on the outcomes, where at the moment when an end-user registers an outcome of an MGR-related activity (publications, producing, patenting) the sharing of monetary or non-monetary benefits would be triggered, has been adopted in Brazil's revised framework for access and benefit-sharing in accordance with the Nagoya Protocol (Brazil, 2019).

These alternative approaches exploring less extensive traceability systems while still supporting the implementation of monetary benefit-sharing provisions influenced the direction of negotiations but were however not codified into the draft text during ICG5. This was arguably because Part II remained a contentious element of the Agreement until the very end of the negotiations which needed to be negotiated and agreed upon as a whole, i.e. as one major political agreement, and not provision by provision. As a consequence, what is now art. 16 on monitoring and transparency is largely a combination of elements from both options in previous draft texts and was largely drafted at the resumed and last IGC 5 in February 2023. This last-minute drafting also may explain why the text only outlines a framework for monitoring the utilization of MGR with many practical aspects remaining to be further defined and clarified.

7.3 Overview of the Monitoring and Transparency System

There are five key components of the Agreement's monitoring and transparency system under Article 16, which are part of a bigger picture of transparency measures outlined in other articles of Part II. The monitoring and transparency system is supported by an institutional infrastructure consisting of an ABS Committee which prepares reports and on Parties' implementation and utilization of MGR based on information from a Clearing-House Mechanism (CHM), which serves as the information and data centre, and finally the CoP which is the decision-making body (Muraki Gottlieb et al., 2025). Parties and these bodies are responsible for implementing the monitoring and transparency system which consists of the following aspects:

1. **Information Sharing Procedures** comprising (Sect. 7.4 below):
   (a) **Notification to the CHM under Article 12**: Parties to the Treaty must implement national measures to ensure that specified information is notified to the CHM at the time of pre-collection of MGR within ABNJ, post-collection and 'utilization' of MGR and DSI on MGR of ABNJ, including modalities for third party access to MGR and DSI (UNGA, 2023a, 2023b Art. 12(1–5) and Art. 12(8), see Humphries et al., 2025). They also must ensure that the ABNJ origin of MGR samples and DSI in repositories and databases in their jurisdiction can be identified (UNGA, 2023a, 2023b Art. 12(6)).

[3] For a detailed discussion of the different options under negotiation at IGC 5 and the following intersessional period, see Langlet and Dunshirn (2023).

(b) **CHM issuance of the BBNJ Identifier to the pre-collection notification under Article 12**; The BBNJ Identifier is automatically issued for a pre-collection notification by the CHM and must also be linked to the post-collection notification and any 'utilization' notification if available (UNGA Art. 12(3) & 12(8)). Information on MGR and DSI linked to the BBNJ Identifier forms the basis of a report from repositories and, where practicable, databases to the ABS Committee about access to MGR and DSI from their repositories and databases (Art. 12(7)).

(c) **Additional procedures for monitoring and compliance:** These are additional procedures that will be adopted by the CoP as recommended by the ABS Committee once the bodies are set up.

2. **Party Periodic Reporting** (Sect. 7.5 below): Parties are obliged to provide regular reports to the ABS Committee. These reports comprehensively detail the implementation of provisions related to activities with respect to MGR and DSI (Article 11) and the equitable sharing of benefits stemming from these activities.
3. **BBNJ Agreement Body Reports** (Sect. 7.6 below)**:** The ABS Committee is responsible for consolidating the information obtained through the CHM. This report is shared with the Parties for review and comments, fostering a collaborative and consultative process. Subsequently, the report, along with any submitted comments, is presented to the CoP for consideration. Finally, the CoP, taking into account the recommendations of the ABS Committee may determine appropriate guidelines for the implementation of the monitoring and transparency system under Article 16, taking into account the national capabilities and circumstances of the Parties.

In essence, this flow establishes a structured process where Parties notify their activities with respect to MGR and DSI, submit reports on their implementation, and engage in a feedback loop through the ABS Committee. The final decisions and guidelines are made by the CoP, ensuring a coordinated and accountable approach to monitoring and transparency in the management of MGR of ABNJ. Parties which are voting members of the CoP are then also obliged to implement the CoP's decisions into national law.

Figure 7.1 provides an overview of the processes and bodies related to monitoring and transparency.

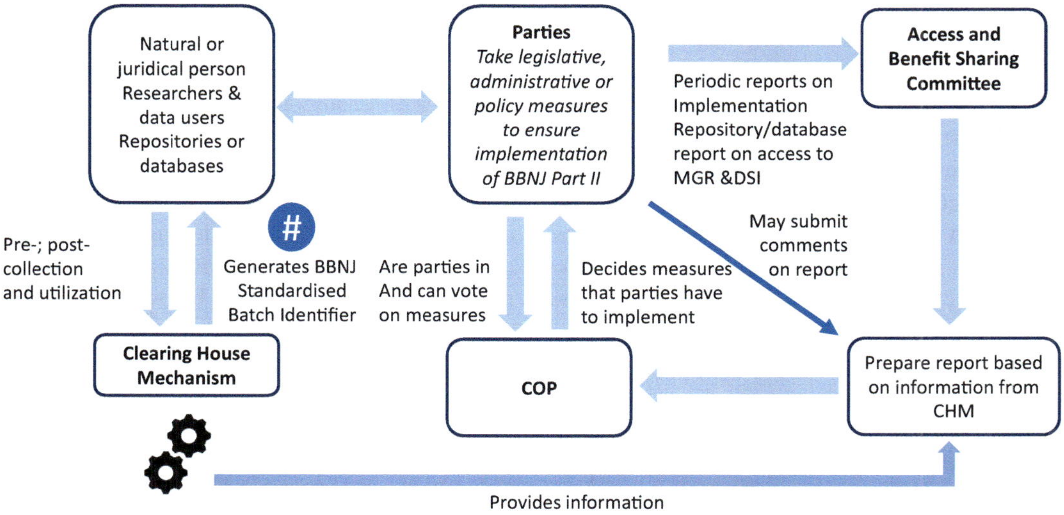

Fig. 7.1 Processes and bodies involved in monitoring and transparency

7.4 Information Sharing Procedures

Article 16(1): Monitoring and transparency of activities with respect to marine genetic resources and digital sequence information on marine genetic resources of areas beyond national jurisdiction shall broadly be achieved through notification to the CHM, through the use of "BBNJ" standardized batch identifiers in accordance with this Part and according to procedures adopted by the CoP as recommended by the ABS Committee.

Monitoring and transparency activities are the responsibility of Parties with the assistance of the CHM and further guidance by the ABS Committee and CoP. They form part of a broader set of obligations under Part II of the BBNJ Agreement which apply to Parties demanding that activities with respect to marine genetic resources of ABNJ by natural or juridical persons under their jurisdiction shall be carried out in accordance with the BBNJ Agreement (Art. 11), leaving it up to the Parties how to transcribe these obligations to the national level.

A key component of transparency is the sharing of information by Parties through the BBNJ Agreement bodies about activities with respect to MGR and DSI that occur by the natural and juridical persons within a Party's jurisdiction and control. Art. 16(1) contains three components to information sharing: notifications of activities with respect to MGR and DSI, the BBNJ identifier system and additional monitoring and compliance procedures. This section outlines the rationale for the three elements and considerations for how they might be implemented in practice.

7.4.1 Notification of MGR and DSI Activities

Monitoring and transparency of activities with respect to MGR and DSI on MGR of ABNJ 'shall be achieved through notification to the Clearing-House Mechanism'. This is not explicitly restricted to the notification system under art. 12 and while it relates to the notification system, it could also include other notification requirements as determined by the CHM (see 7.4.3 below). Chapter 6 of this edited collection contains a detailed interpretation and analysis of the mechanism under art. 12 and how it relates to other articles in the Agreement (Humphries et al., 2025). Art. 12 obliges Parties to take the necessary legislative, administrative or policy measures to ensure that its nationals notify pre-collection, post-collection and utilization activities as well as exchange information about ex situ physical MGR and their DSI within their national jurisdictions. While the whole notification system is relevant to ensuring transparency, there are aspects within art. 12 that play a special role in information sharing and monitoring the location of MGR and DSI of ABNJ as outlined below.

Parties to the Agreement need to arrange for certain information to be notified to the CHM six months or as early as possible prior to the collection in situ of MGR of ABNJ (pre-collection notification) (Art. 12(2)). This includes information about the proposed subject matter and geographical location of the collection and information about the project under which the collection is carried out. Upon pre-collection notification, the CHM will automatically generate a BBNJ Identifier and Parties must take measures to ensure information about the proposed collection is updated within a reasonable period of time prior to collection (Art. 12(3)). The BBNJ Identifier aims to connect the pre- and post-collection information as well as information about utilization of MGR and DSI within national jurisdiction (see 4.2 below).

One year from the MGR collection in ABNJ, Parties must ensure that information about the location of the MGR and DSI, a report detailing the location of collection, and findings from the collection activities are reported to the CHM along with the BBNJ Identifier (Art. 12(5)). Most of the information on MGR in collections/ repositories should be made available through online platforms such as the Global Biodiversity Information Facility,[4] but often collections/repositories are understaffed and the data on their collection/repository is not uploaded in a reasonable timeframe. To meet the information obligation under art. 12, additional resources would need to be made available. From a user, and researcher's perspective, it is recommended that DSI on MGR of ABNJ are deposited with the BBNJ Identifier in a recognized database such as the International Nucleotide Sequence Database Collaboration (INSDC),[5] which is the umbrella body of the three main DSI databases (NCBI, EBI, DDBJ).[6] The INSDC contains most of the world's DSI (Ad Hoc Technical Expert Group on Digital Sequence Information on Genetic Resources, 2020 Para.8) with all other DSI databases drawing their information from it. Scientific good practice is that all sequences are uploaded to the INSDC as soon as possible after they are generated (The Welcome Trust 2003 p. 4). The database will assign upon entry to the database an INSDC Accession Number that needs to be reported in publications with the journals acting as 'gatekeepers'. There is a recent move by scientists to suggest that DSI needs to be made available on generation, without the originator having an embargo on the data to analyse it before depositing it (Amann et al., 2019).

During negotiations, some delegations had proposed obligations for making samples of MGR collected from ABNJ available in open-source platforms so that others could access these materials (UNGA, 2019 Draft art. 10(2)(d)). However, several researchers raised practical challenges for this proposal including questions around who bears the high cost of curation considering millions of organisms may be present in one water or sediment sample and the practicality of sharing a duplicate of a finite sample collected for research (Rabone et al. 2019). The compromise in the final text was to require Parties to ensure that information is shared about where collected MGR and DSI are stored (Art. 12(6) and 12(7)) and that samples that are subject to 'utilization' under their jurisdiction are deposited in publicly repositories/databases within three years of utilization (Art. 14(3)), the access to which may be subject to reasonable conditions including curation costs (Art. 14(4)). This compromise recognizes the impracticability of making all samples freely available to others while promoting transparency about the location of MGR and DSI if others wish to contact the holders for more information. Deposition requirements are crucial for the aggregate report by repositories and databases to the ABS Committee (see Sect. 7.5 below). BBNJ Agreement bodies, however, will need to clarify the procedures and funding arrangements for the deposition of samples subject to utilization.

Thus, state parties shall ensure that MGR/DSI subject to utilization are deposited in repositories/databases under the terms of the Agreement (Art. 14) and that the data follow the FAIR (findable, accessible, interoperable and reusable) principles; collected MGR are subject to a requirement to make the information available about their location (Art. 12). Databases and repositories might wish to be recognized as 'trusted' by the BBNJ Agreement as has been the case for collections under the CBD and Nagoya Protocol. Finally, Parties must ensure that where MGR and (where practicable) DSI are subject to commercial or non-commercial utilization within their jurisdiction, the BBNJ Identifier and certain information is notified to the CHM as soon as the information becomes available (Art. 12(8)). This information primarily relates to the outcomes of research including the results of utilization such as where publications, patents and/or products can be found, where the original sample that is the subject of utilization is held, the modalities for access to MGR and DSI being utilized, a data management plan and if available information on sales of products and further development (Art. 12(8)). While the final notification system has a considerably lighter touch for monitoring the

[4] https://www.gbif.org/.

[5] https://www.insdc.org/.

[6] For the perspectives of traditional knowledge holders, please see Pena-Neira and Coelho (2025).

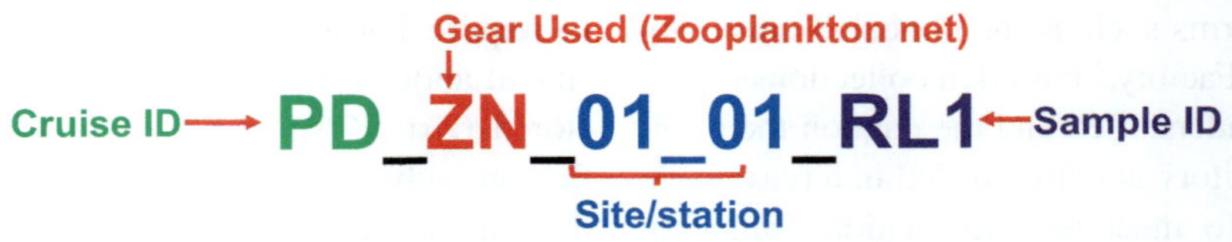

Fig. 7.2 The current use of MSR identifiers giving a range of information on samples collected including the cruise ID, type of gear used for collection, site and station (GPS coordinates will be recorded referencing these) and sample ID. This use, while effective in creating unique identifiers for each sample, reflects the current state of practice and not necessarily the best scientific practice or the foreseen BBNJ batch identifier

location and movement of MGR and DSI compared with previous draft texts, there are a number of gaps in policy and procedure that BBNJ Agreement bodies will need to address over time as outlined in Chap. 5 of this collection (Art. 12(8); Humphries et al., 2025).

7.4.2 BBNJ Standardized Batch Identifier

The BBNJ Identifier will be automatically generated by the CHM upon the pre-collection notification (Art. 12(3)), and Parties are obliged to submit it alongside their post-cruise notification (Art. 12 (6)) and, if available, alongside their utilization notification (Art. 12 (8)). The BBNJ Identifier plays a central role in transparency and information sharing under the BBNJ Agreement framework. The Identifier is mentioned in art. 12 (notification), art. 14 (benefit-sharing), and art. 16 (monitoring and transparency) and is designed as a 'light touch', machine and human-readable identifier linking information about uses of MGR of ABNJ and, where possible, DSI on MGR of ABNJ. While the details about how the BBNJ Identifier will work in practice are yet to be determined by BBNJ Agreement bodies, it is clear that its role in linking information will contribute to benefit-sharing, including through aggregate reports of access to MGR and DSI in repositories and databases (Broggiato et al., 2025). It will not play a direct role in information sharing about the access and use of traditional knowledge associated with MGR in ABNJ, which does not fall within the notification mechanism (Humphries et al., 2025). Instead, the traditional knowledge obligation under art. 13 sets out separate procedures and obligations through a system of prior informed consent and mutually agreed terms (authorization and contractual arrangements) in accordance with national laws (Pena-Neira & Coelho, 2025).

The BBNJ standardized batch identifier follows good practice in marine scientific research (MSR) and can be compared to the use of cruise identifiers. At the moment different national agencies[7] carrying out MSR use different mechanisms to assign cruise identifiers. Such identifiers may not be unique and may vary in the level of information they convey. These cruise identifiers are often used to create sample identifiers as shown in Fig. 7.2. Such identifiers are regarded as scientific good practice, and allow samples to be traced to particular collection events, but are not mandatory and are not legally required. Assignment of a BBNJ standardized batch identifier by the BBNJ CHM will occur on the pre-notification of a research collection event. This may sometimes be synonymous with a cruise identifier (e.g. one identifier for a cruise, autonomous vehicle, seafloor installation etc.), but it is worth noting that negotiators deliberately chose the term 'batch identifier' rather than 'cruise identifier', suggesting that the intent is to allow for, or at the very least not to rule out, finer scale differentiation than 'cruise ID' (e.g. one identifier

[7] For this assessment three national agencies were accessed: IFREMER in France, National Oceanography Centre in the UK and JAMSTEC in Japan.

per project or sample area). Examples of existing batch identifier systems group or aggregate individual sequence accessions together by overall project (BioProject) and individual samples (BioSample) (see also Lawson et al., 2025). A unique identifier as in the BBNJ Agreement needs to obey some fundamental parameters including persistence, discoverability, resolvability and authority. Persistence means the identifier is available in the long term, resolvability that the identifier can be used to directly find the data and discoverability implies that the identifier can be found within and across systems. Authority is important as it refers to curation and standardization of the identifier so that they remain viable in the longer term.

From a researcher's perspective, a system that assigns these BBNJ Identifiers should be automated, following simple rules for how these are used and applied to samples collected during MSR. This could be achieved as a set of good data governance principles as well as terms of use and basic obligations by those carrying out MSR. Once issued, the BBNJ Identifier would attach itself as a 'tag' to all samples collected and upon subsequent work (e.g. DNA extraction, sequencing, chemical extraction and data acquisition) and should be capable of linking to the MGR and DSI on MGR as well as other materials and data (e.g. derivatives, database entries on derivatives). As indicated in Fig. 7.2, a BBNJ Identifier for downstream materials and data may be extended to indicate what has been done with it and what form the data takes. In the example in Fig. 7.2, on DNA extraction the BBNJ Identifier would follow the data in an extended form by adding details on the treatment carried out on it and then deposited into the sequence database to connect it to its origin. The BBNJ Identifier would need to be recognized by other repositories/databases so that it can be incorporated and lead to identification of the original material from which the subsequent material or data was derived. Although this is not specified in the treaty text, the system would then logically require that certain 'gatekeepers' would request the BBNJ Identifier upon submission of a publication or patent for example (see art. 12(8)).

It appears that the implementation of the BBNJ Identifier will be more straightforward when it comes to scientific practices of collecting, sequencing, and publishing materials than for intellectual property-related scenarios. International scientific databases are increasingly promoting standards on the reporting of geographic origins of samples (INSDC, 2021) and it should not be technically complicated to integrate a reporting option for the BBNJ Identifier. However, such transparency practices are much less common in the reporting of genetic resource usage in patent applications (see also Brown, 2025). In 2024, the World Intellectual Property Organization (WIPO) Diplomatic Conference concluded the *WIPO Treaty on Intellectual Property, Genetic Resources And Associated Traditional Knowledge,* (WIPO Treaty) (WIPO, 2024). This requires WIPO members who are Parties to the WIPO Treaty to ensure their domestic patent systems require the disclosure of origin, or the source of, the genetic resources and associated traditional knowledge in patent applications (WIPO Treaty Art. 3). The language in the treaty is broad enough to encompass MGR of ABNJ. While some countries already have disclosure of origin requirements under their national laws (WIPO, 2020), it is unclear how many other countries will implement the WIPO Treaty under their national arrangements and the extent to which there will be a consistent approach. It also remains to be seen how these new WIPO rules will interact with the BBNJ framework.[8]

[8] A mechanism for such an interaction could lie in the possibility to match genetic sequence accession numbers found in patent documents back to their entry on the scientific databases, including to meta-data such as the potentially reported BBNJ identifier. While this increases transparency, it is still not a fully reliable solution as standards are currently also lacking for the reporting of these accession numbers in patent applications (Rohden et al., 2020).

7.4.3 Additional Procedures for Monitoring and Compliance

Next to the notification system (Art. 12) and the BBNJ Identifier, the BBNJ Agreement also allows for the CoP to adopt additional procedures to achieve monitoring and transparency of activities with respect to activities with respect to MGR in ABNJ, based on recommendations by the ABS Committee. This role of the CoP, based on recommendations of the ABS Committee, to provide further guidance on operationalizing the monitoring framework will be important, as there are many elements of the framework and procedures that require further clarification by BBNJ bodies that will come into existence after the Agreement enters into force. There is need for a range of procedures for operationalizing and reviewing the effectiveness of the monitoring framework. The experience of the CBD has demonstrated that effective monitoring requires meaningful goals, actions and indicators for achieving the goals (Conference of the Parties to the Convention on Biological Diversity, 2022a, 2022b) but it remains to be seen whether the BBNJ CoP will develop a similar monitoring framework. Some of the specific issues requiring further clarification outlined in this chapter include exploring how the monitoring system will relate to ABS traceability infrastructure under other information Agreements and national laws; and whether (and how) private entities may voluntarily contribute to the monitoring system directly (instead of going through a Party). Other additional procedures may be determined by BBNJ bodies and require clarification of other provisions of Part II. These are outlined in other chapters of this book and are beyond the scope of this chapter.

7.4.3.1 Relationship with Other ABS Monitoring Infrastructures

Compared with the BBNJ Agreement's monitoring and transparency system, other international ABS frameworks have a range of monitoring infrastructure designed more towards following the movements of biological resources. The CBD and Nagoya Protocol are based on a bilateral system of access authorizations and contractual benefit-sharing on a case-by-case basis (United Nations, 1992 Art. 5; Secretariat of the Convention on Biological Diversity, 2010 Art. 5–7 & Art. 12), whereas the Plant Treaty and PIP Framework have genetic resources within a multilateral mechanism, using standard terms and conditions for ABS (Food and Agriculture Organization of the United Nations, 2001 Art. 12; World Health Organization, 2011 Annex 1). The framework for the DSI multilateral mechanisms is not yet clear, but it will similarly not require ABS negotiations on a case-by-case basis (Kachelriess et al., 2025). It is inevitable, that MGR and DSI on MGR of ABNJ will come into contact with the monitoring systems under these ABS mechanisms, either through MGR inventions or products that incorporate physical resources collected from both within and beyond national jurisdiction or where the metadata is absent or ambiguous for physical resources and DSI. The BBNJ Agreement must be applied in a manner that does not undermine these frameworks (Art. 5(2)) so the CoP will need to determine how to manage the crossovers. Chapter 11 of this collection (Kachelriess et al., 2025) provides a detailed analysis of the other ABS frameworks, whereas this section is an overview of the possible implications for implementation of the monitoring and compliance framework and how stakeholders might manage the overlapping regimes.

Like the BBNJ Agreement, the CBD and Nagoya Protocol have CHMs as an information-sharing and transparency measure. The CBD provides for a general CHM to promote and facilitate technical and scientific cooperation between Parties, including web-based access to official reports and documents on implementation of the CBD generally, rather than ABS specifically (United Nations, 1992 Art. 18(3)). The Nagoya Protocol set up a dedicated ABS Clearing House (ABSCH), which provides access to information from each Party about implementation of the ABS framework under the protocol (Secretariat of the Convention on Biological Diversity, 2010 Art. 14(1)). This includes information about national ABS measures, access permits, codes of conduct, best practices and methods and tools that Parties have developed to monitor genetic resources (Secretariat of the Convention on Biological Diversity, 2010 Art. 14(3)). The monitoring tools at the international level are:

- internationally recognized certificates of compliance (evidence that the genetic resources have been accessed in accordance with prior informed consent and mutually agreed terms if required by the provider country) (Secretariat of the Convention on Biological Diversity, 2010 Art. 17(2–4));
- checkpoints and checkpoint communiques (to collect or receive relevant information related to ABS, such as the source of the genetic resource and the establishment of mutually agreed terms, at designated points along the research and development (R&D) pathway) (Secretariat of the Convention on Biological Diversity, 2010 Art. 17(1));
- user compliance measures (where Parties ensure that genetic resources and traditional knowledge used within their jurisdiction have been accessed in accordance with the provider country's laws) (Secretariat of the Convention on Biological Diversity, 2010 Art. 16); and
- reporting obligations.

The Nagoya Protocol's monitoring infrastructure aims to generate information about genetic resources and Traditional Knowledge to track their use from the original access permit and at designated points along the research and development (R&D) pathway. The ABSCH, where Parties deposit information about these monitoring measures, demonstrates that each of these mechanisms have had a mixed uptake by Parties to the Nagoya Protocol, which creates significant gaps in this track-and-trace style monitoring system (Humphries et al., 2021). However, R&D may contain MGR, associated DSI and traditional knowledge from both within and beyond ABNJ and the Nagoya Protocol monitoring infrastructure may incidentally capture information relevant to transparency under the BBNJ Agreement. The BBNJ bodies will need to clarify the extent to which the BBNJ Agreement's CHM obtains or uses information from this source.

Multilateral ABS frameworks may also have lessons for BBNJ bodies developing information-sharing procedures. The Plant Treaty and PIP Framework ABS mechanisms outlined in Chap. 11 of this collection (Kachelriess et al., 2025) have information systems combined with a Standard Material Transfer Agreement (SMTA) that includes terms and conditions for the use and transfer of materials, the sharing of benefits and information and reporting obligations (Food & Agriculture Organization of the United Nations, 2001 Art. 12; World Health Organization, 2011 Art. 5.4 Annex 1). For example, the Plant Treaty's SMTA contains standard terms and conditions that support Treaty provisions including information and benefit-sharing requirements (Food & Agriculture Organization of the United Nations, 2001 Art. 12), which impose obligations on subsequent users of the genetic material (Governing Body of the International Treaty on Plant Genetic Resources for Food & Agriculture, 2006; Tvedt, 2021). There is no suggestion in the BBNJ Agreement text or the President's reports that the SMTA model has been considered as an implementation tool under the framework. However, from a practical perspective, SMTAs might be an option for transfers of MGR between repositories and between the original collectors and third parties (Tvedt, 2020). The BBNJ bodies may consider issuing guidance on standard terms and conditions for these movements that incorporate the information requirements specified under the BBNJ Agreement (Tvedt, 2020).

The BBNJ bodies may consider procedures for, and lessons learned from, the digital information sharing system under other ABS information systems (Lawson et al., 2019). The purpose of the FAO's Global Information System (GLIS) is to draw from existing information systems to facilitate information exchange on scientific, technical and environmental matters related to plant genetic resources for food and agriculture (PGR) (Food & Agriculture Organization of the United Nations, 2001 Art. 17(1)). It also acts as an early warning system about hazards to PGR and an information source for the FAO Commission on Genetic Resources for Food and Agriculture periodic assessments for the state of the world's PGR (Food & Agriculture Organization of the United Nations, 2001 Art. 17(2) & Art. 17(3)). A specific aim of the GLIS is to bridge the gap in communication between the institutions serving as sources of PGR, those conducting research and added value activities and those using PGR to develop products (Food & Agriculture Organization of the United Nations, 2022). In practice, the GLIS Portal operates as a decentralized network of existing databases and websites, including Genesys,[9] GRIN-Global,[10] the European Search Catalogue for Plant Genetic Resources (EURISCO)[11] and the ABSCH, rather than a curated platform in its own right. It is not the responsibility of the Plant Treaty Governing Body to maintain or verify the accuracy of the information, but rather ensure interoperability with other databases (Lawson et al., 2018). Under this model, there is legal uncertainty for users about the provenance or legal liabilities (such as copyright or proprietary interests) without engaging with the originators of the information (Lawson et al., 2018). The system depends on interoperability between databases but the use of Digital Object Identifiers (DOIs) for the identification of material available in the multilateral system is voluntary (Food & Agriculture Organization of the United Nations, 2023), posing a major challenge for the network. The ongoing challenges with implementing the GLIS, including promoting interoperability and transparency on the rights and obligations of users for accessing, sharing and using the information (Food & Agriculture Organization of the United Nations, 2022), may offer insights for the BBNJ context when setting up the information system.

The information platform under the BBNJ Agreement (the CHM), in contrast, will be a centralized platform (Art. 51(3)). This will require significant oversight and procedures for connecting the information required under the notification mechanism on MGR and DSI, including through other existing databases. It is unclear whether it will have a similar aim to the Plant Treaty's GLIS to bridge the gap in communication between the institutions serving as repositories and databases of MGR and DSI, those conducting research and added value activities and those developing products. Given that the BBNJ Agreement monitoring system does not have a track and trace function, it is likely that the CHM will have more of a passive function for transparency so that users of the centralized database could search through other databases for the origin of a BBNJ Identifier tagged object, if the digital links are not broken. BBNJ bodies will need to clarify the rights and obligations of users of the CHM during implementation of the information platform.

There are, however, significant questions about how the CHM information platform will work in practice. BBNJ bodies will need to develop procedures and policies on interoperability between databases and how to deal with confidential information. Unlike the Plant Treaty's voluntary DOI system, the BBNJ Agreement's transparency system has a compulsory digital identifier (BBNJ Identifier), which will be a considerable advantage for interoperability. However, learning from the Plant Treaty experience, bodies may need to document and inform users about documentation standards for data and metadata, open data standards and common nomenclature and descriptors for MGR of ABNJ (Food & Agriculture Organization of the United Nations, 2022). The Plant Treaty is specific about the types of non-confidential information that Parties must make available to GLIS (Food & Agriculture Organization of the United Nations, 2001 Art. 13(2)(a)), whereas the BBNJ Agreement makes no reference to how it will manage confidential information in accordance with national law.

[9] https://www.genesys-pgr.org/.

[10] http://www.grin-global.org/.

[11] http://www.ecpgr.cgiar.org/resources/germplasm-databases/eurisco-catalogue/.

The BBNJ Agreement's transparency and monitoring framework makes room to accommodate the developments in the multilateral system for DSI under the CBD's Global Biodiversity Framework (Kachelriess et al., 2025). The BBNJ Agreement does not define DSI, which will depend on how the scope and concept evolve in the CoP to the CBD and other ABS international fora (Humphries, 2025). The CoP decision to establish the DSI multilateral mechanism recognizes that 'tracking and tracing of all digital sequence information on genetic resources is not practical' (Conference of the Parties to the Convention on Biological Diversity, 2022a, 2022b para. 5), which also recognizes that there are countries that already attempt to regulate DSI under their bilateral ABS track and trace systems (Scholz et al., 2023). The modalities about how the multilateral mechanism and its information systems will work in practice, including how they will interact with other international and national ABS frameworks are yet to be decided. 'Monitoring and evaluation and review of effectiveness' is one of the items listed for further consideration, along with other measures for transparency including 'principles of data governance' (Conference of the Parties to the Convention on Biological Diversity, 2022a, 2022b Annex). It will be important for policies and procedures on monitoring and information sharing about DSI to be developed consistently between the BBNJ Agreement framework and the DSI multilateral mechanism, which will contain DSI on MGR of ABNJ (DSI Scientific Network, 2023; UNGA, 2023b para. 13).

7.4.3.2 Procedures for Direct Contribution by Private Individuals and Entities

The BBNJ Agreement does not explicitly refer to direct contributions to monitoring and transparency by private individuals and entities. The obligations are on Parties to ensure that the information enters into the Agreement's system. Theoretically, countries may take varying approaches to their obligations, including requiring people and entities under their jurisdiction to report to the country who will pass on the information to the CHM (and ABS Committee in the case of the aggregate data from reports from repositories/databases), or require them to report directly to the BBNJ bodies. In practice, the CoP and other bodies would need to agree on procedures for standardized information to enter the system, otherwise, it will be challenging if not impossible to collate, analyse and report on (see Humphries et al., 2025). Until these procedures and additional information requirements are known, stakeholders can set up their data management systems to collect information required under the Agreement, including under art. 12.

The BBNJ bodies may also encourage more direct contributions by private individuals and entities for monitoring and compliance. An example is the invitation by the CoP to the CBD to 'Parties and relevant organizations to support community-based monitoring and information systems and citizen science and their contributions to the implementation of the monitoring framework for the Kunming-Montreal global biodiversity framework' (Conference of the Parties to the Convention on Biological Diversity, 2022a, 2022b para 6.) Next to this example, there are already a number of citizen science initiatives in ocean conservation that actively involve the public in collecting data and monitoring marine environments (Garcia-Soto et al. 2017). These initiatives not only contribute to scientific research but also enhance public awareness and engagement.

7.5 Reporting by Parties

Article 16(2): Parties shall periodically submit reports to the access and benefit-sharing committee on their implementation of the provisions in this Part on activities with respect to marine genetic resources and digital sequence information on marine genetic resources of areas beyond national jurisdiction and the sharing of benefits therefrom, in accordance with this Part.

Article 16(2) specifies that parties shall periodically submit reports to the ABS Committee on their implementation of provisions related to MGR and the sharing of benefits therefrom. These reports are likely to be transmitted via the CHM, and contain (a) information on legislative, administrative, and policy measures on access and benefit-sharing, (b) contact details on national focal points and (c) other required information (Art 15(4)). This specific report is separate from the biannual aggregate reports on access to BBNJ MGR and DSI, which repositories and databases under Parties' jurisdiction are required to submit directly to the ABS Committee.[12]

National focal points are likely to play an important role in the communication between Parties and the ABS Committee. In this regard, Parties can to some extent draw on experiences or build on existing infrastructures from the CITES, CBD and FAO reporting systems, which equally rely on national focal points that report to a central commission (The State of the World's Aquatic Genetic Resources for Food and Agriculture, 2009). However, it is important to note that in these systems there is always a bilateral country relationship whereas in the BBNJ context, there will be no state of origin. While being helpful points of reference, these existing systems can only be applied to the BBNJ context partially.

Important points for further deliberation are how frequently implementation reports will need to be submitted, and which requirements they will need to fulfill in terms of content and format. An interesting question is how 'implementation of provisions' will be defined, and whether this may also take into account implementation measures at the user or institutional level.

[12] See Chap. 5 Interpreting the Notification System under the High Seas Biodiversity Treaty for further discussion of art. 12.7 (Humphries et al., 2025).

7.6 Treaty Body Reports

Article 16(3): The access and benefit-sharing committee shall prepare a report based on the information received through the Clearing-House Mechanism and make it available to Parties, which may submit comments. The access and benefit-sharing committee shall submit the report, including comments received, for the consideration of the Conference of the Parties. The CoP, taking into account the recommendation of the access and benefit-sharing committee, may determine appropriate guidelines for the implementation of this article, which shall take into account the national capabilities and circumstances of Parties.

The idea of art. 16(3) is that the CoP deliberates not on raw data from the CHM but on a comprehensive report synthesized and pre-processed by experts in the ABS Committee. Upon receiving information entailed in the biannual user report and the periodic implementation report (Art. 16 (2)), and information received through the CHM (Art.16 (3)), the ABS Committee shall prepare a report and, after including comments by Parties, submit the report to the CoP. The idea of this report is to synthesize information received via the CHM and to formulate recommendations concerning the national implementation of the article, to be considered by the CoP. Art. 15(3) describes what these recommendations may encompass, such as guidelines on activities, measures to implement decisions, and matters related to monetary benefit-sharing, financial mechanisms, or information sharing via the CHM.

The ABS Committee will likely play a key role in the implementation of the BBNJ Agreement's MGR provisions, consulting and facilitating the exchange of information with relevant legal instruments, frameworks and bodies (such as the CBD) on activities under its mandate (Art. 15(5)). These activities include benefit sharing, the use of DSI on MGR, 'best practices, tools and methodologies, data governance and lessons learned' (Art. 15(5)). It may make recommendations on this to the CHM, which has a role for providing links to 'relevant global, regional, subregional, national and sectorial clearing-house mechanisms and other gene banks, repositories and databases, including those pertaining to relevant traditional knowledge...and promote where possible, links with publicly available private and non-governmental platforms for the exchange of information' (Art. 51(3)(c)).

While art. 16.1 on Monitoring and Transparency only refers to the CHM in the context of notifications by Parties, Parties could in the future consider its links to other relevant repositories and databases as a means to collect additional information about MGR and DSI on MGR of ABNJ which may be facilitated through the use of the BBNJ Identifier. The language included in art. 16. 3 with regards to information to be used by the ABS Committee for its reports to the CoP could allow for such a broader interpretation (*'information received through the CHM'*). Overall, the ABS Committee, by submitting reports and recommendations to the CoP on the basis of which the CoP will adopt guidance on implementation (Art. 16 (3)), occupies a central role in the monitoring and transparency of activities with regards to MGR from ABNJ.

7.7 Conclusion

This chapter delved into the framework of monitoring and transparency for the governance and regulation of MGR and DSI on MGR of ABNJ under the BBNJ Agreement. It analyzed the three elements of art. 16 discussing the potential implications for different actors in the MGR chain. First, Parties have obligations to establish notification requirements on the collections and uses of people and entities within their jurisdiction, which are to be submitted to the CHM. This shall include the BBNJ identifier for the post-cruise notification and, if available, for the utilization notification. Second, Parties are required to inform the ABS Committee about their implementation of Part II of the BBNJ Agreement concerning activities with respect to MGR and DSI on MGR of ABNJ and the sharing of benefits therefrom. Third, the ABS Committee has the task to compile reports with data from the Parties and information from the CHM to inform decision-making at the CoP.

Monitoring and transparency aspects of MGR in the BBNJ Agreement should be understood in line with developments in other international fora on ABS and their potential impacts on monitoring and transparency under the BBNJ Agreement. Key areas concern the new multilateral mechanism for benefit sharing from DSI under the Global Biodiversity Framework and the WIPO Treaty requiring disclosure of origin of genetic resources and associated traditional knowledge in patent applications. The modalities of both mechanisms are yet to be determined and the extent to which they will affect the operation of the BBNJ Agreement framework is unclear. Regardless, the potential overlaps and conflicts between these systems need careful consideration and cooperation for an effective monitoring and transparency system that has the capacity to inform Parties and the CoP about the extent of MGR and DSI use, informing the benefit-sharing system. As information about Traditional Knowledge is not explicitly included under the reporting arrangements of art. 16, there may need to be additional procedures developed by the CoP to effectively implement the Traditional Knowledge obligation under Article 13 (Pena-Neira & Coelho, 2025).

While art. 16 has undergone significant changes before reaching a compromise that negotiators found acceptable and workable, it is evident that numerous open questions and challenges remain. While the introduction of the BBNJ Identifier as a central tool in the monitoring process has synergies with existing scientific identifier practices and can mark a significant advancement in ensuring traceability, there are many aspects of its operation that are unclear, with implications for the type of data that it will be linked to. It is also yet to be determined, how this identifier will integrate with a range of databases, including patent databases which may potentially pose a challenge for effective implementation. Further, ambiguity about whether the BBNJ Identifier will be issued at the cruise or project level for the purposes of the pre-collection notification will have implications for the type of databases that it will be linked to and the type of information that will filter through to Parties, the ABS Committee, the CHM and the CoP.[13]

The ABS Committee occupies a central role in the system as it compiles regular reports based on information obtained from the CHM and Parties' reports (after Parties have had an opportunity to comment) and shares this with the CoP for discussion. A number of questions regarding these reports, their underlying data and information, their substance and focus and their role in decision-making process remain open and to be addressed before an effective monitoring and transparency system can be implemented. Firstly, while the BBNJ Identifier linked to collections from ABNJ should help the ABS Committee in assessing various information received, important questions remain about the periodic reporting by Parties and the subsequent preparation of comprehensive reports by the ABS Committee. Does information from Party reports to the ABS Committee reach a level of detail such as reporting instances of 'utilization' and the use of MGR in open-access scientific publications? Secondly, regarding data from the CHM, it is unclear if this will demonstrate a timely snapshot of activities under Part II of the Agreement or will there be significant time lags or gaps in data given the time it will take to collect and analyse the data, compile and consult on the reports? This relates to a broader set of questions on the diverse, complex but critical tasks of the CHM (Muraki Gottlieb et al., 2025) and it remains to be seen how this body can fulfil the significant expectations.

While the ABS Committee's reports can greatly contribute to understanding MGR utilization, further questions remain regarding their scope, frequency, detailed requirements, and the time frame in which Parties can submit comments. It is for example foreseeable that the ABS Committee may prepare either very prescriptive reports indicating clear quantities and directions for benefit-sharing or it may prepare very descriptive summaries of the collected data from repositories and the CHM. Considering the scope of information to be processed, it is likely that reports cannot represent all data and information from the whole monitoring and transparency system equally but that the ABS Committee may have to pick and choose elements to focus on in its recommendations. In its reporting, the ABS Committee may for example prioritize among different types of information, such as systematically analyzing individual product-based notices submitted to the CHM while aggregating information related to mere access of DSI labelled with the BBNJ tag. State parties, supported by the CoP and the ABS Committee would do well to clarify these elements and establish certain formats and standards for reporting obligations.

Generally, the role of the monitoring and transparency system and the reports compiled by the ABS Committee for decision-making on the sharing of benefits is a critical area that requires clarification by the CoP. It remains to be decided how data from the CHM and the reports from the ABS Committee will contribute to calculating and deciding on benefit-sharing contributions and how benefits will be allocated (Broggiato et al., 2025). Parties should keep this in mind when moving forward, particularly as it will likely take time for the CoP to develop appropriate guidelines for implementing the monitoring and transparency framework, which is integral to the success of information and benefit sharing. These and other questions leave much work for the BBNJ bodies over the coming years. Nevertheless, the 'light touch' monitoring and transparency under the BBNJ Agreement is an important achievement for ocean governance, conceived through compromise but with many innovative features for effective implementation of Part II of the Agreement.

[13] See Lawson et al. (2025) for further discussion of the BBNJ Identifier.

References

Ad Hoc Technical Expert Group on Digital Sequence Information on Genetic Resources. (2020). Combined Study on Digital Sequence Information in Public and Private Databases and Traceability (CBD/DSI/AHTEG/2020/1/4). Retrieved from https://www.cbd.int/doc/c/1f8f/d793/57cb114ca40cb6468f479584/dsi-ahteg-2020-01-04-en.pdf

Amann, R. I., Baichoo, S., Blencowe, B. J., Bork, P., Borodovsky, M., Brooksbank, C., Xenarios, I., et al. (2019). Toward unrestricted use of public genomic data. *Science, 363*(6425), 350–352. https://doi.org/10.1126/science.aaw1280

Brazil. (2019). Brazil's Position on DSI (Notification 2019-012). Retrieved from https://www.cbd.int/abs/DSI-views/2019/Brazil-DSI.pdf

Broggiato, A., Dunshirn, P., Jaspars, M. & Pena-Neira, S. (2025). Monetary and non-monetary benefit sharing under the BBNJ Agreement. In F. Humphries (Ed.), *Decoding governance marine genetic resource governance under the BBNJ Agreement.* Springer.

Brown, A. E. L. (2025). The place of intellectual property under the BBNJ Agreement. In F. Humphries (Ed.), *Decoding marine genetic resource governance under the BBNJ Agreement.* Springer.

Conference of the Parties to the Convention on Biological Diversity. (2022a). Decision Adopted by the Conference of the Parties to the Convention on Biological Diversity, 15/9 Digital Sequence Information on Genetic Resources (No. CBD/COP/DEC/15/9). Retrieved from https://www.cbd.int/doc/decisions/cop-15/cop-15-dec-09-en.pdf

Conference of the Parties to the Convention on Biological Diversity. (2022b). Monitoring Framework for the Kunming-Montreal Global Biodiversity Framework: Draft Decision Submitted by the President (CBD/COP/15/L.26). Retrieved from https://www.cbd.int/doc/?meeting=cop-15

DSI Scientific Network. (2023). A Harmonized System for Benefit-Sharing from DSI (Policy Brief). Retrieved from https://www.dsiscientificnetwork.org/wp-content/uploads/2023/11/Policy-Brief-Harmonization-of-DSI-BS-systems.pdf

Food and Agriculture Organization of the United Nations. (2001). International Treaty on Plant Genetic Resources for Food and Agriculture. Retrieved from https://www.fao.org/plant-treaty/en/

Food and Agriculture Organization of the United Nations. (2022). Resolution 5/2022: Implementation of the Global Information System. Retrieved from https://www.fao.org/3/nk240en/nk240en.pdf

Food and Agriculture Organization of the United Nations. (2023). Report of the Tenth Session of the Governing Body of the International Treaty on Plant Genetic Resources for Food and Agriculture (IT/GB-10/23/Report). Retrieved from https://www.fao.org/3/nk240en/nk240en.pdf

Garcia-Soto, C., van der Meeren, G. I., Busch, J. A., Delany, J., Domegan, C., Dubsky, K., Fauville, G., Gorsky, G., von Juterzenka, K., Malfatti, F., Mannaerts, G., McHugh, P., Monestiez, P., Seys, J., Węsławski, J. M. & Zielinski, O. (2017). Advancing Citizen Science for Coastal and Ocean Research. French, V., Kellett, P., Delany, J., McDonough, N. [Eds.] Position Paper 23 of the European Marine Board, Ostend, Belgium. 112pp. ISBN: 978-94-92043-30-6

Governing Body of the International Treaty on Plant Genetic Resources for Food and Agriculture. (2006). First Session of the Governing Body of the International Treaty on Plant Genetic Resources for Food and Agriculture (IT/GB-1/06/Report). Retrieved from https://www.fao.org/3/a-y5609e.pdf

Humphries, F., Rabone, M., & Jaspars, M. (2021). Traceability approaches for marine genetic resources under the proposed ocean (BBNJ) treaty. *Frontiers in Marine Science, 8*, 430. https://doi.org/10.3389/fmars.2021.661313

Humphries, F. (2025). Marine genetic resources beyond national jurisdiction: The expansive scope of the BBNJ Agreement. In F. Humphries (Ed.) *Decoding marine genetic resource governance under the BBNJ Agreement.* Springer.

Humphries, F., Jaspars, M., Lavelle, J. & Kachelriess, D. (2025). The novel notification information system for marine genetic resources under the BBNJ Agreement. In F. Humphries (Ed.) *Decoding governance marine genetic resource governance under the BBNJ Agreement.* Springer.

INSDC. (2021). Spatio-temporal annotation policy. INSDC News and Announcements. Retrieved from https://www.insdc.org/news/spatio-temporal-annotation-policy-18-11-2021/

Kachelriess, D. (2023). In M. Epps & C. Chazot (Eds.), *The High Seas Biodiversity Treaty: An Introduction to the Agreement under the United Nations Convention on the Law of the Sea on the Conservation and Sustainable Use of Marine Biological Diversity of Areas Beyond National Jurisdiction.* IUCN.

Kachelriess, D., Dunshirn, P., Langlet, A., Brown, A. E. L. & Scholz, A.H (2025). Benefit sharing provisions of the BBNJ Agreement for MGR and DSI—interlinkages with other ABS frameworks. In F. Humphries (Ed.), *Decoding marine genetic resource governance under the BBNJ Agreement.* Springer.

Laird, S., Johnston, S., Wynberg, R., Lisinge, E., & Lohan, D. (2004). Access and benefit sharing perspectives in protected area management. In *Biodiversity Issues for Consideration in the Planning, Establishment and Management of Protected Area Sites and Networks. Secretariat of the Convention on Biological Diversity.*

Langlet, A., & Dunshirn, P. (2023). Traceability options for marine genetic resource from areas beyond national jurisdiction. HSA Policy Brief.

Langlet, A., Vadrot, A. B. M., Fellinger, S., Dunshirn, P., Ruiz Rodriguez, S. C., & Tessnow-von Wysock, I. (2024). MARIPOLDATAbase (SUF edition). https://doi.org/10.11587/0XXZ0V, AUSSDA, V3.

Lawson, C., Burton, H., & Humphries, F. (2018). The important place of information in the evolving legal and policy framework for the conservation and sustainable use of the world's plant genetic resources for food and agriculture. *European Intellectual Property Review, 40*, 243–259.

Lawson, C., Humphries, F., & Rourke, M. (2019). The future of information under the CBD, Nagoya protocol, plant treaty, and PIP framework. *The Journal of World Intellectual Property, 22*(3–4), 103–119. https://doi.org/10.1111/jwip.12111

Lawson, C., Humphries, F., Jaspars, M. & Rabone, M. (2025). Data management and the 'BBNJ standardized batch identifier' under the BBNJ Agreement. In F. Humphries (Ed.) *Decoding marine genetic resource governance under the BBNJ Agreement.* Springer.

Muraki Gottlieb, H., Ardron, J. & Brown, A. E. L. (2025). BBNJ Agreement: A new infrastructure to foster benefit sharing of marine genetic resources. In F. Humphries (Ed.), *Decoding marine genetic resource governance under the BBNJ Agreement.* Springer.

Pena-Neira, S. & Coelho, L.F. (2025). Traditional knowledge associated with marine genetic resources in areas beyond national jurisdiction. In F. Humphries (Ed.), *Decoding marine genetic resource governance under the BBNJ Agreement.* Springer.

Rabone, M., Harden-Davies, H., Collins, J. E., Zajderman, S., Appeltans, W., Droege, G., Horton, T., et al. (2019). Access to marine genetic resources (MGR): Raising awareness of best-practice through a new agreement for biodiversity beyond national jurisdiction (BBNJ). *Frontiers in Marine Science, 6,* 520. https://doi.org/10.3389/fmars.2019.00520

Rohden, F., Huang, S., Dröge, G., & Hartman Scholz, A. (2020). Combined Study on Digital Sequence Information in Public and Private Databases and Traceability. Retrieved from https://www.cbd.int/abs/DSI-peer/Study-Traceability-databases.pdf

Scholz, A. H., Humphries, F., Vanagt, T., & Jaspars, M. (2023). A new dawn for global benefit-sharing: Capitalizing on the global biodiversity framework for marine genetic resources from areas beyond national jurisdiction. *IUCN Policy Brief.* Retrieved from https://www.iucn.org/sites/default/files/2023-02/bbnj_icg5bis_policy_brief_global_benefit_sharing_1.pdf

Scholz, A. H., Nunez-Vega, G., Weissgold, L., & Wussmann, K. (2024). The future of access and benefit-sharing: What next after the adoption of the global biodiversity framework and decision on digital sequence information? *Diversity, 16*(1), 27. https://doi.org/10.3390/d16010027

Scovazzi, T., Williams, M., Bastmeijer, C. J., Lohan, D., & VanderZwaag, D. (2008). International commons/areas beyond. *Yearbook of International Environmental Law, 18*(1), 267.

Secretariat of the Convention on Biological Diversity. (2010). Nagoya Protocol on Access to Genetic Resources and the Fair and Equitable Sharing of Benefits Arising from their Utilization to the Convention on Biological Diversity. Retrieved from https://www.cbd.int/abs/doc/protocol/nagoya-protocol-en.pdf

Tiller, R., Mendenhall, E., De Santo, E., & Nyman, E. (2023). Shake it off: Negotiations suspended, but hope simmering, after a lack of consensus at the fifth intergovernmental conference on biodiversity beyond national jurisdiction. *Marine Policy, 148*, 105457. https://doi.org/10.1016/j.marpol.2023.105457

Tvedt, M. W. (2020). Marine genetic resources: A practical legal approach to stimulate research, conservation and benefit-sharing. In *The law of the seabed* (pp. 238–254). Brill Nijhoff. https://doi.org/10.1163/9789004391568_013

Tvedt, M. W. (2021). A contract-law analyses of the SMTA of the plant treaty: Can it work as a binding contract? *The Journal of World Intellectual Property, 24*(1–2), 83–99. https://doi.org/10.1111/jwip.12134

UNEP. (2022). *Decision adopted by the conference of the parties to the convention on biological diversity, 15/9 digital sequence information on genetic resources.* In *Conference of the Parties to the Convention on Biological Diversity.* CBD/COP/DEC/15/9. December 19, 2022.

UNGA. (2018). President's aid to negotiations: Intergovernmental Conference on an international legally binding instrument under the United Nations Convention on the Law of the Sea on the conservation and sustainable use of marine biological diversity of areas beyond national jurisdiction, 2nd session, New York, 25 March-5 April 2019. A/CONF.232/2019/1. 3 Dec. 2018. Retrieved from https://www.un.org/depts/los/convention_agreements/texts/BBNJ/A_CONF.232_2019_1_E.pdf

UNGA. (2019). Revised draft text of an agreement under the United Nations Convention on the Law of the Sea on the conservation and sustainable use of marine biological diversity of areas beyond national jurisdiction: Note by the President, Intergovernmental conference on an international legally binding instrument under the United Nations Convention on the Law of the Sea on the conservation and sustainable use of marine biological diversity of areas beyond national jurisdiction. Fourth session. A/CONF.232/2020/3. 18 November 2019. Retrieved from https://www.un.org/depts/los/convention_agreements/texts/BBNJ/A_CONF.232_2020_3_E.pdf

UNGA. (2022a). Further revised draft text of an agreement under the United Nations Convention on the Law of the Sea on the conservation and sustainable use of marine biological diversity of areas beyond national jurisdiction: Note by the President, Intergovernmental conference on an international legally binding instrument under the United Nations Convention on the Law of the Sea on the conservation and sustainable use of marine biological diversity of areas beyond national jurisdiction. Fifth session. A/CONF.232/2022/5. 1 June 2022. Retrieved from https://www.un.org/depts/los/convention_agreements/texts/BBNJ/A_CONF.232_2022_5_E.pdf

UNGA. (2022b). Further refreshed draft text of an agreement under the United Nations Convention on the Law of the Sea on the conservation and sustainable use of marine biological diversity of areas beyond national jurisdiction: Note by the President, Intergovernmental conference on an international legally binding instrument under the United Nations Convention on the Law of the Sea on the conservation and sustainable use of marine biological diversity of areas beyond national jurisdiction. Resumed fifth session. A/CONF.232/2023/2. 12 December 2022. Retrieved from https://www.un.org/depts/los/convention_agreements/texts/BBNJ/A_CONF.232_2023_2_E.pdf

UNGA. (2023a). Agreement under the United Nations Convention on the Law of the Sea on the conservation and sustainable use of marine biological diversity of areas beyond national jurisdiction, Intergovernmental conference on an international legally binding instrument under the United Nations Convention on the Law of the Sea on the conservation and sustainable use of marine biological diversity of areas beyond national jurisdiction. Further resumed fifth session. A/CONF.232/2023/4. 19 June 2023. Retrieved from https://www.un.org/depts/los/convention_agreements/texts/BBNJ/A_CONF.232_2023_4_E.pdf

UNGA. (2023b). Resolution 78/155: [Implementation of the Convention on Biological Diversity and its contribution to sustainable development] (para. 13). Retrieved from https://www.undocs.org/A/RES/78/155

United Nations. (1992). Convention on Biological Diversity. Retrieved from https://www.cbd.int/doc/legal/cbd-en.pdf

Vadrot, A. B., Langlet, A., & Tessnow-von Wysocki, I. (2022). Who owns marine biodiversity? Contesting the world order through the 'common heritage of humankind' principle. *Environmental Politics, 31*(2), 226–250. https://doi.org/10.1080/09644016.2021.1963956

WIPO. (2020). *Key questions on patent disclosure requirements for genetic resources and traditional knowledge* (2nd ed). World Intellectual Property Organization.

WIPO. (2024). Diplomatic Conference to Conclude an International Legal Instrument Relating to Intellectual Property, Genetic Resources and Traditional Knowledge Associated with Genetic Resources, *WIPO Treaty On Intellectual Property, Genetic Resources And Associated Traditional Knowledge* GRATK/DC/7.

World Health Organization. (2011). Pandemic Influenza Preparedness Framework: Annex 1. Retrieved from https://apps.who.int/gb/pip/pdf_files/pandemic-influenza-preparedness-en.p

Arne Langlet holds a PhD in International Relations from the University of Vienna specializing in the marine biodiversity regime complex. His publications address the BBNJ negotiations, questions about the governance of marine genetic resources, the common heritage principle and the use of data in marine governance. He is currently a research associate for the Horizon Europe project MARCO-BOLO, focussing on the use of marine biodiversity data, and a Fisheries Consultant for the FAO, researching ecosystem restoration efforts.

Paul Dunshirn is a PhD researcher at the ERC-funded research project Twin Politics, University of Vienna. His research explores global patent and science landscapes around marine genetic resources and digital sequence information (DSI), aiming to sketch pathways for equitable policymaking. Paul has served as an independent observer and advisor to various UN processes, such as the recently concluded negotiations on the Biodiversity Beyond National Jurisdiction (BBNJ) treaty and the efforts to regulate DSI under the Convention of Biological Diversity (CBD).

Marcel Jaspars is a professor of Organic Chemistry at the University of Aberdeen where he leads the Marine Biodiscovery Centre which focusses on marine resources for novel pharmaceuticals, and to investigate fundamental questions in marine chemical ecology and biosynthesis. Marcel has been active at national and international levels to develop the science, its applications/industrial uptake and associated policy involved in marine biodiscovery and biotechnology. He provides scientific advice to the UK, EU and UN for global policy processes on ocean conservation and digital sequence information via reports, papers and taking part in discussion meetings.

Fran Humphries has specialized in marine and biodiversity law and policy for over two decades in government, academia and consultancies. She is an associate professor at Griffith Law School, Griffith University, Australia, with a background in fisheries management and marine law. She has led large international research consultancies on access and benefit sharing of biological resources for governments, UN organizations and other institutions. She was on the International Council of Environmental Law delegation for the Biodiversity Beyond National Jurisdiction (BBNJ) Agreement negotiations.

Daniel Kachelriess is an expert on oceans, fisheries, wildlife law and policy and followed the negotiations of the BBNJ Agreement as part of the High Seas Alliance and as a member of the IUCN World Commission on Environmental Law. He continues to advise the High Seas Alliance and other organizations on aspects of the BBNJ Agreement, including on Marine Genetic Resources, including the fair and equitable sharing of their benefits. His previous roles include Executive Director of Sea Shepherd Legal, a non-profit law firm, and the Marine Species Officer of the CITES Secretariat.

8 Traditional Knowledge Associated with Marine Genetic Resources in Areas Beyond National Jurisdiction

Sergio Pena-Neira
and Luciana Fernandes Coelho

Abstract

This chapter analyzes and interprets the current international legal discussions and regulations concerning traditional knowledge in the *Agreement Under the United Nations Convention on the Law of the Sea on the Conservation and Sustainable Use of Marine Biological Diversity of Areas Beyond National Jurisdiction*. The analysis focuses on how the agreement addresses traditional knowledge and those producing this kind of knowledge: indigenous peoples and local communities (distinguishing between them) exploring the state of the art in recognizing traditional knowledge across various areas of international law. Additionally, the chapter aims to provide an effective application of these rules by examining the relationship between the agreement and other treaties ruling traditional knowledge, elements of traditional knowledge, and the interpretation and implementation of Article 13 of the Agreement, which serves as the principal source of regulation. A case study is presented to illustrate this implementation. The interpretation is guided by the Vienna Convention on the Law of Treaties and other relevant international agreements concerning genetic resources.

Keywords

Traditional knowledge · Marine genetic resources · Indigenous peoples · Local communities · BBNJ agreement · Recognition in treaties · Regulation

8.1 Introduction

In recent years, there has been an increased recognition of the significance that the traditional knowledge (TK) of indigenous peoples and local communities (IPLCs) plays in natural resource management, biodiversity conservation, and food security, which fostered its integration into the corpus of international law.[1] While the *United Nations Convention on the Law of the Sea* (UNCLOS) (1982/1994), primarily concerned with balancing interstate

S. Pena-Neira (✉)
Universidad Mayor, Santiago, Chile
e-mail: sergio.pena@umayor.cl

L. F. Coelho
Stockholm Environment Institute,
Stockholm, Sweden

[1] Some references on the topic include: Lawson et al. (2022), Mulalap et al. (2020), Radovich (2023), Vierros et al. (2020), Pena-Neira (2017a), Crawford (2019).

F. Humphries (ed.), *Decoding Marine Genetic Resource Governance Under the BBNJ Agreement*,
Sustainable Development Goals Series, https://doi.org/10.1007/978-3-031-72100-7_8

relations, remains silent on dedicated provisions on TK, the *Convention on Biological Diversity* (CBD) (1992/1993), marked a turning point, especially with Article 8, subsection "j", acknowledging its value. The growing awareness that TK of IPLCs may also relate to areas beyond national jurisdiction (ABNJ) through ecological and oceanographic connections (Vierros et al., 2020; Popova et al., 2019) has led to the incorporation of dedicated provisions of TK within the *Agreement Under the United Nations Convention on the Law of the Sea on the Conservation and Sustainable Use of Marine Biological Diversity of Areas Beyond National Jurisdiction* (BBNJ Agreement) (2023).

The BBNJ Agreement's regulation of TK from IPLCs associated with marine genetic resources (MGRs) expands the scope of UNCLOS and signifies the increased influence of biodiversity law within the law of the sea, although further analysis is necessary to fully understand its implications. This is because, while TK may be associated with genetic resources from animals, plants, and microorganisms, not all genetic information necessarily traces its origins to TK, with the elements to evidence the association of TK of IPLCs to the use of genetic resources still disputable. For instance, in the notable case of Basmati rice, the nomenclature was subject to deliberation and dispute (Madhavan, 1998) despite traditional knowledge being meticulously documented (Tata, 2001). In this context, the chapter at hand addresses the dearth of literature and the governance of traditional knowledge associated with MGRs in ABNJ in Article 13 of the BBNJ Agreement. The central inquiry revolves around the effective application of treaty provisions to prospective cases involving traditional knowledge linked to marine genetic resources in these international zones. The primary objective of this chapter is to elucidate the interpretation of this relevant provision.

This chapter advances the critical importance of TK held by IPLCs concerning MGRs in ABNJ, the contemporary implementation of international legal frameworks safeguarding TK and the intricate relationship between TK and other categories of knowledge. It commences by contextualizing the definitions of IPLCs and traditional knowledge and outlining international instruments that addressed this theme prior to the BBNJ Agreement. Subsequently, it provides an overview of key provisions within the BBNJ Agreement referencing to traditional knowledge and IPLCs. Following that, the chapter examines the interpretation of the terms of Article 13, which specifically addresses MGRs associated with TK of IPLCs, as the first element of the interpretation process indicated by the rules of the *Vienna Convention on the Law of Treaties* (VCLT) (1969/1980). It then moves on to analyze significant for the domestic application of these norms, identifying potential gaps that require clarification, such as the definition of "relevant traditional knowledge." It concludes with the discussion of a hypothetical case which illustrates the challenges and opportunities entailed in application of TK within ABNJ.

8.2 The Interface Between the BBNJ Agreement and Other Instruments Governing Traditional Knowledge

Principle 22 of the Rio Declaration on Environment and Development (1992) marked the formal recognition of the nexus between the traditional knowledge of IPLCs and sustainable development in the international arena. After that, considerations pertaining to the access to and use of traditional knowledge held by IPLCs have progressively found incorporation in numerous legal instruments, spanning domains both within and beyond the environmental sector. Notable examples include the CBD, the *International Treaty on Plant Genetic Resources for Food and Agriculture* (PGRFA) (2001/2004), and the *Nagoya Protocol on Access to Genetic Resources and the Fair and Equitable Sharing of Benefits Arising from their Utilization* under the CBD (Nagoya Protocol) (2010/2014), and the establishment of a Traditional Knowledge Division within the World Intellectual Property

Organization (2023). In this context, the BBNJ Agreement builds upon the recognition and experience laid out by these instruments, but it also innovates by explicitly referencing the traditional knowledge of IPLCs within the framework of the law of the sea.

The BBNJ Agreement pertains to the regulation of TK held by IPLCs associated MGRs in areas beyond the sovereignty or jurisdiction of States, hence, the terms should be interpreted in isolation, by in line with Article 31 of the VCLT, as previously referenced. The Agreement does not articulate definitions for "indigenous peoples," "local communities," and "traditional knowledge." This absence is consistent with the approach of the *United Nations Declaration on the Rights of Indigenous Peoples* (2007) (UNDRIP) and the CBD, both of which refrain from endorsing a universal legal definition for IPLCs.However, for the purpose of guiding the interpretation of the BBNJ Agreement, indigenous peoples can be understood as the "descended from a population that inhabited a country at a time of its conquest or colonization by another country, currently consider themselves distinct from other (perhaps more dominant) populations in that country, and retain at least some of their original socioeconomic, cultural, and/or political institutions, which they have rights to enjoy and perpetuate" (Mulalap et al., 2020). Similarly, Mulalap et al. (2020) suggest that local communities can be understood as "a community that has longstanding historic, cultural, and/or political roots in a country and is not typically considered subservient to any other population in the country (although it might have been in the past)."

The glossary of concepts under the scope of Article 8(j), published by the CBD Secretariat, describes traditional knowledge as "the knowledge, innovations, and practices of indigenous and local communities embodying traditional lifestyles relevant for the conservation and sustainable use of biological diversity" (Secretariat of the Convention on Biological Diversity, 2019). This could serve as a basis for interpreting Article 13, with the caveat that it only refers to the traditional knowledge associated with marine genetic resources (MGRs) found in areas beyond national jurisdiction (ABNJ).

8.3 Overview of Key Aspects of the Traditional Knowledge Element in the BBNJ Agreement

The BBNJ Agreement establishes special considerations pertaining to the TK held by IPLCs across virtually all its parts. As an expression of the political desire of the negotiating parties, its preamble affirms that "that nothing in this Agreement shall be construed as diminishing or extinguishing the existing rights of indigenous peoples, including as set out in the United Nations Declaration on the Rights of Indigenous Peoples, or of, as appropriate, local communities."[2] This was an important innovation since, as previously mentioned, UNCLOS does not refer to this topic. Indigenous people and local communities have been ruled together in this international treaty. Indigenous peoples possess collective rights that are distinct from those of local communities, which may comprise both indigenous and non-indigenous members. This legal regulation aligns with Article 8(j) of the CBD and has been addressed in the Nagoya Protocol. However, the rights of indigenous peoples differ significantly from general human rights, particularly in terms of their collective nature and the specific types of knowledge that arise from this collectivity.

Indigenous peoples' rights are often intertwined with minority rights due to their unique interests and legal protections (Sands & Peel, 2018). Their traditional knowledge, which is derived from long-standing cultural practices and observations of natural phenomena, holds a special status. Local communities, while they also develop knowledge through their activities, do not necessarily possess the same collective rights unless they are indigenous.

[2] See Chap. 4.

From a human rights perspective, both indigenous peoples and local communities are protected by rights inherent to their human nature. However, indigenous peoples benefit from additional protections under various international treaties and declarations, such as the *ILO Convention 169 Concerning Indigenous and Tribal Peoples Convention of 1989* (ILO, 1991), UNDRIP of 2007. Local communities do not have access to these specific legal instruments and protections.

Furthermore, states have entered into treaties with indigenous peoples, recognizing their unique status and rights. Such legal relationships are not established with local communities, which are generally considered part of the broader population of each state. The distinctive legal personality and special protections afforded to indigenous peoples are not typically extended to local communities, underscoring the unique position of indigenous peoples within international and domestic legal frameworks.

Furthermore, it is widely recognized, despite some dissenting opinions, that international law distinguishes between biological and genetic resources under the jurisdiction of states (sovereignty or jurisdiction) as per Article 15 of the CBD, and the knowledge associated with those resources, which falls under the rights of groups or individuals (CBD Article 8(j)) (Western Sahara Case, 1975). Knowledge in certain domains may be shared by both indigenous peoples and local communities; however, it differs in its foundations, use, and legal regulation, as outlined in the ILO Convention of 1989. This distinction is further evident in the right to self-determination (Crawford, 2019).

For indigenous peoples, the right to self-determination (recognized in the Western Sahara Case, 1975) carries a unique significance, differing not only from the rights of local communities (e.g., Dixon et al., 2016) but also from subjects of international law. Indigenous peoples' rights include those "attaching to individuals because of their status as members of a group, and rights attaching to the group as such, which individuals can only enjoy in community with others" (Crawford, 2019). In contrast, local communities lack specific legal protections, except in the context of international treaties related to genetic resources, which are primarily concerned with biodiversity conservation (Pena-Neira, 2017b).

Therefore, it is fair to assert that the nature of their rights may differ, even though the Biodiversity Beyond National Jurisdiction (BBNJ) Agreement treats them equivalently from a legal standpoint.

Traditional knowledge also falls within the purview of "General principles and approaches" as outlined in the pertinent legal framework. Accordingly, Article 7, subsection "j" expressly stipulates the "use of relevant traditional knowledge of IPLCs, where available." The term "relevant traditional knowledge" must be taken into account in various contexts, including "Proposals" within the domain of "measures such as area-based management tools." The BBNJ Agreement, as elucidated earlier, incorporates Article 7, subsection 'j' and in Part III, Article 19, subsection 8.3; in proposals related to "identified areas" in Article 19, subsection 8.4, 'j'; in consultations and assessments of proposals in Article 21.2, "c"; in measures that are adopted, as articulated in Article 24.3; and in the context of "Monitoring and Review" undertaken by the Conference of the Parties when making decisions or recommendations, as delineated in Article 26.5. Additionally, the concept of "relevant traditional knowledge" applies in the "Process for environmental impact assessment" encompassing the phases of screening in Article 31.1, "a," "I," "iv"; scoping and impact assessment in Article 31.1 "b" and "c"; in planned activities generating effects on the nature and potential effects on the marine environment, as stipulated in Article 32.3; in impacts of activities in areas beyond national jurisdiction; and in the monitoring of impacts of authorized activities, as articulated in Article 35. In instances where the bearers of this knowledge harbor concerns regarding the measures, as delineated in Article 37.4, "a" the Scientific and Technical Body is mandated to evaluate these concerns utilizing this type of knowledge, pursuant to Article 37.3 "c."

Furthermore, the concept of "relevant traditional knowledge" extends to the involvement of relevant stakeholders in cooperation concerning capacity-building and the transfer of marine technology, as outlined in Article 41.2. It also applies to information dissemination and awareness-raising, as per Article 44.1, "b" and in consultations for public dissemination of information and the facilitation of participation, where due consideration must be given to bearers of this knowledge, in accordance with Article 48.3. These bearers of knowledge may also seek participation as observers in the Conference of the Parties, as provided for in Article 48.4. The expertise to be incorporated into the Scientific and Technical Body, as delineated in Article 49.2, should take into account this concept. Lastly, "relevant traditional knowledge" is instrumental in establishing the "links" to be provided by the Clearing House Mechanism, in accordance with Article 51.3, "c" and in financing conservation and sustainable use programs of these knowledge bearers, as articulated in Article 52.6, "c."

Regrettably, the BBNJ Agreement fails to proffer a precise definition of "relevant traditional knowledge." Nevertheless, it is conceivable to differentiate "traditional knowledge" from "relevant traditional knowledge." In this context, "relevant" should be inextricably linked to areas beyond national jurisdictions, serving as a source of knowledge and tradition for the creators, namely IPLCs. It is a category of knowledge generated inductively by IPLCs based on the effects of MGRs on human beings, animals, and plants within areas beyond national jurisdictions.

8.4 Interpretation of Article 13 of the BBNJ Agreement

Article 13 of the BBNJ Agreement derives its foundation from Article 8, subsection "j" of the CBD, probably also using Article 7 of Nagoya Protocol as a model, for regulating access to TK associated with MGRs in ABNJ. However, it also introduces novel elements, with a particular emphasis on the concept of involvement of IPLCs. Furthermore, as a provision within the framework of UNCLOS and part of an instrument that must be considered in its entirety, the interpretation of Article 13 necessarily differs from its counterparts in the CBD or Nagoya Protocol. This necessitates a nuanced approach that acknowledges both the existing legal framework governing access to TK associated with MGRs and the unique aspects arising from the context of ABNJ.

A close examination of Article 13 reveals a potential ambiguity concerning its scope. While the BBNJ Agreement generally refers to MGRs "of" ABNJ, Article 13 specifically references MGRs "in" ABNJ. This difference in prepositions restricts the geographical and temporal application of Article 13 to TK associated with MGRs physically located within ABNJ at the time of in situ collection.[3] This interpretation raises questions regarding the governance of TK associated with MGRs currently housed in laboratories or gene banks even if collected in ABNJ or MGRs that may have been transported from ABNJ to national waters through natural processes (e.g., ocean currents).[4] Furthermore, it is unclear whether "association" would require a link between the TK and the value or utilization of the MGRs or whether it would encompass a broader range links.

Article 13 of the BBNJ Agreement bestows state parties with three distinct avenues for governing access to TK of IPLCs associated with MGRs in ABNJ: national laws, administrative measures, or policy measures, either individually or in combination. A key aspect of this obligation is the initial determination by each state party of whether it is relevant and appropriate to regulate access to TK associated with MGRs which, ideally, should informed by consultations with relevant stakeholders and relevant studies. A second consideration concerns the potential human rights implications associated with the chosen implementation mechanism. While

[3] See Chaps. 3 and 4.

[4] See Chap. 14.

administrative and policy measures may offer more expeditious enforcement, they also raise the possibility of encroaching upon individual or communal rights, particularly if intellectual property rights are involved if they have been legally recognized (Morgera, 2018; Morgera & Tsioumani, 2010).

In contrast to the general multilateral mechanism for the collection of MGRs under Article 12, which focuses on batch identifiers, notification obligations, and the clearing-house mechanism (CHM),[5] Article 13 prescribes that national measures adopted by state parties shall ensure that access to TK of IPLCs associated with MGRs in ABNJ follows the principles of "free, prior, and informed consent" (FPIC) or "approval and involvement."[6] This distinction creates a dual-mechanism governing the collection of MGRs within Part II of the BBNJ Agreement, with a general mechanism under Article 12 and a special one under Article 13. The latter retains a contractual and private law basis for access to TK of IPLCs, potentially leading to a wide range of approaches across state parties (Leary, 2023). In cases where the state is also a member of the Nagoya Protocol and has enacted legislation regulating the "free, prior, and informed consent," it is likely that the existing legislation might be amended to cover access to MGRs in ABNJ.

The concept of FPIC entails the requirement of obtaining a qualified consent from IPLCs by all parties involved in the creation of this knowledge. While Member States have flexibility in determining the modalities and procedures for obtaining FPIC, these terms must be interpreted in accordance with their ordinary meaning within the context of the BBNJ Agreement and its overall objectives. In the absence of specific guidance within the BBNJ Agreement and from its COP, interpretations of FPIC established under the CBD can serve as a valuable legal source. According to the CBD COP, "free" denotes the absence of coercion, "prior" signifies respect for customary decision-making processes within IPLCs, "informed" requires information sharing in a manner understandable to these communities, and "consent" must be granted by the duly authorized authorities within the communities (CBD/COP/13/25).[7] The interpretation of "approval and involvement," which can be used as a substitute for FPIC, should be subject to similar level of scrutiny. Involvement entails a minimum level of participation and an active role for IPLCs in the decision-making process regarding the utilization of their TK in research, even if the research does not ultimately progress to the utilization phase. However, while the concept of involvement has a historical background, most discussions and legal evolution have centered on terrestrial genetic resources from plants, animals, and microorganisms (Scott, 2018) with reduced precedents to establish a baseline to assess the effectiveness of an involvement process of resources in ABNJ.

Article 13 of the BBNJ Agreement also contemplates a potential role for the CHM established under Article 51 in facilitating access to TK associates to MGRs in ABNJ. The modalities for such facilitation are yet to be determined by the BBNJ COP; however, considering that the CHM will be an open-access platform, two potential avenues for incorporating TK-related information can be envisioned. Under the first option, prior to the conduct of any research activity, the CHM will provide links to existing other clearing-house mechanisms or gene banks containing information concerning national TK associated with MGRs in ABNJ to the CHM or the COP (Article 51.3. "c"). Under the second option, if the existence of TK associated with MGRs in ABNJ comes to light after the in situ collection, the state party conducting the research would then be responsible for sharing this information with the CHM.[8] Another important consideration is that the voluntary language

[5] See Chaps. 6 and 12.

[6] Some commentators have coined the term ´facilitating prior informed consent´ to describe the process, yet it has not been immune to criticism (Perrault 2004; European Union Regulation N. 511/2014).

[7] See Chap. 14.

[8] See Chap. 14.

used in this obligation suggests that parties discretion to opt for alternative tools to promote access to such knowledge, potentially impeding the monitoring of compliance with Article 13.

The last sentence of Article 13 establishes that access to and the use of TK associated with MGRs in ABNJ shall be on mutually agreed terms (MAT), mirroring the system under the CBD and Nagoya Protocol. Notably, the reference to "use" instead of "utilization" (as defined in Article 1.14) suggests a deliberate choice to exclude TK from the multilateral Access and Benefit-Sharing (ABS) mechanism. This exclusion offers potential advantages for IPLCs, as benefits can flow directly to them, bypassing the yet-to-be-determined eligibility criteria for the multilateral ABS mechanism. However, it also necessitates dedicated efforts from state parties to regulate this topic domestically and from IPLCs to negotiate bilateral agreements securing monetary and non-monetary benefits—potentially resulting in a wide range of approaches to the MAT clause (Leary, 2023).

Notably, Article 13 and the BBNJ Agreement are silent on the access and use of TK associated with DSI of MGRs. This omission stands in contrast to the approach under the CBD, where there seems to exist consensus on the imperative to uphold the rights of IPLCs concerning TK associated with MGRs and the data related to them (CBD/WGDSI/1/3). Digital sequence information can establish a direct link with TK, primarily through the "in situ" collection process. When scientists or researchers engage with indigenous people, local communities, or their members regarding the attributes of a biological resource containing genetic components with active elements and the subsequent synthesis of these components, traditional knowledge becomes intertwined with digital sequence information from these genetic resources (Vierros et al., 2020). A case in point is the "Jeevani" case, wherein indigenous people and local communities created the knowledge forming the basis for the eventual "digital sequence information" (Anuradha, 1998).

A distinct issue arises concerning the acknowledgment of a manifestation of willingness to share such knowledge by these indigenous or local communities. While the BBNJ Agreement operates under the premise of common knowledge, it remains unclear whether traditional knowledge qualifies as the shared knowledge of a community. Unlike knowledge produced by scientists, TK arises from accumulated observations regarding the use of genetic resources within biological resources. This raises the crucial question of ownership and delineates the complexities surrounding who possesses what.

8.5 Considerations for How the Article Might Be Implemented in Practice

Effective implementation of Article 13 necessitates a robust national framework. States Parties can utilize legislative, administrative, or policy measures, alone or jointly, to establish a set of fundamental prerequisites for entities seeking access to TK. These prerequisites should include guidance on the application of FPIC and MAT for access and use of such knowledge, to facilitate negotiations among relevant stakeholders to ensure fair and equitable terms. Furthermore, strong mechanisms for review and assessment are crucial to monitor compliance with these prerequisites and ensure the application of this rule.

Additionally, discussions at the levels of the COP, the Access and Benefit-Sharing Committee, and the Scientific and Technical Body should (i) address the criteria for associating TK held by IPLCs with MGRS in ABNJ, (ii) establish a mechanism for exchanging information on the legislative, administrative, and policy measures adopted nationally, (iii) clarify the mandate of the CHM with respect of facilitating access to TK; (iv) provide guidance on the meaning of FPIC and MAT for the purpose of the BBNJ Agreement; and (v) discuss the link of TK associated with DSI on MGRs in ABNJ. Insights gained from experiences in other frameworks may provide useful guidance in this regard.

8.6 The Barossa Gas Project: A Case Study of Traditional Knowledge in Areas Beyond National Jurisdiction

The Barossa Gas Project, spearheaded by energy company Santos, proposes the extraction of natural gas from the Barossa gas field located in the Timor Sea, off the north coast of Australia. The project entails transporting the extracted gas via underwater pipelines to mainland Australia for processing and subsequent export (What is Santos's Barossa Gas Project Proposal? Why is it cast in uncertainty?, 2023).

This large-scale resource extraction initiative has encountered significant opposition. Environmental groups and civil society organizations have voiced concerns regarding the project's potential to generate high levels of greenhouse gas emissions, undermining Australia's climate targets (ibid). Furthermore, the project faced a legal challenge in the Australian Federal Court brought forth by a Tiwi Traditional Owner, who feared the potential for pipeline construction to disrupt submerged cultural heritage sites considered sacred by the Tiwi people (Maxwell, 2023; McClymont, 2023). The claimants requested the project's suspension until the potential damage to the underwater cultural heritage sites and sacred dreaming place were sufficiently assessed.

In early 2024, the Court issued a decision in favor of the Barossa Gas Project, overruling the claims brought forth by the Tiwi Traditional Owner, and paving the way for the project's resumption (Dick, 2024). The Court's reasoning centered on the absence of compelling evidence. The Tiwi claim lacked both archeological proof of cultural heritage sites along the pipeline route and compelling evidence that their songlines, ancestral pathways with spiritual significance, extended into the specific sea country impacted by the project (ibid).

This case exemplifies the complex relationship between TK in ABNJ. While the case underscores the existence of connections between TK and the marine environment in ABNJ, it also highlights the challenges faced by IPLCs in presenting evidence in a manner that aligns with Western legal and scientific frameworks. The burden of proof often falls on these groups to demonstrate the validity and location of their TK, which can be difficult when their knowledge systems rely on oral traditions and spiritual connections to the land and sea, rather than on tangible or quantifiable evidence.

References

Agreement under the United Nations Convention on the Law of the Sea on the Conservation and Sustainable Use of Marine Biological Diversity of Areas beyond National Jurisdiction, New York, 19 of June 2023, C.N.203.2023.TREATIES-XXI.10 of 20 July 2023. https://treaties.un.org/doc/Treaties/2023/06/20230620%2004-28%20PM/Ch_XXI_10.pdf. Accessed 27 Sept 2023

Anuradha, R. V. (1998). *Sharing with the Kanis: A case study from Kerala, India* (pp. 5–7). IUCN.

Convention on Biological Diversity, opened for signature 5 June 1992, 1760 U.N.T.S. 79 (entered into force 29 December 1993).

Convention on Biological Diversity, *Report of the ad hoc open-ended working group on benefit-sharing from the use of digital sequence information on genetic resources on its first meeting*, CBD/WGDSI/1/3, [46].

Convention on Biological Diversity, *Report of the thirteen meeting of the conference of the parties to the convention on biological diversity*, CBD/COP/13/25, [210], Decision XIII/18, Mo'otz Kuxtal Voluntary Guidelines.

Crawford, J. (2019). *Brownlie's principles of international law*. Oxford University Press.

Dick, S. (2024). *Santos wins legal battle against Tiwi Islands elders over $5.7B Barossa gas pipeline*. ABC News. https://www.abc.net.au/news/2024-01-15/santos-wins-barossa-project-battle-against-tiwi-islanders/103320182. Accessed 30 April 2024.

Dixon, M., McCorquodale, R., & Williams, S. (2016). *Cases and materials on international law*. Oxford University Press.

European Union Regulation N. 511/2014 of 16 April 2014 on compliance measures for users from the Nagoya Protocol on Access to Genetic Resources and the Fair and Equitable Sharing of Benefits Arising from their Utilization in the Union, *Official Journal of the European Union*, L 150/159.

ILO Convention 169 Concerning Indigenous and Tribal Peoples Convention of 1989 opened for signature 27 of June 1989, International Labour Conference, (entered into force 5 September, 1991).

International Treaty on Plant Genetic Resources for Food and Agriculture, open for signature 3 November 2001, FAO Res. 3/2003 (entered into force 29 June 2004).

Lawson, C., Rourke, M., & Humphries, F. (Eds.). (2022). *Access and benefit sharing of genetic resources, information and traditional knowledge* (1st ed.). Routledge.

Leary, D. (2023). Mutually agreed terms, arctic genetic resources and private international law. In: *The Routledge handbook of polar law* (pp. 596–612). Routledge.

Madhavan, N. (1998). *India: Biodiversity: Patents, Reuters LTD: New York.* https://wgbis.ces.iisc.ac.in/biodiversity/Environ_sys/doc97html/biodindia313.html. Accessed Feb 2024.

Maxwell, R. (2023). *Tiwi elders ask Plibersek to protect cultural heritage.* The Canberra Times. https://www.canberratimes.com.au/story/8398047/tiwi-elders-ask-plibersek-to-protect-cultural-heritage/. Accessed 30 April 2024.

McClymont, M. (2023). *Tiwi traditional owner files federal court case to protect underwater cultural heritage from Santos' barossa gas pipeline.* Environmental Defenders Office. https://www.edo.org.au/2023/10/31/tiwi-traditional-owner-files-federal-court-case-to-protect-underwater-cultural-heritage-from-santos-barossa-gas-pipeline/. Accessed 30 April 2024.

Morgera, E. (2018). Dawn of new day: The evolving relationship between the convention on biological diversity and international human rights law. *Wake Forest Law Review, 53*(4), 691–712.

Morgera, E., & Tsioumani, E. (2010). The evolution of benefit sharing: Linking biodiversity and community livelihoods. *Review of European, Comparative and International Environmental Law, 19*(2), 150–173.

Mulalap, C. Y., Frere, T., Huffer, E., Hviding, E., Paul, K., Smith, A., & Vierros, M. K. (2020). Traditional knowledge and the BBNJ instrument. *Marine Policy, 122*, 104103.

Nagoya Protocol on Access to Genetic Resources and the Fair and Equitable Sharing of Benefits Arising from their Utilization to the Convention on Biological Diversity, adopted 29 October 2010, 3008 U.N.T.S 3 (entered into force 12 October 2014) UNEP/CBD/COP/DEC/X/1.

Pena-Neira, S. (2017a). Recursos genéticos de animales, plantas y microorganismos y su regulaci6n internacional, Olejnik, Buenos Aires.

Pena-Neira, S. (2017b). Sharing the benefits of marine genetic resources for conservation? *Ocean and Coastal Management, 146*, 132.

Perrault, A. (2004). Facilitating prior informed consent in the context of genetic resources and traditional knowledge. *Sustainable Development Law and Pol'y, 4*, 21.

Popova, E., Vousden, D., Sauer, W. H., Mohammed, E. Y., Allain, V., Downey-Breedt, N., et al. (2019). Ecological connectivity between the areas beyond national jurisdiction and coastal waters: Safeguarding interests of coastal communities in developing countries. *Marine Policy, 104*, 90–102.

Radovich, V. S. (2023). Indigenous peoples and local communities' participation provisions in negotiations on conservation of marine areas beyond national jurisdiction. In: *Sustainability through Participation?* (pp. 406–432). Brill Nijhoff.

Sands, P., & Peel, J. (2018). *Principles of international environmental law.* Cambridge University Press.

Scott, J. (2018). *Overview of relevant work in the context of the convention on biological diversity and the Nagoya protocol concerning prior informed consent or approval and involvement, equitable sharing of benefits and unauthorized access.* OCHR. https://www.ohchr.org/sites/default/files/Documents/Issues/IPeoples/EMRIP/FPIC/ScottJohn.pdf. Accessed Feb 2024.

Secretariat of the Convention on Biological Diversity. (2019). *Glossary of relevant key terms and concepts within the context of Article 8(j) and related provisions*, (p. 9).

Tata, P. (2001). *Basmati victory for India.* New Scientist. https://www.newscientist.com/article/mg17123061-900-basmati-victory-for-india/. Accessed Feb 2024.

United Nations Convention on the Law of the Sea, opened for signature 10 December 1982, 1833 U.N.T.S 3 (entered into force 16 November 1994).

United Nations Declaration on the Rights of Indigenous Peoples, resolution adopted by the General Assembly in 2 October 2007, A/RES/61/295.

Vienna Convention on the Law of Treaties, opened for signature 23 May 1969, 1155 U.N.T.S. 331 (entered into force 27 January 1980).

Vierros, M. K., Harrison, A. L., Sloat, M. R., Ortuño Crespo, G., Moore, J. W., Dunn, D. C., Otq, Y., Cisneros-Montemayor, A. M., Shllinger, G. L., Kehaulani Watson, T., & Govan, H. (2020). Considering indigenous peoples and local communities in governance of the global ocean commons. *Marine Policy, 119*, 104039.

Western Sahara, Advisory Opinion, 1.C.J. Reports 1975, p. 12

What is Santos's Barossa Gas Project Proposal? Why is it cast in uncertainty? (2023) Australian Conservation Foundation. https://www.acf.org.au/what-is-santos-barossa-gas-project. Accessed 30 April 2024.

World Intellectual Property Organization. (2023). *Traditional knowledge division.* https://www.wipo.int/about-wipo/en/activities_by_unit/index.jsp?id=122. Accessed Feb 2024.

Sergio Pena-Neira (Ph.D. in Law, UNIA, Spain, 2014 and Postdocs in King´s College London, YTLCPPL, and University of Cambridge, LRCFIL) is an associate professor at Universidad Mayor, Chile. He has published widely articles, chapters and books on genetic resources, traditional knowledge and climate change.

Luciana Fernandes Coelho is an independent consultant on ocean governance and law of the sea. She has previous experience providing legal advice and consultancy services for several international and civil society organizations. She holds a PhD in maritime Affairs from the World Maritime University, Sweden.

Part II
Treaty Implementation in Practice

The Place of Intellectual Property Under the BBNJ Agreement

9

Abbe E. L. Brown

Abstract

This chapter explores the place of intellectual property (IP) in the negotiations of the BBNJ Agreement and the approach ultimately taken in the BBNJ Agreement. The chapter considers possible steps which could be taken in implementing the BBNJ Agreement in relation to IP regarding access and benefit sharing in respect of marine genetic resources in digital and physical form and to the interaction of the BBNJ Agreement with other international agreements.

Keywords

BBNJ Agreement · Intellectual property · Marine genetic resources · Digital sequence information · Benefit sharing · Biodiversity beyond national jurisdiction

9.1 Introduction

The place of intellectual property (IP) in the Biodiversity Beyond National Jurisdiction (BBNJ) process has been uncertain and controversial from the very start. Indeed, it has been said regarding the marine genetic resources (MGR) negotiations that "the most challenging topics were questions around intellectual property rights (IPR) and monetary benefits at the utilisation stage" (Mendenhall et al., 2023). There were suggested clauses relating to IP in most of the draft texts. Yet when the "ship reached the shore" in March 2023 (Carbon Brief, 2023), the *Agreement Under the United Nations Convention on the Law of the Sea on the Conservation and Sustainable Use of Marine Biological Diversity of Areas Beyond National Jurisdiction* (BBNJ Agreement) did not include a specific IP clause. Yet, there is one reference to "patents granted" in the context of benefit sharing and Clearing House Mechanism notifications [BBNJ Agreement, Article 12.8 (a)]. Further, and reflecting the engagement with IP during the negotiations, it will be suggested in this chapter that IP still has a role in implementing the BBNJ Agreement—on the basis of the Article 12 notifications regarding patents, and also more widely.

This chapter will introduce IP and its relationship with BBNJ, explore the history of the negotiations relating to IP in the BBNJ process, and suggest paths which could be taken to IP in implementation activity.

A. E. L. Brown (✉)
World Commission On Environmental Law, Gland, Switzerland
e-mail: abbe.brown@abdn.ac.uk

F. Humphries (ed.), *Decoding Marine Genetic Resource Governance Under the BBNJ Agreement*, Sustainable Development Goals Series, https://doi.org/10.1007/978-3-031-72100-7_9

9.1.1 An Introduction to IP

IP rights give their owners, who are frequently private entities, the power to control the results of innovation and creativity if threshold requirements are met for the rights to exist. Key IP rights are patents in respect of inventions; trademarks regarding brands, such as logos and words[1]; and database rights and copyright regarding sets of information and words.

From the BBNJ perspective, "inventions" can cover software (in many cases), genetic engineering, genetic vectors, and the use of microorganisms—but in some countries, not gene sequences that replicate genetic information in the DNA of an organism.[2] For a patent to be granted (after a registration process), there must be novelty over the state of the art, inventive step (often termed obviousness), and the invention must be capable of industrial application. So, as an example, there could be a patent in the UK for a cancer drug based on MGR from the areas beyond national jurisdiction (ABNJ), and this can be in force for a maximum period of 20 years. This was the position before the BBNJ Agreement and the Agreement does not change this. When the cancer-related patent is in force, the consent of the patent owner must be obtained by anyone making, disposing of, or offering to do so, using or importing or keeping the product which is the invention—the drug.[3] Also important is that copyright and database rights can be relevant to information platforms and digital sequence information (DSI). Copyright and database rights arise automatically when threshold criteria are met (basically around originality—a lower test than novelty—and labour) and can restrict copying and use of information structures.[4] This type of power can be a significant practical issue for scientists seeking to share and obtain information.

IP regimes do often, however, have specific permitted exceptions to the rights, such that activities which would otherwise infringe will not do so. Continuing the patent example, in the UK, there is an exception in relation to acts done for experimental purposes[5] (which has been interpreted by courts as meaning non-commercial research).[6] The UK also has a Crown Use provision to enable a government department or anyone authorised in writing to do particular acts for the services of the Crown; this is similar to compulsory licencing (forced sharing).[7] Yet the possibility of exceptions which could bring about a more sharing-based approach—one which could be considered more consistent with equity—does not remove a basic fact: IP owners have power through a right, which they can take directly to court, to try to stop some activities of another, if the IP owners choose to so use the right. This issue has received increasing attention as commercial products were being developed based on MGR from ABNJ (Jaspars

[1] See e.g. "Abyssine" patents and trademarks, explored in Humphries et al. (2021). For an introduction to IP, see Brown et al. (2023).

[2] See *European Patent Convention* 1973 art 52 "as such"; Guidelines for Examination in the European Patent Office 2023 Part G-II para 3.1 need for a technical effect for an invention, EU Directive 98/44 on the legal protection of biotechnological inventions OJ L213/13, art 5(2); compare in the *United States Myriad v Association of Molecular Pathology* 569 US 576 (2013) and its relationship with *Mayo v Prometheus* 132 S Ct 1289 (2012) and compare again IP Australia summary https://www.ipaustralia.gov.au/patents/what-are-patents/what-biological-inventions-can-be-patented#:~:text=Patents%20aren't%20available%20for,was%20isolated%20or%20man%2Dmade (accessed 4 February 2024) and *D'Arcy v Myriad Genetics* [2015] HCA 35.

[3] UK Patents Act 1977, s60(1)(a) (UKPA).

[4] Jaspars and Brown (2021), pp. 110–112 and 116–118; *Infopaq International A/S v Danske Dagblades Forening* C-5/08 [2009] ECDR 16 125; *SAS Institute v World Programming* [2103] EWHC 69 (Ch) For database, including extraction, see Directive 96/9 on the legal protection of databases OJ L77/20; *British Horseracing Board v William Hill Organization Ltd* Case C-203/02 [2005] RPC 13; *77M v Ordnance Survey* [2019] EWHC 3007 (Ch); *CV-Online Latvia SIA v Melons* (C-762/19) [2021] ECHR 27; Bernier et al. (2023).

[5] UKPA, s 60(5)(b).

[6] See e.g. *Auchincloss v Agricultural and Veterinary Supplies* [1999] RPC 397.

[7] UKPA, s 55-58; and see article 31(b) TRIPS (permitting compulsory licensing—forced sharing—in some cases).

& Brown, 2021; Lubchenco & Haugan, 2023; Blasiak et al., (2020); Royal Society, 2017), and as patent landscape reviews suggested increasing numbers of potentially relevant patents were owned by a small number of companies from the Global North (Blasiak et al., 2018).

Stepping back, IP rights are relevant to activity which is carried out in ABNJ for two key reasons. Firstly, information enabling technology, DSI, databases, drugs, and the results of genetic engineering will be developed, shared, used, and commercialised in states or regions in states or regions where there are IP rights. Further, alongside IP having a longstanding (and also long challenged) set of theoretical bases (Fisher, 2001; Machlup, 1958; Machlup & Penrose, 1950), IP has since the 1990s been part of the World Trade Organization (WTO) through the TRIPS Agreement. This was controversial (Drahos, 2002), as IP had (and still has) its own international legal framework through the World Intellectual Property Organization (WIPO).[8] Accordingly, WTO members have obligations relating to IP regarding the existence and protection of rights, reflecting those introduced above[9] and these obligations are subject to the robust WTO enforcement provisions.[10] And many parties of the *United Nations Convention on the Law of the Sea* and potential BBNJ parties are also members of the WTO.[11]

The second key issue is the complex relationship between IP, traditional knowledge, genetic resources, and ABNJ. There have been long-running discussions at WIPO towards an instrument on intellectual property, genetic resources, and traditional knowledge associated with genetic resources. These questions build on views that some IP rights have been obtained in circumstances which misappropriate traditional knowledge and genetic resources; and potentially conversely, on arguments (though they are contested) that IP could be a path to protecting traditional knowledge. The WIPO negotiations explored disclosure of origin of genetic resources relating to inventions, which is also important for MGR, BBNJ, and access and benefit sharing (ABS). The place of disclosure, the WIPO Treaty which was agreed in 2024 (after the BBNJ Agreement) in respect of disclosure of origin and ABS, and the potential for fragmentation across regimes, are all considered below.[12]

9.1.2 Evolution of BBNJ's Approaches to IP

Against this backdrop, possible approaches to IP across in the BBNJ process ranged from there being no engagement with it at all, to considering that IP did not belong in the BBNJ rather at WIPO and the WTO, to requiring consistency with other international agreements, to having restrictions on when patents could be granted, and to addressing disclosure of origin of inventions in connection with delivery of benefit sharing in respect of MGR. One commentary noted, in respect of the final text that "one notable change was to completely eliminate the Article on Intellectual Property Rights, because of continued intransigent disagreement on how to address the relationship with other relevant agreements, especially as relates to developments at the World Intellectual Property Organization" (Mendenhall et al., 2023, p. 5). More detail of IP's journey in—and out of—the negotiations is now explored in some more

[8] WIPO website Inside WIPO accessed 3 February 2024.

[9] TRIPS art 9, 27(1) and see also 27(2) and 3(b), 30 regarding exceptions and see article 31(b) above regarding compulsory licences.

[10] TRIPS art 64.

[11] United Nations Convention on the Law of the Sea status of ratifications https://www.un.org/depts/los/convention_agreements/convention_overview_convention.htm and WTO Members and Observers https://www.wto.org/english/thewto_e/whatis_e/tif_e/org6_e.htm accessed 3 February 2024.

[12] An example of limiting IP (albeit in an IP fora) to deliver other goals took place at Marrakesh regarding the Marrakech Treaty to Facilitate Access to Published Works for Persons Who Are Blind, Visually Impaired or Otherwise Print Disabled (2016); see Brown and Waelde (2015) and Thambisetty (2020). See also in this book Kachelriess et al. Chap. 11 considering fragmentation.

depth, and there is also consideration in this book in Humphries et al. (2025a, b).

A 2010 letter[13] from the ad hoc open-ended informal working group[14] refers to having regard to IP,[15] and early BBNJ scholarship engaged with IP (Bonfanti & Trevisanut, 2011; Chiarolla, 2014; Heafey, 2014; Salpin & Germani, 2007). The 2017 Preparatory Committee[16] noted the possible relevance of the relationship with IP including in relation to benefit sharing,[17] and proposals continued to come from scholars (Thambisetty, 2018 compare Broggiato et al., 2018). The President's Aid to Discussions (for IGC 1, 2018) included an IP clause[18]; IGC2 (March 2019)'s President's Aid to Negotiations[19] included as options that there were to be no patents over MGR except where the resources were modified by human intervention in a product capable of industrial application[20]; that states shall take steps to ensure that users of MGR shall disclose the origin of MGR they utilise[21]; that applications which do not comply with that part of the agreement shall not be approved[22]; that states shall take steps to ensure that when applying for patents they propose benefit sharing agreements[23]; that states implement the instrument in a manner consistent with obligations under WIPO and WTO[24]; that states cooperate to ensure IP rights are supportive and not run counter to objectives of instrument[25]; and that there would be "no text".[26]

These clauses, and developments of them, continued to be proposed in formal texts between 2019 and 2023. The IGC3 (August 2019)'s draft text[27]; and IGC4's revised draft text (released early 2020,[28] considered in March 2022), included several IP options. There was a change in approach at IGC5 (August 2022): the further revised draft text had just one IP option[29] and similar approaches were taken during IGC5[30] and at IGC5bis, up to the updated draft text.[31] In final plenaries in March 2023, Palestine said that it saw IP as an important issue, as a proxy for Common Heritage of Mankind.[32] This was in the context of statements in February 2023, by Sierra Leone for the African group, calling for an equitable, fair, and universal approach to BBNJ as a

[13] A/65/68 Letter dated 16 March 2010 from the Co-Chairpersons of the Ad Hoc Open-ended Informal Working Group to the President of the General Assembly (Letter).

[14] Set up by 59/24 (2005) para 73.

[15] Letter, para 77.

[16] 3.2.3. of A/AC.287/2017/PC.4/2.

[17] Section B of A/AC.287/2017/PC.4/2.

[18] A/CONF.232/2018/3, 3.2.3.

[19] A/CONF.232/2019/1 (Aid).

[20] 3.2.3 option 1.1 Aid.

[21] 3.2.3 Option 1,2 Aid.

[22] 3.2.3 Option 1, 3A Aid.

[23] 3.2.3, Option 1, 3B Aid.

[24] 3.2.3 Option II, 1 Aid.

[25] 3.23 Option II, 2 Aid.

[26] 3.2.3, Option III Aid; side event including author; report at https://www.abdn.ac.uk/ncs/departments/chemistry/bbnj/l; Brown (2019); IISD Earth News Bulletin (2019) summarised the breadth of approaches taken to inclusion of IP in BBNJ, including that it was better suited to considered in other fora: EU, Canada, Switzerland, Norway, Holy See, Japan, Korea, Russian Federation, Australia do not support IP rights being included.

[27] A/CONF.232/2019/6 (Draft Text); Art 12.1, 12.2, 12.3, Art 12.4 (a), (b) (c); and see Bengoa Rojas and Yentchare (2023).

[28] A/CONF.232/2020/3, art 12.1, 12.3, 12.4.

[29] A/CONF.232/2022/5, Art 12.

[30] See 21 August 2022 A/CONF.232/2022/CRP1.12 and Add.1 and 26 August 2022 A/CONF.232/2022/CRP.13 and A/CONF.232/2022/CRP.13/Add.1.

[31] A/CONF.232/2023/2 https://www.un.org/bbnj/sites/www.un.org.bbnj/files/aconf-232-2023-2_track_changes_en.pdf and A-CONF-232-2023-CRP1-Add1 (un.org), art 10 (6) (e), art 12.

[32] See recording 1 March 2023 Intergovernmental Conference https://media.un.org/en/asset/k11/k112p0e47q?_gl=1*1hbbsr5*_ga*MjAzOTI0MTI5Mi4xNjg3NDI1MTg3*_ga_TK9BQL5X7Z*MTY5NTEzNDIyMi4xMS4xLjE2OTUxMzQzNzUuMC4wLjA.

whole.[33] Further, through the negotiation process, new suggestions were also put forward by scholars and in policy briefs exploring possible approaches to IP (Humphries et al., 2021; Humphries, 2023; Langlet et al., 2023; Langlet & Dunshirn, 2023; Millicay, 2020; Morris-Sharma, 2020; Oldham et al., 2023; Towards a Package, 2020; Thambisetty, 2022; Thambisetty, 2023) and see reflections in Kanu (2023).

Ultimately, as noted, there was no specific IP clause at all in the final text of the BBNJ Agreement. Yet this has not removed the underpinning issues relating to IP, nor the suggested links between IP, common heritage of humankind, equitable approaches, and equity.[34] Accordingly, there could be a key role for IP in relation to implementation. Three issues are now explored in this respect: disclosure of origin and the BBNJ Agreement, interaction with other regimes in relation to disclosure, and wider interaction with legal fields.

9.2 Implementation

Disclosure of origin can be argued to be important to the patent system and to benefit sharing, in general, for a number of reasons. The sharing of benefits[35] could draw from sales of, for example, pharmaceutical drugs (or the results of any other commercialisation) which are based on resources from a particular place. Benefits could be shared directly as a payment to the community or through for example a school being built in partnership with a community. More details on benefit sharing, particularly in ABNJ and its interface with activities in areas within national jurisdiction, are also considered in this book in Kachelriess et al. (2025) and, in relation to traditional knowledge, in Pena-Neira and Coelho (2025). Importantly here, there are arguments that to enable benefit sharing to come about, it is necessary to establish what genetic resource has been used—and that a path to addressing this can be disclosure, in a patent application, of the origin of the generic resources on which the invention is built (see also e.g. Arnaud-Haond, 2020). Further, there are views that requiring this disclosure of origin would enhance assessment of novelty, which was introduced above, and also play a wider valuable role in establishing trust between communities, the private sector, and the patent system (DOSI, 2020). Several countries have indeed introduced the disclosure of origin requirement in their national patent systems (Castalia, 2018) but until the WIPO Treaty introduced above, there was no international obligation to do so.

The BBNJ Agreement does not include a disclosure of origin requirement. The BBNJ Agreement does, however, provide that parties are to ensure that when MGR of ABNJ, and where practicable DSI on such resources, are utilised, then the resulting patents granted, if available and to the extent possible, shall be notified to the Clearing House Mechanism as soon as information becomes available.[36] For this to be done, some form of engaging with patent offices regarding links between invention and BBNJ, or engagement with patent office databases, is likely to be needed. Such implementing steps in relation to ABNJ could also be aligned with record systems, where they exist, regarding the *Nagoya Protocol on Access to Genetic Resources and the Fair and Equitable Sharing of Benefits Arising From Their Utilization to the Convention on Biological Diversity* 2011 (Nagoya Protocol); with the more formal patent disclosure of origin systems if countries have them, as noted above; and also, in due course with obligations under

[33] See recording labelled 24 February 2023 Intergovernmental Conference https://media.un.org/en/asset/k1b/k1bd5e0htu?_gl=1*1moq9a3*_ga*MjAzOTI0MTI5Mi4xNjg3NDI1MTg3*_ga_TK9BQL5X7Z*MTY5NTEzNDIyMi4xMS4xLjE2OTUxMzQzMzUuMC4wLjA.

[34] See consideration in Thambisetty et al. (2023), pp. 48–49, 76–77.

[35] See also in this book Broggiato et al. (2025).

[36] See also in this book Muraki Gottlieb et al., (2025a).

the new WIPO Treaty.[37] Patents are likely, therefore, to be an important practical part of BBNJ Agreement implementation regarding disclosure of origin and benefit sharing and the building of new practice regarding ABNJ activity (see also in this book Lawson et al., 2025). This discussion raises the second issue: interaction with other legal regimes in relation to disclosure.

There have been negotiations at the WTO regarding TRIPS' interface with the *Convention on Biological Diversity* (CBD) and disclosure of origin since 2003, in which period there has been the CBD's Nagoya Protocol. These negotiations have not progressed (Chiarolla, 2019; Pavoni, 2014).[38] Until 2022, negotiations at WIPO under the Intergovernmental Committee on Intellectual Property and Genetic Resources, Traditional Knowledge and Folklore, the work of which included disclosure of origin, were not progressing to consensus.[39] These impasses made it clear that the issue of disclosure is contested in itself; and also that if disclosure of origin was considered to be important from the ABNJ perspective, it would have been valuable for a solution to be advanced in the BBNJ discussions.

This position changed. In 2022, there was a WIPO Chair' text of a Draft International Legal Instrument Relating to Intellectual Property, Genetic Resources, and Traditional Knowledge Associated with Genetic Resources,[40] accompanied by explanatory notes. These notes referred to ABNJ—in the context of being an example to which the suggested arrangements of disclosure (of a country of origin, people, or community)—would not be workable, and that in such cases, the source shall be provided in the patent application. In 2023, the Intergovernmental Committee and Preparatory Committee met and approved a draft text as a Basic Proposal to be considered at a Diplomatic Conference; and in May 2024, after 25 years of negotiation, the WIPO Treaty on Intellectual Property, Genetic Resources, and Associated Traditional Knowledge was adopted.[41] The Treaty does not refer to the Ocean or to ABNJ at all. Rather, it provides that when the claimed invention in a patent application is based on genetic resources or when the claimed invention is based on traditional knowledge associated with genetic resources, contracting parties shall require applicants to disclose the country of origin of the genetic resource or the indigenous peoples of local community who provided the knowledge; however, it goes on to provide that if this information is not known or there is no country of origin, the source of the genetic resources shall be disclosed.[42] Importantly, "source of genetic sources" is defined[43] as any source from which the applicant obtained genetic resources, and there are some non-exhaustive examples. The definition would cover ANBJ or a particular location in ABNJ,

[37] See in this book Broggiato et al. (2025), Humphries et al. (2025b), Kachelriess et al. (2025).

[38] See WTO Review 'Art 27.3 (b), traditional knowledge, biodiversity https://www.wto.org/english/tratop_e/trips_e/art27_3b_e.htm, and proposals for new article 29bis TN/C/W/59 of 2011.

[39] Intergovernmental Committee on Intellectual Property and Genetic Resources, Traditional Knowledge and Folklore https://www.wipo.int/tk/en/igc/ accessed 3 February 2024; genetic resources from ABNJ were in square brackets (indicating particular lack of agreement) regarding whether they would or would not be exceptions to any obligations WIPO/GRTKF/IC/28/4 (2014), art 4.1(e); WIPO/GRTKF/IC/30/4 2016, art 3.1(e); WIPO/GRTKF/IC/34/4 2017, art 4(1)(e).

[40] 2022 Chair Text WIPO/GRTKF/IC/43/5 and notes p 10 para 6 (b); draft report prepared by Chair of WIPO IGC on IPGRTKF 30 April 2019 https://www.wipo.int/edocs/mdocs/tk/en/wipo_grtkf_ic_40/wipo_grtkf_ic_40_chair_text.pdf.

[41] WIPO/GRTKF/IC/SS/GE/23/4 (Decisions 2023) and Basic Proposal for an International Legal Instrument Relating to Intellectual Property, Genetic Resources and Traditional Knowledge Associated with Genetic Resources GRATK/DC/3 (2023); Diplomatic Conference to Conclude an International Legal Instrument Relating to Intellectual Property, Genetic Resources and Traditional Knowledge Associated with Genetic Resources—https://www.wipo.int/diplomatic-conferences/en/genetic-resources/index.html (accessed 3 February 2024) and WIPO Treaty GRATK/DC/7.

[42] WIPO Treaty, arts 3.1 and 3.2.

[43] WIPO Treaty, art 2.

for example, identified by a global positioning system.

There is, therefore, an obligation in the WIPO Treaty to disclose origin, by source, in patent applications in respect of innovation based on[44] ABNJ. This will enhance transparency.[45] Yet the obligation in respect of source applies to "genetic resources". These are defined as "genetic material of actual or potential value"[46] and genetic material is defined as "any material of plant, animal, microbial, or other origin containing functional units of heredity".[47] It is suggested that this does not cover DSI—and noted above, DSI is increasingly important at a practical level in scientific research based on ABNJ. There is to be a review four years after entry into force including issues arising from new and emerging technology (WIPO Treaty, Article 8), and this may lead to DSI being considered. For now, however, there are significant limits on the scope of the disclosure obligation.

This links with the third issue for consideration regarding IP and BBNJ—wider interaction with legal fields. Reflecting the approach taken to DSI at WIPO, the South Centre closing statement said that it was "imperative to address remaining gaps such as the regulation of Digital Sequence Information (DSI) to prevent further exploitation without benefit sharing" (South Centre, 2024)—although this statement did not refer specifically to ABNJ and the Ocean. There is no substantive provision in the WIPO Treaty in relation to ABS. The Preamble (WIPO Treaty, Preamble para 5) provides, however, that the "Treaty and other international instruments related to genetic resources and traditional knowledge associated with genetic resources should be mutually supportive"; and the "Treaty shall be implemented in a mutually supportive manner with other international agreements relevant to this Treaty", with a footnote that "nothing in this Treaty shall derogate from or modify any other international agreement" (WIPO Treaty Article 7 and note 5).

The interaction between the BBNJ Agreement, IP, and other fields in relation to ABS, and looking beyond disclosure of origin, is considered elsewhere in this book.[48] IP also has a place in BBNJ Agreement implementation in wider interaction with legal fields. Firstly, this is through the "not undermining" provisions in the BBNJ Agreement. This covers "relevant legal instruments and frameworks and relevant global, regional, subregional, and sectoral bodies and that promotes coherence and coordination with those instruments, frameworks, and bodies" [BBNJ Agreement, Article 5(2)]. Further, the BBNJ Agreement provides that "parties shall endeavour to promote... the objectives of this Agreement when participating in decision-making under other relevant legal instruments and frameworks" [BBNJ Agreement, Article 8(2)].

The search for "coherence" could be a base for countries to choose to embrace flexibilities to patent and copyright through TRIPS and to create some, TRIPS-permitted, limits on IP rights. This could bring about a more sharing-based approach to the results of innovation and creativity. Such an approach could in turn be consistent with the BBNJ Agreement's provision that to achieve its objectives, parties shall be guided by inter alia,[49] the principle of the common heritage of humankind which is set out in the Convention [BBNJ Agreement, Article 7(b)], the principle of equity [BBNJ Agreement, Article 7(d)][50] and the use of the

[44] See WIPO Treaty, art 2 "means that the genetic resources and/or traditional knowledge associated with genetic resources must have been necessary for the claimed invention, and that the claimed invention must depend on the specific properties of the genetic resources and/or on the traditional knowledge associated with genetic resources".

[45] See WIPO Treaty, art 1(a) objective.

[46] WIPO Treaty, art 2.

[47] WIPO Treaty, art 2.

[48] See also in this book Kachelriess et al. (2025).

[49] Importantly, see also BBNJ Agreement art 7(c) freedoms of the high seas and see in this book Muraki Gottlieb et al., (2025a).

[50] These also include the fair and equitable sharing of benefits.

best available science and scientific information [BBNJ Agreement, Article 7(i)]. Of importance here is that the reference to common heritage of humankind can be argued, and by some is clearly hoped—see closing statements of the G77 and China (2023 Compilation 2023, p. 5) and African group (Compilation 2023, p. 7 and Kanu, 2023) including regarding inequities in approaches to negotiations)[51]—to be a means to bring about a more sharing, equitable approach to BBNJ (Carbon Brief, 2023). In contrast, the USA stated that "[w]e do not interpret anything in this Agreement as authorizing or permitting any waiver or undermining of existing intellectual property rights and obligations under international or national law; …. requiring mandatory disclosure in patent applications of the origin or source of marine genetic resources; or requiring compulsory licenses" (Compilation 2023, p. 113). The proposals made in this chapter do not go as far as this—but the comments of the USA are an important reminder of the diverging reviews which remain.

A path to explore possible new approaches to IP, if countries wished to do this, could be through the new ABS Committee established by the BBNJ Agreement (BBNJ Agreement, Article 15).[52] This committee is to comprise experts from a range of fields. It could explore new guidelines about forms of engaging with TRIPS, such as through encouraging greater use of existing voluntary IP licencing opportunities (e.g. those building on Creative Commons, Malaria Vaccine, and CAMBIA models).[53] There could also be exploration of the different approaches which can be taken to the concept of "openness" in the BBNJ Agreement: the term "open access" appears (without clarification) regarding non-monetary benefit sharing [BBNJ Agreement, Article 14(2)(c)]; there are references to publicly searchable and accessible forms [BBNJ Agreement, Article 14(2)(d)]; there are requirements that BBNJ standard identifiers are to be posted in publicly accessible repositories and databases (BBNJ Agreement, Article 15.3); and the Clearing House is to consist primarily of an open-access platform (BBNJ Agreement, Article 51). Yet as the author has explored, "open access" can mean so many different things (Jaspars & Brown, 2023). If one does not engage with IP rights (and their limits, new and potential), then the private (closed) nature of IP rights could be used to limit openness. This is an important area for creativity and practical support in the implementation process.

9.3 Conclusion

The relationship between and approaches taken to private power, Global North dominance, and a sharing and collective approach to addressing societal issues, run through the BBNJ Agreement negotiations. The limited engagement with IP in the final BBNJ Agreement has not removed the central importance of IP to these questions. One should not (be one activist, diplomat, lawyer, or scientist) adhere too readily to the view that one cannot interfere with IP rights or indeed engage with them. There is the potential for balance and exception within IP law itself; and this can assist in delivering outcomes which are more consistent with a common heritage of humankind-based approach to MGR and DSI. Yet the possible benefits of

[51] 5 March and 19 June 2023 Tweets Jeremy Raguain @Columbia_ESP and Tweet 5 March 2023 from Martin Kimani @KenyaMissionUN 5 March "The Africa group ran an amazing negotiation. There was intimidation, bullying, desperation, pleading, reasoning and at some point, tears were drawn from some quarters."

[52] And see in this book Muraki Gottlieb et al. (2025b).

[53] When we share, everyone wins—Creative Commons When we share, everyone wins - Creative Commons https://creativecommons.org/share-everyone wins/#:~:text=CC%20is%20an%20international%20nonprofit,a%20brighter%20future%20for%20all.; PATH's Malaria Vaccine Initiative | Home https://www.path.org/our-impact/resources/the-pathmalaria-vaccine-initiative/; Microsoft Word - BiOS License V1_5.doc (cambia.org) https://cambia.org/wp-content/uploads/2017/10/BiOS-License-and-Tech-Support-Agreement-version-1_5.pdf.

IP mean that it cannot readily be discounted or wholly attacked when considering, for example, how one might support new product supply chains relating to MGR and DSI from ABNJ—which, of course, may give rise to a benefit to be shared.

Establishing the place of IP will be a continuing journey towards new shores. The variety of views will continue and it should not be assumed parties reaching consensus means that the BBNJ Agreement necessarily involves a commitment to a common heritage of humankind focussed approach. The closing statements also saw the USA (Compilation 2023, p. 113) stating "[w]e support this Agreement creating a system for the fair and equitable sharing of benefits related to MGR of ABNJ, even though these resources are not the common heritage of humankind'.[54] In contrast, Palestine stated "[t]he international community must strive for further progress through the instrument´s governance and implementation, including benefit sharing applied across the entire instrument, disclosure of information on use of MGRs, and engagement on intellectual property rights in alignment with other established biodiversity instruments and the common heritage of humankind" (Compilation 2023, p. 98).

Nonetheless, the momentum at the end of the BBNJ Agreement negotiations could provide a different willingness from all to partner in new ways. The lack of an IP clause could be seen as a platform for practical action in implementation and as a valuable opportunity for the BBNJ community—policy maker, scientist, established industry, and spin-out company—to make IP work, fairly, to the benefit of all.[55]

Acknowledgements The author has sought to influence regard to IP in the BBNJ negotiations: see REF2021 'Embedding Intellectual Property Rights in the UN Marine Biodiversity Beyond National Jurisdiction Process'. In Impact Case Study Database) https://results2021.ref.ac.uk/impact/submissions/d78a9c70-995e-412d-8eaa-f758873a4f2a/impact and the' Song of the Ocean' Sharing the benefits of the ocean (abdn.ac.uk) (both accessed 3 February 2024); and she was a member of the IUCN delegation to the BBNJ negotiations. Alongside IP rights there are trade secrets; for reasons of space, trade secrets will not be explored separately in this chapter. This chapter is part of a project which has received funding from the European Union's Horizon 2020 research and innovation programme under Grant Agreement no. 101000392 (MARBLES). This output reflects only the author's view and the European Research Executive Agency (REA) cannot be held responsible for any use that may be made of the information contained. This chapter is also supported by UK Research and Innovation under the UK Government's Horizon Europe funding guarantee Grant No IFS 1007167 (University of Aberdeen). Views and opinions expressed are however those of the author(s) only and do not necessarily reflect those of the European Union or the European Research Executive Agency (REA) or UKRI. Neither the European Union nor the granting authority can be held responsible for them.

[54] See also Vadrot et al. (2022), regarding the US position that MGR negotiations undermine the current Intellectual Property Rights regime.

[55] See Collective Statement (2023) referring to equity and coherence.

References

A/CONF.232/2023/INF.5 3 August 23 compilation of statements (Compilation 2023)

Agreement Under the United Nations Convention on the Law of the Sea on the Conservation and Sustainable Use of Marine Biological Diversity of Areas Beyond National Jurisdiction open for signature 20 September 2023 C.N.203.2023 TREATIES-XXI.10 of 20 July 2023 A/CONF.232/2023/4* ("BBNJ Agreement").

Arnaud-Haond, S. (2020). Mind the gap between biological samples and marine genetic resources in areas beyond national jurisdiction: Lessons from land. In T. Heidar (Ed.), *BBNJ and MGRs: Practical solutions for benefit-sharing. New knowledge and changing circumstances in the law of the sea* (pp. 29–39). Brill.

Bengoa Rojas, C., & Yentchare, P.-Y. M. (2023). Multilateral matters #10: A sea of possibilities: Intellectual property considerations in the BBNJ negotiations (part two) UCT IP unit. http://ip-unit.org/2021/multilateral-matters-10-a-sea-of-possibilities-intellectual-property-considerations-in-the-bbnj-negotiations-part-two/. Accessed 3 Feb 2024.

Bernier, A., Busse, C., & Bubela, T. (2023). Public biological databases and the *Sui Generis* database right. *IIC, 43*, 1316–1358.

Blasiak, R., Wynberg, R., Grorud-Colvert, K., Thambisetty, S., et al. (2020). *The ocean genome: Conservation and the fair, equitable and sustainable use of marine genetic resources*. Washington, DC: World Resources Institute. Available online at www.oceanpanel.org/blue-papers/

ocean-genome-conservation-andfair-equitable-and-sustainable-use-marine-genetic

Blasiak, R., Jouffrey, J.-P., & Wabnitz, C. C. C. (2018). Corporate control and global governance of marine genetic resources. *Science Advances, 4*(6), 527. https://doi.org/10.1126/sciadv.aar5237

Bonfanti, A., & Trevisanut, S. (2011). TRIPS on the high seas: Intellectual property rights on marine genetic resources. *Brooklyn Journal of International Law, 37*, 215–220.

Broggiato, A., Dunshirn, P., Jaspars, M., & Pena-Neira, S. (2025). Monetary and non-monetary benefit sharing under the BBNJ Agreement. In F. Humphries (Ed.), *Decoding governance marine genetic resource governance under the BBNJ Agreement.* Springer.

Broggiato, A., Vanagt, R., Lallie, L., et al. (2018). *Mare Geneticum*: Balancing governance of marine genetic resources in international waters. *International Journal of Marine and Coastal Law, 12*, 3–33.

Brown, A. (2019). *Global challenges, interdisciplinary solutions*. https://www.abdn.ac.uk/law/blog/global-challenges-interdisciplinary-solutions-law-chemistry-and-biodiversity-beyond-national-jurisdiction-/

Brown, A., Kheria, S., Cornwell, J., & Iljadica, M. (2023). *Contemporary intellectual property: Law and policy*. Oxford University Press.

Brown, A., & Waelde, C. (2015). IP, disability, culture and exceptionalism: Does copyright law deal with difference? In G. Dinwoodie (Ed.), *Intellectual property and general legal principles* (pp. 215–245). Edward Elgar.

Carbon Brief. (2023). *Q and A: What does the high seas treaty mean for climate change and biodiversity*. https://www.carbonbrief.org/qa-what-does-the-high-seas-treaty-mean-for-climate-change-and-biodiversity/. Accessed 3 Feb 2024.

Castalia. (2018). Economic evaluation of disclosure of origin requirements. 3252-castalia-economic-assessment-evaluation-disclosure-origin-requirements-pdf (mbie.govt.nz) Report on the Compilation of Materials on Disclosure Regimes Relating to Genetic Resources and Associated Traditional Knowledge (wipo.int). Accessed 3 Feb 2024.

Chiarolla, C. (2014). Intellectual property rights and benefit sharing from marine genetic resources in areas beyond national jurisdiction: Current discussions and regulatory options. *Queen Mary Journal of Intellectual Property, 4*(3), 171–194.

Chiarolla, C. (2019). Intellectual property from a global environmental law perspective. Key lessons from the implementation of patent disclosure requirements for genetic resources and traditional knowledge. *Transnational Environmental Law, 8*(3), 503–521.

Collective Statement. (2023). The High Seas Treaty from negotiation to implementation" 6–7 October 2023. https://www.iatlantic.eu/bbnjsymposium/#:~:text=ENDORSE%20THE%202023%20SYMPOSIUM%20COLLECTIVE%20STATEMENT&text=This%20landmark%20'BBNJ%20Agreement'%20represents,thirds%20of%20the%20world's%20ocean

Convention on Biological Diversity, opened for signature 5 June 1992, 1760 UNTS 79 (entered into force 29 December 1993).

DOSI. (2020). *Intellectual property rights: Implications for deep-ocean stewardship*. https://www.dosi-project.org/wp-content/uploads/070-DOSI-Policy-brief-Intellectual-Property-Rights-V2-web1.pdf. Accessed 3 Feb 2024.

Drahos, P. (2002). Negotiating intellectual property rights: Between coercion and dialogue. In P. Drahos, & R. Mayne (Eds.), *Global intellectual property rights. Knowledge, access and development* (pp. 161–182). Palgrave Macmillan.

European Patent Convention 1973 1065 UNTS 199 (entered into force 7 October 1977).

Fisher, W. (2001). Theories of intellectual property. In S. R. Muinzer (Ed.), *New essays in the legal and political theory of property*. Cambridge University Press.

Heafey, E. (2014). Access and benefit sharing of marine genetic resources from areas beyond national jurisdiction: Intellectual property—friend. *Not Foe, 14*(2), 5469.

Humphries, F., Berry, T., & Muraki Gottlieb, H. (2025a). Bridging divides: The evolution of marine genetic resources governance beyond national jurisdiction. In F. Humphries (Ed.), *Decoding marine genetic resource governance under the BBNJ Agreement.* Springer.

Humphries, F., Jaspars, M., Lavelle, J., & Kachelriess, D. (2025b). The novel notification information system for marine genetic resources under the BBNJ Agreement. In F. Humphries (Ed.), *Decoding governance marine genetic resource governance under the BBNJ Agreement.* Springer.

Humphries, F. (2023). Commentary on the August 2022 draft text marine genetic resource element of the treaty on marine biodiversity beyond national jurisdiction. *SSRN*. https://doi.org/10.2139/ssrn.4337883

Humphries, F., Rabone, M., & Jaspars, M. (2021). Traceability approaches for marine genetic resources under the proposed ocean (BBNJ) treaty. *Frontiers in Marine Science, 8*, 661313. https://doi.org/10.3389/fmars.2021.661313

IISD Earth News Bulletin. (2019). Summary report 25 March to 5 April 2019. https://enb.iisd.org/events/2nd-session-intergovernmental-conference-igc-conservation-and-sustainable-use-marine/summary. Accessed 3 Feb 2024.

Intergovernmental Committee on Intellectual Property and Genetic Resources, Traditional Knowledge and Folklore *CONSOLIDATED DOCUMENT RELATING TO INTELLECTUAL PROPERTY AND GENETIC RESOURCES* WIPO/GRTKF/IC/28/4 (2014).

Intergovernmental Committee on Intellectual Property and Genetic Resources, Traditional Knowledge and Folklore *CONSOLIDATED DOCUMENT RELATING*

TO INTELLECTUAL PROPERTY AND GENETIC RESOURCES WIPO/GRTKF/IC/30/4 (2016).

Intergovernmental Committee on Intellectual Property and Genetic Resources, Traditional Knowledge and Folklore *CONSOLIDATED DOCUMENT RELATING TO INTELLECTUAL PROPERTY AND GENETIC RESOURCES* WIPO/GRTKF/IC/34/4 (2017).

Intergovernmental Committee on Intellectual Property and Genetic Resources, Traditional Knowledge and Folklore, *Chair's Text of a Draft International Legal Instrument Relating to Intellectual Property, Genetic Resources and Traditional Knowledge Associated with Genetic Resources* WIPO/GRTKF/IC/43/5 (2022).

Intergovernmental Committee on Intellectual Property and Genetic Resources, Traditional Knowledge and Folklore *Basic Proposal for an International Legal Instrument Relating to Intellectual Property, Genetic Resources and Traditional Knowledge Associated with Genetic Resources* GRATK/DC/3 (2023).

Intergovernmental Committee on Intellectual Property and Genetic Resources, Traditional Knowledge and Folklore *Decisions of Special Session* WIPO/GRTKF/IC/SS/GE/23/4 (2023).

Jaspars, M., & Brown, A. (2021). Benefit sharing: Combining intellectual property, trade secrets, science and an ecosystem-focused approach. In M. Nordquist & R. Long (Eds.), *Marine biodiversity of areas beyond national jurisdiction* (pp. 97–130). Brill Nijhoff.

Jaspars, M., & Brown, A. (2023). What should we mean by 'open access.' In C. Lawson, M. Rourke, & F. Humphries (Eds.), *Access and benefit sharing of genetic resources, information and traditional knowledge* (pp. 89–111). Routledge.

Kachelriess, D., Dunshirn, P., Langlet, A., Brown, A. E. L., & Scholz, A.H. (2025). Marine genetic resources and digital sequence information under the BBNJ Agreement: Interlinkages with other access and benefit sharing frameworks. In F. Humphries (Ed.), *Decoding marine genetic resource governance under the BBNJ Agreement.* Springer.

Kanu, M. I. (2023). The Ship has reached the shore and the contribution of the African Group. https://www.ftbchambers.co.uk/elblog/view/the-ship-has-reached-the-shore-the-bbnj-treaty-and-contributions-of-the-african-group. Accessed 25 April 2023

Langlet, A., & Dunshirn, P. (2023). Traceability options for marine genetic resources for areas beyond national jurisdiction. https://www.highseasalliance.org/wp-content/uploads/2023/02/traceability-options-paper-1.pdf. Accessed 3 Feb 2024.

Langlet, A., Dunshirn, P., Kachelriess, D., & Currie, D. (2023). Non-paper: Marine genetic resources, including questions on the sharing of benefits—a brief on broad options for benefit sharing. https://highseasalliance.org/resources/non-paper-marine-genetic-resources-including-questions-on-the-sharing-of-benefits-a-brief-on-broad-options-for-benefit-sharing/. Accessed 3 Feb 2024

Lawson, C., Humphries, F., Jaspars, M., & Rabone, M. (2025). Data management and the 'BBNJ Standardized Batch Identifier' under the BBNJ Agreement. In F. Humphries (Ed.), *Decoding marine genetic resource governance under the BBNJ Agreement.* Springer.

Lubchenco, L., & Haugan, H. (2023). The ocean genome: Conservation and the fair, equitable and sustainable use of marine genetic resources the blue compendium.

Machlup, F. (1958). An economic review of the patent system. US Senate, 85 Congress Second Session No 15, 80.

Machlup, F., & Penrose, E. (1950). The patent controversy in the nineteenth century. *Journal of Economic History, 10*(1), 4155.

Marrakech Treaty to Facilitate Access to Published Works for Persons Who Are Blind, Visually Impaired or Otherwise Print Disabled (entered into force 30 September 2016).

Marrakesh Agreement Establishing the World Trade Organization. (1994). 1867 UNTS 154, Art. II and Annex IC *Agreement on Trade-Related Aspects of Intellectual Property Rights* (TRIPS Agreement) (entered into force 1 January 1995).

Mendenhall, E., Tiller, R., & Nyman, E. (2023). The ship has reached the shore: The final session of the 'Biodiversity beyond National Jurisdiction' negotiations. *Marine Policy, 155*, 105686.

Microsoft Word—BiOS License V1_5.doc (cambia.org) (Cambia)

Millicay, F. (2020). Marine genetic resources of areas beyond national jurisdiction and intellectual property rights. In T. Heidar (Ed.), *BBNJ and MGRs: Practical solutions for benefit-sharing new knowledge and changing circumstances in the law of the sea* (pp. 65–78). Brill.

Morris-Sharma, N. (2020). BBNJ and MGRs: Practical solutions for benefit-sharing. In T. Heidar (Ed.), *New knowledge and changing circumstances in the law of the sea* (pp. 79–98). Brill.

Muraki Gottlieb, H., Kachelriess, D., & Slobodian, L. (2025a). Understanding the preamble, principles and objectives of the BBNJ Agreement: A focus on the fair and equitable sharing of benefits of marine genetic resources. In F. Humphries (Ed.), *Decoding marine genetic resource governance under the BBNJ Agreement.* Springer.

Muraki Gottlieb, H., Ardron, J., & Brown, A. E. L. (2025b). BBNJ agreement: a new infrastructure to foster benefit sharing of marine genetic resources. In F. Humphries (Ed.), *Decoding governance marine genetic resource governance under the BBNJ Agreement.* Springer.

Nagoya Protocol on Access to Genetic Resources and the Fair and Equitable Sharing of Benefits Arising From Their Utilization to the Convention on Biological Diversity opened for signature 2 February 2011 3008 UNTS 3 (entered into force 12 October 2014).

Oldham, P., Chiarolla, C., & Thambisetty, S. (2023). Digital sequence information in the UN high seas treaty: Insights from the global biodiversity framework-related decisions. In *LSE law: Policy briefing paper No. 53*. https://doi.org/10.2139/ssrn.4343130

PATH's Malaria Vaccine Initiative | Home (Malaria Vaccine).

Pavoni, R. (2014). The Nagoya protocol and WTO law. In E. Morgera, M. Buck, & E. Tsioumani (Eds.), *UnRavelling the Nagoya protocol*. Brill.

Pena-Neira, S., & Coelho, L. F. (2025). Traditional Knowledge associated with marine genetic resources in Areas Beyond National Jurisdiction. In F. Humphries (Ed.), *Decoding governance marine genetic resource governance under the BBNJ Agreement*. Springer.

Royal Society. (2017). Future of oceans: Metal-rich minerals and genetics evidence pack 2017. https://royalsociety.org/-/media/policy/projects/future-oceans-resources/future-of-oceans-evidence-pack.pdf. Accessed 3 Feb 2024

Salpin, C., & Germani, V. (2007). Patenting of research results relating to genetic resources from areas beyond national jurisdiction: The crossroads of the law of the sea and intellectual property law. *Review of European Community and International Environmental Law, 16*(1), 12–23.

South Centre. (2024). Statement to the WIPO diplomatic conference on a treaty on intellectual property, genetic resources and associated traditional knowledge. https://www.southcentre.int/wp-content/uploads/2024/05/SC-Statement-WIPO-Treaty_24-May.pdf. Accessed 4 June 2024

Thambisetty, S. (2018). Marine genetic resources beyond national jurisdiction: Elements of a new international legally binding instrument. In *LSE law: Policy briefing paper No. 32*. https://doi.org/10.2139/ssrn.3219995

Thambisetty, S. (2022). Marine genetic resources beyond national jurisdictions: Negotiating options in intellectual property. In *South Centre 148 Research Paper*. https://www.southcentre.int/wp-content/uploads/2022/03/RP148_MARINE-GENETIC-RESOURCES-BEYOND-NATIONAL-JURISDICTIONS_EN.pdf. Accessed 3 Feb 2024

Thambisetty, S. (2023). The unfree commons: Freedom of marine scientific research and the status of marine genetic resources beyond national jurisdiction. In LSE Working Paper 24/2003 forthcoming Modern Law Review 2024.

Thambisetty, S., Oldham, P., & Chiarolla, C. (2023) Developing state positions in the making of the BBNJ treaty: An expert briefing document on marine genetic resources. https://ssrn.com/abstract=

Thambisetty, S. (2020). Biodiversity beyond national jurisdiction: (Intellectual) property heuristics. In M. Nordquist & R. Long (Eds.), *Marine biodiversity of areas beyond national jurisdiction* (pp. 131–145). Brill Nijhoff.

"Towards a Package" January 2020 BBNJ Edinburgh event summary and report. https://www.abdn.ac.uk/ncs/documents/workshop-report.pdf. Accessed 3 Feb 2024.

United Nations Convention on the Law of the Sea opened for signature 10 December 1982 1833 UNTS 3 (entered into force 16 November 1994).

Vadrot, A. B. M., Langlet, I., & von Wysocki, T. I. (2022). Who owns marine biodiversity? Contesting the world order through the 'common heritage of humankind' principle. *Environmental Politics, 31*(2), 226–250.

When we share, everyone wins—Creative Commons.

WIPO Treaty on Intellectual Property, Genetic Resources and Associated Traditional Knowledge GRATK/DC/7 open for signature 24 May 2024

Abbe Brown is a professor in Intellectual Property Law at the University of Aberdeen. Before returning to academia, she practised as an intellectual property and commercial litigator at leading firms in London, Melbourne and Edinburgh. Abbe has a strong interest in the ocean and in interdisciplinary research and is a member of the World Commission on Environmental Law and the Deep Ocean Stewardship Initiative.

Considerations Concerning State Ratification of the BBNJ Agreement

10

Jeff A. Ardron, Daniel Kachelriess, Christopher H. C. Lyal, Chilenye Nwapi, Muriel Rabone, Aysegul Sirakaya, and Alison Swaddling

Abstract

This paper presents procedural, policy, and legal considerations that states may encounter before and after ratifying the BBNJ Agreement, with a focus on its marine genetic resources (MGR) provisions. It briefly examines the behaviours of parties to the other two previously ratified implementing agreements to the United Nations Convention on the Law of the Sea, as well as some ratification lessons learnt from other relevant treaties. The other three 'pillars' of the BBNJ Agreement sit on well-established foundations of national and international law. However, as discussed here, the legal foundation of the fourth pillar, MGR, is less well established nationally or internationally, and much remains to be determined. Therein, the BBNJ Agreement represents a significant advancement of international law. However, the reality is that most parties will not adopt all the necessary new legislation prior to their ratification, nor can they, because many of the MGR provisions are as yet still unclear. States are therefore likely to choose a progressive approach, ratifying the Agreement to signal commitment while gradually developing the legal framework necessary for full compliance, as well as policies to guide its implementation.

Keywords

BBNJ ratification · MGR · Marine genetic resources · Nagoya protocol · CITES · ISA · DSM · UNFSA · Plant treaty

J. A. Ardron (✉)
Africa Oceans, The Nature Conservancy, Mombasa, Kenya
e-mail: jeff.ardron@tnc.org

D. Kachelriess
IUCN World Commission On Environmental Law, High Seas Alliance (Advisor), Vienna, Austria

C. H. C. Lyal · M. Rabone · A. Sirakaya
Department of Life Sciences, Natural History Museum, London, UK

C. Nwapi · A. Swaddling
The Commonwealth Secretariat London, London, UK

10.1 Introduction

In June 2023, after nearly 20 years of preparations, open-ended discussions, and negotiations, the text of the *Agreement under the United Nations Convention on the Law of the Sea on the conservation and sustainable use of marine biological diversity of areas beyond national jurisdiction* (BBNJ Agreement) was adopted by consensus (A/CONF.232/2023/4. See also: Druel and Gjerde, 2014). Opened for signature

F. Humphries (ed.), *Decoding Marine Genetic Resource Governance Under the BBNJ Agreement*, Sustainable Development Goals Series, https://doi.org/10.1007/978-3-031-72100-7_10

on 20 Sept. 2023, the BBNJ Agreement will enter into force 120 days after the date of deposit of the sixtieth ratification, approval, acceptance, or accession [henceforth, 'ratification' will be used to capture all of these possibilities, see Article 68(1)]. As of October 2024, more than 100 states and the European Union have *signed* the Agreement; and of these, 14 have also ratified it.[1]

This chapter presents considerations that states may encounter before and after ratifying the Agreement. (Note that much of the discussion here may also apply to regional economic integration organisations.) Regarding the adoption of national legislation necessary to implement the Agreement, the chapter briefly examines the behaviours of parties to the other two previously ratified implementing agreements to the United Nations Convention on the Law of the Sea (UNCLOS). Finally, some ratification lessons learnt from other treaties outside of the UNCLOS framework, and relevant to marine genetic resources (MGR) of areas beyond national jurisdiction (ABNJ), are outlined.

10.1.1 General Procedures for Ratification

Once the terms of a treaty have been agreed, the negotiating parties can sign it. By signing a treaty, the parties declare their willingness or intention to be bound by the treaty terms. The signing itself is, however, not binding on the signatories, although it obligates the signatories to refrain from taking steps likely to frustrate the treaty purpose (Moore, 2012). To be bound, a signatory state must take a further step to ratify the treaty, i.e. to express its consent to be bound by the treaty. Once a treaty is in force, which is determined when the conditions specified in the treaty are met (BBNJ Article 68), ratification of the treaty binds the ratifying state in its relations with other ratifiers as well as with third parties, insofar as those third parties are impacted by the provisions of the treaty. Ratification can influence domestic respect for the terms of the treaty through policy changes even before the treaty is incorporated into domestic law (Baccini and Urpelainen, 2014; Elkins et al., 2013). For instance, ratification can create a sense of obligation among domestic institutions to respect the terms of the treaty; domestic courts may view the ratified treaty as an interpretive guide to ensure that domestic judicial decisions align with the state's international commitments; and civil society, including non-governmental organisations, (NGOs) can apply the ratified treaty to put pressure on governmental institutions to uphold the obligations enacted even before the treaty is fully integrated into domestic law (Elkins et al., 2013; Von Stein, 2016).

The ratification process however varies from state to state. In some cases, ratification is an executive act requiring no parliamentary involvement while in others, parliamentary approval is required (an example being the USA where the President can ratify but only with the 'advice and consent' of the Senate (see Taylor, 2019; note there is however a category of international agreements known in the USA as 'executive agreements', which scholars have argued does not require the Senate's advice and consent. Two examples are the Paris Agreement and the Minamata Convention on Mercury; see Bodansky & Spiro, 2016; Wirth, 2015, 2017). In yet others, ratification is purely a parliamentary act (Maluwa, 2012). Furthermore, while treaty ratification is generally within the constitutional power of the national government, some states have the convention of consulting with their subnational governments before ratification, especially where the subject matter of the treaty touches on matters within the legislative jurisdiction of subnational governments under the state's constitution (Nwapi, 2011; Paquin, 2010). For example, in Canada, although the federal government has sole constitutional

[1] https://treaties.un.org/Pages/ViewDetails.aspx?src=TREATY&mtdsg_no=XXI-10&chapter=21&clang=_en&_gl=1*54w05o*_ga*MTIwNTg5ODExNS4xNjk1NjkwNzM1*_ga_TK9BQL5X7Z*MTY5NTY5MDczNC4xLjEuMTY5NTY5MjE2OS4wLjAuMA.

authority to negotiate and ratify international treaties, as a matter of policy it consults with the provinces and territories before ratifying treaties that touch on provincial or territorial jurisdiction (Barnett, 2021). Subnational involvement in the treaty ratification process facilitates effective domestic implementation and enforcement of the treaty at the subnational level (Barnett, 2021; Paquin, 2010). Generally, however, the ratification process is at the national level and follows the following steps:

1. **Initial review:** After the negotiating states have agreed the terms of the treaty, but before signing, the treaty undergoes a review process by relevant government departments of the state to indicate whether ratifying the treaty is plausibly beneficial to the state.
2. **Signature:** The state formally indicates its intention to sign the treaty, sometimes followed by a signing ceremony.
3. **Further legal scrutiny**: The treaty undergoes a second more extensive review process to ensure the treaty aligns with the national constitution and existing domestic policies and laws. While incompatibility with existing policies and laws may not lead to a refusal to ratify—as these may be revised (if necessary) to align with the treaty—where the terms of the treaty are incompatible with the constitution, ratification may be unlikely due to the fundamental nature of constitutional provisions and the consequent difficulty of amending the constitution.
4. **Consultation with subnational governments**: In federal states, where the state's practice includes consultation with subnational governments, the state submits the text of the treaty to its subnational governments to obtain their views before ratification. This process may go hand in glove with step 3 above.
5. **Legislative approval**: In states where legislative involvement is required before ratification, the treaty is submitted to the national legislature for approval to ratify. In other states where ratification is purely an executive act, the President or Prime Minister will simply ratify the treaty if satisfied with its provisions.
6. **Domestic legislation**: In some instances, states may need to amend existing legislation or develop new legislation to meet the treaty obligations. This is where substantial divergence occurs, as discussed below. Some states might not need to enact new legislation if existing laws already enable them to generally meet treaty obligations. However, for the MGR components of the treaty, it is likely that states' existing laws (if any) will be insufficient to address the extensively negotiated and innovative BBNJ MGR provisions.
7. **Instrument of ratification:** The state then submits an 'instrument of ratification' to the United Nations treaty depository.
8. **Implementation, monitoring, enforcement, and reporting**: After the treaty comes into force, there is ongoing monitoring, control, enforcement, and reporting. Potentially, domestic legislative updates will be required to ensure continued compliance.

10.1.2 General Considerations for States Before Ratification

As part of a state's legal scrutiny (step 3), prior to ratification, the following points, *inter alia,* should be considered, some of which may bring political and legal hurdles, briefly outlined below.

- **States parties and non-parties to UNCLOS** may sign and ratify this Agreement, along with regional economic integration organisations, as made clear in the BBNJ Agreement's Articles 65 and 66. Article 6 stipulates that the legal status of non-parties to UNCLOS or any other related agreements is not affected by ratifying this Agreement. Article 44, paragraphs 5 and 6, specify the dispute settlement options available to parties of this Agreement that are not parties to UNCLOS. Thus, non-UNCLOS states can be parties to this Agreement and have mechanisms for resolving disputes under it.

- **Jurisdictional issues**: The Agreement pertains to areas beyond national jurisdiction (ABNJ). The ability of a state to enforce its domestic laws on its entities operating in international waters will need further elaboration and agreements by all parties to the Treaty, especially if there are various nationalities involved in a given (research) cruise—as is typically the case.
- **Marine genetic resources (MGR)**: Of interest here and discussed in more detail below, the significance of digital sequence information on marine genetic resources of ABNJ is acknowledged in the Preamble and further detailed in Part II of the treaty, but operational details remain to be addressed (see Humphries, 2025; Rabone et al., 2025).
- **Benefit sharing from MGRs**: The fair and equitable sharing of benefits arising from MGRs with developing states, considering the special circumstances of small island developing states and least developed countries, also remains to be addressed (see Broggiato et al., 2025; Lavelle & Wynberg, 2025).
- **Common heritage of humankind principle** (Article 7(b)), suggests that certain resources in ABNJ are the collective inheritance of humanity. Although not explicitly voiced in the Treaty, there is the possibility that this principle could be taken into consideration in the implementation of MGR provisions, though much will depend on future interpretations and decisions taken by the Conference of Parties (COP). That MGRs in ABNJ could be the common heritage of humankind might challenge national norms of property and intellectual rights. However, given that this principle is also in UNCLOS Part XI, regarding mineral resources of the Area, states may already have had some experience reconciling this principle with domestic legislation.
- **Freedom of marine scientific research** (Article 7(c)) as applied to ABNJ, should align with states' national legislation. UNCLOS Part XII also speaks to this freedom, and therefore, it should not be particularly challenging.
- **Data management**: With several articles of the treaty stipulating open sharing of data, states will need to build robust data management systems (see Lawson et al., 2025).
- **Intellectual property**: The Agreement is silent on the implications for intellectual property rights and commercial interests more generally (see Brown, 2025).
- **Polluter-Pays principle:** Article 7(a) requires a state to be able to attribute liability and costs to those responsible for pollution, a task that will require international cooperation in waters outside of states' jurisdictions and normal maritime domain awareness.
- **Military exclusions**: Article 10(3) excludes military activities from the MGR provisions related to utilisation, which raises certain questions for domestic jurisdictions where the line between military and government activities can be blurry.
- **Biosecurity**: Article 11(7) stipulates that activities should be conducted solely for peaceful purposes. Domestic legislation would have to set forth stringent measures to ensure biosecurity and prevent potential misuse of MGR.
- **Rights of Indigenous Peoples** are acknowledged in the Preamble and Article 13. The treaty ensures that it does not diminish or extinguish the existing rights of Indigenous Peoples. States with indigenous communities might need to update their domestic laws, particularly if rights of Indigenous Peoples are not already fully recognised.
- **Capacity building and technology transfer**: The treaty recognises the need for capacity building, development, and transfer of marine technology to support developing states, both coastal and landlocked. Article 9(b) emphasises the importance of building capacity in countries where that capacity is lacking. Domestic policies would need to facilitate technology transfer according to Part V of the Agreement, as well as skill development, which might be viewed by some players as relinquishing a competitive and economic advantage.

- **Burdens on small island and developing states**: Article 25(3) recognises the need to not impose disproportionate burdens on small island developing states or least developed countries in implementing the treaty's area-based measures. How this will be operationalised remains to be determined.
- **Environmental impact assessments**: Part VI covers environmental impact assessments including the participation of potentially most affected states and stakeholders, including local communities and Indigenous Peoples (Article 31(1)(a)(iv)). However, the criteria by which such participation is to be evaluated remain to be determined. Article 29 considers the relationship between the Agreement and environmental impact assessment processes under relevant legal instruments and frameworks and relevant global, regional, subregional, and sectoral bodies but does not consider its relationship with national legislation. EIA legislation is common in most countries but also varies widely. States' harmonisation of domestic legislation with the treaty's (yet to be established) EIA requirements could in some instances take time.
- **Financial mechanism**: Article 52(3) establishes a financial mechanism that includes a voluntary trust fund (Article 54(4a)), a 'special fund' (Article 54(4b)), and a Global Environment Facility trust fund (Article 54(4c)). The COP may also consider establishment of additional funds, as part of the financial mechanism, to support the conservation and sustainable use of marine biological diversity of areas beyond national jurisdiction, to finance rehabilitation and ecological restoration of marine biological diversity of areas beyond national jurisdiction (Article 54(5)), which suggests that there is some acknowledgement that the three funds already established under the Agreement may not fully meet these needs. Eligibility for access to funding shall be open to developing States Parties on the basis of need (Article 52(12)). How 'need' will be established, as well as the priorities of the funds more generally will be on the agendas of the early COPs. It will in the interests of many states to be party to those early discussions; i.e. to have ratified the Agreement.

10.2 Domestic Obligations Regarding Marine Genetic Resources

Below is a brief summary of states' obligations regarding MGRs in ABNJ, as outlined in the articles of the Agreement, and discussed elsewhere in this volume. Readers are reminded that the signature (but not yet ratification) nonetheless obligates states to refrain from acts which would defeat the object and purpose of the treaty (Vienna Convention on the Law of Treaties, Article 18).

- **Collection and use of MGRs**: Article 11(4 and 7) states that the collection of MGRs should respect the rights and interests of coastal states in accordance with UNCLOS, and no state can claim sovereignty over MGRs in ABNJ.
- **Benefit humanity**: Article 11(6) further states that activities with respect to MGR of areas beyond national jurisdiction are for the benefit of all humanity, particularly for the benefit of advancing the scientific knowledge and promoting the conservation and sustainable use of marine biological diversity.
- **Notification and reporting**: Article 12 requires parties to notify the clearing house mechanism about collection of MGRs and provide reports detailing the geographical area of collection and other relevant data and information on the utilisation of MGR, including the BBNJ standardised batch identifier (see Chap. 14, this volume, Rabone et al. (forthcoming)).
- **Access to traditional knowledge**: Article 13 stipulates that traditional knowledge associated with MGRs held by Indigenous Peoples and local communities shall only be accessed with their free, prior, and informed consent or approval and involvement.

- **Benefit sharing**: Article 14(1) stipulates that all benefits arising from MGRs shall be shared in a fair and equitable manner and contribute to the conservation and sustainable use of marine biological diversity of areas beyond national jurisdiction (see Broggiato et al., 2025).
- **Monetary benefits**: Article 14(5) stipulates that monetary benefits shall be shared through the financial mechanism (Article 52), again for the conservation and sustainable use of marine biological diversity of areas beyond national jurisdiction. After the Agreement's entry into force, developed parties shall make annual contributions to the special fund referred to in Article 52. However, the COP must decide on the modalities for the sharing of these monetary benefits, taking into account the recommendations of the access and benefit-sharing committee (Article 14(7)). These discussions are likely to be of broad interest to States Parties with a range of views.
- **Review of monetary benefits**: Article 14(10) further specifies that parties must assess monetary benefits from the utilisation of MGRs on a biennial basis. The first review must take place no later than five years after the entry into force of the Agreement.

These provisions collectively underscore States Parties' commitment to responsible management of MGRs, sharing of information, emphasising peaceful use, respect for indigenous rights, international cooperation, and fair and equitable benefit sharing.

10.2.1 Timelines for Domestic Legislation

The treaty does not specify a timeline for when states must take legislative, administrative, or policy measures to implement the treaty. However, several articles imply that these measures should be implemented in a way that supports the objectives of the Agreement and facilitates compliance:

- **Article 12 (Notification on Activities)**: States are required to take measures to ensure that information is notified to the clearing house mechanism in accordance with the treaty. This would necessitate having the appropriate legislative and administrative systems in place by the time the clearing house mechanism is established.
- **Article 13 (Traditional Knowledge):** Parties should take measures to ensure that traditional knowledge associated with MGRs is accessed only with the free, prior, and informed consent or approval of Indigenous Peoples and local communities. This implies an immediate obligation, starting after ratification and continuing as relevant situations arise.
- **Article 14 (Fair and Equitable Sharing of Benefits):** Parties shall take the necessary measures to ensure that benefits arising from activities with respect to MGRs by physical and juridical persons under their jurisdiction are shared in accordance with the Agreement. Assuming the state wishes to engage in MGR-related activities, an ongoing obligation to establish or adjust legislative or policy frameworks is required, which will be largely dictated by the timeliness of COP decisions on modalities and other details—a process which could take some years.

The above examples indicate that the Agreement anticipates States Parties to take the necessary legislative and policy measures in a manner and timeframe that supports its objectives, but that some of these changes will only become clear after the Conference of Parties (COP) starts to meet; i.e. after the Agreement enters into force. Ideally, states would begin preparing or adjusting their legislative and administrative frameworks as they move towards ratification and continue to adapt these frameworks as necessary to remain in compliance with the treaty's evolving modalities. However, as discussed further below, the reality is that most parties will not adopt all the necessary new legislation prior to their ratification, nor can they, because many of the MGR provisions are as yet still unclear. Notwithstanding that many countries have constitutional or legal requirements that necessitate bringing domestic laws in line with international agreements before ratification, these requirements appear to be interpreted in a variety of ways, as outlined below. States might update existing laws where they align already with the Agreement's provisions before ratification, and work on new legislation and/or more comprehensive revisions afterwards. States might choose a progressive approach, ratifying the Agreement to signal commitment while gradually developing the legal framework necessary for full compliance, as well as policies to guide implementation. For developing countries, legislative development might benefit from international development assistance prior to ratification, but post-ratification could further benefit from access to the capacity building opportunities provided under the Agreement itself, suggesting that legislative work will continue post-ratification as well.

Deliberately waiting to change or significantly amend laws provides a state time to respond to developments and clarifications that might emerge in the initial stages of the Agreement's implementation, such as those agreed upon in the early COPs. However, waiting too long may send the message that a state is not committed to fully implement the Agreement, leaving legal uncertainties that could send MGR-related research and development (R&D), and associated investments and business elsewhere. On the other hand, acting proactively could establish a state as a world leader, help shape critical decisions of the early COPs, as well as encouraging the establishment of MGR R&D within its jurisdiction.

10.3 The Two UNCLOS Implementing Agreements

To get a sense of how BBNJ ratification may be carried out in practice by states, one can look to the two UNCLOS implementing agreements currently in force: the Part XI Agreement and the Fish Stocks Agreement. However, it is worth noting that both of these are almost thirty years old (1994 and 1995, respectively) and that international norms and expectations have evolved, particularly concerning public participation and transparency (Ardron et al., 2014, 2023), which could take longer than in the past.

10.3.1 Part XI Agreement (1994)

In the Part XI Agreement, concerning deep-seabed mining (DSM), there are substantive legislative obligations for states, particularly if they are interested in sponsoring or carrying out DSM, including: environmental safeguards, financial contributions, benefit sharing, reporting, and compliance.

The International Seabed Authority (ISA), which regulates DSM, requests States Parties to report on the status of their domestic legislation. In practice, however, reporting has been variable and most States Parties had no legislation in place at the time of ratification, or indeed may still not, decades later. However, for sponsoring states involved in exploration, the situation is different; they generally passed specific domestic legislation after the ISA approved exploration regulations, which was after they ratified the Agreement (Chen, 2020). Note the International Seabed Authority's DSM *exploration* regulations were first created for polymetallic nodules in 2000, polymetallic sulphides in 2010, and cobalt-rich ferromanganese crusts in 2012. The *exploitation* regulations are still under development, as of 2024 (https://www.isa.org.jm/the-mining-code/draft-exploitation-regulations/). For example:

- China consented to be bound by the Agreement in 1996 and passed the *Law of the People's Republic of China on Exploration for and Exploitation of Resources in the Deep Seabed Area* in 2016 (Chen, 2020);
- Cook Islands' accession to the Agreement was in 1996, and they passed the *Seabed Minerals Act* in 2009, which was revised in 2019 and further amended in 2020 (https://www.sbma.gov.ck/laws);

- Kiribati consented to be bound by the Agreement in 2003 and passed the *Seabed Minerals Act* in 2017 (https://www.fao.org/faolex/results/details/en/c/LEX-FAOC177489);
- Nauru consented to be bound by the Agreement in 1996 and passed the International Seabed Minerals Act in 2015 (http://ronlaw.gov.nr/nauru_lpms/files/acts/0ca467c92b6467be3d58fa7ebc53ad8b.pdf);
- Singapore consented to be bound by the Agreement in 1994 and passed the *Deep Seabed Mining Act in 2015* (https://sso.agc.gov.sg/Act/DSMA2015);
- Tonga consented to be bound by the Agreement in 1995 and passed the Seabed Minerals Act in 2014 (https://ago.gov.to/cms/images/LEGISLATION/PRINCIPAL/2014/2014-0008/SeabedMineralsAct2014_3.pdf);
- The UK was an something of an exception in that it had previous legislation; it ratified the Agreement in 1996 and amended their existing *Deep Sea Mining Act* of 1981 with the *Deep Sea Mining Act 2014* to give effect to the obligations under the Agreement (https://www.legislation.gov.uk/ukpga/1981/53/contents and https://www.legislation.gov.uk/ukpga/2014/15/section/1).

Therefore, at first glance, ratification of the Part XI Agreement did not push states towards passing new legislation right away. Rather, the creation of domestic deep-seabed mining legislation generally hinged on (a) whether the state was interested in becoming active under the Agreement, and (b) the readiness of the international regulatory regime. It is also worth noting that some of the above states also made amendments to other existing legislation, i.e. sections of their Environment, Tax, and other Acts as necessary to accommodate DSM components.

10.3.2 United Nations Straddling Fish Stocks Agreement (1995)

Unlike DSM, fisheries have been operating in one form or another for millennia, with legislative roots sometimes more than a century old. Yet, like DSM, many states modified domestic legislation several years after ratifying the United Nations Fish Stocks Agreement (UNFSA). Examples of these include:

Australia (ratified 1999): Fisheries Management Act of 1991 (https://www.legislation.gov.au/Series/C2004A04237) was provided with ministerial direction in 2005 to, *inter alia*, revise catch limits in accordance with the requirements of the UNFSA (https://www.agriculture.gov.au/agriculture-land/fisheries/domestic/harvest_strategy_policy/2005_ministerial_direction_to_afma).

Canada (ratified 1999): Fisheries Act (1985) has been revised numerous times, but recognition of the application of a precautionary approach and an ecosystem approach, consistent with UNFSA, did not occur until 2019 (https://laws-lois.justice.gc.ca/eng/acts/F-14/page-1.html?txthl=ecosystem#s-2.5).

New Zealand (ratified 2001): Fisheries Act of 1996 governs the management of fisheries resources, including straddling and highly migratory fish stocks. As part of several revisions, five years later in 2001 the Act directly incorporated the UNFSA (as Schedule 1A; https://www.legislation.govt.nz/act/public/1996/0088/latest/DLM401101.html).

Norway (ratified 1996): has adapted its fisheries management and inspection systems to align with UNFSA provisions. The country's rewritten Marine Resources Act (2009; https://www.fiskeridir.no/English/Fisheries/Regulations/The-marine-resources-act) represents a paradigm shift in the management of Norwegian fisheries consistent with the objectives of the UNFSA (Gullestad et al., 2017).

USA (ratified 1996): While the USA has not ratified UNCLOS, it is a party to the UNFSA. The reauthorised *Magnuson-Stevens Fishery Conservation and Management Act* (2007; https://media.fisheries.noaa.gov/dam-migration/msa-amended-2007.pdf) serves as the principal federal law for managing fisheries and was amended to better align with the UNFSA principle of ecosystem-based management (though the precautionary approach was not incorporated).

10.3.3 Possible Reasons for Slow Legislative Implementation of UNCLOS Agreements

As demonstrated above, the domestic legislation specifically aimed at meeting aspects of the other two UNCLOS implementing agreements usually was enacted post-ratification—sometimes several years later. This phenomenon could be due to a variety of overlapping factors, all of which may take longer than anticipated. For example, comprehensive legal scrutiny and public stakeholder consultations, which often precede legislative enactment, can be lengthy. Furthermore, drafting domestic legislation may require the involvement of multiple governmental departments and possibly even the formation of new administrative bodies.

In the meantime, it is not uncommon for states to use existing legislation, in instances where it does not explicitly contradict the Agreement, as a placeholder to temporarily meet obligations while new, more explicit revisions are being developed, acting in good faith to bring its laws into full compliance with the Agreement over time. However, these 'interim' measures can stay in place longer than anticipated. As outlined above, there are always many unanswered questions about how a new treaty will be implemented. States could be reluctant to put efforts into detailed legislation too soon. The time lag between ratification and domestic legislation can offer a state the flexibility to adapt to emerging best practices and to learn from the experiences of others. Finally, some states may simply see implementation of provisions related to activities in ANBJ (e.g. mining, fishing, or MGR discovery) as a low priority, until such time that the benefits, economic, or otherwise to the state are clearer.

10.4 Relevant Examples Outside of UNCLOS

As is further discussed in Chap. 11 of this volume (Kachelriess et al., 2025), the BBNJ Agreement was negotiated in the context of existing agreements and frameworks. Article 5.2 states that '*the Agreement shall be interpreted and applied in a manner that does not undermine relevant legal instruments and frameworks and relevant global, regional, subregional and sectoral bodies and that promotes coherence and coordination with those instruments, frameworks and bodies*'. Here, we briefly explore three relevant treaties already in force, noting key mechanics of their operation which must not be undermined by the BBNJ Agreement when it enters into force and which furthermore offer lessons learned regarding implementation of MGR provisions.

10.4.1 The Convention on International Trade in Endangered Species (CITES)

The Convention on International Trade in Endangered Species (CITES; 1973, entered into force 1975) is among the most widely ratified multilateral agreements with 184 parties at the time of writing. Its objective is to regulate international trade in species of wild fauna and flora to ensure that it is not detrimental to their survival in the wild. It does so by putting in place a system of permits and certificates that national authorities designated by parties are required to issue at certain steps along the international trade chain (https://cites.org/eng/disc/how.php) when trading species listed on CITES Appendices, which includes many marine species (Pavitt et al., 2021).

Relevant to BBNJ, CITES' definition of international trade includes 'introduction from the sea', which in turn is defined as 'transportation into a state of specimens of any species which were taken in the marine environment not under the jurisdiction of any state' (CITES Convention, Article I (c) and (e)). 'Specimen' is defined as any animal or plant, alive or dead, and for animals is further defined as 'any readily recognisable part or derivative thereof' (CITES Article I (b)). Further guidance on 'readily recognisable' adopted by the CITES COP clarifies that this 'include[s] any specimen which appears from an accompanying document, the packaging or a mark or label, or from any other circumstances, to be a part or derivative of an animal or plant of a species included in the Appendices' (CITES Resolution Conf. 9.6 (Rev. CoP19). This means that the collection of physical samples, including for marine genetic research, of CITES-listed species from areas beyond national jurisdiction already needs to follow CITES obligations.

CITES' Conference of the Parties at its 16th meeting in 2013 adopted guidance (https://sdg.iisd.org/commentary/guest-articles/cites-growing-role-in-international-shark-conservation/) in Resolution Conf. 14.6 (Rev.CoP16) that outlines different scenarios for chains of custody that involve introduction from the sea and how the issuance of CITES documents would work for each case. A subsequent survey by the CITES Secretariat in 2018 found that a limited number of parties were implementing the new guidance, while many reported still not fully implementing the provisions (https://cites.org/sites/default/files/eng/com/sc/70/E-SC70-34.pdf).

The patchy implementation and limited experience with CITES provisions applicable to specimens collected in areas beyond national jurisdiction have had unintended negative knock-on effects on work under other relevant arrangements, e.g. scientific assessments for the management of CITES-listed species conducted under Regional Fisheries Management organisations (RFMOs; see e.g. https://www.iccat.int/com2023/ENG/PLE_120_ENG.pdf) and has been identified as a priority challenge to resolve in order to improve coordination and synergies between CITES and RFMOs. (https://www.bmuv.de/fileadmin/Daten_BMU/Download_PDF/Artenschutz/cites-rfmo_workshop_3_bf.pdf).

CITES' experience however also highlights opportunities for complementarity and mutually reinforcing mandates. For example, the Food and Agriculture Organisation (FAO) and CITES Secretariat jointly produced technical resources and capacity building material to support parties in translating CITES obligations in parallel and complementary to sectoral (fisheries) legislation Nakamura and Kuemlangan (2020). Also, the FAO Port State Measures Agreement (https://www.fao.org/port-state-measures/en/) and related training material explicitly incorporate CITES requirements in its Port State inspection procedures (see FAO PSMA, Annex B, paragraph d).

When transcribing the obligations from the BBNJ Treaty, in particular those of Part II, into national legislation, prospective parties can learn from the experience from the CITES community with the implementation of 'introduction from the Sea' in multiple ways:

- Include CITES when assessing existing relevant domestic regulations or frameworks, and build on/learn from existing experience in implementing CITES' Introduction from the Sea.
- Consider likely chains of custody scenarios when operationalising, in particular, the notification provisions, and consider when, how, and by whom information should be shared. (https://www.iddri.org/en/publications-and-events/blog-post/trading-experiences-what-can-global-ocean-treaty-learn-1970s).
- Ensure early consultation, awareness, and capacity building of relevant sectors and authorities to avoid unintended negative knock-on effects.
- Consider synergetic or mutually reinforcing instruments and their national implementation, such as the FAO Port State Measures Agreement.

10.4.2 The Nagoya Protocol, Under the Convention on Biological Diversity (CBD)

Regarding MGRs, the parallel legal framework for national jurisdiction is the Nagoya Protocol on Access to Genetic Resources and the Fair and Equitable Sharing of Benefits Arising from their Utilisation (Nagoya Protocol, 2010, entered into force 2014) under the 1992 Convention on Biological Diversity (CBD). This Agreement, currently with 142 ratifications, covers genetic resources within national jurisdictions, including marine genetic resources (https://www.cbd.int/abs/nagoya-protocol/signatories/). While many species' ranges traverse national boundaries, it is where individual genetic resources are accessed that determines the access and benefit-sharing (ABS) legislation that applies. Because many marine scientific research activities collect across areas within and beyond national jurisdiction (Rabone et al., 2019), it would be beneficial for states to harmonise the administrative frameworks regulating such collections, to the extent possible, in order to minimise duplication and confusion due to similar obligations of the BBNJ Agreement, the Nagoya Protocol, and provisions under UNCLOS (e.g. Article 248, *Duty to provide information to the coastal state*). However, the legal distinction between bilateral arrangements made under national jurisdiction of the coastal state (e.g. benefit-sharing agreements) and BBNJ would remain distinct.

Similar to the implementing agreements under UNCLOS discussed above, the Nagoya Protocol is also experiencing significant periods of time between ratification by states and the adoption of processes and procedures. In many cases, new legislation is required. Translating an international treaty into national legislation requires understanding and managing inconsistencies, as well as clarifying terms with uncertain meanings. An example of this is the term 'utilisation', the meaning of which is ambiguous (the EU, for example, understands research in the absence of development to be classified as 'R&D' while others might consider both elements to be necessary to qualify an action as 'utilisation').

While the Nagoya Protocol's implementation is of relevance for MGR found within national jurisdictions, related developments under the CBD are of further pertinence for areas beyond national jurisdictions as well; for example, the ongoing discussions on digital sequence information (DSI) include mechanisms on benefit-sharing and data governance deriving from the use of DSI that might be extended to DSI from MGR (draft summary of the 1st Meeting of the Open-Ended Working Group on DSI: https://www.cbd.int/doc/c/b3c5/e301/e4cdc9663fb0001e5196ef8e/wgdsi-01-l-02-en.pdf). Harmonising practices under the Nagoya Protocol and the BBNJ Agreement could encourage open and responsible data governance as well as clarify benefit-sharing options and modalities within and beyond national jurisdictions. Whatever is ultimately decided, one can expect new national legislation for implementation of the CBD DSI provisions, which would set a precedent and could presumably be adapted for use under the BBNJ Agreement, when in force.

The Kunming-Montreal Global Biodiversity Framework (GBF; https://www.cbd.int/gbf), adopted in 2022, also falls under the CBD. The GBF Goal C Targets 13 and 15 deal with benefit sharing from the use of genetic resources and DSI that are found within national jurisdiction. Given that scientific surveys often collect samples from areas both within and beyond national jurisdiction within one survey, the implementation of the GBF also has implications for the implementation of the BBNJ Treaty, where administrative harmonisation would again be beneficial.

Academia and ABS practitioners have long criticised the ABS system under the Convention on Biological Diversity and the Nagoya Protocol on its effectiveness for adhering to its objectives (Laird et al., 2020; Prathapan et al., 2018). Implementation of the Nagoya Protocol by some states, for example Brazil, has hindered biodiversity research activities, which has necessitated further revision of national legislation. The literature has been debating whether ABS, as an international legal framework aiming at building trust-based relationships under the principles of fairness and equity between the developing and developed countries, is currently achieving any more than the bureaucratisation of rights over genetic resources and procedural rights that should allow for a smooth and meaningful exercise of these rights (Ruiz Muller, 2018; Pauchard, 2017; Sirakaya, 2022; Wynberg, 2023). The inability to quantify or fully recognise or even ascribe relevance to the non-monetary benefits also poses challenges (Pauchard, 2017; Prathapan et al., 2018; Rabone et al., 2019). The implementation of the BBNJ Agreement could learn from these experiences. Awareness of the requirements and active engagement by the scientific community alongside stakeholders is a key first step (see Chap. 14, this volume, Rabone et al.).

10.4.3 The Plant Treaty

International Treaty for Plant Genetic Resources for Food and Agriculture (Plant Treaty) is relevant to the BBNJ Agreement in that its objectives include the conservation and sustainable use of (plant) genetic resources and the fair and equitable sharing of the benefits arising out of their use (Plant Treaty, Article 1.1). Both treaties include benefit sharing through a centralised institution and a benefit-sharing fund (BBNJ Agreement, Articles 11 and 51; Plant Treaty, Article 13). To date, the Plant Treaty has been brought into effect via the establishment of a multilateral legal framework among providers and users (Plant Treaty, Article 10.2.). Through the use of specialised contracts (Standard Material Transfer Agreements or SMTAs), the treaty relies on already existing international and domestic legal frameworks for the interpretation and enforcement of the SMTAs. However, the SMTA model has received criticisms related to its design and effectiveness in generating benefits, as well as efficacy as a binding contract (e.g. Tvedt, 2021). The Governing Body of the Plant Treaty has for the past ten years been considering an enhancement of the multilateral SMTA system, with the aim of effectively responding to these critiques (https://www.fao.org/3/nn605en/nn605en.pdf). The ongoing work under the Plant Treaty to develop a ten-year capacity building strategy also illustrates the growing shared need across various international instruments dealing with the utilisation of genetic resources to build internal (i.e. secretariat) as well as external capacity. Similar to approach to benefit-sharing mechanisms above, this also illustrates the need for harmonisation of approach and alignment across different UN fora, as far as is practical.

10.5 Concluding Remarks

Ratification of the BBNJ Agreement signals a Party's willingness to comply with its provisions, including those on MGR, but does not require that all its national legislation is readied prior to ratifying (though this may be a requisite of the state itself). Indeed, looking towards the other two UNCLOS implementing agreements, few states had relevant legislation prepared at the time of their ratifications. Often a lag occurred of about a decade between the date of ratification and significant legislative revisions. The speed of legislative development can depend on many factors, largely driven by the state's interest (or lack thereof) in becoming active under the treaty, and the maturity of the accompanying international legal and policy regime, upon which national legislation can be built. Given that most of the procedures and modalities associated with MGRs (discussed in Part I in this volume) are yet to be established, states may have little choice but to develop their laws progressively, starting with enabling legislation and working down into regulatory detail when implementation becomes clearer—perhaps several years later. In the meantime, review of existing legislation that might be impacted can be performed, and drafting of enabling legislation can still be undertaken, including provisions that allow for, *inter alia*:

- Recognition of the Polluter-Pays principle.
- Recognition of the common heritage of humankind principle.
- Peaceful use of MGR from ABNJ, without sovereign claim.

- Principles for open and responsible data governance.
- Free prior and informed consent by Indigenous Peoples.
- Notification and reporting, including an MGR traceability system, and support for a batch digital identifier.
- Capacity building and technology transfer –to be determined by parties to the treaty.
- Fair and equitable sharing of benefits, including monetary benefits –to be determined by parties to the treaty in the second phase.

Ultimately, parties will need to consider synergetic or mutually reinforcing instruments in their national implementation of the BBNJ Agreement with other international agreements they may be signatories to; for example, the FAO's Port State Measures Agreement, and CITES. The Nagoya Protocol under the CBD, which applies to national jurisdictions, highlights the need for global cooperation in benefit sharing arising from genetic resources, wherever they may occur.

States will have an obligation to implement these rules in relation to activities within their jurisdiction or control, which includes state enterprises, activities of flag vessels, activities of nationals, and activities that take place within their jurisdiction. Multiple states could have jurisdiction over the same activity. Notwithstanding that no state shall claim or exercise sovereign rights over MGRs from ABNJ (Article 11(4)), it is nonetheless the States Parties who will decide what rules, regulations, and/or laws apply once MGRs arrive within their jurisdiction, so long as they are consistent with the BBNJ Agreement (Preamble and Article 6). Therefore, the BBNJ Agreement, through its COP and committees, could facilitate a collaborative approach among states to address these shared multifaceted challenges.

The other three 'pillars' of the BBNJ Agreement sit on well-established foundations of national and international law regarding area-based marine protection (e.g. various CBD commitments), environmental impact assessments (Song, 2022), capacity building and transfer of marine technology (e.g. UNCLOS Article 144). However, as discussed here, the legal foundation of the fourth pillar, MGR, is less well established nationally or internationally. As evidenced by the other chapters in this volume, much remains to be determined. Therein, the BBNJ Agreement represents a significant advancement of international law, and arguably one of the strongest reasons for states to ratify it; or alternatively, for states to shy away! Regardless, future States Parties understandably will be cautious to commit their national laws and policies with regard to MGRs, until certain details become clearer. Parties will need to apply an incremental and adaptive approach, which may begin before their ratification of the Agreement, and will continue beyond.

References

Ardron, J. A., Lily, H., & Jaeckel, A. (2023). Public participation in the governance of deep-seabed mining in the Area. In R. Rayfuse, & A. Jaeckel (Eds.), *Chapter 16, research handbook on international marine environmental law* (2nd ed.).

Ardron, J. A., Clark, N., Seto, K., Brooks, C., Currie, D., & Gilman, E. (2014). Tracking 24 years of discussions about transparency in international marine governance: Where do we stand? *Stanford Environmental Law Journal, 33*(2), 167–190.

Baccini, L., & Urpelainen, J. (2014). Before ratification: Understanding the timing of international treaty effects on domestic policies. *International Studies Quarterly, 58*(1), 29–43.

Barnett, L. (2021). Canada's approach to the treaty-making process. Parliament of Canada, Publication No. 2008-45-E. https://lop.parl.ca/staticfiles/PublicWebsite/Home/ResearchPublications/HillStudies/2008-45-e.pdf.

Bodansky, D., & Spiro, P. (2016). Executive agreements+. *Vanderbilt Journal of Transnational Law, 49*(4), 885–903.

Broggiato, A., Dunshirn, P., Jaspars, M., & Pena-Neira, S. (2025). Monetary and non-monetary benefit sharing under the BBNJ Agreement. In F. Humphries (Ed.), *Decoding governance of marine genetic resources under the BBNJ Agreement.* Springer.

Brown, A. E. L. (2025). The place of intellectual property under the BBNJ Agreement. In F. Humphries (Ed.), *Decoding governance of marine genetic resources under the BBNJ Agreement.* Springer.

Chen, C. Z. (2020). China's domestic law on the exploration and development of resources in deep seabed areas. In C. Banet (Ed.), *The law of the Seabed: Access, uses, and protection of seabed resources* (pp. 335–370). Brill Nijhoff.

Druel, E., & Gjerde, K. M. (2014). Sustaining marine life beyond boundaries: Options for an implementing agreement for marine biodiversity beyond national jurisdiction under the United Nations Convention on the Law of the Sea. *Marine Policy, 49*, 90–97.

Elkins, Z., Ginsburg, T., & Simmons, B. (2013). Getting to rights: Treaty ratification, constitutional convergence, and human rights practice. *Harvard International Law Journal, 54*, 61.

Gullestad, P., Abotnes, A. M., Bakke, G., Skern-Mauritzen, M., Nedreaas, K., & Søvik, G. (2017). Towards ecosystem-based fisheries management in Norway–practical tools for keeping track of relevant issues and prioritising management efforts. *Marine Policy, 77*, 104–110.

Humphries, F. (2025). Marine genetic resources beyond national jurisdiction: The expansive scope of the BBNJ Agreement. In F. Humphries (Ed.), *Decoding governance of marine genetic resources under the BBNJ Agreement*. Springer.

Kachelriess, D., Dunshirn, P., Langlet, A., Brown, A. E. L., & Scholz, A.H. (2025). Marine genetic resources and digital sequence information under the BBNJ Agreement: Interlinkages with other access and benefit sharing frameworks. In F. Humphries (Ed.), *Decoding governance of marine genetic resources under the BBNJ Agreement*. Springer.

Laird, S., Wynberg, R., Rourke, M., Humphries, F., Muller, M. R., & Lawson, C. (2020). Rethink the expansion of access and benefit sharing. *Science, 367*(6483), 1200–1202.

Lavelle, J., & Wynberg, R. (2025). Benefit sharing under the BBNJ agreement in practice. In F. Humphries (Ed.) *Decoding governance of marine genetic resources under the BBNJ Agreement.* Springer.

Lawson, C., Humphries, F., Jaspars, M., & Rabone, M. (2025). Data management and the 'BBNJ standardized batch identifier' under the BBNJ Agreement. In F. Humphries (Ed.), *Decoding governance of marine genetic resources under the BBNJ Agreement.* Springer.

Maluwa, T. (2012). Ratification of African Union Treaties by member states: Law, policy and practice. *Melbourn Journal of International Law, 13*, 636.

Moore, D. H. (2012). The president's unconstitutional treatymaking. *UCLA Law Review, 59*, 598–675.

Nakamura, J. N., & Kuemlangan, B. (2020). *Implementing the convention on international trade in endangered species of wild fauna and flora through national fisheries legal frameworks: A study and a guide* (Vol. 4). Food & Agriculture Organisation of the United Nations.

Nwapi, C. (2011). International treaties in Nigerian and Canadian courts. *African Journal of International and Comparative Law, 19*(1), 38–65.

Paquin, S. (2010). Federalism and compliance with international agreements: Belgium and Canada compared. *The Hague Journal of Diplomacy, 5*(1–2), 173–197.

Pauchard, N. (2017). Access and benefit sharing under the convention on biological diversity and its protocol: What can numbers tell us about the effectiveness of the regulatory regime. *Resources, 6*(1), 11.

Pavitt, A., Malsch, K., King, E., Chevalier, A., Kachelriess, D., Vannuccini, S., & Friedman, K. (2021). *CITES and the sea trade in commercially exploited CITES-listed marine species*. FAO Fisheries and Aquaculture Technical Papers No 666.

Prathapan, K. D., Pethiyagoda, R., Bawa, K. S., Raven, P. H., Rajan, P. D., et al. (2018). When the cure kills—CBD limits biodiversity research. *Science, 360*(6396), 1405–1406.

Rabone, M., Horton, T., Humphries, F., Lyal, C., Muraki Gottlieb, H., Scholz, A. H., Vanagt, T., & Jaspars, M. (2025). BBNJ agreement: Considerations for scientists and commercial end users of MGR at research, development and commercialisation stages. In F. Humphries (Ed.), *Decoding governance of marine genetic resources under the BBNJ Agreement.* Springer.

Rabone, M., Harden-Davies, H., Collins, J. E., Zajderman, S., Appeltans, W., Droege, G., Brandt, A., Pardo-Lopez, L., Dahlgren, T. G., Glover, A. G., & Horton, T. (2019). Access to marine genetic resources (MGR): Raising awareness of best-practice through a new agreement for Biodiversity Beyond National Jurisdiction (BBNJ). *Frontiers in Marine Science, 6*, 520.

Ruiz Muller, M. (2018). *Access to genetic resources and benefit sharing 25 years on: Progress and challenges*. International Centre for Trade and Sustainable Development (ICTSD).

Sirakaya, A. (2022). Is the Nagoya Protocol designed to conserve biodiversity? *Plants, People, Planet, 4*(1), 68–75. https://doi.org/10.1002/ppp3.10221

Song, Y. (2022). The obligation of EIA in the international jurisprudence and its impact on the BBNJ negotiations. *Sustainability, 15*(1), 487.

Taylor, M. L. (2019). Senate treatment of select international agreements, 2013–2018. *American Journal of International Law, 113*, 184–188.

Tvedt, M. W. (2021). A contract-law analyses of the SMTA of the plant treaty: Can it work as a binding contract? *The Journal of World Intellectual Property, 24*(1–2), 83–99.

United Nations. (1969). Vienna convention on the law of treaties, United Nations, treaty series, entered into force 27 January 1980 (p. 331).

Von Stein, J. (2016). Making promises, keeping promises: Democracy, ratification and compliance in international human rights law. *British Journal of Political Science, 46*(3), 655–679.

Wirth, D. A. (2015). An international and domestic law of climate change: A binding international agreement without the senate or congress. *Harvard Environmental Law Review, 39*, 515–566.

Wirth, D. A. (2017). Executive agreements relying on implied statutory authority: A response to Bodansky and Spiro. *Vanderbilt Journal of Transnational Law, 50*, 741–755.

Wynberg, R. (2023). Biopiracy: Crying Wolf or a lever for equity and conservation? *Research Policy, 52*(2), 104674. https://doi.org/10.1016/j.respol.2022.104674

Jeff A. Ardron has more than thirty years of experience in marine policy and conservation science. Having worked for governments, intergovernmental organizations, academia and non-governmental organizations, he is currently Africa Oceans Director for The Nature Conservancy and is based in Mombasa, Kenya. He holds a PhD in ocean sciences from University of Southampton, UK, and an MSc in environmental management from Royal Roads University, Canada.

Daniel Kachelriess is an expert on oceans, fisheries, wildlife law and policy and followed the negotiations of the BBNJ Agreement as part of the High Seas Alliance and as a member of the IUCN World Commission on Environmental Law. He continues to advise the High Seas Alliance and other organizations on aspects of the BBNJ Agreement, including on Marine Genetic Resources, including the fair and equitable sharing of their benefits. His previous roles include Executive Director of Sea Shepherd Legal, a non-profit law firm, and the Marine Species Officer of the CITES Secretariat.

Christopher H. C. Lyal is a taxonomist with a strong level of policy engagement with the Convention on Biological Diversity, particularly regarding the Nagoya Protocol and Digital Sequence Information. He has undertaken international consultancies and been a delegate in CBD COPs representing the UK, and participated a several CBD expert groups. He has also advised on the implementation of the BBNJ agreement.

Chilenye Nwapi is a legal adviser in the Oceans and Natural Resources Section of the Commonwealth Secretariat, London. He holds a PhD in law from the University of British Columbia, Canada.

Muriel Rabone is a researcher based in the deep-sea ecology and systematics group of the Natural History Museum, London. She also studies the neglected tropical diseases: schistosomiasis and paragonimiasis.

Aysegul Sirakaya is a legal expert on access and benefit sharing at Abyss, and the ABS Officer at the Natural History Museum of London. She teaches ABS at various research institutions and advises actors on ABS compliance.

Alison Swaddling is an ocean governance adviser at the Commonwealth Secretariat, providing technical advice and support to governments on various aspects of ocean governance within national jurisdiction, and in their engagement with international agreements covering the areas beyond national jurisdiction. She leads the Secretariat's BBNJ Project. She has a Master of Environmental Law and a Bachelor of Marine Science. Her policy development experience is complimented by offshore industry experience.

Marine Genetic Resources and Digital Sequence Information Under the BBNJ Agreement—Interlinkages with Other Access and Benefit-Sharing Frameworks

11

Daniel Kachelriess, Paul Dunshirn, Arne Langlet, Abbe E. L. Brown, and Amber H. Scholz

Abstract

The BBNJ Agreement, including its access and benefit-sharing provisions for marine genetic resources and digital sequence information, was negotiated against the backdrop of an existing tapestry of international law. Considerations of potential future interactions with other relevant instruments, frameworks, and bodies have shaped parts of the BBNJ Agreement and such interactions will play an important role in its future successful implementation. This chapter discusses the BBNJ Agreement's general approach to regime interaction, highlights several instruments, frameworks, and bodies of particular relevance to the BBNJ negotiations and the future implementation of the BBNJ Agreement, and explores concrete scenarios and possible challenges of future regime interactions.

Keywords

Biodiversity · Marine genetic resources · Access and benefit sharing · Digital sequence information · Regime complex · Global biodiversity framework

11.1 Introduction

On 19 June 2023, UN Member States adopted the new *Agreement under the United Nations Convention on the Law of the Sea on the conservation and sustainable use of marine biological diversity of areas beyond national jurisdiction* (BBNJ Agreement), following over two decades of discussions, including over five years in formal negotiations. Through the BBNJ Agreement, UN Member States sought to address gaps in the *United Nations Convention on the Law of the Sea* (UNCLOS), specifically in four areas (see UN General Assembly document A/66/119, 2011):

- Marine genetic resources, including questions on the sharing of benefits;

D. Kachelriess (✉)
IUCN World Commission on Environmental Law, High Seas Alliance (Advisor), Vienna, Austria

P. Dunshirn · A. Langlet
Department of Political Science, University of Vienna, Vienna, Austria

A. E. L. Brown
World Commission On Environmental Law, Gland, Switzerland

A. H. Scholz
Department of Science Policy and Internationalization, Leibniz Institute DSMZ-German Collection of Microorganisms and Cell Cultures, Braunschweig, Germany

F. Humphries (ed.), *Decoding Marine Genetic Resource Governance Under the BBNJ Agreement*, Sustainable Development Goals Series, https://doi.org/10.1007/978-3-031-72100-7_11

- measures such as area-based management tools, including marine protected areas;
- environmental impact assessments; and
- capacity-building and the transfer of marine technology.

which are now reflected as Parts II-V of the BBNJ Agreement.

This chapter will focus on Part II: "Marine genetic resources (MGRs), including the fair and equitable sharing of benefits" and will discuss past and (likely) future interactions between the BBNJ Agreement and other access and benefit-sharing (ABS) systems. It will highlight several of the other instruments and bodies that States considered when negotiating the BBNJ Agreement and that will be relevant for its future implementation and discuss specific likely interactions. This is particularly important in the context of monetary benefit sharing for digital sequence information (DSI). Incompatibilities between different benefit-sharing systems for DSI could cause significant challenges and potentially high costs for implementation as well as create the risk of double payment for users that process large amounts of DSI during R&D potentially falling under multiple instruments (Halewood et al., 2023). It will close with some considerations for future discussions in BBNJ and other fora.

11.2 The BBNJ Agreement's General Provisions on Regime Interaction

The BBNJ Agreement was negotiated against the backdrop of an existing tapestry of international agreements and other negotiation processes (Langlet & Vadrot, 2023a), some of which have the potential of overlapping mandates or scope. It is therefore no surprise that the question of how the BBNJ Agreement will interact with other such frameworks was very present in the lead up to and during the final BBNJ negotiation session. In the context of the provisions on marine genetic resources, DSI, and the fair and equitable sharing of benefits thereof, a lot of attention was paid, in particular, to future interactions with the Convention on Biological Diversity (CBD) framework (Scholz et al., 2021).

The fact that the BBNJ regime will interact with other regimes is not a new phenomenon to international governance nor unique to the BBNJ case. Literature has dealt with issues of regime overlap and interplay largely under the term "regime complexity", which refers to the landscape of overlapping and interlinked legal and institutional arrangements that govern specific issue areas (Alter & Raustiala, 2018). Regime complexity in international governance can create specific outcomes and presents both challenges and opportunities for global governance, with literature suggesting both positive and negative implications. On the one hand, the division of governance across various institutions can lead to regulatory conflicts, reduced effectiveness in global cooperation, and inequalities due to forum-shopping, a practice where actors strategically choose advantageous forums (Gehring & Faude, 2013, Gomez-Mera et al., 2020). Legal studies emphasize potential conflicts in legal rules and inconsistencies (Davis, 2009; Raustiala & Victor, 2004). On the other hand, political science perspectives highlight the benefits of such complexity, including increased flexibility, adaptability, and pooling of expertise (Keohane & Victor, 2011; Lesage & van de Graaf, 2013). Institutional literature suggests synergies and functional divisions of labour between regimes, fostering competition that can lead to structured responsibilities and enhanced cooperation effectiveness (Conca, 2007; Pratt, 2018). Ultimately, the impact of regime complexity hinges on its dynamic evolution and the strategic interactions of states and non-state actors across institutions, potentially signalling normative progress in global governance (Faude & Große-Kreul, 2020; Kelley, 2009).

In the BBNJ Agreement, States set out the general approach to regime interactions in Article 5.2, which provides that "*the Agreement shall be interpreted and applied in a manner that does not undermine relevant legal instruments and frameworks and relevant global, regional, subregional, and sectoral bodies and that promotes coherence and coordination with those instruments, frameworks and bodies".* This text presents a compromise between the proposals championed by different groups of States: "does not undermine the competencies of"—which only looks at competencies for the provision to apply—and "does not undermine the effectiveness of"—which would have shifted the test to the actual functioning and implementation of the relevant provisions. The constructive ambiguity of the compromise language makes it likely that both ideas will remain influential when Parties discuss the implementation of the provisions at future BBNJ CoPs, which could then take pragmatic decisions on a case-by-case basis.

This more or less soft—depending on future practice—guidance on the applicability of the BBNJ Agreement is then counter-balanced in Article 8.2 which sets out that *"Parties shall endeavour to promote, as appropriate, the objectives of this agreement when participating in decision-making under other relevant legal instruments, frameworks, or global, regional, subregional or sectoral bodies".*

11.3 Relevant Other Instruments and Bodies, with a Focus on Other ABS Frameworks

11.3.1 United Nations Fish Stocks Agreement and Regional Fisheries Bodies

The general approach to regime interactions in the BBNJ Agreement was to a significant degree shaped by States' considerations of how the BBNJ Agreement would or would not interact with its sibling UNCLOS implementing agreement, the *Agreement for the Implementation of the Provisions of the United Nations Convention on the Law of the Sea of 10 December 1982 relating to the Conservation and Management of Straddling Fish Stocks and Highly Migratory Fish Stocks* (UN Fish Stocks Agreement) and the related regional fisheries bodies, in particular in the context of Part III of the BBNJ Agreement on area-based management tools, including marine protected areas (MPAs) (Langlet & Vadrot, 2023a, 2023b).

However, the concern that the provisions of Part II of the Agreement on MGRs could also negatively affect fishing or fishing-regulated activities is evident in the explicit exclusion of "fishing-regulated under relevant international law and fishing-related activities" and "Fish or other living marine resource known to have been taken in fishing or fishing-related activities from ABNJ, except where such fish or other marine resources are regulated as utilization under this part" in Article 10.2 a) and b) of the BBNJ Agreement respectively. While the intent of this exclusion was clear to the negotiators, potential ambiguities and loopholes have already been highlighted and additional guidance by the BBNJ Conference of the Parties (CoP) may be necessary to achieve the original intent (Humphries, 2025). The second half of the second exclusion ("except where such fish or other marine resources are regulated as utilization under this part") explicitly opens up the possibility that the provisions of Part II of the BBNJ Agreement, including the notification and monitoring provisions, could apply to fish or other marine resources under specific circumstances. Where this would be the case, the implementation of the provisions would need to be done in a manner that "does not undermine" regulations of relevant regional fisheries bodies.

11.3.2 World Trade Organization—Agreement on Trade-Related Aspects of Intellectual Property Rights (TRIPS)

The link between intellectual property (IP) and BBNJ and its combination of private rights over the results of innovation and creativity and the potential public benefit from using these results is explored in Chap. 9 of this book (Brown, 2025). Chapter 9 also considers the BBNJ negotiations' engagement with the availability of patents over MGRs and ongoing discussions on the disclosure of origin in patents at World Intellectual Property Organization (WIPO) in the context of genetic resources and traditional knowledge and previous discussions in respect of disclosure of origin at WTO. The position under the Agreement on Trade-Related Aspects of Intellectual Property Rights (TRIPS), which is part of the WTO Agreement, is particularly important in relation to access and benefit sharing. WTO members are to require that patent applications shall disclose the invention "in a manner sufficient clear and complete" so that it can be carried out by a person skilled in the art (TRIPS art 29). This obligation can be a very practical form of non-monetary benefit sharing: once the patent has expired, everyone can benefit from and use the information in the patent, but even during the patent term that information can be used as a base for other work, provided such work does not infringe on the IP of the patent owner. There are, however, frequent disputes about whether there has actually been sufficient disclosure in a patent (see e.g. Kirin-Amgen V Hoechst, 2005, R Regeneron Pharmaceuticals Inc v Kymab Ltd UKSC 27, 2020). In a similar vein, there are debates regarding whether or not scientists actually do engage with patents, including because of challenges in carrying out searches of patent databases and because patents are written using language more geared to lawyers and the scope of the patent, than to benefiting other scientists (Oulette, 2017).

Further, even during the patent term, some national laws have restricted exceptions to the rights conferred, enabling others to make use of the invention (say for research) during the patent term. The opportunity to increase benefit sharing, however, depends on countries' choice (TRIPS art 30, Brown, 2025). A base upon which States might choose to do this lies in TRIPS' objectives and principles provisions, which engage with the fact that IP rights sit in a wider landscape. This refers to the mutual advantage of producers and users of technological knowledge, social and economic welfare, balancing rights and obligations, and adopting measures necessary to protect public health and nutrition; there must, however, be consistency with the TRIPS agreement in respect of the steps taken by States (TRIPS arts 7, 8).

11.3.3 CBD and Nagoya Protocol

Both past and recent discussions under the framework of the 1992 CBD were highly important in shaping the BBNJ Agreement and potential future interactions, in particular with regard to DSI, and will remain an important issue in the operationalization of the BBNJ Agreement.

The Nagoya Protocol on Access to Genetic Resources and the fair and equitable sharing of benefits arising from their utilization (Nagoya Protocol)[1] is a 2010 supplementary agreement to the CBD which aims to provide a transparent legal framework for the fair and equitable sharing of benefits arising out of the utilization of genetic resources. Critics have argued that the Nagoya Protocol's bureaucratic complexity has prevented the fair and equitable utilization of genetic resources and rather hindered scientific research, innovation, and exchange, particularly noting the complexity caused by each country having its own unique system for benefit sharing (Prathapan et al., 2018). While the mandate of the CBD is distinct from the BBNJ mandate in terms of its geographical scope, States and stakeholders have frequently referenced developments within the CBD, highlighting the advantage and the opportunity to harmonize both systems to a certain degree. References largely expressed the need to "learn from Nagoya", e.g. avoiding the repetition of unintended consequences of the CBD's Nagoya Protocol for the BBNJ ABS provisions or referred to the issue of DSI. Furthermore, the same genetic resources can often be found both within and beyond national jurisdiction, which means that having two entirely separate systems may create undesirable incentives to "forum shop".

[1] The Nagoya Protocol was adopted by CBD CoP10 on 29 October in Nagoya, Japan. It entered into force on 12 October 2013. At the time of writing of this manuscript (June 2024) it has 141 Parties.

Although it may be difficult to approximate the exact influence that the Nagoya Protocol experience exerted on the BBNJ negotiations, it appeared to have strongly informed some of the options and arguments that were put forward by States or discussed informally. During the 5 sessions of the Intergovernmental Conference (IGC) and online dialogues, States made reference to the Nagoya Protocol on 88 occasions, oftentimes highlighting its original intent, e.g. the fair and equitable sharing of genetic resources while also raising sensitivity to potential unintended consequences (Langlet & Vadrot, 2023a, 2023b, Langlet et al., 2024). Such reference to the Nagoya Protocol was often made by proponents of a bureaucratically light ABS system noting that the high seas were not under the sovereignty of any State that would necessitate granular approaches to ABS as is necessary under the CBD, in which each individual country can exercise sovereignty as it sees fit. In February 2024 during a seminar, Dr. Thambisetty, who served as a technical advisor to the G77 chair, remarked that regrets about the design and regulatory burden of the Nagoya Protocol were "a ghost" in the BBNJ negotiations and that the desire to "not go there again was very strong" (Thambisetty, 2024).

More recently, DSI has emerged as a critical discussion in various international fora, in particular in the CBD and also BBNJ. One of the DSI-related concerns during the BBNJ negotiations was how DSI could or should be defined, e.g. what sort of biological or genetic data and information should be classified as DSI and how arising benefits should be shared, particularly when DSI can easily be shared and accessed globally without any physical transfer of genetic material (see also Broggiato et al., 2025). For some stakeholders, the idea of tracking and tracing DSI through the value chain was perceived to likely generate either more costs than benefits (Scholz et al., 2022, Langlet & Dunshirn, 2023) or to trigger avoidance behaviour if BBNJ compliance or benefit-sharing mechanisms are more complex than under other instruments.

At its 15th Conference of the Parties, the CBD made significant advances in its discussions in decision 15/9 in December 2022 (CBD/COP/DEC/15/9). Among other things, this decision:

- Affirmed the need for a fair and equitable sharing of benefits derived from the utilization of DSI on genetic resources (Art. 2).
- Acknowledged the impracticality of tracking and tracing all DSI on genetic resources (Art. 5).
- Listed a set of principles, inter alia, including that the solution should generate more costs than benefits, not hinder research and innovation, and support open science principles, which a benefit-sharing mechanism should meet (Art.9).
- Identified that the monetary and non-monetary benefits [...] should principally support the conservation and sustainable use of biological diversity, benefiting, among others, Indigenous peoples and local communities (Art. 10).
- Resolved to create [...] a multilateral benefit-sharing mechanism concerning the use of DSI on genetic resources, including of a global fund (Art. 16).

A deepening understanding of the technical and logistical complexity of R&D with DSI, made possible through several commissioned studies and pandemic-induced informational webinars, played a role in facilitating debates on multilateral or hybrid approaches for DSI, in contrast to purely bilateral approaches (see e.g. CBD, 2022b and reports referenced therein). The broader context of benefit sharing and the complexity of R&D processes when using DSI also served as a justification for the exploration of decoupled payment approaches.

In the BBNJ context, the timing of CBD decision 15/9 (CBD, 2022a) both caused and solved problems. When the first session of IGC5 was suspended in August 2020, negotiators had achieved a breakthrough in principle with regard to monetary benefit sharing, but the inclusion of DSI in that regime was unclear (IISD/ENB, 2022). While acknowledging that decisions made in other fora did not automatically translate to the BBNJ context, many Parties and stakeholders argued that not explicitly referencing DSI in the BBNJ Agreement following the CBD decision would be regressive (see e.g. Oldham et al., 2023).

On the other hand, the CBD decision alleviated significant negotiation burden for BBNJ negotiators and supplied some clarifications. It was also significant that the decisions recognize that a strict track and trace system, as had been proposed in the BBNJ context prior to IGC5, may not be the most practical way to achieve traceability of all DSI (see Langlet & Dunshirn, 2023). The decision also included the idea that monetary benefits "should, in particular, be used to support the conservation and sustainable use" of biodiversity (CBD/COP/DEC/15/9, paragraph 10), an important compromise that BBNJ Agreement Art. 14 (5) codifies in stricter form ("shall be used"). This opened the door for many developed countries to agree to monetary benefit sharing for a specific purpose, namely the implementation of the BBNJ Agreement itself. Further, the CBD decision contained elements that the BBNJ negotiators could consider to draw from; deciding on the matter of benefit sharing without first defining DSI (a "non-definition approach") (Thambisetty et al., 2023), which the BBNJ Agreement also followed, and a list of principles.

While the actual functioning of the CBD's multilateral benefit-sharing framework for DSI is yet to be substantiated, with work currently ongoing (see e.g. CBD, 2023), the BBNJ Agreement includes the option to align with the mechanism to potentially be established for DSI under the CBD at a later stage in Article 14.7 and 14.9. These provisions specify that monetary benefit-sharing provisions currently in the text will be reviewed biannually, with the first review to take place no later than five years after the entry into force. The CoP can then adopt other modalities for the benefit-sharing mechanism (see also Broggiato et al., 2025). In doing so, the text specifies, the CoP should be mutually supportive of and adaptable to other ABS mechanisms, a clear link to CBD's multilateral benefit-sharing framework for DSI, and an opportunity to align the two systems. A harmonized system might have the substantial benefit of not having to distinguish between CBD and BBNJ origins when accessing and sharing DSI while leaving the option for international bodies to evaluate monetary benefit sharing for different types of DSI separately.

One area in which the BBNJ Agreement gets out "ahead" of the CBD is by requiring a notification system from the utilization of DSI—without clarifying what DSI is, what kind of DSI utilization would trigger notification (beyond publications) and whether if these notifications on DSI connect to benefit sharing itself. During the 2022–24 intersessional period, CBD Parties begin to develop the new multilateral mechanism and might or might not look to the BBNJ system to evaluate compatibility and harmonization between the two systems.

11.3.4 FAO Plant Treaty and WHO Pandemic Influenza Preparedness Framework

Besides the CBD and the Nagoya Protocol, the BBNJ framework would ideally also align with other international ABS systems on specific types of genetic resources, namely under the International Treaty on Plant Genetic Resources for Food and Agriculture ('Plant Treaty') and the WHO's Pandemic Influenza Preparedness (PIP) framework in order to ensure legal certainty and reduce avoidance behaviour or competition for users among these various frameworks. These specialized frameworks were originally developed to address physical genetic resources with potential overlaps to genetic resources being sourced from ABNJ. Current discussions about extending these frameworks to DSI also have important implications for BBNJ and, more generally, for ideas to establish harmonized ABS systems on DSI across legal contexts.

The ABS instrument under the Plant Treaty, adopted in 2001, was originally designed to protect farmers' intellectual property rights about conserving, improving, and making available plant genetic resources for food and agriculture

(FAO, 2019). This instrument is designed as a multilateral framework and built around pre-negotiated (i.e. non-editable by either users or providers) standard material transfer agreements that define access and benefits to be shared upon use of such resources (Rabitz, 2017). Even though the Commission on Genetic Resources for Food and Agriculture (CGRFA) has published guidelines in regard to aquatic genetic resources (e.g. aquacultures), no legal obligations similar to the farmer's rights provision have been decided on so far (FAO, 2019). Interactions between the BBNJ treaty and the FAO system may nevertheless come to matter if genetic resources from ABNJ are used for agricultural purposes (aquatic or terrestrial such as kelp farming or aquaculture). Additionally, interactions between these two frameworks may also take the form of information sharing—the FAO is specifically mentioned under BBNJ Article 51 as one of the relevant international cooperation partners for its clearing-house mechanism.

Similarly to the Plant Treaty, the WHO introduced a multilateral ABS mechanism based on standard material transfer agreements for pandemic influenza virus samples in 2011 (Rourke, 2019). This framework ensures the global sharing of virus material through the WHO's Global Influenza Surveillance and Response System while guaranteeing the return of benefits derived from such sharing, including vaccines and antivirals. However, it remains disputed to what extent the framework actually fulfils its purpose, particularly on the benefit-sharing aspects which essentially trade access to pathogens as a bargaining chip for medical countermeasures which for many seems fundamentally flawed (Rourke, 2019). Furthermore, as the PIP framework is limited to pandemic influenza strains only, it has a very narrow scope. However, for BBNJ discussions, the PIP model is the basis for broader discussions on Pathogen Access and Benefit Sharing under the currently negotiated WHO Pandemic Preparedness CA+Agreement, which could have a much broader material scope, theoretically overlapping with microbial organisms found also in ABNJ.

When it comes to interrelations between the existing ABS frameworks, the Nagoya Protocol does not apply to specific types of genetic resources for which other international ABS instruments exist (Nagoya Protocol art 4.4.; Humphries et al., 2021a, 2021b). This article likely applies to the Plant Treaty and PIP framework and indeed the EU Regulation 511/2014 implementing the Nagoya Protocol explicitly exempts its users from the Nagoya Protocol if accessing GR through these instruments (Lawson et al., 2020.). It is unclear how the BBNJ treaty will be integrated into this system of general (CBD and Nagoya Protocol) and specialized (Plant Treaty and PIP framework) ABS frameworks. It may be argued that BBNJ cannot and need not fall under the exemption for specialized frameworks of the Nagoya Protocol, as no similar sovereign rights exist in the BBNJ context. In any case, it will be important to consider how to avoid "duplicate obligations" for access and benefit sharing across different legal contexts.

Maybe the most important overlaps between the BBNJ Agreement, the Plant Treaty, and the PIP framework are again the governance of DSI (referred to as "genetic sequence data" under the latter) (Aubry et al., 2022; Lawson et al., 2020). As previously discussed for DSI benefit sharing under the future CBD multilateral framework, relevant DSI is to a large extent stored in the same international databases, making it a relevant yet complicated task to design a harmonized system across those various legal contexts.

11.3.5 International Seabed Authority

Within the "family" of UNCLOS and its implementing agreements, to which the BBNJ Agreement belongs, another existing example of a benefit-sharing system is the International Seabed Authority (ISA). This was set up in Art. 156 of UNCLOS, to organize and control activities in the area (UNCLOS Art. 157). The Area is defined as "the seabed and ocean floor and subsoil thereof beyond the limits of national jurisdiction" (UNCLOS Art 1.1 (1)). UNCLOS sets out that the area and its resources are the common heritage of [hu]mankind (UNCLOS Art.

136) and that activities in the Area are to benefit [hu]mankind as a whole (UNCLOS Art. 140.1). This includes the equitable sharing of financial and other economic benefits derived from such activities (UNCLOS Art 140.2).

The provisions of UNCLOS for the Area (Part XI) and the corresponding implementing agreement of 1994 clearly focus on the mining of resources, defined as "all solid, liquid or gaseous mineral resources in situ in the Area at or beneath the seabed, including polymetallic nodules" (UNCLOS Art. 133), but there has been a longstanding argument which regime would apply to resources of the Area more broadly, including marine genetic resources (see Humphries, 2025; Milicay, 2007). Regardless of a future resolution of that debate, it is well recognized that even if the regime of the Area were to apply, neither UNCLOS nor the 1994 implementing agreement included any provisions that would give the ISA a specific role in the operationalization of such an interpretation.

Given that, contrary to the above, the BBNJ Agreement's scope explicitly includes MGR in ABNJ and an operationalization of fair and equitable benefit sharing for such resources, it seems logical and likely that the benefit-sharing systems of the ISA and BBNJ would not overlap—with the former covering benefits from mineral resources from the Area and the latter covering marine genetic resources from both the high seas and the Area.

11.4 Specific Scenarios and Possible Challenges

Concrete scenarios in which genetic resources traverse various jurisdictions and legal contexts should be considered when assessing interactions between the BBNJ Agreement and other ABS frameworks. This includes, for instance, cases when researchers collect MGRs from ABNJ but use them (e.g. analysing, storing, or processing) in a country that regulates genetic resource use under its jurisdiction independently of resource origin. While the CBD's jurisdictional scope does not apply to MGRs in ABNJ, Humphries et al. (2021a, 2021b) discuss examples where national ABS laws may nevertheless cover MGRs from ABNJ. The BBNJ Agreement does not provide a specific mechanism to address such potential overlaps at the outset other than the generic guidance on its relationship with other agreements. The CoP in deciding on future modalities of benefit sharing (Article 14.7), and the access and benefit-sharing committee in making recommendations, are to take into account that the benefit-sharing provisions "*should be mutually supportive of and adaptable to other access and benefit-sharing instruments*". (Art 14.9).

It is worth noting that (most) non-monetary benefits are not exhausted by being shared via multiple obligations and the monetary benefit sharing for the initial period after the entry into force will be decoupled from individual access and utilization events, rendering the practical implications of an overlap minimal. This would change if different modalities for monetary benefit sharing that were in some way coupled to access and utilization events were to be adopted under Article 14.7 in the future. In this context, it will also be important to consider any potential loophole arising from the fact that countries may be party to the BBNJ but not to other international agreements.

Another relevant scenario occurs when organisms are cosmopolitan across multiple legal contexts—for instance, if specifically valuable bacteria would be found in a deep-sea hydrothermal vent ecosystem in ABNJ and similar properties would be later found in another jurisdiction either during the same cruise or a completely separate cruise. Would there be any requirements for users to stick to the original ABS procedures under BBNJ or be free to pick and choose the arrangements that are most suitable for them? If the former, how would Parties monitor compliance with this provision by natural or juridical persons under their jurisdiction? This problem has also been discussed in the CBD context, and so far no clear solutions have been found (Jaspars et al., 2021).

Complications may also emerge in the (rather likely) scenario where researchers “mix” DSI from both ABNJ and other legal contexts when developing a product or biotechnology. It will need to be considered how existing database infrastructures should address such cases, especially as it relates to questions about “weighing” the importance of individual sequence data to end products or patents to assess monetary benefits. For instance, how should the traceability and benefit-sharing systems deal with cases where researchers analyse thousands of sequences from national jurisdiction and one sequence from ABNJ?

11.5 Considerations on Balance Reached Within the BBNJ Agreement While Open to Coordination with Others

On the one hand, many States expressed an appetite to build on the principles in the CBD DSI decision and not to create incompatible systems particularly as scientists and technical advisors stressed that scientific practice does not differentiate between different sources of DSI, uses the same databases and that currently less than 1% of sequences are from areas from national jurisdiction (Scholz et al., 2021). There were also major concerns about the type of notification system that would be necessary to implement some of the ideas on the table for BBNJ that could lead to disincentives to study DSI from ABNJ (see Humphries et al., 2025).

On the other hand, it is important to acknowledge the different starting points for the discussion. In the CBD negotiations, it was clear that MGRs and DSI fall under national jurisdiction. In the BBNJ context, some delegations held the view that MGRs and DSI from ABNJ would fall under freedom of the high seas and others that they would fall under common heritage of [hu]mankind under UNCLOS. The discussion on those principles was so controversial that it was the last issue to be addressed during the negotiations. For a detailed discussion of these principles and how they were reflected in the BBNJ Agreement, see Chap. 4 (Muraki Gottlieb et al., 2025).

The BBNJ compromise on benefit sharing can therefore be interpreted as a hard-won, careful balance and many delegations and stakeholders will likely want to take great care not to upset that balance going forward when considering the interrelation with other bodies and agreements.

References

Alter, K. J., & Raustiala, K. (2018). The rise of international regime complexity. *Annual Review of Law and Social Science, 14*(1), 329–349.

Aubry, S., Frison, C., Medaglia, J. C., Frison, E., Jaspars, M., Rabone, M., Sirakaya, A., Saxena, D., & van Zimmeren, E. (2022). Bringing access and benefit sharing into the digital age. Plants People Planet, 1–8.

Broggiato, A., Dunshirn, P., Jaspars, M., & Pena-Neira, S. (2025). Monetary and non-monetary benefit sharing under the BBNJ Agreement. In F. Humphries (Ed.) *Decoding governance of marine genetic resources under the BBNJ Agreement*. Springer.

Brown, A. E. L. (2025). The place of intellectual property under the BBNJ Agreement. In F. Humphries (Ed.), *Decoding governance of marine genetic resources under the BBNJ Agreement*. Springer.

Conca, K. (2007). Institutional interaction in global environmental governance: Synergy and conflict among international and EU policies. *Perspectives on Political Science, 36*(3), 181.

Convention on Biological Diversity. (2022a). Decision adopted by the Conference of the Parties to the Convention on Biological diversity 15/9. Digital sequence information on genetic resources. CBD/COP/DEC/15/9.

Convention on Biological Diversity. (2022b). Report of the Ad Hoc Technical Expert Group on Digital Sequence Information on Genetic Resources. CBD/DSI/AHTEG/2020/1/7.

Convention on Biological Diversity. (2023). Report of the Ad Hoc Open-ended Working Group on Benefit-sharing from the Use of Digital Sequence Information on Genetic Resources on its first meeting. CBD/WGDSI/1/3*.

FAO. (2019). *The state of the world's aquatic genetic resources for food and agriculture*. FAO Commission on Genetic Resources for Food and Agriculture Assessments.

Faude, B., & Große-Kreul, F. (2020). Let's justify! How regime complexes enhance the normative legitimacy of global governance. *International Studies Quarterly, 64*(2), 431–439. https://doi.org/10.1093/isq/sqaa024

Gehring, T., & Faude, B. (2013). The dynamics of regime complexes: Microfoundations and systemic effects. *Global Governance, 19*(1), 119–130.

Gómez-Mera, L., Morin, J. F., & Van de Graaf, T. (2020). Regime complexes. In *Architectures of earth system governance: Institutional complexity and structural transformation* (pp. 137–157).

Halewood, M., Bagley, M. A., Wyss, M., & Hartmann Scholz, A. (2023). New benefit-sharing principles for digital sequence information. *Science, 382*(6670), 520–522.

Humphries, F. (2025). Marine genetic resources beyond national jurisdiction: The expansive scope of the BBNJ Agreement. In F. Humphries (Ed.), *Decoding governance of marine genetic resources under the BBNJ Agreement.* Springer.

Humphries, F., Jaspars, M., Lavelle, J., & Kachelriess, D. (2025). The novel notification information system for marine genetic resources under the BBNJ Agreement. In F. Humphries (Ed.), *Decoding governance of marine genetic resources under the BBNJ Agreement.* Springer.

Humphries, F., Muriel, R., & Marcel, J. (2021a). Traceability approaches for marine genetic resources under the proposed ocean (BBNJ) treaty. *Frontiers in Marine Science, 8*, Article 661313. https://doi.org/10.3389/fmars.2021.661313

Humphries, F., Laird, S., Wynberg, R., Morrison, C., Lawson, C., & Kolisnikova, A. (2021b). *Survey of access and benefit-sharing country measures accommodating the distinctive features of genetic resources for food and agriculture and associated traditional knowledge*. Food and Agriculture Organization of the United Nations (FAO).

International Institute for Sustainable Development (IISD)/Earth Negotiations Bulletin (ENB). (2022). Summary of the fifth session of the Intergovernmental Conference on an International Legally Binding Instrument under the UN Convention on the Law of the Sea on the Conservation and Sustainable Use of Marine Biodiversity of Areas Beyond National Jurisdiction: 15–26 August 2022. Retrieved May 5, 2024, from https://enb.iisd.org/sites/default/files/2022-08/enb25240e.pdf

Jaspars, M., Humphries, F., & Rabone, M. (2021). Tracing options for marine genetic resources from within national jurisdictions. Report prepared for Commonwealth Secretariat.

Kelley, J. (2009). The more the merrier? The effects of having multiple international election monitoring organizations. *Perspectives on Politics, 7*(1), 59–64.

Keohane, R. O., & Victor, D. G. (2011). The regime complex for climate change. *Perspectives on Politics, 9*(1), 7–23. https://doi.org/10.1017/s1537592710004068

Kirin-Amgen Inc V Hoechst Marion Roussel Ltd. (2005). *Reports of Patent, Design and Trade Mark Cases, 122*(6), 169–207. https://doi.org/10.1093/rpc/2005rpc9

Langlet, A., & Dunshirn, P. (2023). Traceability options for marine genetic resource from areas beyond national jurisdiction. Non-paper commissioned by the High Seas Alliance. Retrieved May 3, 2024, from https://www.highseasalliance.org/wp-content/uploads/2023/02/traceability-options-paper-1.pdf

Langlet, A., & Vadrot, A. (2023a). Not 'undermining' who? Unpacking the emerging BBNJ regime complex. *Marine Policy, 147*, 105372. https://doi.org/10.1016/j.marpol.2022.105372

Langlet, A., & Vadrot, A. (2023b). IOs in the BBNJ regime complex—the dataset. *Data in Brief, 48*, 109153. https://doi.org/10.1016/j.dib.2023.109153

Langlet, A., Vadrot, A.B.M., Fellinger, S., Dunshirn, P., Ruiz R., Silvia C,, & Tessnow-von Wysocki, I. (2024). 'MARIPOLDATAbase (SUF edition)'. https://doi.org/10.11587/0XXZ0V, AUSSDA, V4

Lawson, C., Rourke, M., & Humphries, F. (2020). Information as the latest site of conflict in the ongoing contests about access to and sharing the benefits from exploiting genetic resources. *Queen Mary Journal of Intellectual Property, 10*(1), 7–33. https://doi.org/10.4337/qmjip.2020.01.01

Lesage, D., & van de Graaf, T. (2013). Thriving in complexity? The OECD system's role in energy and taxation. *Global Governance, 19*, 92.

Milicay, F. (2007). A legal regime for the biodiversity of the area. *Ocean Yearbook, 36*, 188–236.

Muraki Gottlieb, H., Kachelriess, D., & Slobodian, L. (2025). Understanding the preamble, principles and objectives of the BBNJ Agreement: A focus on the fair and equitable sharing of benefits of marine genetic resources. In F. Humphries (Ed.) *Decoding governance of marine genetic resources under the BBNJ Agreement.* Springer.

Ouellette, L. (2017). Who reads patents? *Nature Biotechnology, 35*, 421–424. https://doi.org/10.1038/nbt.3864

Prathapan, K., Divakaran, R., Pethiyagoda, R., Bawa, K. S., Raven, P. H., & Rajan, P. D. (2018). When the cure kills—CBD limits biodiversity research. *Science, 360*(6396), 1405–1406. https://doi.org/10.1126/science.aat9844

Pratt, T. (2018). Deference and hierarchy in international regime complexes. *International Organization, 72*(3), 561–590. https://doi.org/10.1017/S0020818318000164

Rabitz, F. (2017). *The global governance of genetic resources: Institutional change and structural constraints*. Routledge.

Raustiala, K., & Victor, D. G. (2004). The regime complex for plant genetic resources. *International Organization, 58*(2), 277–309.

Regeneron Pharmaceuticals Inc v Kymab Ltd. (2020). UKSC 27 (24 June 2020). Retrieved from http://www.bailii.org/uk/cases/UKSC/2020/27.html

Rourke, M. (2019). Access by design, benefits if convenient: A closer look at the Pandemic Influenza Preparedness Framework's standard material transfer agreements. *The Milbank Quarterly*. https://doi.org/10.1111/1468-0009.12364

Scholz, A. H., Lange, M., Habekost, P., Oldham, P., Cancio, I., Cochrane, G., & Freitag, J. (2021).

Myth-busting the provider-user relationship for digital sequence information. *GigaScience, 10*(12), giab085. https://doi.org/10.1093/gigascience/giab085
Scholz, A. H., Freitag, J., Lyal, C. H. C., et al. (2022). Multilateral benefit-sharing from digital sequence information will support both science and biodiversity conservation. *Nature Communications, 13*, 1086. https://doi.org/10.1038/s41467-022-28594-0
Thambisetty, S. (2024). The oceans treaty as a win for multilateralism: What lies ahead. *LSE Law School*. Retrieved February 6, 2024, from https://www.youtube.com/watch?v=D-bpoPnfk28&list=PLK4elntcUEy0ptu7NwzFVXLwI6OrPQl4U&index=50
Thambisetty, S., Oldham, P., & Chiarollo, C. (2023). The expert briefing document: A developing country perspective on the making of the BBNJ Treaty. LSE Law, Society and Economy Working Papers, 30/2023. London School of Economics and Political Science. Law School.
UN General Assembly. (2011). Letter dated 30 June 2011 from the Co-Chairs of the Ad Hoc Open-ended Informal Working Group to the President of the General Assembly, A/66/119.

Daniel Kachelriess is an expert on oceans, fisheries, wildlife law and policy and followed the negotiations of the BBNJ Agreement as part of the High Seas Alliance and as a member of the IUCN World Commission on Environmental Law. He continues to advise the High Seas Alliance and other organizations on aspects of the BBNJ Agreement, including on Marine Genetic Resources, including the fair and equitable sharing of their benefits. His previous roles include Executive Director of Sea Shepherd Legal, a non-profit law firm, and the Marine Species Officer of the CITES Secretariat.

Paul Dunshirn is a PhD researcher at the ERC-funded research project Twin Politics, University of Vienna. His research explores global patent and science landscapes around marine genetic resources and digital sequence information (DSI), aiming to sketch pathways for equitable policymaking. Paul has served as an independent observer and advisor to various UN processes, such as the recently concluded negotiations on the Biodiversity Beyond National Jurisdiction (BBNJ) treaty and the efforts to regulate DSI under the Convention of Biological Diversity (CBD).

Arne Langlet holds a PhD in International Relations from the University of Vienna specializing in the marine biodiversity regime complex. His publications address the BBNJ negotiations, questions about the governance of marine genetic resources, the common heritage principle and the use of data in marine governance. He is currently a research associate for the Horizon Europe project MARCO-BOLO, focussing on the use of marine biodiversity data, and a Fisheries Consultant for the FAO, researching ecosystem restoration efforts.

Abbe Brown is a professor in Intellectual Property Law at the University of Aberdeen. Before returning to academia, she practised as an intellectual property and commercial litigator at leading firms in London, Melbourne and Edinburgh. Abbe has a strong interest in the ocean and in interdisciplinary research and is a member of the World Commission on Environmental Law and the Deep Ocean Stewardship Initiative.

Amber H. Scholz is a microbiologist and head of the Science Policy and Internationalization Department at the Leibniz Institute DSMZ in Braunschweig, Germany. She leads projects on international science policy especially on access and benefit sharing and digital sequence information and founded the DSI Scientific Network and Germany Nagoya Protocol HuB.

Data Management and the 'BBNJ Standardized Batch Identifier' Under the BBNJ Agreement

12

Charles Lawson, Fran Humphries, Marcel Jaspars, and Muriel Rabone

Abstract

This chapter addresses two innovations in the BBNJ Agreement—the '"BBNJ" standardized batch identifier' (BBNJ Identifier) and the data management plan (DMP). The BBNJ Identifier is a means to link information about the subject matter of the BBNJ Agreement—marine genetic resources (MGRs) and digital sequence information on MGRs (DSI)—back to reporting on usage, to enable transparency and equitable benefit sharing. DMPs outline processes and standards for data creation, access, ownership, management and storage, and the roles and responsibilities of stakeholders where data is derived from MGRs and associated DSI. Both requirements are important to fulfilling the BBNJ Agreement obligation that MGR data are Findable Accessible Interoperable and Reusable, or FAIR. This chapter outlines the BBNJ Agreement obligations and the areas that will require further input as the agreement develops into practice, with direction from the subsidiary bodies: the Conference of the Parties (COP); the Scientific and Technical Body (STB), and the Access and Benefit-Sharing Committee (ABSC). It provides legal perspectives and context on the data requirements in relation to other relevant legal frameworks. The chapter concludes that the BBNJ Identifier and DMPs can contribute to modalities for the sharing of the benefits arising from the use of MGRs and DSI that are mutually supportive of, and adaptable to other access and benefit-sharing instruments.

Keywords

BBNJ agreement · Marine biodiversity beyond national jurisdiction · BBNJ standardized batch identifier · Data management plans · Marine genetic resources · Digital sequence information

12.1 Introduction

The *Agreement under the United Nations Convention on the Law of the Sea on the Conservation and Sustainable Use of Marine Biological Diversity of Areas Beyond National*

C. Lawson (✉) · F. Humphries
Griffith Law School, Griffith University, Nathan, Queensland, Australia
e-mail: c.lawson@griffith.edu.au

M. Jaspars
Department of Chemistry, Marine Biodiscovery Centre, University of Aberdeen, Aberdeen, UK

M. Rabone
Department of Life Sciences, The Natural History Museum, London, UK

F. Humphries (ed.), *Decoding Marine Genetic Resource Governance Under the BBNJ Agreement*, Sustainable Development Goals Series, https://doi.org/10.1007/978-3-031-72100-7_12

Jurisdiction (BBNJ Agreement) establishes a regime for notification and benefit sharing of marine genetic resources (MGRs) and digital sequence information on MGRs (DSI) of Areas Beyond National Jurisdiction (ABNJ) (A/CONF.232/2023/4 articles 9–16; see also UNCLOS, parts VII and IX). This chapter focuses on two requirements in the BBNJ Agreement that represent legal innovations that are generally not seen in other treaties—specific reference to the use of a unique identifier and to data management plans (DMP). A DMP is 'a living document for a research project, which outlines data creation, data policies, access and ownership rules, management practices, management facilities, and equipment, and who will be responsible for' data management (see, e.g., Australian Research Data Commons). The information from the DMP must be included in the pre-collection notification and updated where necessary in the post-collection notification (article 12(2)(j)); 12(5)(d)). A DMP is also required for third-party access to MGRs and DSI under the 'utilization' notification (article 12(8)(d)). Monitoring and transparency of 'activities' is achieved through notifications as aggregate reporting to the Clearing-House Mechanism (CHM) using '"BBNJ" standardized batch identifiers' (BBNJ Identifier hereafter; articles 12(7) and 16(1)). These identifiers are to be 'automatically generated' upon prior notification to the CHM (article 12(3)). Little is known about how the (undefined) BBNJ Identifier will work in practice, but it will be used 'according to procedures adopted by the Conference of the Parties (COP) as recommended by the access and benefit-sharing committee' (article 16(1)).

The BBNJ Identifier is essentially a tool for linking information about the subject matter of the BBNJ Agreement—MGRs and DSI—to reporting mechanisms via the CHM. Contracting Parties are required to implement the necessary legislative, administrative, or policy measures to ensure that information is notified to the CHM upon three triggering events—pre-collection of in situ MGRs (when the CHM issues the BBNJ Identifier), post-collection, and 'utilization' of MGRs, as well as reporting by repositories and databases (where practicable) on access to MGR and DSI (articles 12(1), (2), (5), (7) and (8)). The BBNJ Identifier in the BBNJ Agreement has a range of functions, including providing a digital link between the notification information about MGRs and DSI and details about where they are housed—the publicly accessible repositories and databases (article 14(3)). Additional and/or alternative future modalities for the sharing of monetary benefits from the utilization of MGRs (and DSI) will be determined by the COP upon recommendations by the Access and Benefit-Sharing Committee (ABSC). The BBNJ Agreement text does not explicitly assign a role for the BBNJ Identifier to link information concerning the use of traditional knowledge associated with MGRs (or DSI) in ABNJ (article 13), although it is likely to be a role that the ABS Committee, the STB and BBNJ Agreement COP will investigate.

The aim of this chapter is to offer legal interpretations and perspectives on the DMP and BBNJ Identifier obligations under the BBNJ Agreement. This chapter provides legal context and background to the later Chap. 14 (Rabone et al., 2025) that explores practical considerations for the holders of MGRs and DSI of ABNJ (such as scientists and research institutions) to align their practices with the intent of the DMP and BBNJ Identifier obligations under the BBNJ Agreement. It builds on the interpretation of information-sharing requirements outlined in Chap. 5 (notification system) (Humphries et al., 2025), Chaps. 6 and 13 (benefit-sharing obligations) (Broggiato et al., 2025; Lavelle & Wynberg, 2025), and Chap. 7 (monitoring and transparency obligations) (Langlet et al., 2025). These jurisdictional issues will include consideration and compliance with the overlapping obligations under the United Nations' *Convention of Biodiversity* (CBD) and its *Nagoya Protocol on Access to Genetic Resources and the Fair and Equitable Sharing of Benefits Arising from their Utilization to the Convention on Biological Diversity* (Nagoya Protocol) that provide a framework for access and benefit sharing (ABS) of MGRs within national jurisdiction (Kachelriess et al., 2025).

It draws from lessons learned for the data ecosystem from other agreements like the Food and Agriculture Organization of the United Nations' (FAO) *International Treaty for Plant Genetic Resources for Food and Agriculture* (Plant Treaty) and the World Health Organization of the United Nations' (WHO) *Pandemic Influenza Preparedness Framework* (PIP Framework).

The chapter is structured as follows: Sect. 12.2 outlines the subject matter and activities that the DMP/BBNJ Identifier is seeking to capture. Section 12.3 discusses the BBNJ Identifier in context of other relevant legal frameworks with reference to terminology and definitional considerations, and Sect. 12.4 suggests how the BBNJ Identifier system requirements may be operationalized from an international governance point of view. Section 12.5 offers some insights into what the BBNJ Agreement envisages as the purpose and content of DMPs. The chapter concludes that the BBNJ Identifier and DMPs can contribute to modalities for the sharing of the benefits arising from the use of MGRs and DSI on MGR that are mutually supportive of, and adaptable to other access and benefit-sharing instruments.

12.2 What is Being Identified, How and Why? The Function of the BBNJ Identifier as Intended by Negotiators

The digitization of biodiversity data, including for MGRs and associated data including DSI, is progressing rapidly, and there is an urgent need to preserve linkages between these data to improve the chain of provenance (origin) and the assembly of large datasets for biodiversity research (Guralnick et al., 2015; Rabone et al., 2023a, 2023b). A resolution is to use recognizable, persistent, globally unique, and stable identifiers to link the data (Guralnick et al., 2015 p. 134). The BBNJ Identifier (and DMP) requirements are important for fulfilling the BBNJ Agreement obligation that MGR data are Findable, Accessible, Interoperable, and Reusable, or FAIR (Wilkinson et al., 2016).

Labelling or tagging objects (e.g., a sample, a document) is critical for scientific and management (administration) endeavours so that objects can be traced, and associations can be made between objects. The purpose of the BBNJ Identifier is to associate the collected samples/materials from ABNJ with aggregate reports and the outputs of utilization as one information point for the benefit-sharing regime (see articles 12(8), 14(7) and 16(1); Humphries, 2025). For this purpose, the BBNJ Identifier is a label that distinguishes a collection of (or a batch of) samples (or materials) collected on a particular cruise from other cruises and links related information. There are two elements to this purpose. First, a unique identifier that distinguishes the object from all other objects by a unique sign or symbol. Secondly, a description about the uniquely identified object that facilitates other aspects of the uniquely identified object to be searched so that it can be identified and grouped, and importantly, linked with all the related objects/associated information, e.g., ecological data, sequence results, patents, or licences. The key data requirement for a physical MGR sample and its associated information such as DSI is that they remain linked or associated with each other (see Rabone et al., 2019). MGR and information including DSI are distinct, but intrinsically linked entities.

First, what is being identified? The BBNJ Agreement defines MGRs very broadly as 'any material of marine plant, animal, microbial, or other origin containing functional units of heredity of actual or potential value' (article 1(8)). It does not define DSI, nor does it define 'sample' (consistent with the CBD, see Sect. 12.3; Chap. 3, Humphries, 2025). The definition indicates that samples are physical materials as opposed to intangible materials, but this is a question for resolution by the COP. The adoption of a distinct DSI terminology, 'marine genetic resources *and* digital sequence information on marine genetic resources' (emphasis added) (articles 9–12 and 14–16) confirms that information (DSI) is distinct to the physical materials (MGRs). The phrase 'digital sequence information' (DSI) arose in discussions in the

CBD and Nagoya Protocol forum (UNEP/CBD/COP/13/25 para. 321 and Decision XIII/16 (para. 1); see also CBD/COP/14/14 p. 172 (Decision 14/20)). DSI has also been discussed at forums including the Plant Treaty and the PIP Framework (see, e.g., Lawson et al., 2020). The 2022 Conference of the Parties to the CBD adopted the Kunming-Montreal Global Biodiversity Framework (GBF) (CBD/COP/DEC/15/4 para. 1 and Annex) where the phrase has been reframed as 'digital sequence information on genetic resources' (CBD/COP/DEC/15/9; see also CBD/COP/DEC/15/4 para. 2(e)). However, the term DSI was undefined in the CBD forum as a 'placeholder' (CBD/DSI/AHTEG/2018/1/4 Annex, para 1; see also CBD/DSI/AHTEG/2020/1/3). The usage of this term in the BBNJ Agreement (and non-definition) was undoubtedly necessary to reach agreement as there is still no consensus about what DSI is, including DSI, and whether there is a consistent meaning across different forums with the CBD and Nagoya Protocol forum using the terminology 'digital sequence information on genetic resources' (see, e.g., CBD/COP/DEC/15/9). Both the BBNJ Agreement and GBF now appear to use this term as a 'definition' and no longer as a 'placeholder' even though there is recognition of 'divergent views' about the scope of DSI (CBD/COP/DEC/15/9 Preamble).

There are several clues in the evolution of the BBNJ Agreement text to indicate that the intent of drafters for the BBNJ Identifier was a light touch, centrally issued machine-readable identifier capable of linking information about uses of MGR of ABNJ, including end uses, for a range of purposes including aggregate reports. However, the form it may take could evolve as technologies develop, for example, the use of artificial intelligence technologies. During the preparatory committee work and the Intergovernmental Committee meetings, there was considerable discussion about whether traceability or track and trace mechanisms might be applied to MGRs linking the collection to all the subsequent research and development (R&D) and commercialization (see Chap. 6 Broggiato et al., 2025). Tracking and tracing means following the sample through each step from collection to research, development, and commercialization, while traceability means being able to find the origin of the sample from a later step (see Jaspars et al., 2021). Blockchain was proposed as a possible pathway for implementing track and trace (Morgera et al., 2020) although the feasibility, desirability, and scientific necessity for such a scheme have been questioned (Oldham et al., 2023; Scholz et al., 2022). The Plant Treaty's approach to monitoring attempts to track biological resources through the R&D process (see Sect. 12.3). Yet, time and financial resources for true track and trace infrastructure are disproportionate to the expected financial benefits from their use (Humphries et al., 2021a; Oldham et al., 2023; Scholz et al., 2022). A clear complexity is the use of DSI, which is not often subject matter under national ABS laws and can often be obtained through open access databases and used independently from the physical resources (see Lawson et al., 2020 pp. 19–26). The Conference of the Parties to the CBD (CBD COP) recognized that 'tracking and tracing of all digital sequence information on genetic resources is not practical' (CBD/COP/DEC/15/9 2022 para. 5). The CBD COP has established an Ad Hoc Open-ended Working Group on Benefit Sharing from the Use of Digital Sequence Information on Genetic Resources to advice on these matters (CBD/COP/DEC/15/9 2022 para. 18).

The requirement for aggregate reports indicates the BBNJ Identifier is intended to be a traceability end-user system, rather than track and trace. Parties 'must ensure that repositories, to the extent practicable, and databases under their jurisdiction prepare, on a biennial basis, an aggregate report on access to' MGR and DSI 'linked to their "BBNJ" standardized batch identifier, and make the report available to the' ABS Committee (article 12(7)). This aggregate report is linked to the sharing of monetary payments, the additional and/or alternative future modalities of which will be decided by the BBNJ Agreement COP, but, for example, may include a tiered fee 'based on a diversified set of indicators measuring the aggregate level of activities by a Party' (article 14(7)(c)). This means that an aggregate report will not report on the use that has been made of a specific MGR, but only provides some indication on how MGRs and, where practicable, DSI have been 'accessed' from a repository or database within jurisdiction of a Party (article 12(7)), which might then be used to determine potential financial contributions to the multilateral fund.

The intention for a light touch system is indicated by the decision to accommodate existing scientific practices. Earlier drafts of the BBNJ Agreement text appeared to use the term identifier in the sense of a legal identifier used for administrative purposes, rather than a unique identifier associated with a specimen generally used for scientific purposes (Rabone et al., 2019). Broggiato et al. (2018) proposed a notification system requiring data already collected by scientists and a unique identifier to be associated with each notification to keep track of cruise information and samples collected. Early drafts of the BBNJ Agreement text suggested the use of a unique identifier for individual MGRs (e.g., A/CONF.232/2019/6 draft article 13(3)(a)). Humphries et al. (2021a) suggested that the cruise code as a bulk collection identifier would align better with scientific practice than usage of a (separate) legal identifier and be more practical than identifiers for each MGR. As an illustration, one water sample may contain thousands of organisms, and all may eventually receive identifiers as they are identified/worked on; however, each MGR will be linked back to the collection information, i.e., a 'one-to-any' relationship whereby the sample origin can always be traced. Therefore, it is the collection identifier that is relevant/needed in terms of the BBNJ requirements. The success and credibility of the BBNJ Identifier depends on the ability therefore to identify samples and associated information, including DSI, that are derived from the original cruise collections without having to track each iteration of the MGRs and their associated information.

The language and system of the 'batch' identifier adopted during ICG5 appears to have been based on a model proposed by Oldham and Thambisetty (2023) (see Tur de la Concepción, 2024 p. 4). In this model:

1. A cruise leader makes a straightforward notification under the Treaty.
2. A batch identifier ... is automatically generated to accompany samples and is linked to appropriate use under the Treaty.
3. The batch identifier is included in existing biodiversity information systems.
4. The batch identifier is included in the outputs of scientific research, and data on its use can be retrieved through automated means in scientific, taxonomic, publication, and patent databases.
5. Companies gain legal certainty under the Treaty and may use the identifier in support of marketing and advertising.
6. It becomes possible to automate the development of indicators on marine genetic resources and digital sequence information on genetic resources under the Treaty.
7. A range of flexible monetary benefit-sharing measures organized around payment tiers are enabled by the use of the identifier' (Oldham & Thambisetty, 2023 pp. 1–2).

Whether these features will be adopted by the BBNJ Agreement remains unclear as the batch identifiers are to be automatically generated by the CHM (article 12.3) according to the mechanism that will be determined by the COP (see A/CONF.232/2023/4, articles 15.3(d) and 51). What is established is that the BBNJ Identifier will be issued as part of the pre-collection notification, which includes information about where the materials and data will be deposited (article 14.3). In this way, the BBNJ Identifier is findable and accessible and could be used to generate aggregate reports. Different datasets will have the identifier associated with them so that they can be linked. There will be no need for the metadata from pre-collection, post-collection, and 'utilization' events to be recorded with every data deposition if it is already available elsewhere—the BBNJ Identifier could connect them. For example, a record in the Global Biodiversity Information Facility (GBIF) can contain data on samples and their taxonomy, while the International Nucleotide Sequence Database Collaboration (INSDC) can contain the DSI. If both have the same BBNJ Identifier, it will be the link that connects this information. The INSDC Bioproject system works in a similar way (see Humphries et al., 2021a pp. 12–14). The identifiers need to be unique so that the object associated with that identifier can be located and 'links to relevant global, regional, subregional, national and sectoral clearing-house mechanisms and other gene banks, repositories, and databases' (article 51.3(c)). Table 12.1 shows the type of information on MGR activities that will be associated with the BBNJ Identifier. One unknown is to whom the BBNJ Identifier will be issued. This will hinge on who is responsible for the pre-collection notification. While this could be any number of actors, from the Party to the BBNJ Agreement, to the project funders, it is likely to be the principal investigator, i.e., the cruise or autonomous uncrewed vessel lead (see Chap. 5 Humphries et al., 2025).

12.3 The BBNJ Identifier in the Context of Relevant Legal Frameworks

Identifiers are also present in other forums for a governing access and benefit sharing of biological resources. The CBD and Nagoya Protocol have a monitoring and compliance framework that includes an Internationally Recognized Certificate of Compliance (IRCC) system (Nagoya Protocol, article 17), the ABS Clearing House (ABSHCH) (CBD, article 18(3); Nagoya Protocol, article 14) and checkpoints (Nagoya Protocol, article 17(1)). The IRCC includes a unique identifier ('ABSCH Unique Identifier') for the certificate associated with the subject matter or genetic resources covered by the certificate (Nagoya Protocol, article 17(4)). The main purpose of the IRCC, however, is to confirm satisfaction of the requirements for prior informed consent and mutually agreed terms (Nagoya Protocol, article 17(4)), and not necessarily the physical materials (see Humphries et al., 2021a).

The Plant Treaty in contrast has developed a Digital Object Identifier (DOIs) scheme to address the need for unique identifiers that facilitate links between the physical plant samples (genetic resources) to information about those samples, with agreed identifiers and descriptive metadata (see Alercia et al., 2018) coordinated through a Global Information System (GLIS) (see Plant Treaty, article 17; see Lawson, 2015; Lawson et al., 2019). The key elements of the GLIS are that it facilitates linkages between existing systems neither replacing nor duplicating their functionality, provides for DOIs applicable to all types of plant samples/materials and DOIs created for other systems, and its use is voluntary (Alercia et al., 2018 pp. 1–2). The functionality of the GLIS is facilitated through the data forms recommended (Alercia et al., 2018 pp. 14–25) and guidelines on the use of DOIs (Plant Treaty Secretariat 2017).

In a further contrast, the PIP Framework specifically provides for a traceability mechanism to 'track in real time' the receipt and transfer of materials within the WHO system ('Influenza Virus Traceability Mechanism') (PIP Framework, article 5.3.1). However, this does not follow the materials outside the WHO system (such as private pharmaceutical companies) where there is only a requirement to report material transfers to WHO (PIP Framework, article 5.4.2 and Annex 2 (para 4.4)). These outside entities are encouraged to deposit 'genetic sequences' and associated data in public domain or public access databases such as GenBank of the International Nucleotide Sequence Database Collaboration (INSDC; PIP Framework, article 5.2.2), where they would necessarily be assigned an accession number that is an unique identifier for each deposited sequence (Lawson & Rourke, 2016; Rohden et al., 2020).

These systems also vary in terms of uptake. Multiple issues have been identified with usage of the CBD's IRCC and ABSCH Unique Identifier system, including lack of uptake and challenges in traceability given the identifier is not machine readable (see Rabone et al., 2019; Rohden et al., 2020; Scholz et al., 2022). To date, there have only been 5065 IRCCs with

Table 12.1 Information to be provided to the clearing-house mechanism that is associated with the '"BBNJ" standardized batch identifier'

Article	Information obligation
12(3)	Unique '"BBNJ" standardized batch identifier' (automatically generated)
12(2)(a)	The nature and objectives under which the collection is carried out, including, as appropriate, any programme(s) of which it forms part
12(2)(b)	The subject matter of the research or, if known, the marine genetic resources to be targeted or collected, and the purposes for which such resources will be collected
12(2)(c)	The geographical areas in which the collection is to be undertaken
12(2)(d)	A summary of the method and means to be used for collection, including the name, tonnage, type and class of vessels, scientific equipment and/or study methods employed
12(2)(e)	Information concerning any other contributions to proposed major programmes
12(2)(f)	The expected date of first appearance and final departure of the research vessels, or deployment of the equipment and its removal, as appropriate
12(2)(g)	The name(s) of the sponsoring institution(s) and the person in charge of the project
12(2)(h)	Opportunities for scientists of all States, in particular scientists from developing States, to be involved in or associated with the project
12(2)(i)	The extent to which it is considered that States that may need and request technical assistance, in particular developing States, should be able to participate or to be represented in the project
12(2)(j)	A data management plan prepared according to open and responsible data governance, taking into account current international practice
12(5)(a)	The repository or database where digital sequence information on marine genetic resources is or will be deposited
12(5)(b)	Where all marine genetic resources collected in situ are or will be deposited or held
12(5)(c)	A report detailing the geographical area from which marine genetic resources were collected, including information on the latitude, longitude, and depth of collection, and, to the extent available, the findings from the activity undertaken
12(5)(d)	Any necessary updates to the data management plan
12(6)	Identifying information that samples and DSI on MGR in repositories or databases under their jurisdiction can be identified as originating from Areas Beyond National Jurisdiction (to the extent practicable)
12(7)	Information on access to MGRs and DSI in repositories, to the extent practicable, and databases for the purpose of the aggregate report to the access and benefit-sharing committee
12(8)(a)	Where the results of the utilization, such as publications, patents granted, if available and to the extent possible, and products developed, can be found
12(8)(b)	Where available, details of the post-collection notification to the Clearing-House Mechanism related to the marine genetic resources that were the subject of utilization
12(8(c)	Where the original sample that is the subject of utilization is held
12(8)(d)	The modalities envisaged for access to marine genetic resources and digital sequence information on marine genetic resources being utilized, and a data management plan for the same
12(8)(e)	Once marketed, information, if available, on sales of relevant products and any further development
14(2)(d)	Information contained in the notifications and '"BBNJ" standardized batch identifiers' in publicly searchable and accessible forms
15(4)(c)	Any information required by the decisions taken by the Conference of the Parties
16	Reporting requirements under the monitoring and transparency mechanism may include information associated with the BBNJ Identifier

ABSCH Unique Identifiers recorded by the ABSHCH (as of May 2024), with most involving non-commercial research and representing records from very few of the CBD's and Nagoya Protocol's Contracting Parties (see Avilés-Polanco et al., 2019). In contrast, the Plant Treaty adopted DOIs under a voluntary arrangement encouraging the use of these identifiers with newly minted DOIs of each new iteration of plant material, such as the project seed from a cross, and all coordinated through a GLIS facilitating linkages between existing database systems (see, e.g., Plant Treaty Secretariat 2017; see also Alercia et al., 2018). Exploiting plant materials has a well-understood pipeline from the seedbank through R&D (plant breeding) to an economic product (an improved plant variety) that can in most cases be easily tracked and traced, and yet the monetary benefits are meagre, even though the non-monetary benefits are considerable (see, e.g., López Noriega et al. 2019 pp. 823–829). This probably reflects the voluntary nature of monetary benefit sharing where the commercialized improved plant varieties only have monetary benefits when their further uses are restricted, such as intellectual property protected. The DOIs, however, have attracted popular support among plant breeders where, to date, 1.4 million DOIs have been assigned and the GLIS provides information about those records including links to related information in other databases, publications, or projects (IT/GB-10/23/11 para. 7).

There is no doubt that utilizing genetic resources through R&D has delivered all sorts of benefits, although it is also clear that these benefits have not been evenly distributed with global gaps in development, capacity, and technology (see Laird et al., 2020 p. 1202; Prathapan et al., 2018 p. 1405). What remains uncertain, however, is whether benefit sharing can be harnessed in a way that *actually* captures benefits downstream from something that was collected upstream. Lessons from the Plant Treaty and PIP Framework for monitoring and compliance have prompted an emphasis on capacity building, technology transfer, and other non-monetary benefits under the BBNJ Agreement (see Chaps. 7 and 11 of this collection—Langlet et al., 2025, Kachelriess et al., 2025). From a monetary benefit perspective, it focuses on aggregate reports to show that X (MGR) and Y (DSI on MGR) came from an activity Z (cruise/autonomous underwater vehicle or AUV) for the purpose of benefit sharing and transparency, rather than tracking and tracing through the pipeline. The aggregate report does not allow for direct traceability (product Y came from MGR X). However, data that has the BBNJ Identifier can be traced back to the collection activity and the full cruise report will give exact location and environmental data for each MGR collected (Table 12.2).

12.4 How Could the BBNJ Identifier System Requirements Be Operationalized from an International Governance Viewpoint?

Much of the detail about how BBNJ Identifiers will work in practice is yet to be determined by the input of the subsidiary bodies. The decision to adopt BBNJ Identifiers is the first step. The next steps will require the ABSC and STB to develop guidelines, modalities, and protocols for BBNJ Identifiers and their uses to deliver on their objectives and adopted by the COP (articles 15 and 16(1)). The BBNJ Agreement presents a linear vision of science—from collection to utilization and commercialization—but the reality is vastly different (as outlined in Chap. 14 Rabone et al., 2025) and input from these bodies will be crucial for practical implementation. In terms of operation, the idea is that the BBNJ Identifier will be automatically generated by the CHM during the pre-collection notification (article 12(3)). Then it will link the pre-collection notification to the post-collection update about the precise location of the original collection and current location of the MGRs and DSI, i.e., where they are housed (articles 12(5) and (6)). It will link this information to any subsequent notification concerning 'utilization' of

Table 12.2 Monitoring and compliance measures in the Nagoya Protocol compared to those in the BBNJ Agreement, noting that the procedures to be adopted by the Conference of the Parties as recommended by the access and benefit-sharing committee have not yet been formulated

Nagoya protocol		BBNJ agreement	
Article	Measure	Article	Measure
13.1	National focal points	–	–
13.1	Competent national authorities	–	–
13.2	Access requirements satisfied	–	–
15.1	PIC and MAT satisfied (including imports)	–	–
15.2	Measures for non-compliance with PIC and MAT	–	–
16.1	TK used has indigenous peoples' PIC and MAT	13	FPIC and MAT
16.2	Measures for non-compliance with indigenous peoples' PIC and MAT	–	–
17.1(a)	Checkpoints	12(7)	Aggregate reports from repositories where practicable and databases
17.1(b)	MAT addresses reporting requirements	–	–
18.1	MAT addresses dispute resolution	–	–
17.1(c)	Monitoring and compliance communications tools and systems	–	–
17.3 and 17.4	Internationally recognized certificates of compliance confirming PIC and MAT, including: (a) Issuing authority (b) Date of issuance (c) The provider (d) Unique identifier of the certificate (e) The person or entity to whom prior informed consent was granted (f) Subject–matter or genetic resources covered by the certificate (g) Confirmation that mutually agreed terms were established (h) Confirmation that prior informed consent was obtained (i) Commercial and/or non-commercial use	16(1)	'BBNJ' standardized batch identifiers
6.3(e) and 17.2	Notifying the clearing-house about access permits and compliance certificates	–	–
18.2 and 18.3	Legal avenues for resolving disputes	–	–
19.1	Model MAT contract clauses	–	–
20.1	Codes of conduct, guidelines, and best practice or best standards	–	–
21	Awareness raising	–	–
29	Country-level monitoring and reporting	16(2)	Country-level monitoring and reporting
30	Conference of the Parties-level reporting	16(3)	Conference of the Parties-level reporting

PIC—Prior informed consent; FPIC—Free, prior, and informed consent; MAT—Mutually agreed; TK—Traditional knowledge

MGRs, and where practicable DSI, including the results of the utilization, the location of the original sample, modalities about how others might access the MGRs and DSI under the utilization notification and sales of products if applicable (article 12(8)). Within three years of the utilization, the MGR and DSI (with the BBNJ Identifier) are to be deposited in publicly accessible repositories and databases as a form of benefit sharing, but access to these may be subject to reasonable conditions (articles 14(3) and (4)). The assumption here is that the MGRs and the associated information including DSI can be associated with the results of 'utilization' that is identified in the 'utilization' notification, like publications, patents, registration of products (article 12(8)). Importantly, the BBNJ Identifier should be able to link different datasets where the common element is the same MGR or DSI on MGR (see Tur de la Concepción, 2024 p. 4).

One area of ambiguity for reporting (but less so for the BBNJ Identifier) relates to the definition of the triggering event. Chapter 4 analyses the temporal challenges with definitions of the collection and 'utilization' triggers (Humphries et al., 2025). The identifier is only issued once and is not updated as new information becomes available. Where there is a material change to the information in the pre-collection notification, any updates are to be made prior to the collection (article 12(4)), but the BBNJ Identifier is unchanged. Like the identifier associated with the CBD and Nagoya Protocol's IRCC system, the BBNJ Identifier is attached to the proof of the activity and not assigned to the genetic resources individually. However, the BBNJ Identifier is not updated when records are amended unlike the CBD, which has revision numbers added for amendments to records (Humphries et al., 2021a p. 5). The notification may be amended and updated, but the identifier is simply a tag that is resolvable, persistent, authoritative, and unique.

The BBNJ Agreement decision to form an ABS Committee (article 15) and for the Conference of the Parties to decide on the modalities for monetary benefit sharing (article 7) will ultimately require Contracting Parties to make definitional decisions about 'DSI', 'samples' and 'utilization', probably also with the involvement of the STB. This was not necessary for the Nagoya Protocol as that agreement only required Contracting Parties to adopt appropriate legislative, administrative, and policy measures in their domestic laws (Nagoya Protocol, articles 5.2, 6.2, and 7), and there they could apply the definitional requirements differently—which they have (see, e.g., Humphries et al., 2021b; some Contracting Parties have included 'digital sequence information' in their domestic laws: see CBD/DSI/AHTEG/2020/1/5 Annex (pp. 13–16)). The BBNJ Agreement, in contrast, is seeking a multilateral approach with its benefit-sharing fund (see articles 14(7) and 52) and will require a common understanding among the Contracting Parties so that those applying the scheme conducting 'activities' and collecting samples (articles 11(1) and 11(3)) will know whether they are a part of the scheme, and require a BBNJ Identifier to be generated by the Clearing-House Mechanism for their pre-collection notification (article 12(3)).

The Plant Treaty's recommendations (Alercia et al., 2018 pp. 14–25) and guidelines (Plant Treaty Secretariat 2017) about the uses of (voluntary) DOIs should provide useful guidance. Significant work was conducted to develop an understanding of the data forms required together with recommendations about the main features and benefits of DOIs and a set of basic principles for users to determine when to assign them. A set of descriptors was selected by consulting experts in the relevant disciplines: *Mandatory* descriptors essential to the identification of the material that had to be provided in order to assign a DOI (e.g., the name and address of the holding institution); *Highly recommended* descriptors that should be provided if known in order to enrich the description of the material and facilitate the discovery function (e.g., the biological status of the material; and *Context* descriptors that should be provided, if available, to assist in appreciating the material (e.g., information on the collecting mission) (see IT/GB7/SAC-1/16/3 para. 15). These are essentially the metadata that will facilitate to uses of the unique identifiers and impact how efficient and effective the mechanisms will work. Unlike the voluntary Plant Treaty DOI system, however, the BBNJ Identifiers are to be established through legislative, administrative, or policy measures and the CHM (articles 12(1) and (3)), and inputs from the ABSC and STB.

The Plant Treaty also provides some guidance about the uses of unique identifiers like DOIs. Recall that the higher purpose of the BBNJ Identifier is about facilitating fair and equitable benefit sharing 'arising from activities' with MGRs and DSI (articles 9(a) and 14(1)). The legislative, administrative, or policy measures and the CHM-mandated uses of the BBNJ Identifiers will be important for benefit sharing. Here, the Plant Treaty has issued guidelines on the use of DOIs and assists in clarifying the uses of DOIs to maintain links with the development of information about the materials and the later uses to develop new materials (Plant Treaty Secretariat 2017). To deliver on its function of tracing, the BBNJ Identifier will need to be connecting records linking the MGRs and DSI to their subsequent activities that can then deliver fair and equitable benefits. A key element of this will be the data management plans (DMPs) (article 12(2)(j)).

12.5 Data Management Plans

The BBNJ Agreement imposes obligations on Parties to ensure that DMPs are carried out by various actors. The first is the actor responsible for the pre-collection notification who must prepare a DMP 'according to open and responsible data governance, taking into account current international practice', who is also responsible for making any updates to the DMP as part of the post-collection notification (article 12(2)(j) and 12(5)(d)). The second is the Party where the 'utilization' occurs, who is required to notify the CHM about 'the modalities envisaged for access [to MGRs and DSI] being utilized, and a data management plan for the same' (article 12(8)(d)). It is unclear whether the DMP must be unique to each utilization, or whether this obligation requires Parties to set our requirements for DMPs generally, which will presumably be clarified by the BBNJ Agreement COP (see Chap. 4 Humphries et al., 2025). The BBNJ Agreement contemplates a DMP per se and some limits on its form (article 12(2)(j)). The BBNJ Agreement already provides strict limits on how (remaining) MGRs and DSI are to be handled: that is to say they have to be deposited in publicly accessible repositories and databases, maintained either nationally or internationally, no later than three years from the start of the utilization (article 14(3)).

First, what is a DMP per se? Put simply, data management is about the collection, organization, storage and documentation of data and information about something (see, e.g., Hudson-Vitale & Moulaison-Sandy, 2019). The ideal is that a DMP will be a formal document describing the roles, responsibilities, and activities for data collected and used from the initial research stages through the lifecycle of the data, and include three components: data, management, and a plan (see Gajbe et al., 2021). There is a lot of guidance available in a number of jurisdictions about what a DMP might include and how it might be evaluated (see, e.g., Science Europe: see also, e.g., Gajbe et al., 2021; Miksa et al., 2019a) and a number of available tools, such as DMPOnline, DMPTool, ezDMP, and so on (see, e.g., Stodden et al., 2019). As set out above, the substance and role of DMPs will presumably be clarified by the BBNJ Agreement COP. A good example of some of the relevant elements of a DMP could or should include (Australian Research Data Commons):

- a back-up strategy;
- an existing data survey;
- outline of data to be created;
- file format guidance;
- data/metadata instructions;
- ownership, access and security information;
- data organization and naming conventions;
- information on managing data transfers and synchronization between machines;
- guidelines for collaborative writing with colleagues;

- version control;
- data storage locations;
- hardware budget and management information;
- bibliography management tools;
- data sharing, publishing, archiving and licencing instructions;
- data destruction rules;
- responsibility allocations; and
- a budget for the overall DMP.

This listing is not comprehensive and the specific context of DMPs for the BBNJ Agreement will need to be determined, perhaps starting with the requirement that they include the BBNJ Identifier (article 12(3)). The BBNJ Agreement COP will need to set out the relevant elements and this might be expected to be an output from the ABS Committee (article 16(1)), with inputs from the STB. While DMPs are now a common requirement for researchers, they have proven to generally be problematic because researchers are largely ambivalent to their compliance with their DMPs, and they have been found to be largely ineffectual (Hudson-Vitale & Moulaison-Sandy, 2019 pp. 323–324). The likely important elements of every DMP going forward are that they be a living document that can be updated (as envisioned by the BBNJ Agreement) (article 12(5)(d)), that they are machine readable (Miksa et al., 2019b), and that there are appropriate metadata frameworks to make the data and information available (as addressed above) (see Singh and Madalli et al. 2023).

Next its form, what is a DMP 'made in accordance with open and responsible data governance, taking into account current international practice' (article 12(2)(j))? Current data and information practices in science should, as a matter of good and best practices, favour open access only limited by intellectual property claims (see OECD/LEGAL/0463 pp. 6 and 9–11; UNESCO pp. 3–4 and 11; see also Lawson et al., 2024a) and comply with normative standards such as the FAIR Data Principles that promotes simplifying the discovery, evaluation and reuse of information (Wilkinson et al., 2016). These data and information practices will also need to address the concerns of Indigenous Peoples reflected in expressions such as Indigenous Data Sovereignty (Kukutai & Taylor, 2016) and the CARE Principles for Indigenous Data Governance (Research Data Alliance International Indigenous Data Sovereignty Interest Group, 2019; see also Lawson et al., 2024a). According to these principles, Indigenous Peoples are likely to claim ownership and control over some forms of data and information including data and information about them, their territories, and their ways of life (Carroll et al., 2019). As data and information changes, especially with the introduction of artificial intelligence, these normative standards can be expected to change and the BBNJ Agreement is open to the evolution of these expectations. The principles articulated in the GBF (CBD/COP/DEC/15/9 paras. 4, 5 and 9) also provide guidance both for DMPs and the BBNJ Identifier under the agreement:

1. Encourages the depositing of more digital sequence information on genetic resources, with appropriate information on geographical origin and other relevant metadata, in public databases;
2. Recognizes that tracking and tracing of all digital sequence information on genetic resources is not practical …
3. Also agrees that a solution for fair and equitable benefit sharing on digital sequence information on genetic resources should, inter alia:
   (a) Be efficient, feasible, and practical.
   (b) Generate more benefits, including both monetary and non-monetary, than costs.
   (c) Be effective.
   (d) Provide certainty and legal clarity for providers and users of digital sequence information on genetic resources.
   (e) Not hinder research and innovation.
   (f) Be consistent with open access to data.
   (g) Not be incompatible with international legal obligations.
   (h) Be mutually supportive of other access and benefit-sharing instruments.
   (i) Take into account the rights of Indigenous Peoples and local communities, including

with respect to the traditional knowledge associated with genetic resources that they hold'.

12.6 Conclusions

The BBNJ Agreement is the latest in a series of agreements regulating genetic resources in international law (Lawson, 2012). Up to the BBNJ Agreement, these agreements have maintained a distinct separation between the physical materials and information about those materials. They essentially provided for the information obligations about the ABS legislative, administrative and policy requirements applied by a country (see Humphries et al., 2021b pp. 17–19) and then the information about the resources such as DSI to be free, open, and accessible (Lawson et al., 2019 pp. 107–111). There are some exceptions where countries have included information like DSI in their ABS schemes as a genetic resources or derivatives (CBD/DSI/AHTEG/2020/1/5 Annex, pp. 13–16). Put slightly differently, there has been a conflict between the general obligations dealing with information as a resource derivative within the ABS transaction and the requirements to disclose and exchange information. This conflict has been directly engaged by the GBF (CBD/COP/DEC/15/4 para. 1 and Annex) to establish a distinct multilateral mechanism to share the benefits from accessing DSI (CBD/COP/DEC/15/9 (2022) paras. 3 and 9; see also CBD/COP/DEC/15/4 para. 2(e)).

The BBNJ Agreement maintains this distinction between the physical materials and information and establishes a common multilateral fund mechanism for sharing benefits from both MGRs and DSI (see Chap. 6; Broggiato et al., 2025). The BBNJ Agreement has, however, given to the COP the possibility to decide on additional and/or alternative modalities for sharing the monetary benefits from the use of MGR and DSI that anticipates information being located across a series of repositories and databases, mediated through the unique BBNJ Identifier and a centrally available CHM. This has the potential to create multiple separate mechanisms dealing with DSI under different United Nations agreements (CBD, Plant Treaty, PIP Framework), but conversely it could lead the way towards a harmonized DSI benefit-sharing system (see DSI Scientific Network: see also Aubry et al., 2022), especially because the modalities to be established by the COP for the sharing of the benefits (arising from the use of DSI on MGR) should be mutually supportive of and adaptable to other access and benefit-sharing instruments (article 14(9)). This would meet the principles delineated in the GBF with benefits flowing back into each of these agreements based on the type of aggregate report specified in the BBNJ Agreement. Given these sound principles and the indication that the GBF and BBNJ Agreement will be mutually supportive of other ABS instruments, there is every hope that a harmonized solution to benefit sharing for MGRs and DSI on (M)GR is possible as it will be greatly beneficial to basic research and biodiversity conservation, as well as biotechnology among other applications.

It is unusual for a treaty to include the level of detail as the BBNJ Agreement about identifier and data management mechanisms. However, these were seen as essential inclusions in the text by negotiators, particularly the G77 and China, for effective implementation of the notification, transparency and benefit-sharing systems (see Tur de la Concepción, 2024). It is important to remember that the form of these innovations may evolve as technologies advance, for example, through the use of artificial intelligence and future tools for information exchange. In the treaty text, there is no link between the traditional knowledge obligation and the BBNJ Identifier or the DMP, nor is there explicit reference to the CARE principles for indigenous data management that complement the FAIR principles articulated in the treaty text (see Lawson et al., 2024b). The ABS Committee, SBT, and COP might discuss these linkages in future for a more integrated approach to the governance of MGR, DSI, and associated traditional knowledge under Part II of the BBNJ Agreement.

A longer-term aspiration might be that if such a system is successful and benefits accrue to be used for biodiversity conservation and fundamental biodiversity research, that Article 10 in the Nagoya Protocol might be revisited and a global multilateral benefit-sharing mechanism developed for genetic resources, following in the footsteps of the proposed system for MGRs and DSI under the BBNJ Agreement. The bare details are addressed in the BBNJ Agreement, and more work will be needed by the proposed ABSC and the BBNJ Agreement COP decision(s) on the modalities for monetary benefit sharing, such as suitable guidelines. The BBNJ Identifier was considered the lowest cost option, in effect a 'tag' that labels MGRs and DSI as well as associated data as arising from ABNJ. However, the ideal of the BBNJ Identifier although simple, will require careful implementation, alignment with existing data systems and considerable curation to work effectively.

References

A/CONF.232/2019/6—UN General Assembly. (2019). *Draft Text of an Agreement under the United Nations Convention on the Law of the Sea on the Conservation and Sustainable use of Marine Biological Diversity of Areas Beyond National Jurisdiction*, A/CONF.232/2019/6. UN.

A/CONF.232/2023/4—Intergovernmental Conference on an International Legally Binding Instrument under the United Nations Convention on the Law of the Sea on the Conservation and Sustainable Use of Marine Biological Diversity of Areas Beyond National Jurisdiction. (2023). *Agreement under the United Nations Convention on the Law of the Sea on the Conservation and Sustainable Use of Marine Biological Diversity of Areas Beyond National Jurisdiction* A/CONF.232/2023/4. UN.

ABSHCH—CBD Access and Benefit-sharing Clearing House from https://absch.cbd.int/en

Alercia, A., López, F. M., Sackville Hamilton, N. R. & Marsella, M. (2018). *Digital object identifiers for food crops: Descriptors and guidelines of the global information system*. FAO.

Aubry, S., Frison, C., Medaglia, J. C., Frison, E., Jaspars, M., Rabone, M., Sirakaya, A., Saxena, D., & Van Zimmeren, E. (2022). Bringing access and benefit sharing into the digital age. *Plants, People, Planet, 4*(1), 5–12.

Australian Research Data Commons, *Data Management Plans* from https://ardc.edu.au/resource/data-management-plans

Avilés-Polanco, G., Jefferson, D. J., Almendarez-Hernández, M. A., & Beltrán-Morales, L. F. (2019). Factors that explain the utilization of the Nagoya Protocol framework for access and benefit sharing. *Sustainability, 11*(20), 5550.

Broggiato, A., Vanagt, T., Lallier, L. E., Jaspars, M., Burton, G., & Muyldermans, D. (2018). Mare Geneticum: Balancing governance of marine genetic resources in international waters. *The International Journal of Marine and Coastal Law, 33*(1), 3–33.

Broggiato, A., Dunshirn, P., Jaspars, M. & Pena-Neira, S. (2025). Monetary and non-monetary benefit sharing under the BBNJ Agreement. In F. Humphries (Ed.), *Decoding governance of marine genetic resource governance under the BBNJ Agreement.* Springer.

Carroll, S. R., Rodriguez-Lonebear, D., & Martinez, A. (2019). Indigenous data governance: Strategies from United States native nations. *Data Science Journal, 18*, 31.

CBD—*Convention on Biological Diversity*, opened for signature 5 June 1992, 1760 U.N.T.S. 79 (entered into force 29 December 1993) (CBD).

CBD/COP/14/14—Conference of the Parties to the Convention on Biological Diversity. (2019). *Report of the Conference of the Parties to the Convention on Biological Diversity on its Fourteenth Meeting.* UNEP.

CBD/COP/DEC/15/4—Conference of the Parties to the Convention on Biological Diversity. (2022). *Kunming-Montreal Global Biodiversity Framework* CBD/COP/DEC/15/4. UNEP.

CBD/COP/DEC/15/9—Conference of the Parties to the Convention on Biological Diversity. (2022). *Digital sequence information on genetic resources* CBD/COP/DEC/15/9. UNEP.

CBD/DSI/AHTEG/2018/1/4—*Ad Hoc* Technical Expert Group on Digital Sequence Information on Genetic Resources. (2018). *Report of the Ad Hoc Technical Expert Group on Digital Sequence Information on Genetic Resources* CBD/DSI/AHTEG/2018/1/4.

CBD/DSI/AHTEG/2020/1/3—*Ad Hoc* Technical Expert Group on Digital Sequence Information on Genetic Resources. (2020). *Digital sequence information on genetic resources: Concept, scope and current use* CBD/DSI/AHTEG/2020/1/3.

CBD/DSI/AHTEG/2020/1/5—*Ad Hoc* Technical Expert Group on Digital Sequence Information on Genetic Resources. (2020). *Fact-finding study on how domestic measures address benefit-sharing arising from commercial and non-commercial use of digital sequence information on genetic resources and address the use of digital sequence information on genetic resources for research and development* CBD/DSI/AHTEG/2020/1/5. UNEP.

DSI Scientific Network, *A harmonized system for benefit-sharing from DSI*, Policy Brief from https://www.dsiscientificnetwork.org/wp-content/uploads/2023/11/Policy-Brief-Harmonization-of-DSI-BS-systems.pdf

Gajbe, S. B., Tiwari, A., & Singh, R. K. (2021). Evaluation and analysis of data management plan tools: A parametric approach. *Information Processing & Management, 58*(3), 102480.

Guralnick, R. P., Cellinese, N., Deck, J., Pyle, R. L., Kunze, J., Penev, L., Walls, R., Hagedorn, G., Agosti, D., Wieczorek, J., & Catapano, T. (2015). Community next steps for making globally unique identifiers work for biocollections data. *ZooKeys, 494*, 133–154.

Rabone, M., Horton, T., Humphries, F., Lyal, C., Muraki Gottlieb, H., Scholz, A.H., Vanagt, T. & Jaspars, M. (2025). BBNJ Agreement: Considerations for scientists and commercial end users of MGR at research, development and commercialisation stages. In F. Humphries (Ed.), *Decoding marine genetic resource governance under the BBNJ Agreement.* Springer.

Hudson-Vitale, C., & Moulaison-Sandy, H. (2019). Data management plans: A review'. *DESIDOC Journal of Library & Information Technology, 39*(6), 322–328.

Humphries, F. (2025). Marine genetic resources beyond national jurisdiction: The expansive scope of the BBNJ Agreement. In F. Humphries (Ed.), *Decoding marine genetic resource governance under the BBNJ Agreement.* Springer.

Humphries, F., Gottlieb, H. M., Laird, S., Wynberg, R., Lawson, C., Rourke, M., Tvedt, M. W., Oliva, M. J., & Jaspars, M. (2020). A tiered approach to the marine genetic resource governance framework under the proposed UNCLOS agreement for biodiversity beyond national jurisdiction (BBNJ). *Marine Policy, 122*, 103910.

Humphries, F., Rabone, M., & Jaspars, M. (2021a). Traceability approaches for marine genetic resources under the proposed ocean (BBNJ) treaty'. *Frontiers in Marine Science, 8*, 661313.

Humphries, F., Laird, S., Wynberg, R., Morrison, C., Lawson, C. & Kolesnikova, A. (2021b). *Survey of access and benefit-sharing country measures accommodating the distinctive features of genetic resources for food and agriculture and associated traditional knowledge*, Background Study Paper 70. FAO.

Humphries, F., Jaspars, M., Lavelle, J. & Kachelriess, D. (2025). The novel notification information system for marine genetic resources under the BBNJ Agreement. In F. Humphries (Ed.), *Decoding marine genetic resource governance under the BBNJ Agreement.* Springer.

IT/GB-10/23/11—Governing Body of the International Treaty for Plant Genetic Resources for Food and Agriculture. (2023). *Report on the implementation of the global information system* IT/GB-10/23/11. FAO.

Jaspars, M., Humphries, F. & Rabone, M. (2021). *Tracing options for marine genetic resources from within national jurisdiction.* Commonwealth Secretariat.

Kachelriess, D., Dunshirn, P., Langlet, A., Brown, A. E. L. & Scholz, A.H. (2025). Marine genetic resources and digital sequence information under the BBNJ Agreement: Interlinkages with other access and benefit sharing frameworks. In F. Humphries (Ed.), *Decoding marine genetic resource governance under the BBNJ Agreement.* Springer.

Kukutai, T., & Taylor, J. (2016). Data sovereignty for Indigenous Peoples: Current practice and future needs. In T. Kukutai & J. Taylor (Eds.), *Indigenous data sovereignty: Toward an agenda* (pp. 1–22). ANU Press.

Laird, S., Wynberg, R., Rourke, M., Humphries, F., Muller, M. R., & Lawson, C. (2020). Rethink the expansion of access and benefit sharing. *Science, 367*(6483), 1200–1202.

Langlet, A. & Dunshirn, P. (2022). *Traceability options for marine genetic resource from areas beyond national jurisdiction.* High Seas Alliance.

Langlet, A., Dunshirn, P., Jaspars, M., Humphries, F., & Kachelriess, D. (2025). Monitoring and transparency aspects of MGR-utilization under the BBNJ Agreement. In F. Humphries (Ed.), *Decoding marine genetic resource governance under the BBNJ Agreement.* Springer.

Lavelle, J. & Wynberg, R. (2025). Benefit sharing under the BBNJ Agreement in practice. In F. Humphries (Ed.), *Decoding marine genetic resource governance under the BBNJ Agreement.* Springer.

Lawson, C. (2010). Taxonomic conceptions of algae, animals, fungi and plants in granting intellectual property privileges. *Griffith Law Review, 19*(3), 472–503.

Lawson, C. (2012). *Regulating genetic resources: Access and benefit sharing in international law.* Edward Elgar.

Lawson, C. (2015). Information intellectual property and the global information system for plant genetic resources for food and agriculture. *Australian Intellectual Property Journal, 26*(1), 27–38.

Lawson, C., & Rourke, M. (2016). Open access DNA, RNA and amino acid sequences: The consequences and solutions for the international regulation of access and benefit sharing. *Journal of Law and Medicine, 24*(1), 96–118.

Lawson, C., Humphries, F., & Rourke, M. (2019). The future of information under the CBD, Nagoya protocol, plant treaty and PIP framework. *Journal of World Intellectual Property, 22*(3–4), 103–119.

Lawson, C., Rourke, M., & Humphries, F. (2020). Information as the latest site of conflict in the ongoing contests about access to and sharing the benefits from exploiting genetic resources. *Queen Mary Journal of Intellectual Property, 10*(1), 7–33.

Lawson, C., Humphries, F., & Rourke, M. (2024a). Challenging the existing order of knowledge sharing governance with digital sequence information on genetic resources. *Journal of Intellectual Property Law and Practice, 19*(4), 337–357.

Lawson, C., Humphries, F. & Rourke, M. (2024b). Genetic resources as culture and heritage: Repatriation and benefit sharing. *Melbourne Journal of International Law* 24(1): in press.

López, I., Halewood, M., Abberton, M. T., Amri, A., Angarawai, I. I., Anglin, N. L., Blümmel, M., Bouman, B., Campos, H., Costich, D., Ellis, D., Gaur, P. M., Guarino, L., Hanson, J., Kommerell, V., Kumar, L., Lusty, C., Ndjiondjop, M.-N., Payne, T., … Wenzl, P. (2019). CGIAR operations under the plant treaty framework. *Crop Science, 59*(3), 819–832.

Miksa, T., Walk, P., & Neish, P. (2019a). *RDA DMP Common Standard for Machine-actionable Data Management Plans* from https://www.rd-alliance.org/group/dmp-common-standards-wg/outcomes/rda-dmp-common-standard-machine-actionable-data-management.

Miksa, T., Simms, S., Mietchen, D., & Jones, S. (2019b). Ten principles for machine-actionable data management plans. *PLoS Computer Biology, 15*(3), e1006750.

Morgera, E., Switzer, S. & Geelhoed, M. (2020) *Study for the European Commission on 'Possible ways to address digital sequence information—legal and policy aspects'*. ENV.F.3/SER/2019/6175145. European Commission.

Nagoya Protocol—Conference of the Parties to the Convention on Biological Diversity. (2010). *Report of the Tenth Meeting of the Conference of the Parties to the Convention on Biological Diversity* (2010) UNEP/CBD/COP/10/27, [103] and Annex (Decision X/1, Annex 1 (Nagoya Protocol), pp. 89–109). UNEP.

OECD/LEGAL/0463—Organisation for Economic Co-operation and Development. (2021). *Recommendation of the council on enhancing access to and sharing of data*, OECD/LEGAL/0463. OECD.

Oldham, P., & Thambisetty, S. (2023). *ONEST: The middle way for monetary benefit sharing in BBNJ Negotiations*. https://doi.org/10.5281/zenodo.7573699

Oldham, P., Chiarolla, C., & Thambisetty, S. (2023). *Digital sequence information in the UN High Seas Treaty: Insights from the Global Biodiversity Framework-related decisions*, LSE Law-Policy Briefing Paper 53. LSE Law School.

PIP Framework—World Health Organisation. (2011). *Pandemic Influenza Preparedness: Sharing of Influenza Viruses and Access to Vaccines and Other Benefits*, Sixty-Fourth World Health Assembly WHA64.5. WHO.

Plant Treaty—*International Treaty on Plant Genetic Resources for Food and Agriculture*, 3 November 2001, 2400 U.N.T.S. 303 (entered into force 29 June 2004).

Plant Treaty Secretariat—Secretariat of the International Treaty for Plant Genetic Resources for Food and Agriculture. (2017). *Guidelines for the optimal use of digital object identifiers as permanent unique identifiers for plant genetic resources for food and agriculture—v.2*. FAO.

Prathapan, K. D., Pethiyagoda, R., Bawa, K. S., Raven, P. H., Rajan, P. D., et al. (2018). When the cure kills—CBD limits biodiversity research. *Science,* 360(6396): 1405–1406.

Rabone, M., Harden-Davies, H., Collins, J. E., Zajderman, S., Appeltans, W., Droege, G., Brandt, A., Pardo-Lopez, L., Dahlgren, T. G., Glover, A. G., & Horton, T. (2019). Access to marine genetic resources (MGR): Raising awareness of best-practice through a new agreement for biodiversity beyond national jurisdiction (BBNJ). *Frontiers in Marine Science, 6*, 520.

Rabone, M., Wiethase, J. H., Simon-Lledó, E., Emery, A. M., Jones, D. O., Dahlgren, T. G., Bribiesca-Contreras, G., Wiklund, H., Horton, T., & Glover, A. G. (2023a). How many metazoan species live in the world's largest mineral exploration region? *Current Biology, 33*(12), 2383–2396.

Rabone, M., Horton, T., Jones, D. O. B., Simon-Lledó, E., & Glover, A. G. (2023b). A review of the International seabed authority database DeepData from a biological perspective: Challenges and opportunities in the UN Ocean Decade. *Database* 2023, p.baad013.

Research Data Alliance International Indigenous Data Sovereignty Interest Group. (2019). *CARE principles for indigenous data governance.* Global Indigenous Data Alliance from https://www.gida-global.org/care

Rohden, F., Huang, S., Dröge, G., & Hartman-Sholz, A. (2020). Combined study on Digital Sequence Information (DSI) in public and private databases and traceability. Ad Hoc Technical Expert Group on Digital Sequence Information on Genetic Resources to the CBD (Report No. CBD/DSI/AHTEG/2020/1/4).

Scholz, A. H., Freitag, J., Lyal, C. H., Sara, R., Cepeda, M. L., Cancio, I., Sett, S., Hufton, A. L., Abebaw, Y., Bansal, K., & Benbouza, H. (2022). Multilateral benefit-sharing from digital sequence information will support both science and biodiversity conservation. *Nature Communications, 13*(1), 1086.

Science Europe, *Research Data Management* from https://scienceeurope.org/our-priorities/research-data/research-data-management

Singh, R. K., & Madalli, D. P. (2023). DMPFrame: A conceptual metadata framework for data management plans. *Journal of Library Metadata, 23*(3–4), 121–160.

Stodden, V., Ferrini, V., Gabanyi, M., Lehnert, K., Morton, J., & Berman, H. (2019). Open access to research artifacts: Implementing the next generation data management plan. *Proceedings of the Association for Information Science and Technology, 56*(1), 481–485.

Tur de la Concepción, R. (2024). Negotiating fair and equitable sharing of benefits in the BBNJ Agreement: Role of the Group of 77 and China. *Marine Policy, 163*, 106085.

UNCLOS—*United Nations Convention on the Law of the Sea* opened for signature 10 December 1982, 1833 U.N.T.S 397 (entered into force 16 November 1994).

UNEP/CBD/COP/10/27—Conference of the Parties to the Convention on Biological Diversity. (2011). *Report of the Tenth Meeting of the Conference of the Parties to the Convention on Biological Diversity*, UNEP/CBD/COP/10/27, [103] and Annex 1 (Decision X/1, pp. 89–109) (Nagoya Protocol). UNEP.

UNEP/CBD/COP/13/25—Conference of the Parties to the Convention on Biological Diversity. (2016). *Report of the Conference of the Parties to the Convention on Biological Diversity on its Thirteenth Meeting*, UNEP/CBD/COP/13/25. UNEP.

UNESCO—United Nations Educational, Scientific and Cultural Organization. (2021). *UNESCO recommendation on open science*. UNESCO.

Wilkinson, M. D., Dumontier, M., Aalbersberg, I. J., Appleton, G., Axton, M., Baak, A., Blomberg, N., Boiten, J. W., da Silva Santos, L. B., Bourne, P. E., & Bouwman, J. (2016). The FAIR guiding principles for scientific data management and stewardship. *Scientific Data, 3*(1), 1–9.

Charles Lawson is a professor at the Griffith Law School, Griffith University, Australia, with interests in intellectual property law, public finance law and the interactions between law and science including access and benefit sharing biological resources.

Fran Humphries has specialized in marine and biodiversity law and policy for over two decades in government, academia and consultancies. She is an associate professor at Griffith Law School, Griffith University, Australia, with a background in fisheries management and marine law. She has led large international research consultancies on access and benefit sharing of biological resources for governments, UN organizations and other institutions. She was on the International Council of Environmental Law delegation for the Biodiversity Beyond National Jurisdiction (BBNJ) Agreement negotiations.

Marcel Jaspars is a professor of Organic Chemistry at the University of Aberdeen where he leads the Marine Biodiscovery Centre which focusses on marine resources for novel pharmaceuticals, and to investigate fundamental questions in marine chemical ecology and biosynthesis. Marcel has been active at national and international levels to develop the science, its applications/industrial uptake and associated policy involved in marine biodiscovery and biotechnology. He provides scientific advice to the UK, EU and UN for global policy processes on ocean conservation and digital sequence information via reports, papers and taking part in discussion meetings.

Muriel Rabone is a researcher based on the deep-sea ecology and systematics group of the Natural History Museum, London. She also studies the neglected tropical diseases: schistosomiasis and paragonimiasis.

Benefit Sharing Under the BBNJ Agreement in Practice

13

Jessica Lavelle and Rachel Wynberg

Abstract

This chapter examines different models and examples of benefit sharing to explore how article 14 of the BBNJ Agreement may be implemented in practice. It describes monetary and non-monetary benefits, as differentiated in the BBNJ Agreement, and considers the different modalities that may be employed for their distribution. Special attention is given to existing access and benefit sharing (ABS) frameworks to highlight possibilities for streamlining and alignment. A focus is placed on approaches to strengthen the capacity of low- and middle-income countries and enhance the conservation and sustainable use of biodiversity of areas beyond national jurisdiction.

Keywords

Access and benefit sharing · Digital sequence information · Marine genetic resources · Nagoya protocol · UNCLOS

J. Lavelle (✉) · R. Wynberg
Department of Environmental and Geographical Science, University of Cape Town, Private Bag, Rondebosch 7701, South Africa
e-mail: Jessica-Jane.Lavelle@uct.ac.za

R. Wynberg
e-mail: Rachel.Wynberg@uct.ac.za

13.1 Introduction

The *Agreement under the United Nations Convention on the Law of the Sea on the Conservation and Sustainable use of Marine Biological Diversity of Areas Beyond National Jurisdiction* (BBNJ Agreement) (UNGA, 2023) is the first internationally binding global legal instrument for the conservation and sustainable use of marine biodiversity of areas beyond national jurisdiction (ABNJ). Adopted in June 2023, the BBNJ Agreement sets a framework for the fair and equitable sharing of benefits from marine genetic resources (MGRs) and associated digital sequence information (DSI), capacity-building and transfer of marine technologies, and area-based management tools, including marine protected areas and environmental impact assessments. Article 14 sets out the benefit sharing obligations Parties have to fulfil, including the fair and equitable sharing of monetary and non-monetary benefits arising from the use and commercialisation of MGRs and DSI in ABNJ. This article also stipulates that benefit sharing should focus on the conservation and sustainable use of biodiversity in ABNJ. In Chap. 6, Broggiato et al. (2025) provide a detailed interpretation of the treaty text, while this chapter shares ideas for the practical implementation of article 14. These conceptions are drawn from different models and examples of benefit sharing across diverse sectors,

F. Humphries (ed.), *Decoding Marine Genetic Resource Governance Under the BBNJ Agreement*, Sustainable Development Goals Series, https://doi.org/10.1007/978-3-031-72100-7_13

partnerships and national experiences to illuminate how best to strengthen benefits for the conservation and sustainable use of marine biodiversity while simultaneously enhancing the capacity of low- and middle-income countries (LMICs).

The concept of benefit sharing arose from the 1992 *Convention on Biological Diversity* (CBD) (United Nations, 1992) which has as one of its three objectives the fair and equitable sharing of benefits from the use of genetic resources. In this context, benefit sharing refers to the act of giving a portion of profits derived from the use of genetic resources and associated traditional knowledge, as well as non-monetary benefits, to the providers of those resources and/or knowledge to achieve fairness (Schroeder, 2007). Benefit sharing as conceptualised in the CBD was intended to incentivise biodiversity conservation, although no direct link between benefit sharing and biodiversity conservation was made explicit, and the extent to which this has happened is limited (Sirakaya, 2021; Wynberg & Laird, 2023). With the 2014 adoption of the *Nagoya Protocol on Access to Genetic Resources and the Fair and Equitable Sharing of Benefits Arising from their Utilisation to the Convention on Biological Diversity* (Nagoya Protocol) (Secretariat of the Convention on Biological Diversity, 2010), fair and equitable benefit sharing has become central to access and utilisation of genetic resources within national jurisdictions. However, in practice, it has been difficult to achieve. These difficulties have been attributed partly to fragmented and overly bureaucratic processes and an unwillingness on the part of users to acknowledge historical and ongoing inequities in research and innovation between high and LMICs. Moreover, little consideration has been given to procedural fairness and to ways in which the power imbalances between parties, and the limited capacities of LMICs, can be addressed to negotiate equitable benefits. In the context of these challenges, the inclusion of MGRs of ABNJ in benefit sharing measures is of interest, linked in part to increased research interest in MGRs and associated DSI.

The inclusion of monetary benefits in the BBNJ Agreement was highly contested between Parties, linked to concerns about the burdensome and costly 'track and trace' processes that could hinder research and innovation. Those arguing for the inclusion of monetary benefits in the agreement raised concerns around the exclusion of Parties from accruing benefits from shared resources, the vast differences in capacity among Parties to access and use MGRs and DSI in ABNJ, and the need for funds to contribute to the conservation and sustainable use of biodiversity in ABNJ. A compromise was ultimately reached, including the establishment of a multilateral funding mechanism (the 'Special Fund') to allocate benefits (art 52). This is in alignment with Decision 15/9 of the Kunming-Montreal Global Biodiversity Framework (GBF) (Conference of the Parties to the Convention on Biological Diversity, 2022) adopted in December 2022, which also established a multilateral funding mechanism to allocate monetary benefits from the utilisation of DSI, thus streamlining the approach to monetary benefit sharing across these legal instruments. Both the BBNJ Agreement and the GBF emphasise the use of monetary benefits towards the conservation and sustainable use of biodiversity. This chapter draws on different models and examples of benefit sharing and considers how the implementation of article 14 may work in practice within the unique context of the BBNJ Agreement. We place a particular focus on benefits for the conservation and sustainable use of marine biodiversity. The chapter begins with describing approaches for sharing monetary benefits and is then followed by section describing those for sharing non-monetary benefits. A final section discusses how research programmes can be better aligned to conservation priorities.

13.2 Monetary Benefits

Article 14, Paragraph 5 sets out the obligation for the fair and equitable sharing of monetary benefits from the utilisation of MGRs and DSI in ABNJ. Monetary benefits are to be managed through a Special Fund and Paragraph 6 requires high-income Parties to make an annual contribution to the Special Fund until additional and/or alternative payment modalities are agreed upon by the Conference of the Parties (CoP). The novel inclusion of an annual contribution by high-income Parties serves to provide regular and immediately available funds for capacity building and support to LMICs even in the absence of successfully commercialising MGRs which may take many years. Calculated at 50 percent of a Party's assessed contribution to the annual budget of the BBNJ Agreement, this innovative way to provide for monetary benefit sharing is decoupled from any specific access or use of MGRs. It therefore does not need a burdensome track and trace system (see also Chapter 6, Broggiato et al., 2025). The additional and/or alternative payment modalities are to then be determined based on the advice of an ABS Committee established under article 15. While multiple modalities may be employed, these are required to be in alignment with other ABS instruments and can be accrued from milestone payments, a percentage of the revenue from the sale of commercialised products, or a tiered fee, among others.

Milestone payments are tied to the application or exploitation of intellectual property rights in cases where the payment is triggered by an activity or occurrence of an event. While milestone payments may enable the capturing of benefits from the early stages of commercialisation, they require the ongoing tracking of research and development which may not be feasible within the unique geographical, political and jurisdictional context of ABNJ and given the complex and fragmented use of MGRs and DSI in the development of products (Humphries et al., 2021). As highlighted in Chap 7 (Langlet et al., 2025), the traceability mechanism employed by the Clearing-House Mechanism needs to be of a light touch, practically feasible for implementation, and not impose additional burdens on scientific users. Given the end user approach adopted in the BBNJ Agreement, whereby research outcomes can be traced to the original collection notification via the 'BBNJ Standardised Batch Identifier' (BBNJ Identifier), the payment of royalties or a levy upon commercialisation may offer a simpler solution. Under the Nagoya Protocol, royalties are usage-based payments made by the user of a genetic resource to the provider and are often agreed to as percentages of gross or net revenue. Percentages of royalty payments have been used by both Brazil and South Africa in national ABS agreements. For example, in South Africa a 'traditional knowledge levy' calculated at 1.5% of the farm-gate price, is paid by users to institutions representing holders of traditional knowledge about rooibos (Wynberg et al., 2023). In Brazil, the framework regulating ABS sets out conservation priorities and lists options for users of genetic resources to support high biodiversity areas, promote sustainable use and/or support Indigenous peoples in protected areas. The law creates a model with two payment options from which users can choose: (i) the first requires companies or users to pay 1% of the annual net revenue derived from the product(s) using genetic resources directly into a national fund ringfenced ad infinitum for conservation and sustainable use; (ii) the second option enables companies or users to perform their own projects or hire NGOs to execute projects. Vetted at a national level to facilitate maximum impact, these projects are required to promote biodiversity conservation and sustainable use and a 'bank' of projects has been created by the Ministry of Environment that companies can choose from, and third parties such as NGOs can be employed to support implementation (Wynberg et al., 2022).

Parties are required to introduce the necessary legislative, administrative and policy measures to comply with benefit sharing provisions under the BBNJ Agreement. Therefore, Parties must ensure that there is a mechanism for those utilising MGRs and DSI of ABNJ within their jurisdiction to notify the state to fulfil the necessary notification and benefit sharing obligations of the BBNJ Agreement. Use and commercialisation may be from MGRs and/or DSI collected or accessed via a repository or database. While not suggested in the BBNJ Agreement or a

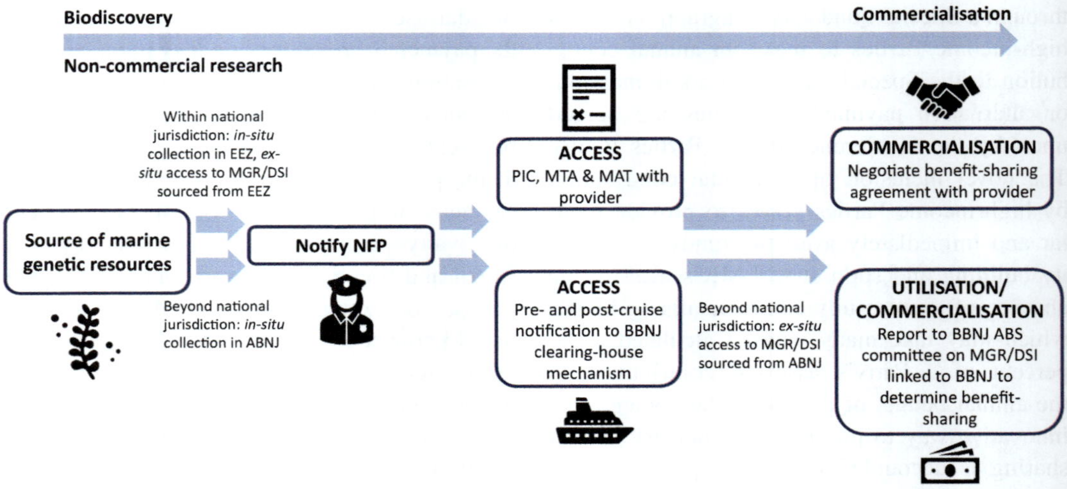

Fig. 13.1 Optional streamlining of national ABS mechanism with BBNJ notification and benefit sharing obligations

requirement, one option could be for Parties to streamline ABS instruments and facilitate notification and fulfilment of BBNJ Agreement benefit sharing obligations through existing National Focal Points.[1] However, the notification and benefit sharing modalities would differ given the Nagoya Protocol uses a bilateral mechanism and the BBNJ Agreement a multilateral mechanism (see Fig. 13.1).

Article 14 articulates that the sharing of all benefits should focus on the conservation and sustainable use of marine biodiversity. Article 52 provides further details on monetary benefit sharing specifying that the Special Fund shall be utilised *inter alia* to fund capacity building projects under the BBNJ Agreement, assist LMICs to implement the BBNJ Agreement, and support conservation and sustainable use programmes by Indigenous Peoples and local communities. Capacity building projects include projects on the conservation and sustainable use of marine biodiversity and activities and programmes, including training related to the transfer of marine technology. While the distinction between monetary and non-monetary benefits can be considered arbitrary given that all benefits come at a financial cost, the allocation of funds to specified capacity building projects enables the channelling of funds where most needed as self-identified by LMICs. Importantly, these projects will strengthen capacity within national jurisdictions, which over time will help to facilitate the full participation of LMICs in the conservation and sustainable use of ABNJ. Capacity building could include strengthening foundational biodiversity research, including taxonomy and the development of environmental monitoring baselines, and support for marine protected areas including their development, management, monitoring and enforcement.

13.3 Non-monetary Benefits

13.3.1 Access to Samples, Sample Collections, DSI and Data in Databases and Repositories

A key focus of non-monetary benefit sharing is to facilitate access to samples, sample collections and DSI in accessible repositories and databases. This stems from the uneven geography of rapidly progressing genomics, alongside

[1] https://chm.cbd.int.

disparities in ocean science, including access to ABNJ and deep-sea research (Blasiak et al., 2020; Sink et al., 2021). In this way, scientists in LMICs can be supported to conduct research on MGRs and DSI in ABNJ for scientific discovery, biodiversity research and conservation purposes. MGR research allows scientists to identify and describe species and communities and to explore how they are interrelated, thus developing foundational biodiversity knowledge. Knowledge of this genetic variation between organisms and across jurisdictions is critical to understanding ecological connectivity and enabling effective management approaches for biodiversity conservation, including the implementation of area-based management tools and environmental impact assessments under the BBNJ Agreement.

The storage of and access to MGR samples varies significantly across states. For example, the United States National Cancer Institute's Natural Products Repository houses extracts collected from the Exclusive Economic Zone of countries around the world, and a range of other specialised collections exist that are housed at museums, universities and private research institutes. This is often not the case in less well-resourced countries. Prior to the CBD and Nagoya Protocol, samples were often freely collected and exported, creating disparities in repositories between provider and user countries. Such disparities also impacted the capacity of provider countries to undertake further research, with little to no ready access to samples and limited capacity and funding to undertake and maintain their own collections. Enhancing existing or developing new repositories for samples and sample collections, particularly in regions of high biodiversity but with limited resources, presents an ideal opportunity to develop the capacity of LMICs for biodiversity conservation research, both within and beyond national jurisdictions (see Box 13.1) (Collins et al., 2020a).

The increasing shift towards the use of genetic sequence data and databases for their storage also raises critical issues relating to their access and use given the internet bandwidth and technology required to do so. The deposit of DSI in public databases is already standard international scientific practice and there exist multiple online open-access or open-source databases. For example, the International Nucleotide Sequence Database Collaboration (INSDC) brings together the DNA Databanks of Japan (DDBJ), based at the National Institute of Genetics, Japan; GenBank based at the National Center for Biotechnology Information (NCBI), the USA; and the European Nucleotide Archive based at the European Bioinformatics Institute, European Molecular Biology Laboratory (EMBL-EBI), United Kingdom. Because countries hosting databases (e.g. the USA, Europe and Japan) provide funds, expertise and technological capacity to store, analyse and manage data within these public databases, their development and maintenance is a form of benefit sharing. However, others consider access to databases and technology an insufficient benefit for genetic sequence information unless paired with complementary capacity building skills to use and access those databases. This is generally the case for countries rich in biodiversity but which lack sufficient molecular research capacity or biotechnology infrastructure to make use of these global database systems, thus a key area needing support.

In this regard, article 12 requires that notification should be given of MGRs collected in situ. Such information should include the geographical area to be accessed, and the assignment of a unique BBNJ Identifier to enable samples to be linked to their origins. Depositing samples in the regions closest to collection areas, and supporting the establishment and management of such repositories, could help to build scientific and technical capacity in biodiversity-rich countries (see Box 13.2). Paragraph 4(d) notes that if Parties house samples in repositories within their own jurisdiction, access to these samples should be at preferential rates or no-cost to developing countries.

Box 13.1: Development of Repositories for Samples and Sample Collections in Less Well-Resourced Countries—The Case of South Africa

In South Africa, a biodiscovery collaboration between Rhodes University, Scripps Institution of Oceanography and SmithKline Beecham (now GlaxoSmithKline) resulted in the collection of 336 different marine invertebrates over the period 1994–1995. While no patents or commercial products emerged from the collaboration, the extracts were returned to Rhodes University in 1996 and housed at the South African Institute for Aquatic Biodiversity, a National Collection Facility searchable via *the Global Biodiversity Information Facility (GBIF) and accessible to local and international researchers. This collection continues to be analysed, furthering taxonomic knowledge of invertebrate diversity in southern Africa, including the discovery of new species.*

Source Lavelle and Wynberg (2022)

13.3.2 Capacity Building, Transfer of Marine Technology and Scientific Cooperation

Capacity building is an important determinant of the successful implementation of the BBNJ Agreement given that countries have different levels of knowledge, resources and capacities to understand, value and make decisions on marine biodiversity, particularly in the deep seas (Hoareau et al., 2022). As highlighted in Broggiato et al. (2025), access to samples, DSI and data needs to be paired with relevant scientific research capabilities to be of value as a benefit. The well-documented imbalance in research capacity, technology, finances and intellectual property rights in relation to MGRs influences the capacity of LMICs to conserve and sustainably manage marine areas within national jurisdiction (Morgera, 2022). This also has direct implications for their capacity to contribute to the conservation of ABNJ given the ecological connectivity between areas (Popova et al., 2019).

Increasing efforts have focused on improving the effectiveness of capacity building. These include a departure from short-term training opportunities or time onboard cruises to long-term partnerships that foster mutual exchange and collaborative research (Collins et al., 2019). Short-term exchanges can be of value to the early development of research capacity among LMIC researchers to provide the foundation for longer-term collaborations that enable researchers to be part of strategic networks that bring together complementary expertise (Hoareau et al., 2022) (see Box 13.3). The AOSIS Declaration on Marine Science highlights the need for capacity building and technology transfer initiatives to be monitored, reviewed and adjusted according to the self-identified needs of LMICs, and hence, the need for sufficient flexibility and a focus on mutual learning among all partners.

Box 13.2: Examples of Non-monetary Benefits that Support Conservation and Sustainable Use of Marine Biodiversity

- *Research collaborations that bring together specialised expertise, resources and infrastructure;*
- *scientific and environmental education and training with institutions undertaking biodiversity and conservation research, conservation agencies, traditional knowledge holders and resource custodians (bioinformatics training is particularly needed);*
- *capacity building that can be applied to fundamental biodiversity research including taxonomy and phylogenetics through knowledge exchange and training;*
- *capacity building among biodiversity institutions, conservation agencies and other relevant groups to develop*

specialised expertise, equipment and infrastructure that has application to biodiversity research and conservation;

- *the acquisition of specialised deep-sea sampling equipment and infrastructure that may not be readily available;*
- *support for biodiversity and conservation research in funding proposals;*
- *sharing data and information in national and/or public databases with accurate metadata for ease of access to avoid duplication of research and to enable interoperability;*
- *alignment of research programmes to conservation priorities;*
- *partnerships between industry, biodiversity scientists and conservationists;*
- *sharing data and information with conservation agencies about resources that are accessed to inform resource management decision making;*
- *depositing samples in repositories including biobanks, genebanks, chemical extract libraries and museums that contribute to the conservation of threatened and rare species and for use in other research;*
- *access to and transfer of technology related to marine resources or applicable to biodiversity research and conservation, this might include omics technologies that can be used to generate genetic data for taxonomy or analytical technologies for molecular networking of species to investigate the effect of environmental factors; and,*
- *including conservation benefits in collection notifications.*

Deep-sea sampling is inaccessible for many states, particularly LMICs. Barriers to entry include training deficiencies, inadequate finances, limited access to vessels and technology, and inadequate access and engagement in international research (Collins et al., 2020b). Limited field experience in deep-sea research and difficulty obtaining visas also precludes researchers in developing countries from participating in deep-sea research conferences, thereby making it difficult for these researchers to 'catch up with global standards' (Sink et al., 2021). These barriers to sampling and deep-sea research limit the potential for the widespread adoption and application of genetic and other approaches to biodiversity conservation and undermine the objectives of the BBNJ Agreement.

An important aspect of research partnerships is the need for mutual capacity building opportunities and genuine scientific cooperation to facilitate appropriate technology transfer/co-development of technology. Mutual capacity building recognises that while high-income countries may have greater scientific and technical capacities, LMICs, in addition to scientific capacity, also have capacity in other knowledges relevant to the ocean. Bringing these knowledges together through innovative partnerships can result in novel understandings of marine biodiversity. Similarly, the co-development of technology allows for the input and innovation of all partners with their nuanced needs and expertise. Under article 12, the BBNJ Agreement provides an opportunity for biodiversity scientists and conservationists from LMICs to be supported to participate in sampling expeditions and research cruises that collect mutually beneficial data and jointly analyse results with joint research outputs that strengthen transdisciplinary science; therefore, benefit sharing starts already at the notification phase. Early involvement of scientists from LMICs in the planning stages of research is critical to embedding conservation benefits at the onset. Such initiatives also help emerging scientists to gain exposure to international research and are thus fundamental to capacity building. Collaborations in ABNJ can also feed back into conservation priorities within national jurisdictions, thus supporting approaches for basin-scale management.

Equally important is for capacity building and technology transfer initiatives to develop collaborations of diverse expertise that give attention to research for biodiversity conservation. For example, molecular approaches used in biodiscovery have direct application to ecosystem assessments and area-based management tools, while taxonomic and phylogenetic capacity are foundational to both conservation and biodiscovery. This requires a shift of focus given that attention to date has largely hinged on industrial applications.

Box 13.3: Long-Term Research Partnerships for Capacity Building

*PharmaSea (*www.pharma-sea.eu*) was a large-scale, five-year (2012–2017) collaborative project to collect and screen samples to discover novel products for the treatment of infections, inflammation and neurodegenerative diseases. The project was backed by more than €9.5 million of EU funding and brought together 24 partners from 13 countries from industry, academia and non-profit organisations. The project focused on the biodiscovery, development and commercialisation of new compounds from marine organisms with a focus on underexploited marine phyla of cultivable microorganisms. It aimed to achieve optimised and sustainable production of relevant biomass and high added-value compounds for pharmaceutical, nutraceutical and cosmeceutical applications, and to overcome some of the major bottlenecks in the drug discovery pipeline.*

PharmaSea, led by Aberdeen University, partnered with the Institute for Microbial Biotechnology and Metagenomics at the University of the Western Cape, South Africa with this partnership resulting in long-term collaboration beyond the time frame of the original project with multiple beneficial outcomes. For example, through PharmaSea, funding was obtained by the Institute to develop a large collection of microorganisms isolated from South African marine invertebrates. This collection continues to be analysed with five to ten additional research projects having been initiated further to the PharmaSea project. Further, the collection has been approved as part of the national microbial biobank by the Department of Science and Innovation housed by the South African National Biodiversity Institute, thus enabling access to other researchers including those working in biodiversity conservation. Another aspect of the collaboration included exchange programmes in which the Institute was the training centre for all project partners for genomics and genetic engineering. Together with opportunities to lead research and publish as lead authors, and recognition through multiple awards, this has facilitated internationalisation of the Institute, opening doors to further funding and research partnership opportunities. Another key outcome of the partnership was the building of a critical mass in marine biodiscovery research in South Africa through training postgraduate students and the Institute being invited by the EU on a number of occasions to advise on incorporating South African researchers in EU programmes.

[1]*While not specific to the development of capacity for biodiversity conservation this example highlights the importance of long-term collaborations.*

Source Lavelle and Wynberg (2022) *and pers. comm. with UWC 7th December 2023*

13.4 Alignment of Research Programmes to Conservation Priorities

Article 12 of the BBNJ Agreement requires notification of all planned in situ collection of MGRs to the Clearing-House. While article

12 does not stipulate the need to align research programmes to biodiversity conservation priorities, marine biodiscovery research often targets areas of high marine biodiversity that may be of conservation priority. Thus, synergistic opportunities could exist for biodiscovery research to contribute to area-based management tools and marine spatial planning, including marine protected areas (see Box 13.4) and to environmental impact assessments. There could also be opportunities to offset the high costs of deep-sea sampling and to support genetic approaches to marine conservation. For example, biodiscovery sampling expeditions could collaborate with conservation scientists to strategically contribute to improved baseline studies of the state of marine biodiversity in ABNJ.

Box 13.4: Embedding Conservation Benefits into Biodiscovery Research: The Case of Algoa Bay

The Marine Natural Products Chair at Rhodes University in South Africa has an active programme in biodiscovery research for pharmaceuticals. The use of 'omic' technologies forms the basis of the Chair's research resulting in large-scale genomic and metabolomic datasets for the discovery pipeline. Recently, the Chair formed the Algoa Bay Community of Practice with three other universities and three national research institutions to create a transdisciplinary team in marine and social science with the objective to make technology and data accessible beyond the initial focus of research. In this way, the scope of data being collected and applied is dramatically widened. For example, technology applications used on sponges for biodiscovery are being applied to marine ecology and conservation. The conservation application of such a discovery is that by understanding the molecular foundation of the ecosystem which is very sensitive to change and establishing a baseline, data and technology can be used to monitor ecosystem health. Modelling can also be used to predict the impact of potential activities on the ecosystem which can inform economic and environmental decision-making. Thus, by making knowledge and information more accessible through the Community of Practice, it can be used more broadly.

Source Wynberg et al. (2022)

13.5 Conclusion

This chapter set out to describe what benefit sharing might look like in practice and to describe approaches that might contribute to the conservation and sustainable use of marine biodiversity within the context of the BBNJ Agreement. There remains a lot of uncertainty, however, regarding the exact modalities of how monetary benefits will accrue and be disbursed under the BBNJ Agreement. Incentivising the payment of monetary benefits will require a simple modality that is not burdensome to users, and that is transparent and accountable in the allocation of those benefits. Given that the commercialisation of MGRs is difficult to achieve, benefit sharing will largely depend on non-monetary benefits. What is clear is that long-term collaborations, built on the co-development of needs that are self-identified by LMICs, and that fulfil capacity building and technology transfer needs, will be key to the fair and equitable co-production of ocean science and shared responsibility for the conservation and sustainable use of ABNJ.

References

Blasiak, R., Wynberg, R., Grorud-Colvert, K., Thambisetty, S., Bandarra, N. M., Canário, A. V. M., da Silva, J., Duarte, C. M., Jaspars, M., Rogers, A., Sink, K., & Wabnitz, C. C. C. (2020). The ocean genome and future prospects for conservation and equity. *Nature Sustainability, 3*(8), 588.

Broggiato, A., Dunshirn, P., Jaspars, M. & Pena-Neira, S. (2025). Monetary and non-monetary benefit sharing under the BBNJ Agreement. In F. Humphries (Ed.), *Decoding governance of marine genetic resources under the BBNJ Agreement.* Springer.

Collins, J. E., Harden-Davies, H., Jaspars, M., Thiele, T., Vanagt, T., & Huys, I. (2019). Inclusive innovation: Enhancing global participation in and benefit sharing linked to the utilization of marine genetic resources from areas beyond national jurisdiction. *Marine Policy, 109*(2019), 103696.

Collins, J. E., Rabone, M., Vanagt, T., Amon, D. J., Gobin, J., & Huys, I. (2020a). Strengthening the global network for sharing of marine biological collections: recommendations for a new agreement for biodiversity beyond national jurisdiction. *ICES Journal of Marine Science*. https://doi.org/10.1093/icesjms/fsaa227

Collins J.E., Vanagt, T. & Huys, I. (2020b). Stakeholder perspectives on access and benefit-sharing for areas beyond national jurisdiction. *Frontiers in Marine Science*, *7*, 265. https://doi.org/10.3389/fmars.2020.00265

Conference of the Parties to the Convention on Biological Diversity. (2022). Kunming-Montreal Global Biodiversity Framework, United Nations Biodiversity Conference COP15/CP-MOP10/NP-MOP4. December 2022.

Hoareau, K., Pouponneau, A., Morgera, E., Lavelle, J. & Wynberg, R. (2022). Mutual learning through capacity building on marine biological diversity of areas beyond national jurisdiction. *One Ocean Hub Policy Brief.* https://oneoceanhub.org/wp-content/uploads/2022/08/Policy-brief_6_R1_Mutual-learning-_OOH-FINAL.pdf

Humphries, F., Rabone, M., & Jaspars, M. (2021). Traceability approaches for marine genetic resources under the proposed ocean (BBNJ) treaty. *Frontiers in Marine Science*, *8*. https://doi.org/10.3389/fmars.2021.661313.

Langlet, A., Dunshirn, P., Jaspars, M., Humphries, F., & Kachelriess, D. (2025) Monitoring and transparency aspects of MGR-utilization under the BBNJ Agreement. In F. Humphries (Ed.), *Decoding governance of marine genetic resources under the BBNJ Agreement.* Springer.

Lavelle, J., & Wynberg, R. (2022). Marine Biodiscovery in South Africa. Science, Conservation, Governance and Equity. One Ocean Hub and University of Cape Town. https://oneoceanhub.org/wp-content/uploads/2022/09/Lavelle-Wynberg-2022_Marine-Biodiscovery-in-South-Africa.pdf

Morgera, E. (2022). The relevance of the human right to science for the conservation and sustainable use of marine biodiversity of areas beyond national jurisdiction: A new legally binding instrument to support co-production of ocean knowledge across scales. In De Lucia et al. (Eds.), *International law and marine areas beyond national jurisdiction: Reflections on justice, space, knowledge and power* (pp. 242–274). Brill.

Popova, E., Vousden, D., Sauer, W., Mohammed, E., Allain, V., Downey-Breedt, N., Fletcher, R., Gjerde, K., Halpin, P., Kelly, S., Obura, D., Pecl, G., Roberts, M., Raitsos, D., Rogers, A., Samoilys, M., Sumaila, U., Tracey, S., & Yool, A. (2019). Ecological connectivity between the areas beyond national jurisdiction and coastal waters: safeguarding interests of coastal communities in developing countries. *Marine Policy, 104*, 90–102. https://doi.org/10.1016/j.marpol.2019.02.050

Schroeder, D. (2007). Benefit sharing: It's time for a definition. *Journal of Medical Ethics*, *33*(4), 205–209. https://doi.org/10.1136/jme.2006.016790

Secretariat of the Convention on Biological Diversity. (2010). *Nagoya Protocol on Access to Genetic Resources and the Fair and Equitable Sharing of Benefits Arising from their Utilisation,* opened for signature 29 October 2010, [2012] ATNIF 3 (entered into force 12 October 2014).

Sink, K., McQuaid, K., Atkinson, L., Palmer, R., Van der Heever, G., Majiedt, P., Dunga, L., Currie, J., Adams, R., Wahome, M., Howell, K. & Patterson, A. (2021). *Challenges and Solutions to Develop Capacity for Deep-sea Research and Management in South Africa* (35 pp.). South African National Biodiversity Institute.

Sirakaya, A. (2021). Is the Nagoya Protocol designed to conserve biodiversity? *Plants, People, Planet,* 4(4). https://doi.org/10.1002/ppp3.10221

UNGA. (2023). *Agreement under the United Nations Convention on the Law of the Sea on the conservation and sustainable use of marine biological diversity of areas beyond national jurisdiction*, Intergovernmental conference on an international legally binding instrument under the United Nations Convention on the Law of the Sea on the conservation and sustainable use of marine biological diversity of areas beyond national jurisdiction. Further resumed fifth session. A/CONF.232/2023/4. 19 June 2023.

United Nations. (1992). *Convention on Biological Diversity*, opened for signature 5 June 1992, 1760 UNTS 69 (entered into force 29 December 1993).

Wynberg, R., & Laird, R. (2023). Access and benefit sharing and biodiversity conservation: The unrealised connection. In C. Lawson, M. Rourke, & F. Humphries (Eds.), *Access and benefit sharing of genetic resources, information and traditional knowledge* (pp. 50–70). Routledge.

Wynberg, R., Ives, S., & Bam, J. (2023). How access and benefit sharing entrenches inequity: The case of Rooibos. *Journal of Southern African Studies, 49*(4), 589–610. https://doi.org/10.1080/03057070.2023.2301640

Wynberg, R., Crouch, N., Lavelle, J., van Niekerk, J., & Ndwandwe, S. (2022). Guidelines for integrating the conservation and sustainable use of biodiversity in access and benefit-sharing approaches in South Africa. Department of Forestry, Fisheries and the Environment, University of Cape Town, South African National Biodiversity Institute, GIZ and Bioinnovation Africa.

Jessica Lavelle holds a PhD in environmental governance and is a research associate of the Bio-economy Research Chair at the University of Cape Town.

Rachel Wynberg is a professor in the Department of Environmental and Geographical Science at the University of Cape Town in South Africa where she holds a government-funded Research Chair focussed on Environmental and Social Dimensions of the Bio-economy. With a background in both the natural and social sciences, she has a strong interest in inter- and transdisciplinarity and policy engagement across the humanities, arts and sciences. Her research spans topics relating to just, ethical and biodiverse bio-economies; seeds, farmers' rights and agrobiodiversity; knowledge politics; agroecology and food sovereignty; the governance of wild species; and emerging technologies and equity in science.

BBNJ Agreement: Considerations for Scientists and Commercial End Users of MGR at Research, Development and Commercialization Stages

14

Muriel Rabone, Tammy Horton, Fran Humphries, Christopher H. C. Lyal, Hiroko Muraki Gottlieb, Amber H. Scholz, Thomas Vanagt, and Marcel Jaspars

Abstract

The research, development and commercialization pipeline for accessing, using and sharing marine genetic resources (MGR) of areas beyond national jurisdiction (ABNJ) is highly varied and complex. Equally complex is the governance framework under the 2023 agreement on the conservation and sustainable use of marine biological diversity of ABNJ, for which many practical details, including procedures, are yet to be decided by treaty Parties. This chapter draws from real world examples to analyse ways in which current scientific practice is supported or challenged by framework elements, including notification, monitoring and benefit sharing systems and associated infrastructure such as the BBNJ Standardized Batch Identifier and data management plans. It compares how the elements and infrastructure may work in practice using six R&D scenarios ranging from an idealized linear pathway to more complex pathways involving automation, sequence information and traditional knowledge associated with MGR in different geographical and temporal scales. For an efficient and 'future proofed' framework that supports innovation and fulfils treaty objectives, it is proposed that treaty bodies and policy makers need to look beyond the idealized R&D pathways envisaged in the treaty and engage directly with scientists and commercial end users when designing the practical details of implementation.

Muriel Rabone and Fran Humphries: These authors contributed equally to this chapter.

M. Rabone · C. H. C. Lyal
Department of Life Sciences, The Natural History Museum, London, UK

T. Horton
Ocean BioGeosciences, National Oceanography Centre, Southampton, UK

F. Humphries
Griffith Law School, Griffith University, Nathan, QLD, Australia

H. Muraki Gottlieb
Department of Organismic and Evolutionary Biology, Harvard University, Cambridge, MA, USA

A. H. Scholz
Department of Science Policy and Internationalization, Leibniz Institute DSMZ-German Collection of Microorganisms and Cell Cultures, Braunschweig, Germany

T. Vanagt
3BIO, Brussels, Belgium

M. Jaspars (✉)
Marine Biodiscovery Centre, Department of Chemistry, University of Aberdeen, Aberdeen, UK
e-mail: m.jaspars@abdn.ac.uk

F. Humphries (ed.), *Decoding Marine Genetic Resource Governance Under the BBNJ Agreement*, Sustainable Development Goals Series, https://doi.org/10.1007/978-3-031-72100-7_14

Keywords

Marine genetic resources (MGR) · Digital sequence information (DSI) · Access · Utilization · Benefit sharing · FAIR · BBNJ Standardized Batch Identifier · Reporting requirements · BBNJ agreement

Abbreviations

ABSC	Access and benefit sharing committee
AI	Artificial intelligence
AUV	Autonomous underwater vehicle
API	Application programming interface
BS	Benefit sharing
BBNJ identifier	BBNJ standardized batch identifier
CHM	Clearing house mechanism
COP	Conference of the parties
DES	Digital extended specimen
DMP	Data management plan
DSI	Digital sequence information
FPIC	Free and prior informed consent
IPLC	Indigenous peoples and local communities
ITPGRFA	International treaty on plant genetic resources for food and agriculture
MAT	Mutually agreed terms
MGR	Marine genetic resource
STB	Scientific and technical body

14.1 Introduction

Following almost two decades of negotiations, the adoption by consensus of the *Agreement under the United Nations Convention on the Law of the Sea on the Conservation and Sustainable Use of Marine Biological Diversity of Areas Beyond National Jurisdiction* (BBNJ Agreement (UNGA, 2023)) marks a new phase in marine biodiversity governance in roughly two-thirds of the world's oceans known as areas beyond national jurisdiction (ABNJ). No States have sovereignty or sovereign rights to ABNJ, which encompass the water column of the high seas and the deep seabed below (UNCLOS[1] Parts VII and XI). In response to alarming marine biodiversity decline (Díaz, 2019), the BBNJ Agreement is a treaty[2] that was designed to fill a gap in biodiversity governance and address questions of equity in the exploration of marine genetic resources (MGR) of ABNJ. This chapter provides insights into how Part II of the BBNJ Agreement, MGR governance, may be applied in practice. Questions from an operational perspective remain as modalities and clarification of the MGR framework will be decided in future by the Conference of the Parties to the BBNJ Agreement (CoP), supported by treaty (subsidiary) bodies and infrastructure. These include the Access and Benefit Sharing Committee (ABSC), the Scientific and Technical Body (STB) and the Clearing House Mechanism (CHM). Building on the textual treaty interpretation of Chaps. 2–8 of this edited collection,[3] this chapter analyses key practical considerations for implementation of treaty obligations for scientists and commercial end users, and more broadly for the entities (public and private) that collect, hold and utilize MGRs of ABNJ and associated digital sequence information (DSI) and traditional knowledge (TK) at the pre/post-collection, research and development (R&D) and commercialization stages.

The BBNJ Agreement creates a framework for Parties to cooperate on marine biodiversity governance in accordance with treaty objectives. The overall objective is 'to ensure the conservation and sustainable use of marine BBNJ, for

[1] *United Nations Convention on the Law of the Sea*, opened for signature 10 December 1982, 1833 UNTS 397 (entered into force 16 November 1994).

[2] Which will come into force 120 days after 60 states become Parties to the treaty.

[3] Broggiato et al. (2025), Humphries (2025), Humphries et al., (2025a, b), Langlet et al. (2025), Muraki Gottlieb et al., (2025a, b), Pena-Neira and Coelho (2025).

the present and in the long term, through effective implementation of the relevant provisions of the Convention[4] and further international cooperation and coordination (art 2). The BBNJ Agreement has four elements—Part II (MGR, including the fair and equitable sharing of benefits), Part III (Area-Based Management Tools, including Marine-Protected Areas), Part IV (Environmental Impact Assessment) and Part V (Capacity Building and the Transfer of Marine Technology) plus other provisions such as Part VI (institutional arrangements). The objectives for Part II are

(a) The 'fair and equitable sharing of benefits arising from activities with respect to' MGR and DSI on MGR of ABNJ 'for the conservation and sustainable use of marine biological diversity of ABNJ';
(b) The 'building and development of the capacity of Parties', 'particularly developing States Parties' and other categories listed, to carry out these activities;
(c) 'the generation of knowledge, scientific understanding and technological innovation, including through the development and conduct of marine scientific research, as fundamental contributions to the implementation of this Agreement'; and
(d) 'the development and transfer of marine technology in accordance with this Agreement' (art 9).

Part II provides a framework for MGR governance. Under the treaty, MGR means 'any material of marine plant, animal, microbial or other origin containing functional units of heredity of actual or potential value' (art 1(8)). DSI is undefined but is a placeholder term that is increasingly used in international fora to denote information associated with genetic resources such as DNA, RNA, proteins and possibly metabolites (CBD/COP/DEC/15/9, CBD/DSI/AHTEG/2020/1/3). 'Traditional Knowledge' is also undefined, but its scope is likely to be determined under national laws by governments or Indigenous Peoples and local communities (IPLCs) (Humphries, 2025; Pena-Neira & Coelho, 2025). Key elements of the framework are:

(a) A notification system for users of MGR and DSI encompassing:
    a. pre- and post-collection notifications;
    b. 'utilization' notification; and
    c. reporting on 'access' to MGR and DSI in repositories and databases;
(b) A system for the fair and equitable sharing of benefits from the use of MGR that contributes to the conservation and sustainable use of marine biological diversity in ABNJ;
(c) A monitoring and transparency system, including a BBNJ Standardized Batch Identifier (BBNJ Identifier); and
(d) Provisions on access and use of TK of IPLCs associated with MGR in ABNJ.

Regarding (c), the CHM will automatically generate a BBNJ Identifier upon receipt of a pre-collection notification (see Sect. 2.1). This is a unique identifier that tags the whole collection (the 'batch') to provide a stable link between information about the collection event (including the location of collection) and any MGR or DSI that is subsequently held or deposited in a repository or database. The idea is that the original collection will be linked to any subsequent unique identifiers for the MGR and DSI to help ascertain provenance (i.e. original location where they were collected) of the MGR and DSI that will be the subject of R&D and aggregate reports to the CHM (see Sect. 2.4). It is equivalent to an identifier for a deep-sea research cruise (as in current usage) but would meet certain characteristics of being 'persistent' or stable over time and globally unique, resolvable and authoritative (Guralnick et al., 2015; Page, 2023). Usage of persistent identifiers is key to database interoperability and to making data FAIR, or Findable, Accessible, Interoperable and Reusable (Islam et al., 2023; Juty et al., 2020; Rabone et al., 2023a, 2023b;

[4] *United Nations Convention on the Law of the Sea*, opened for signature 10 December 1982, 1833 UNTS 397 (entered into force 16 November 1994).

Wilkinson et al., 2016). The importance of FAIR and persistent identifiers for MGR traceability and the treaty has been discussed previously in the literature (Humphries et al., 2021; Rabone et al., 2019). FAIRness of MGR data is now a requirement of the BBNJ Agreement (art 14). Utilization of MGR from ABNJ (e.g. publication of papers or patents, or development of products) necessitates notification to the CHM when this information is available, to (1) allow transparency and (2) determine the level of benefit sharing; key objectives of Part II. The BBNJ Identifier is intended to be integrated into existing databases and embedded in the outputs of scientific research (including publications and patents), facilitating automated retrieval (Oldham & Thambisetty, 2023). However, many questions remain on how this could be implemented in practice.

Similarly, many of the practical details such as procedures and guidance on interpretation and scope are yet to be determined by treaty bodies and Parties. Scientists, repositories, commercial end users and other stakeholders however can already start thinking about how the treaty framework will affect them when implemented under national law. These stakeholders may already have aligned their practices and procedures with access and benefit sharing (ABS) procedures under national laws that implement the *Convention on Biological Diversity* (CBD) and the *Nagoya Protocol on Access to Genetic Resources and the Fair and Equitable Sharing of Benefits Arising from their utilization to the Convention on Biological Diversity* (Nagoya Protocol) and other international ABS frameworks (Kachelriess et al., 2025). The task is now to consider how to also to align their practices with the new BBNJ Agreement framework as it unfolds.

It must be noted that if MGR is collected from areas within national jurisdiction (AWNJ), it may be subject to national ABS laws that implement the international frameworks of the CBD (Kachelriess et al., 2025). Analysis of other legal frameworks governing MGR in AWNJ is beyond the scope of this chapter, but this highlights how the R&D pipeline for MGR may be subject to different governance regimes for the same research project, depending on where the MGR was originally collected.

The aim of this chapter is to analyse how the BBNJ Agreement infrastructure, procedures and processes might apply under a series of scenarios of R&D and commercialization pathways, to better understand the effects of treaty implementation. Section 14.2 outlines the key requirements of Part II of the BBNJ Agreement—notifications (pre- and post-collection and utilization), 'accessing' MGR and DSI from repositories and databases, reporting requirements and benefit sharing. It outlines ways in which current scientific practice is both supported and challenged by key elements of the MGR governance framework, with examples. Section 14.3 outlines six scenarios of R&D and commercialization ranging from an idealized linear approach to more complex scenarios including those involving the use of DSI and TK under different temporal and spatial scales. These scenarios highlight areas of ambiguity in the treaty obligations, which can vary significantly depending on the R&D pathway. It argues that this variety highlights the need for the CoP and other treaty bodies to think beyond the idealized linear R&D pathway when developing policies and guidance to Parties on implementation of Part II. Engaging directly with scientists, commercial end users, repositories and other stakeholders during implementation can 'help future proof' the treaty and ensure its objectives are met, including the generation of knowledge, scientific understanding and technological innovation.

14.2 Implications of the MGR Framework for Stakeholders

The other chapters in this edited collection provide a detailed analysis and interpretation of Part II provisions (see Humphries et al., 2025a for an overview of chapters). The purpose of this section is to highlight elements of the MGR governance framework that are relevant for demonstrating the extent to which current R&D

practices are supported or challenged by these elements. This analysis includes

- an overview of notification, benefit sharing and transparency requirements and consistency with current practice;
- real world examples to illustrate complexities of research processes including examining the UK as a case study;
- perspectives on practicalities for implementation and how current practice can be best adapted for requirements of the BBNJ Agreement; and
- identification of areas for input by the treaty bodies.

The focus of this section concerns the points in time where the R&D pipeline interacts with the requirements for notification (art 12), transparency (art 16) and benefit sharing (art 14). A key focus of the analysis is how the BBNJ Identifier connects these elements. The BBNJ Identifier is a key innovation in the framework, which is an administrative tag or identifier automatically issued by the CHM upon pre-collection notification that can link the CHM with other scientific or administrative identifiers and databases to assist with information gathering for treaty obligations as above (see also Lawson et al., 2025). Scenario 6 in Sect. 14.3 also relates to how R&D might interact with article 13 on TK associated with MGR in ABNJ, which has been analysed in other chapters of this collection, including Pena-Neira and Coelho (2025). To assist with understanding the elements in this section and Sect. 14.3, Fig. 14.1 outlines a graphical representation of the notification requirements in article 12, showing timelines for pre- and post-collection and utilization notifications and links between the BBNJ Identifier and downstream MGR and DSI unique identifiers. The numbers refer to relevant articles in the BBNJ Agreement.

14.2.1 Pre-collection Notification

The first step in the notification process is the pre-collection notification to the CHM, which Parties are required to ensure is completed six months or as early as possible prior to the collection or sampling of MGR from ABNJ (art 12(2)). The practicalities for the infrastructure and procedures are yet to be determined by the CoP at the time of writing. This obligation is on Parties but in practice, under national law,

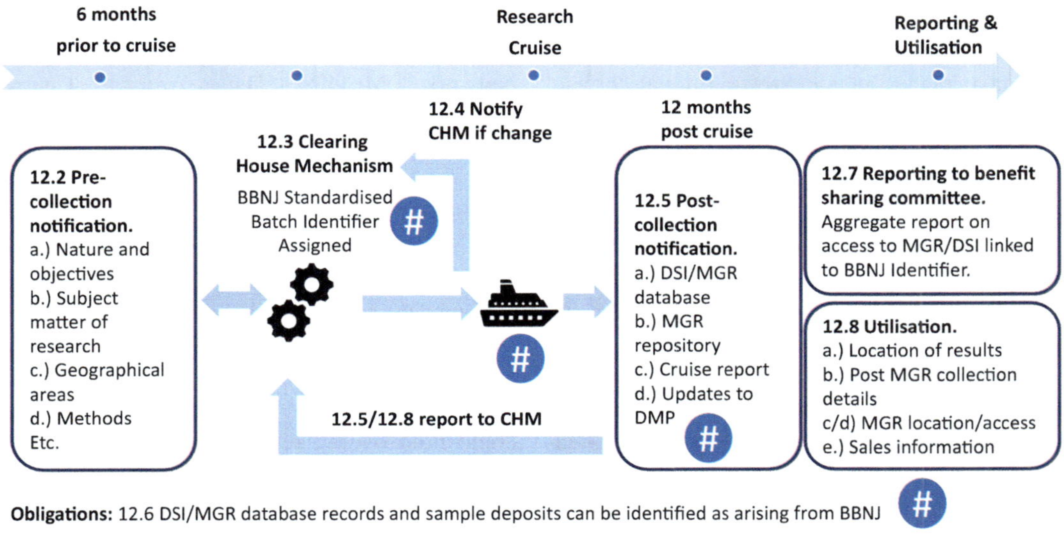

Fig. 14.1 Article 12 notification requirements under the BBNJ agreement

Parties may require their governments to act as a conduit for all notifications or may require their nationals to notify the CHM directly.

Table 14.1 shows a comparison of pre-collection notification in practice and the BBNJ Agreement's requirements in article 12(2). This shows that most of the notification requirements are already met under existing scientific good practice. However, providing opportunities for researchers from developing States to take part in the proposed research (which is also referenced in Part V of the BBNJ Agreement on Capacity Building and the Transfer of Marine Technology) could be an area that the CoP may benefit from input of the treaty bodies.

Within the UK marine scientific research community, these pre-collection notification requirements are already standard practice, managed by the National Marine Facilities at the National Oceanography Centre, through the Marine Facilities Planning portal (MFP). For example, the JC263 cruise to the Porcupine Abyssal Plain in 2024 is listed with the dates, sampling equipment, and planned data and samples to be collected.[5] In the UK, this planning procedure is currently in usage only for national research vessels owned or administered by the Natural Environment Research Council (NERC). The European institutes, the Royal Netherlands Institute for Sea Research (NIOZ) and Institute of Marine Research (IMR, Norway) also use the same system. Sampling may take place outside the traditional 'cruise' pattern, ie from autonomous vessels (see scenario 4 in Sect. 3.4), or private/philanthropic-owned vessels, which could also potentially use this system or, if not, equivalent compliance with BBNJ Agreement requirements will need to be ensured. This portal could also inform a global model for scientific vessels given the need for a more harmonized approach.

The MGR that fall within scope of the BBNJ Agreement is broad, but the term "sample" is undefined, which may need further clarification from the treaty bodies (Humphries, 2025). A collection can encompass a wide range of sample types, from: environmental samples of water, ice or sediment that (may) contain whole or partial organisms; through to whole organisms, e.g. single identified specimens, or mixed samples of specimens; to samples derived from any of these, such as extracted DNA or tissue preparations (Rabone et al., 2019). Samples may also be collected for other purposes, but later utilized for MGR research. Collection of any physical sample that may contain MGR (water, sediment, fauna) could be included under the BBNJ Agreement, regardless of the intent of use (commercial or non-commercial). This is because the trigger for the treaty is a collection event in ABNJ and appears not be limited to collections for the purpose of investigating the genetic attributes of the organisms, unlike the CBD framework (Humphries et al., 2024b). When developing the modalities of the BBNJ Agreement, it will be important for treaty bodies and Parties to consider exclusions for samples which are collected but not intended to be used for MGR research and not stored (such as water samples collected for physical oceanography data).

The importance of data management plans (DMPs) and data archiving are well recognized by science funders. In the UK, oceanographic data are archived (British Oceanographic Data Centre; BODC) as is marine biodiversity data (Marine Environmental Data and Information Network, MEDIN, Data Archive Centre/ DASSH, The Archive for Marine Species and Habitats Data). A DMP is provided for each cruise as standard practice in the UK, but further clarification of the sampling may be needed, covering what is current compliance and what is needed for implementation. If there are several independent scientists and research programmes on board, this may necessitate adapted DMPs. Here a set of protocols could be developed with input from the STB and the cruise lead could hold overall responsibility. For example, DMPs could be made available on the treaty's CHM. There are also opportunities within the DMP requirements to apply FAIR data formats and to provide suggestions as to the databases

[5] https://nerc.marinefacilitiesplanning.com/programme.

Table 14.1 Information to be notified to the CHM 6 months or as early as possible prior to the collection in situ of MGRs of ABNJ

Article 12.2	Current good practice and opportunities for improvement
(a) The nature and objectives under which the collection is carried out, including, as appropriate, any programme(s) of which it forms part	Provided as part of a cruise plan
(b) The subject matter of the research or, if known, the marine genetic resources to be targeted or collected and the purposes for which such resources will be collected	Provided as part of a cruise plan
(c) The geographical areas in which the collection is to be undertaken	Provided as part of a cruise plan. Additional detail is frequently provided as route of vessel defined before cruise departure. Some countries, such as the United States, have security concerns that may not make it possible to provide precise information prior to departure
(d) A summary of the method and means to be used for collection, including the name, tonnage, type and class of vessels, scientific equipment and/or study methods employed	Vessel information is available via the operator (e.g. national oceanographic agency, private operator or charitable organization). Scientific equipment equipment/methods are defined in the cruise plan but subject to change
(e) Information concerning any other contributions to proposed major programmes	Whether there will be contributions will depend on the funder and research programme
(f) The expected date of first appearance and final departure of the research vessels, or deployment of the equipment and its removal, as appropriate	The relevant dates are provided as part of the cruise plan
(g) The name(s) of the sponsoring institution(s) and the person in charge of the project	The information is provided as part of the cruise plan. The sponsoring institution could be a research funder, national oceanographic institution, or charitable organization. Person in charge is usually the cruise leader or the principal investigator on the cruise application
(h) Opportunities for scientists of all States, in particular scientists from developing States, to be involved in or associated with the project	Opportunities to be part of the cruise may be available. However, information on opportunities is not kept in one platform
(i) The extent to which it is considered that States that may need and request technical assistance, in particular developing States, should be able to participate or to be represented in the project	Participation of States with needs for technical assistance that request it may be able to participate or be represented in a project. However, information is not kept in one platform
(j) A data management plan prepared according to open and responsible data governance, taking into account current international practice	A data management plan is provided as part of a cruise plan. However, there may be opportunities to harmonize data formats and reporting. Further, best practices for where certain types of data should be deposited could be considered

that could be used for particular types of data (Lawson et al., 2025).

While pre-collection notification requirements in Table 14.1 are consistent with current research practice as covered in scenario 1 (see Sect. 3.1), there are several challenges for understanding how all scenarios will be governed by the treaty and the national laws supporting its implementation. The BBNJ Agreement is silent on the responsible entity that would submit the required information to the CHM, but it may be the home institution of the cruise principal investigator. Clarification on who is responsible for notification could be provided by the treaty bodies. Further, the CHM is to be notified of any "material change" to the cruise planning. Many variables can affect cruise planning and operations. It is unclear where the threshold lies for reporting these changes to the CHM, but pragmatism is

Table 14.2 Information to be notified along with the BBNJ identifier to the CHM as soon as it becomes available, but no later than 1 year from the collection in situ of marine genetic resources of areas beyond national jurisdiction

Article 12(5)	Current good practice and opportunities for improvement
(a) The repository or database where digital sequence information on marine genetic resources is or will be deposited	This information is provided as part of the data management plan, which forms part of the cruise plan. Similar to the data management plan outlined in Table 14.1, there may be an opportunity to harmonize the consistency and requirements on reporting the relevant information
(b) Where all marine genetic resources collected in situ are or will be deposited or held	The information about where the physical materials collected in ABNJ is recorded as part of the cruise database, which forms part of the cruise report
(c) A report detailing the geographical area from which marine genetic resources were collected, including information on the latitude, longitude and depth of collection, and, to the extent available, the findings from the activity undertaken	The geographical sampling location is recorded as part of the cruise database, which forms part of the cruise report. For operational reasons, some data may be missing from the database. The result of the collection activities may take some time depending on a few factors: the number of samples collected, the number of personnel available, and financial and other resources
(d) Any necessary updates to the data management plan provided under paragraph (2) (j) above	This is a new requirement for the DMP. Depending on the extent of updates, the new requirement may be relatively easy for the researchers to fulfil

needed to avoid overloading both researchers and operations of the CHM itself with unnecessary reporting and data (see Scenario 1). Reporting guidelines would facilitate compliance with the notification requirement. Here the treaty bodies such as the STB can play a significant role in providing recommendations to the CoP.

14.2.2 Post-Collection Notification

After a vessel returns to shore with MGR of ABNJ, post-collection notification requirements to the CHM are expected no later than a year following the MGR collection (art 12(5) (a–d)). This includes information on the repository where the MGR samples are held, the databases where the DSI are, or will be, deposited, and a report detailing what was collected and where and a general summary of findings. Most of the notification requirements can be fulfilled by providing the cruise report and the relevant cruise database/s (Table 14.2). For UK marine scientific research, this information is usually captured in a Cruise Summary Report and/or the later Cruise Report.[6] The capture of the post-expedition requirements could be incorporated into this existing process in the UK, which can be revised to ensure compliance with the BBNJ Agreement.

Table 14.2 shows a comparison of post-collection notification in practice and the BBNJ Agreement's requirements in article 12(5). As with the pre-collection notification requirements in Table 14.1, most of the requirements align with existing scientific practices, but there may be opportunities for further harmonization, such as reporting the repository or database where DSI on MGRs is or will be deposited. Guidance by treaty bodies on practicalities is important given the complexities throughout the R&D pathways described in this chapter.

The research process initiated on the cruise may take many months to years to complete (Engel et al., 2021; Humphries et al., 2021). Subsequent research may result in MGR sample deposits in repositories similar to those

[6] https://www.ukri.org/councils/nerc/facilities-and-resources/find-A-nerc-facility-or-resource/marine-facilities-policy-and-guidance/.

listed here, associated records in taxonomic biodiversity and DSI databases, and research publications, which would, in theory, be linked to a BBNJ Identifier. The requirement in BBNJ Agreement's article 12(5)(c) to detail "findings from the activity undertaken" within one year from collection could be viewed as challenging in some cases due to the number of samples obtained and lack of personnel available to carry out the work during the term of the research funding (i.e. "findings" are yet to be ascertained). Follow-on research often involves additional researchers not involved in the original cruise and associated research project. Cruise funding is often time limited and restricted to vessel time itself, mobilizing/demobilizing research project teams, consumables needed for research and sample shipment. Even if downstream research is funded, it may be time limited and often ceases within a few years of the cruise. This means that there is little support for follow-on research and additional reporting unless additional funding is obtained. In any case, research timescales may be lengthy especially when large collections have been obtained that require curation and analysis. This highlights inherent issues in science funding, primarily the disconnect between short term timescales of grants and long timescales for research and maintenance of collections and databases (Rabone et al., 2019). The BBNJ Agreement does qualify such requirement by stating, "in accordance with current international practice and *to the extent practicable*" (emphasis added) to ensure that the requirements can be implemented.

The requirement under article 12(5)(d) to update the DMP in article 12(2)(j) is not a current scientific practice, but with adequate resources (e.g. financial, personnel, etc.), such work may be possible. Considering article 12(8)(d) on modalities envisaged for access to MGR and DSI being utilized and "*"a data management plan for the same"*, it is not clear what the role of a DMP is here or why it is required if the earlier parts of article 12 are complied with, and good scientific practice is followed. The treaty bodies could also consider the question of how a researcher may handle results from the original samples with the BBNJ Identifier that is identified and has value for research years or even decades after the DMP was submitted. That said, the determination of practicability could be better determined at the Party level to ensure that the requirement can be implemented.

14.2.3 Utilization Notification and the "BBNJ" Standardized Batch Identifier

The utilization notification focuses on reporting the outcomes of R&D (commercial or non-commercial) on MGR and associated DSI. Article 12(8) requires that where MGR of ABNJ and where practicable the DSI "on such resources are subject to utilization, including commercialization, by natural or juridical persons under their jurisdiction, Parties shall ensure that the following information, including the 'BBNJ' standardized batch identifier, if available, be notified to the Clearing-House Mechanism as soon as such information becomes available:

(a) Where the results of the utilization, such as publications, patents granted, if available and to the extent possible, and products developed, can be found;
(b) Where available, details of the post-collection notification to the Clearing-House Mechanism related to the marine genetic resources that were the subject of utilization;
(c) Where the original sample that is the subject of utilization is held;
(d) The modalities envisaged for access to [MGR and DSI on MGR] being utilized, and a DMP for the same;
(e) Once marketed, information, if available, on sales of relevant products and any further development".

The term "utilization" is defined as "to conduct research and development on the genetic and/or biochemical composition of MGRs, including through the application of biotechnology" (art 1(14)). This means that there need only be an investigation into the genetic or biochemical composition of the MGR or associated DSI for the activity to fall within scope of utilization—it does not require some form of genetic manipulation or human intervention. "Biotechnology" may be one form of utilization, but "utilization" can encompass a much broader range of activities including taxonomic and conservation research, subject to confirmation from the CoP. The CoP may need to review the operation of the utilization notification and any future benefit sharing arrangements associated with utilization to ensure the broad "utilization" trigger supports conservation objectives of the treaty (Humphries, 2025). The MGR definition does not explicitly include derivatives, which are instead brought into the treaty through the definition of biotechnology. "Biotechnology" means "any technological application that uses biological systems, living organisms, or derivatives thereof, to make or modify products or processes for specific use" (art 1(3)). Arguably this means that the utilization notification may apply to derivatives only if they are used to make or modify products or processes for specific use, rather than simply investigating them for their genetic or biochemical composition (Humphries, 2025).

Article 12(8) on reporting utilization of MGR and associated DSI could be interpreted to mean that any DNA sequencing conducted as part of research activities constitutes utilization of MGR, i.e. not limited to commercialization routes. Under the CBD and Nagoya Protocol approach to ABS, countries have the discretion to define the scope of activities that trigger obligations. In the ABNJ context, the framework will only be effective if there is a common understanding of key definitions and the geographical, temporal and subject matter scope of the obligations (Humphries, 2025). Guidance on the types of activities that fall within the scope of the "utilization" trigger will be important for the BBNJ Agreement given the implications for basic research if the reporting requirements are impractical. For example, in the EU implementation of the Nagoya Protocol, usage of DNA sequencing data (i.e. DSI) for species identification is excluded from "utilisation" (Regulation (EU) No 511/2014)).

One of the requirements is to report where the original sample that is the subject of utilization is held. Since the BBNJ Identifier potentially covers many different MGR (specimens, sample types and species) from the pre-collection notification, and these may be held in many different repositories, this requirement may have challenges for implementation. MGR samples are likely to be transferred from the original repository (e.g. that reported in 12(5)(b)) to another as work continues; therefore, due diligence is needed to ensure the BBNJ Identifier accompanies samples and data. A new data standard Latimer Core[7] for biological collections captures information at the level of the collection, rather than of the individual specimen or sample like Darwin Core and could facilitate this (e.g. by recording the BBNJ Identifier). Given how R&D and commercialization can progress in a non-linear way (e.g. scenarios 2–6), due diligence will be necessary between actors throughout the process to ensure that the BBNJ Identifier is maintained with the MGR and DSI on MGR and downstream materials/products/data so that these can be recorded as required in article 12(8)(a).

Implementing the requirements for global databases, such as those holding biodiversity records and DSI may be achievable, but there may also be unforeseen challenges. For biodiversity databases like Ocean Biodiversity Information System (OBIS) and Global Biodiversity Information Facility (GBIF), MGR records, e.g. relating to specimens/samples held in collections, the BBNJ Identifier could be captured by existing data fields in global data standards such as Darwin Core, and the BBNJ Identifier incorporated into the occurrence record in the database. More development may be required for DSI records in INSDC, for example, addition of a field on a DSI page for the BBNJ Identifier. These global databases will

[7] https://tdwg.github.io/ltc/index.html.

likely need to engage with the treaty bodies to provide technical details about the functions and limitations (e.g. funding and technology) to ensure future compliance with article 12(8)(c) and article 12(6) outlined in Sect. 2.4 below (see also Muraki Gottlieb et al., 2025a).

There are significant potential informatics requirements for the CHM itself to meet the obligation to generate and resolve BBNJ Identifiers and handle all the notifications. It may need to interface with external databases and meet confidentiality and security requirements. In designing, implementing and maintaining the architecture of the CHM, a non-trivial operation, integration with existing data systems and awareness of emerging practices is essential. It is important that the BBNJ process, via bodies like the STB, are agile to latest developments in data science. Whatever the configuration of the CHM, the BBNJ Identifier will need to be robust to changes in technology and transcend the architecture, as data and web infrastructures may change considerably over time.

As explored in the scenarios in Sect. 14.3, collections without a BBNJ Identifier utilized in R&D may need to be identified or tagged, either as part of the BBNJ Identifier system or in some other way. This would capture scenarios such as legacy MGR collected prior to the BBNJ Agreement entering into force (scenario 2) or automation that may not trigger a pre-collection notification (scenario 4) or MGR utilized in a product but originally collected from harvest fisheries that are outside the scope of Part II (scenario 5). How this could work in practice requires input from repositories and other stakeholders and could be informed by existing and developing approaches. For example, Latimer Core as above, and GGBN has developed data standards for directly linking specimen records with Nagoya Protocol permit requirements (Droege et al., 2016; Schiller et al., 2024). The emerging digital specimen identifier concept, or Digital Extended Specimen (DES), which is based on a key principle that each object has a globally unique, persistent, authoritative and actionable identifier is also relevant to the BBNJ Identifier system in general (Hardisty et al., 2022; Islam et al., 2023; Page, 2023). Overall, the system should be light touch, embedded in community practice through broad consultation and not require "null" reporting where no utilization has occurred.

(1) *Publications*

The BBNJ Agreement requires Parties to ensure that the location of the results of utilization (publications, patents granted, and products developed) are reported to the CHM (art 12(8)(a)). Regarding publications, if supported by journals, the BBNJ Identifier could be recorded, for example as part of the methods, so that publications can be automatically retrieved using text mining e.g. via an application programming interface or API (Oldham & Thambisetty, 2023). Scientific journals sometimes support compliance with international agreements such as the Nagoya Protocol and scientific good practice, such as the requirement to provide accession numbers for DSI sequenced in the research reported (Humphries et al., 2021). Asking journals to add the BBNJ Identifier to the list of required information may be feasible. Implementation of the BBNJ Identifier would therefore benefit from cooperation between journals, databases (e.g., INSDC) funding organizations and the treaty bodies.

Challenges may arise during text mining such as the need for access to the full text of a publication which is possible for open access publications but may be problematic for publications behind a paywall. Further, text mining currently often requires scrutiny by a human operator. For example, results may give false positives and these need to be removed manually, potentially a very labour-intensive process. Given that a Party is responsible for the reporting obligation, text mining would have to be modified to include only the Party in question. One interpretation of article 12(8) is that the Party where the utilization occurs is responsible for the "utilization" notification (art 12(8)), but this poses practical challenges when the results of research are in countries outside of where the utilization occurs (Humphries et al., 2025b).

Also, many publications have authors from multiple countries, and this may lead to double counting of outputs as the publication will be reported in the aggregate report for each country, unless there is agreement that, for example, the Party where the principal investigator (often the last author) resides has responsibility, in a way possible under the EU regulations implementing the Nagoya protocol (Regulation (EU) No 511/2024). Further, States that are not Parties to the BBNJ Agreement do not have notification, monitoring or benefit sharing obligations under the treaty, which is likely to create gaps and loopholes in the BBNJ Identifier reporting system for data about DSI "access" and "utilization". The reporting requirements in article 12(8) are qualified by stating that the required information including the BBNJ identifier should be notified to the CHM "if available". Another reason for this qualification is to address utilization of MGR (or its DSI) that was collected from ABNJ prior to the treaty or relevant law entering into force if a Party does not elect to override the retroactivity provision under article 10(1) (see Sect. 3.2). As with other provisions, the inclusion of practicability is important to maintain flexibility so that the requirements can be effectively implemented by Parties.

(2) *Patents and Products*

The requirement to report on the location of patents granted and if available products developed may now be achievable following recent international developments. The 2024 World Intellectual Property Organization *WIPO Treaty on Intellectual Property, Genetic Resources and Associated Traditional Knowledge* (WIPO, 2024) requires the disclosure of origin or source of genetic resources and traditional knowledge in patent applications where a product or process is based on the resources or knowledge (Brown, 2025). For commercial products there is no centralized registry for origin of genetic resources included in patents. It may be possible for certain regulated patented products (e.g. pharmaceuticals, food/feed) to obtain disclosure of origin information. However, for others (e.g. industrial enzymes) obtaining information about the origin of genetic resources may be more challenging. It may still be possible to find out, if good scientific practice has been followed and due diligence applied i.e. that the BBNJ Identifier is associated with a product. It should also be noted that some industries do not patent, which may impact compliance with article 12(8) notification for "utilization". The challenge will be for a Party to obtain information on products containing MGR linked with a BBNJ Identifier where the R&D was carried out by a multinational corporation and different elements of R&D carried out in different countries, including under the jurisdiction of non-Parties. Also, a product may have been developed using multiple MGR with different BBNJ Identifiers, which may necessitate complex record keeping with unintended non-compliance (see also scenario 3 in Sect. 3.3 below).

Requirements for information on any products developed and sales of such products (art 12(8)(e)) may be the most challenging requirement to meet. Potentially use of the BBNJ Identifier throughout the R&D process and due diligence if research outputs are published will allow the reporting of data on products, but accessibility of these data are very rare once MGR enters the commercialization phase (Humphries et al., 2021). This may require reliance on self-reporting by industry, potentially as part of corporate social responsibility or a system similar to that applied for ethical biotrade. Companies could benefit from legal certainty in products developed, and even utilize the BBNJ Identifier in marketing of products (Oldham & Thambisetty, 2023). Parties could potentially regulate for this within national jurisdictions. The second challenge will be to determine which Party reports on the activities of a multinational corporation and its products. A further, unaddressed, question is what the implications are for a product that uses DSI sourced from multiple origins with different international frameworks including CBD and the BBNJ Agreement and/or DSI outside the scope of these two frameworks. In these cases of "mixed

DSI use" which legal requirements will prevail in the event of inconsistency, or will all apply equally? What if the ABNJ portion's contribution to a product is a small minority or not part of the claim but "only" part of the reference material of the patent application? Will BBNJ agreement reporting requirements still apply? These practical questions will require further consideration (see also scenario 3).

The BBNJ Agreement has an initial monetary benefit sharing scheme that is decoupled from the MGR R&D processes discussed here. That said, once the BBNJ Agreement enters into force, the CoP will decide on the modalities for a monetary benefit sharing scheme arising from the utilization of MGR and DSI, requiring an understanding of current practices of scientists and commercial end users. In that regard, there is a lack of baseline data on products from BBNJ. Private sector entities could provide records of products developed/costs/profits directly to the CHM since such information may be requested by a Party that has jurisdiction over the entity. This would support the work of the ABS Committee to make guidelines or a code of conduct for activities with respect to MGR and DSI (art 15(3)(a)). While details associated with the future monetary benefit sharing tied to samples from BBNJ may not be decided for a year or more, the private sector could explore potential implications of the requirements and prepare avenues for data sharing on commercial uses.

14.2.4 Access to MGR and DSI in Repositories and Databases

Both the notification and benefit sharing systems contain obligations about access to MGR and DSI in repositories and databases. Parties must take measures to ensure that MGR and DSI on MGR of ABNJ (together with their BBNJ Identifier) that are subject to "utilization" are "deposited in publicly accessible repositories and databases, maintained either nationally or internationally, no later than three years from the start of such utilization, or as soon as they become available, taking into account current international practice" (art 14(3)). Access to MGR and DSI may be subject to the reasonable conditions of:

(a) The need to preserve the physical integrity of MGR;
(b) "Reasonable costs associated with maintaining the relevant gene bank, biorepository or database in which the sample, data or information is held";
(c) Reasonable costs associated with providing access; and
(d) Other reasonable conditions in line with the objectives of the BBNJ Agreement.

The treaty goes on to say that opportunities "for access on fair and most favourable terms, including on concessional and preferential terms, may be provided to researchers and research institutions from developing States" (art 14(4)).

As soon as information becomes available, Parties must notify to the CHM the modalities envisaged for access to MGR and DSI that are subject to "utilization" (art 12(8)(d)). These "access" provisions are distinct from the concept of "utilization", although the term "access" is undefined in the treaty (Humphries, 2025).

Several UK institutes collect, house and conduct work on MGR from ABNJ including publicly funded research institutes, universities, museums, and private consultancies. Relevant organizations include the Marine Biological Association (MBA), the National Oceanography Centre (NOC), the British Antarctic Survey (BAS), the Natural History Museum (NHM), Plymouth Marine Laboratory (PML), and a numerous universities and consultancies. These include both "official" national repositories of the NHM for example, which fall under existing legislation (the British Museum Act) and more general scientific repositories, housed in university, government, or commercial research institutes and laboratories (e.g. biotechnology companies, private consultancy). These entities and sectors will have different degrees of formalization for collections management. Depending on interpretation of the

BBNJ Agreement, some may comprise entities responsible for reporting. These organizations currently house MGR, which would fall under the BBNJ Agreement only if the retroactive application of provisions under article 10(1) are accepted by the UK when ratifying. But whether retroactivity is applied or not, all MGR collected and housed in these repositories *following* ratification would be in scope. Retroactivity also has implications for handling and storage of MGR and DSI. If applied, then ensuring that MGR are identified as originating from ABNJ could be required for existing collections (art 12(6)) which may necessitate significant additional curation and databasing (with cost implications, as outlined below).

Article 14(4) as above covers what the negotiators considered were reasonable conditions for facilitating access. Scientific sample repositories in the UK (including but not restricted to museum collections) are available for research as standard practice globally, any limitations on access are likely to arise for the reasons stated in the text, "preserving the physical integrity of samples". This recognizes that samples as physical entities are finite by nature and undergo attrition, i.e. can be "used up" in the research process. Other reasonable conditions could include existing research agreements, for example samples may be embargoed during the research project phase until the project is completed. "Reasonable costs" will arise through provision of access to samples (e.g. sample processing, staff time) to account for the time-intensive process of curation of biological collections and associated data. These potential fees could be waived or reduced for developing States, as is current practice for some museums and collections. Potentially the benefit sharing fund (see Sect. 2.5 below) could subsidize the costs of collection material for users from developing States. Input on "reasonable conditions" and "reasonable costs" could also be provided by the STB. It is important to recognize also that while most collections of MGR are currently housed in developed States (Collins et al., 2021), whatever the challenges for these institutes to meet the treaty requirements, those based in developing States will be far greater.

The requirement for repositories and databases to ensure MGR/DSI can be identified as originating from ABNJ in article 12(6) could be challenging as outlined in 2.3 above. The onus is on the Party that hosts these entities to ensure and monitor compliance which may be possible for national repositories with adequate resources. The BBNJ Identifier will need to be captured by in-house databases of marine research institutes/repositories, which as for global databases in Sect. 2.3 above would require development or field modification for the relevant database and associated documentation. Automating reporting will be important for larger organizations to minimize administrative burden.

Article 12(7) requires Parties to ensure that repositories, to the extent practicable, and databases under their jurisdiction prepare a biennial aggregate report on "access" to MGR and DSI linked to their BBNJ Identifier and make the report available to the ABS Committee. The language "to the extent practicable" recognizes that it may not be possible to require all repositories to compile the report, but at least the major repositories holding MGR may have the capacity to compile such a report. One key issue for understanding what to report is that the term "access" is undefined. For example, a scientist may borrow an MGR from a collection simply to compare it physically with another MGR. Such action would not be considered utilization, but potentially considered "access" by default. Practicalities of such an interpretation would need to be carefully considered given common practice of exchange of material across institutions, and scientists visiting institutions to study their collections. Monitoring "access" at such granularity would be of little value to the benefit sharing system but could have massive cost implications. Similarly, what defines "access" for DSI will need to be carefully considered (see scenario 3 in Sect. 3.3). It is important that disproportionate and burdensome reporting requirements are avoided, such as equating a BLAST search with access. Monitoring access to DSI in any case will not be feasible as stated in the 2022 Kunming-Montreal Global Biodiversity Framework (GBF) decision on DSI that "Recognizes that tracking and tracing of all digital sequence information on genetic resources is not practical" (CBD/COP/DEC/15/9; see also Rohden et al., 2020; Scholz et al., 2022). Further, while the aggregate report in theory would be light touch, there would be a need to collate information in totality to be able to report in aggregate. Ideally reporting processes would be automated as far as possible (see Sect. 2.3). National legislation implementing the treaty obligations will need to address what "access" means, and guidance from the CoP will be important to ensure consistency across Parties.

14.2.5 Benefit Sharing

Article 14 provides a framework for the fair and equitable sharing of monetary and non-monetary benefits from activities with respect to MGR and DSI on MGR of ABNJ (Broggiato et al., 2025; Lavelle & Wynberg, 2025). The treaty text provides an inclusive list of non-monetary benefits, including

- Access to samples, sample collections and DSI;
- Open access to FAIR (findable, accessible, interoperable, reusable) scientific data;
- Information contained in the notifications and BBNJ Identifier in publicly searchable and accessible forms;
- Transfer of marine technology and capacity building;
- Increased technical and scientific cooperation, in particular with those in developing States; and
- Other forms of benefits as determined by the CoP (art 14(2)).

It establishes a special fund that will be funded through annual Party contributions, additional contributions from Parties and private entities and payments in accordance with article 14(7) monetary benefit system (art 52(4)). Under this system, the CoP will decide on the modalities for the sharing of monetary benefits from the utilization of MGR and DSI, taking into account the recommendations of the ABS Committee. Modalities may include milestone payments, payments related to the commercialization of products, a tiered fee based on aggregate level of activities by a Party or other forms the CoP decides.

Article 14(3) provides that one form of non-monetary benefit sharing is open access to FAIR data in national databases "in accordance with current international practice and open and responsible data governance". Existing principles of FAIR data management are well established in the scientific community (Page, 2023) and open access to DSI, required for peer-reviewed publication, is consistent with these principles. Standard practice for UK-funded research are data policies to ensure data are openly accessible within a two-year window, with some UK institutes mandating publication solely in open access journals, contributing to non-monetary benefit sharing (art 14(2)(a–h)).

Article 14(9) indicates that benefit sharing modalities under the BBNJ Agreement "should be mutually supportive of and adaptable to other access and benefit-sharing instruments". Article 15(5) sets out how the ABS Committee "may consult and facilitate the exchange of information with relevant legal instruments and frameworks ... including benefit-sharing, the use of DSI on MGR, best practices, tools and methodologies, data governance and lessons learned". Current discussions under the CBD and the GBF regarding how benefits from the use of DSI should be shared recognize the complexity of the situation and are developing a framework for the instruments that deal with DSI to work together towards a common solution. Guiding principles are listed in the "DSI Decision" under the GBF and are based on sound scientific and pragmatic principles (CBD/COP/DEC/15/9). If such an overarching global benefit sharing mechanism could be agreed for all DSI falling under different UN instruments, then a multilateral DSI fund could disburse funds for conservation and sustainable use of biodiversity both within and beyond national jurisdiction.

This section demonstrates that many of the requirements under the MGR framework are already part of scientific practice, but there is a long way to go in incorporating practices by commercial end users into the framework. It is important for implementing the objectives of both Part II and Part V of the BBNJ Agreement to go beyond these contributions, including the instances of contributions to capacity building and transfer of marine technology (CB/TT) listed in Article 14(2)(e–h) (Harden-Davies et al., 2022). Funding for capacity building could be scaled up through existing practices, where opportunities are promoted and encouraged through science programme funding calls. Alignment with capacity building initiatives under the Nagoya Protocol is also important. Mandatory funding for CB/TT could be considered by research funding bodies, with careful input on ensuring sustainability. There are opportunities for greater harmonization in CB/TT efforts. Policies should be checked at national levels and revised to ensure compliance with BBNJ Agreement requirements and ensure that reporting can be collated centrally from a range of sources including repositories, databases, ABS clearing house mechanisms and Party implementation infrastructure, to share with the ABS Committee and the CoP for ongoing decision-making on notifications and benefit sharing.

14.3 Scenarios About How the BBNJ Agreement May Apply to R&D and Commercialization

This section explores scenarios based on existing scientific practices associated with MGR research, which follow the pathway of a collected sample from ABNJ through the R&D pipeline (including potential commercialization): in effect, a "day in the life" of an MGR. Scenario 1 is a simple linear example followed by more complex, non-linear scenarios concerning MGR, DSI and TK. Scenarios cover the following areas/activities:

1. MGR collected from a research cruise using a national research vessel (Sect. 3.1);
2. the use of MGR collected prior to the BBNJ Agreement/national laws coming into force (Sect. 3.2);
3. the use of ABNJ-sourced DSI (Sect. 3.3);
4. automation in collection and R&D (Sect. 3.4),
5. MGR for R&D sourced from fish harvest-related activities (Sect. 3.5), and
6. the use of TK associated with MGR in ABNJ (Sect. 3.6).

14.3.1 Scenario 1—Simple Linear Scenario—Collecting MGR from ABNJ with a Research Vessel (Cruise)

This scenario describes a linear example where MGR is collected on a research cruise using a national research vessel carrying out biological/biodiscovery research in ABNJ (see, e.g. Alcock, 2014; Clark et al., 2016; Rabone et al., 2019). Research cruises to ABNJ are commonly funded by national research funding bodies through a grant application (which would typically also include a cruise proposal and cruise application). Such a grant application may include plans for sample and data collection and utilization, and a DMP, which are requested by many funders at this stage. To enable efficient use of valuable cruise time, many research cruises involve multiple teams, often from different countries, carrying out distinct research projects. This requires careful cruise planning, including use of equipment and on-board facilities and may involve compromises among the teams. A plan for training early career researchers including from developing States may be included, but is not necessarily required, which is a potential area for improvement in the treaty process. Once a grant for a deep-sea expedition is awarded, the next stage may involve confirming technical feasibility of the proposed work and availability of sampling and on-board facilities. The cruise path may cross regions under special designation such as marine protected areas and associated permits or environmental impact assessments would need to be in place. During the research cruise, the sampling sites and deployments, and even intended cruise path may change due to weather conditions. The functionality of sampling gear and discoveries made while on the cruise may also alter the research aims and objectives (Clark et al., 2016). Once collected, samples may be preserved on board or analysed immediately, and data generated. As described in Sect. 14.2, any physical samples collected in ABNJ could be in scope as (a) may contain MGR, and (b) the collection event in ABNJ rather than the intent of use triggers requirements (Humphries 2025).

The BBNJ Agreement requires a pre-collection notification to be submitted to the CHM six months or as early as possible prior to the collection or sampling of MGR from ABNJ (art 12). Once submitted, the CHM will automatically issue a BBNJ identifier to be linked to the pre-collection notification. The issued BBNJ Identifier would link the sample/organism and associated DSI that are subsequently identified from the collection if it is included in the metadata or other records associated with the samples or DSI. In other words, the user of the MGR or DSI may be able to trace the organism or data back to the original collection if the BBNJ Identifier maintains its link within repositories or databases. This requirement can only be fulfilled once the CHM and its BBNJ Identifier function are operational, the timeframe of which is currently unknown. After the BBNJ Identifier is issued, there is a requirement that "updated information shall be notified to the CHM within a reasonable period of time and no later than the start of collection in situ, when practicable" if there is a "material change to the information provided to the CHM". It is not clear what constitutes a "material change" and what entity would determine whether the changes exceed the threshold (see Sect. 2.1 above).

During operations, a database will be populated with information about cruise operations, potentially including deployments and sample collections. This information forms part of the subsequent cruise report that presents the results obtained within a set time from demobilization, often within a year. Once the cruise returns to port, samples may be shipped to the institutions of respective project teams. This means that the same MGR, and different MGR with the same BBNJ Identifier, may end up in multiple repositories in various countries, with implications for reporting (see Sects. 2.2 and 2.3 above). The final cruise report may contain a narrative including a summary, study sites and scientific sampling. Several national oceanographic institutions have online portals for depositing the cruise plan and cruise reports and have sophisticated mechanisms to interrogate the sampling data generated, but this capacity is limited to developed States with resources (see Sect. 2.1 above).

After the research activities are complete and the vessel returns to shore, there is the requirement in the BBNJ Agreement's article 12(5) to notify the CHM with required post-collection information "as soon as it becomes available, but no later than one year from the collection in situ of marine genetic resources of areas beyond national jurisdiction". Most of the notification requirements can be fulfilled by providing the cruise report and the relevant cruise database (see Sect. 2.2 above). Article 12(7) requires that Parties to prepare an aggregate report on MGR and DSI linked to the BBNJ Identifier for the ABS Committee every two years "to the extent practicable". In principle, each sample record in the cruise database would be associated with the same BBNJ Identifier, which will stay linked to MGR/DSI from the original collection. If a research programme has multiple project leads, there is likely to be one database per project. There is an opportunity for treaty bodies to encourage the consistent application of data and metadata standards to ensure that all data complies with FAIR principles (see Sect. 14.2). Much of the detail around how the process will work remains to be determined by the CoP,

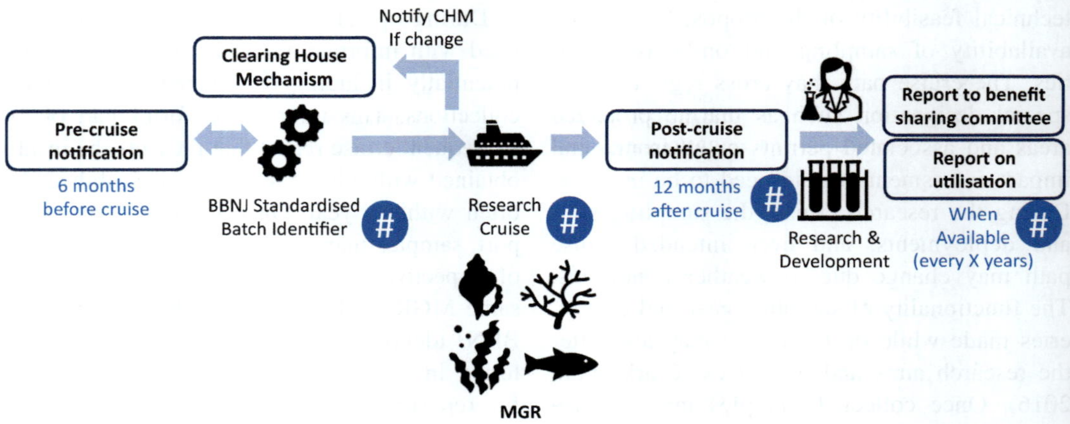

Fig. 14.2 Requirements of article 12 applied to a simple linear scenario involving a planned cruise by a national research vessel

with the support of its subsidiary bodies, but the intent appears to be to keep the system as simple as possible using the BBNJ Identifier (art 12(7), 14(7)).

In summary, cruise plans and reports in current practice already fulfil many of the BBNJ requirements (see also Tables 14.1 and 14.2) but even a linear scenario can have many variables. Further, while this scenario describes a cruise in ABNJ, cruises may collect both within and beyond national jurisdictions (Rabone et al., 2019) (Fig. 14.2).

14.3.2 Scenario 2—The Use of MGR and DSI Collected or Generated Prior to the BBNJ Agreement Coming into Force

Article 10(1) provides that the BBNJ Agreement applies to activities with respect to MGR and DSI on MGR of ABNJ collected and generated after the entry into force of the treaty for the respective Party. In other words, a State that has ratified the treaty by implementing their obligations under national laws (i.e. become a Party to the treaty), will state the date from which their laws will apply to the collection and generation of MGR and DSI for their nationals who undertake these activities. This follows the ordinary rules of international law that treaties and laws are not normally retrospective (see Humphries, 2025). However, the article goes on to say that the treaty obligations apply to "utilization" of MGR and DSI on MGR of ABNJ that were collected or generated *before* entry into force, unless a Party makes and exception when ratifying. In other words, while the provisions for collection, including pre- and post-collection notifications and the provisions for "access" to MGR and DSI in repositories are not retrospective, depending on the way each Party deals with temporal scope under their national laws, the provisions for "utilization" may be retrospective and cover MGR from ABNJ collected before the legislation came into force.

If a Party does not opt out of the retrospective effect for utilization, MGR collected prior to the law coming into force that is the subject of "utilization" (e.g. legacy MGR) will not have a BBNJ Identifier as there will have been no pre-collection notification (Fig. 14.3). This MGR may still need to be included in any report to ABS Committee and subsequent reports on utilization, however. The BBNJ Agreement does not specify how relevant information about MGR without a BBNJ Identifier should be supplied to the CHM. In the first instance, it may not even be possible to determine what is in scope given legacy collections may not have locality data to ascertain whether they were collected

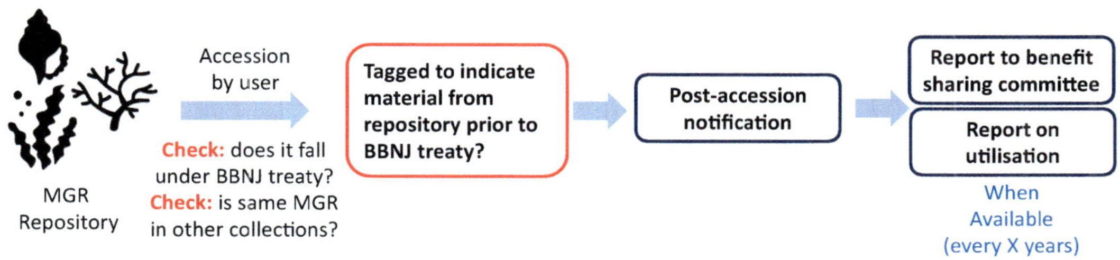

Fig. 14.3 Potential workflow for MGR collected prior to a treaty housed in an MGR repository, accounting for potential retroactivity of article 10

from ABNJ. Retroactivity for MGR under the BBNJ Agreement raises other complexities. The same MGR from the same pre-BBNJ cruise may be housed by institutes in different States, some that are Parties have opted out of retroactivity and some that have not and some States that have not become Parties and are not under BBNJ Agreement obligations, which would complicate reporting for ongoing R&D and commercialization of MGR and associated DSI. It may also promote jurisdiction shopping where "utilization" is conducted in States with the least regulatory and reporting burden.

Legacy MGR potentially could be "tagged" in some way to indicate it was obtained from ABNJ before the BBNJ Agreement and relevant Party's law entered into force (Fig. 14.3). Such a tag could both differentiate the pre-BBNJ MGR and also act as a flag that not all data will be available and therefore may not meet the notification requirements for utilization (e.g. location where the original sample is held, details of the post-collection notification and the modalities of access by third parties (art 12(8)). The data outlined in the pre-collection and post-collection notifications may not have been collected originally, or there may be significant costs for the repository in finding the information (see Sect. 14.2), although the only reporting requirements in this retroactive scenario concern those relating to "utilization". However, the technical feasibility of implementing such a tag, or how it may align with the BBNJ Identifier, is unclear, and whether it would clarify, or further complicate matters requires consultation with repositories, databases and other stakeholders (Sect. 14.2).

14.3.3 Scenario 3—Use of ABNJ-Sourced DSI

For DSI on MGR of ABNJ that has been deposited in a database after the BBNJ Agreement enters into force, with the source MGR having been notified under a pre-collection notification (Fig. 14.4), a BBNJ Identifier will be associated with the MGR and connected to the resulting DSI (art 12(5)(a), 12(6)) to enable downstream reporting on "access" and "utilization" (arts 12(7), 12(8)). These requirements raise the question of what comprises "access" and "utilization" for DSI. For example, does using DSI in a comparative search (e.g. BLAST) or in a phylogenetic tree constitute access? Or would such activities need to be more substantial than simply comparative? And, if so, can and will this distinction be made? It is important interpretations of these terms by treaty bodies are grounded in practicalities and researchers are cognizant of what activities may apply (see Sect. 14.2).

The requirement for post-collection notification to the CHM on the original MGR and the repository where the sample is kept (arts 12(8)(b–c)) could be met using the BBNJ Identifier associated with the DSI on MGR, allowing it to be traced to the relevant notification (although not necessarily to the relevant repository—see Sect. 14.2). If research yields a patentable discovery and a patent is applied for and granted (i.e. if claims are being made on the DSI on MGR or if the DSI is needed "in order to enable a practitioner skilled in the art to reproduce

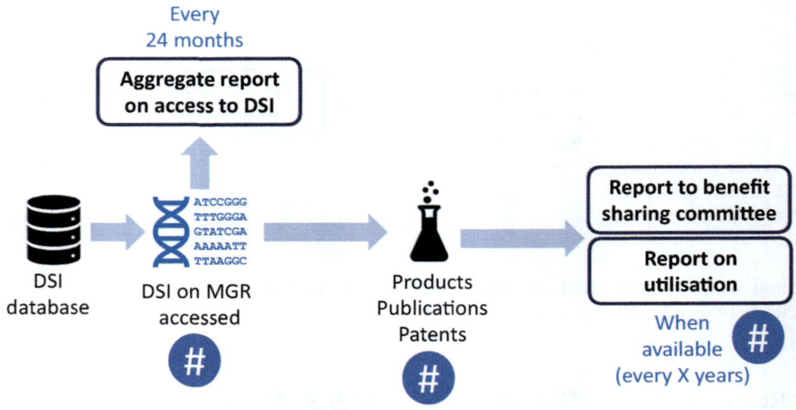

Fig. 14.4 A "simple linear" example of the use of DSI on MGR of ABNJ where the DSI on MGR is used directly from a database with no modification to generate a product

the invention"; TRIPS agreement,[8] article 27), then article 12(8) also requires notification to the CHM of where this patent can be found. The WIPO recently revised policies and now requires disclosure of origin so this should allow compliance (WIPO, 2024, see Sect. 2.3 above). Similarly, INSDC now requires spatiotemporal information for all DSI published on the database, with some exceptions. Issues may also arise if the sequence queried is not unique but shared, for example in members of a single species that has been collected both in ABNJ and in AWNJ.

In reality, the use of DSI is far more complex than the above simple linear scenario suggests. DSI used in a product may be derived from multiple MGR from both ABNJ and AWNJ, potentially also collected before the BBNJ Agreement entered into force (Fig. 14.5). At the start of this process, DSI on MGR is queried against a DSI database containing millions of sequences returning potentially 10–1000 s of sequences from different organisms, including e.g. terrestrial microorganisms (under CBD), plants [under the *International Treaty on Plant Genetic Resources for Food and Agriculture*[9]] and deep-sea fish (MGR collected prior to the BBNJ Agreement). Sequences from these three sources may be combined with the original DSI on MGR used in the query leading to a "hybrid" consensus sequence (a synthetic sequence based on many sequences). This new hybrid may be subject to even further modification based on additional sequences or via targeted mutations or directed evolution. After a research process taking months to years, a final sequence is arrived at which could be used in a final product that may be patented and commercialized. It will be challenging to determine the individual contribution of each original sequence was to the final product. Further modifications may take the product DSI far from the original sequence so that it is likely impossible to trace it back to the original DSI on the source MGR. Also, only the DSI derived from MGR collected after the BBNJ Agreement for the respective Party came into force will have a BBNJ Identifier. It will be very challenging therefore to calculate which benefits should be

[8] *Marrakesh Agreement Establishing the World Trade Organization*, Annex 1C (*Agreement on Trade-Related Aspects of Intellectual Property Rights*), opened for signature 15 April 1994, 1869 UNTS 299 (entered into force 1 January 1995) (TRIPS Agreement).

[9] *International Treaty on Plant Genetic Resources for Food and Agriculture*, opened for signature 3 November 2001, 2400 UNTS 303 (entered into force 29 June 2004).

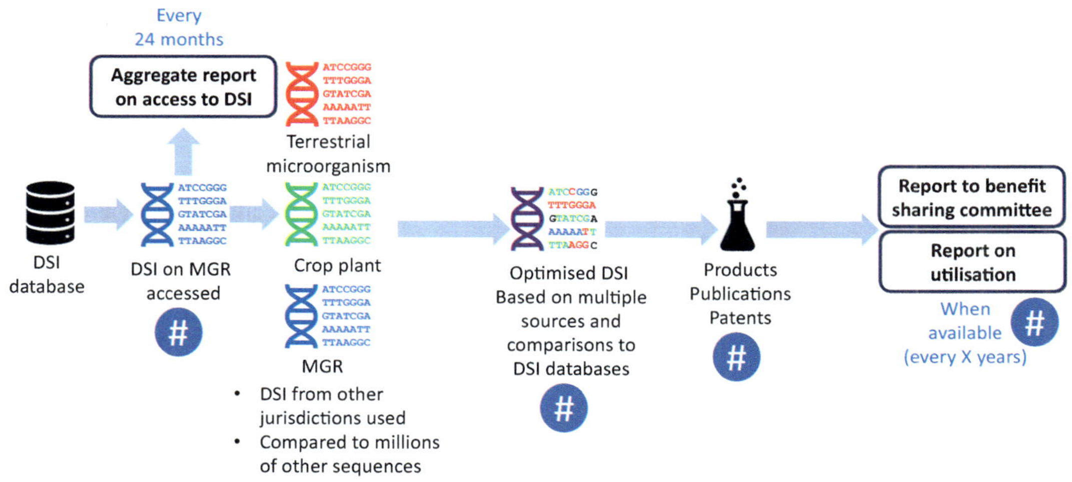

Fig. 14.5 A more complex and realistic example of how DSI on MGR might be used to create a product. DSI on MGR is compared to other sequences in the database, and the eventual sequence contains elements of DSI from other types of organisms that fall under different regulations/regimes. Additional optimization may be carried out to derive the eventual DSI that is used to create a product

shared with which UN ABS fora for each of the DSI utilized. In any case, there are currently no mechanisms in place to deal with benefit sharing from DSI under any UN instrument dealing with DSI (UNEP, 2022).

If a DSI on MGR that is used was deposited prior to the BBNJ Agreement entered into force (legacy DSI), then article 10(1) allows a Party to opt out of retroactivity (scenario 2). Some data held in databases prior to the BBNJ Agreement will not have spatiotemporal information that makes it easy to identify as originating from ABNJ, and therefore in scope. Even if the DSI can be identified as coming from MGR of ABNJ, it will not have a BBNJ Identifier, or even the required associated data, so it may not be feasible to comply with article 12. The application of retroactivity on DSI may require the "tagging" or identification of pre-BBNJ DSI records in databases. How this tag could (or even should) be implemented requires further input by treaty bodies and Parties. Further, given the main DSI databases (INSDC) are multinational collaborations, the question who will implement and report on this data remains.

14.3.4 Scenario 4—Automation in Collection, Research and Development

"Collection in situ" includes the activities of collection and sampling (art 1(4)). Autonomous underwater vehicles (AUVs), also called Uncrewed Marine Vessels, that collect data and samples from ABNJ can have different levels of autonomy, ranging from remotely operated vehicles (ROV) with some human operation to fully autonomous vessels with machine learning and artificial intelligence (AI) capabilities (IMO, 2021). AUVs can take samples of water and sediment that contain MGR or samples of live biological specimens. They can collect, integrate and transmit information related to the physics, chemistry and biology of the ocean (Chai et al., 2020). Techniques for automated collection and analysis are rapidly evolving. For example, sequencing can now be carried out in situ (in the field), using portable sequencing technologies, and combined with environmental DNA (eDNA) sampling, which allows sequencing of DNA filtered directly from water or sediment samples to analyse genetic material in the cells and identify species present (Harrison et al., 2019).

Parties have an obligation to ensure information is notified to the CHM when MGR and, where practicable, DSI are subject to utilization, including commercialization "by natural or juridical persons under their jurisdiction" (art 12(8)). The legal status of AI and whether it can be a "juridical person" varies between States (see Humphries, 2025). Whereas ROVs for example have some human involvement, it is likely that the wording of the "utilization" notification for MGR and DSI will have a loophole for fully autonomous entities without legal personhood. In contrast, the pre- and post-collection notifications and obligations relating to "access" to MGR and DSI in repositories and databases have no such limitations and can be triggered by activities carried out by remotely operated and fully autonomous entities. However, the pre- and post-collection notifications are confined to collection in situ of MGR of ABNJ, which means the physical samples and not the information components such as DSI, which is subject to obligations as distinct subject matter (see Humphries, 2025). It is only collection of the physical MGR that is the trigger for the pre-collection notification and the automatic assignment of the BBNJ Identifier, meaning that those activities that collect data directly from ABNJ as described above may not be captured.

Further, the nature of autonomous collections and sampling do not necessarily fit within the expected timeframes of the pre- and post-collection notifications. Without humans on board, the patterns and timing of collection can vary significantly from the cruise envisaged in scenario 1. Autonomous collections and sampling may be from permanent moorings, AUVs or floats that can conduct uninterrupted missions for months if not years (Chai et al., 2020). The information may be transmitted to a research facility within national jurisdiction in real time or AUVs may carry out in situ analysis in ABNJ (Chai et al., 2020). The research projects might change during the uninterrupted (possibly indefinite in future) deployment but notification of material changes to the pre-collection must be done before the vessel leaves shore for the first time (art 12(4)). It may be impractical for the post-collection notification to be carried out no later than one year from collection in situ if the collection is ongoing.

Whereas the "utilization" obligation was envisaged by the negotiators as being an activity that comes after the collection activity in ABNJ under the linear scenario 1 with advances in technology, "utilization" of MGR and DSI may occur within ABNJ at the time of collection as above (Chai et al., 2020). There are rapid advances in employing machine learning/AI techniques for the utilization of DSI, but if these are considered to be fully autonomous, they may fall outside scope of the utilization notification and information sharing obligations. In practice, it may be impossible for a Party to distinguish between those utilization activities carried out within their jurisdictions by natural and juridical persons and those that are not.

Currently it is unclear what activities fall under obligations concerning "access" to MGR and DSI in repositories and databases (see Sect. 2.4 above). If access extends to a BLAST search or similar, then it would not matter whether the activity was generated through AI or not because the "access" provisions are not similarly constrained as the "utilization" provisions to those undertaken by natural or juridical persons. Figure 14.6 outlines a common-sense approach to notification of activities with respect to MGR and DSI on MGR of ABNJ following the linear scenario for manned cruises. One option is post-deployment notification after 12 months as shown (Fig. 14.6) but an AUV may collect data for a full year or beyond, therefore another option could be 12 months after final data upload/mission completion.

By including different subject matter triggers for "collection" and "utilization" activities, the BBNJ Agreement creates possible loopholes for AI-related research that is increasingly being used in each of these activities. MGR and DSI as physical and digital entities are distinct but intrinsically linked (Rabone et al., 2019). The artificial separation of collection for physical

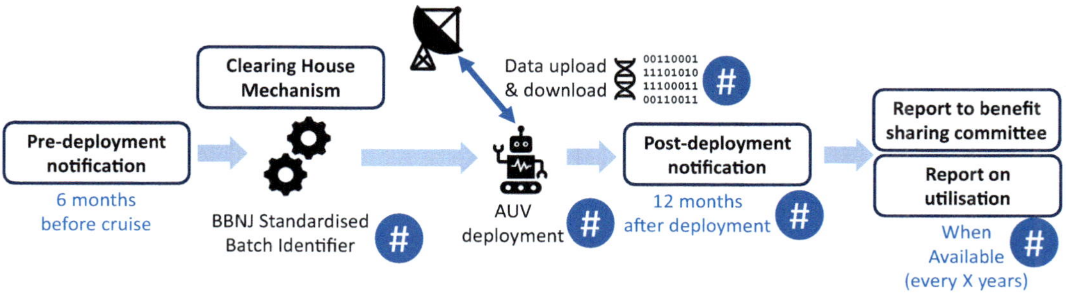

Fig. 14.6 Potential pathway for MGR collected by unmanned underwater vessels, using artificial intelligence in the process

samples and for the information components creates loopholes and complexities in the notification system. Until the BBNJ Agreement bodies clarify the situation for AUVs, applying a common-sense approach, Parties should ensure that the pre-notification is carried out prior to the first deployment of the AUV to ABNJ if the original project intends to collect samples and data. The notification mechanism is largely tailored to the linear ideal of collection, sampling and utilization of MGR from ABNJ via manned cruises rather than automation in these activities. In reality, BBNJ Agreement bodies will need to provide specific guidance for accommodating the distinguishing features of fully autonomous activities, such as the length of deployment and the geographical location of subject matter including activities that do not fit neatly within the linear scenario.

14.3.5 Scenario 5—Fish and Fishing-Related Activities

Article 10(2) of the BBNJ Agreement provides two related exemptions to the scope of application for Part II MGR. These are (1) fishing regulated under relevant international law and fishing-related activities (art 10(2)(a); and (2) fish or other living marine resources known to have been taken in fishing and fishing-related activities from ABNJ, except where such fish or other living marine resources are regulated as utilization under part II on MGR (art 10(2)(b)). Together, these provide a broad exclusion of both fisheries harvest activities and living marine resources, including fish, caught as a commodity in ABNJ. The exclusion not only covers the activity of fishing but also the resources that are the result of such activities. Like the Nagoya Protocol, the determining factor is the presence/absence of utilization. In terms of fishing activities, the exclusion means that such activities will not have to comply with any of the collection-related requirements of Part II MGR, such as a pre-cruise notification, which might pose some practical challenges.

Regarding living marine resources, many different scenarios could be described in which these are collected with other purposes than utilization as an MGR, yet later the resource, parts of it, or associated organisms are utilized as MGR. Examples include specimens acquired from recreational fishing, or from commercial fishing vessels (including bycatch) via observers, or acquired later, e.g. from fish markets, and/or subsamples thereof, including ecto- or endo-parasites (Koepper et al., 2022), gut microbiomes, or tissue samples. In these cases, the MGR will enter the BBNJ MGR pathway at a later stage (Fig. 14.7). All the post utilization steps, such as notification on use and reports to the ABS Committee, will be similar to standard MGR collections.

A simple case can illustrate how the exemption in article 10(2)(b) may apply. An unusual fish is harvested as part of a catch from a

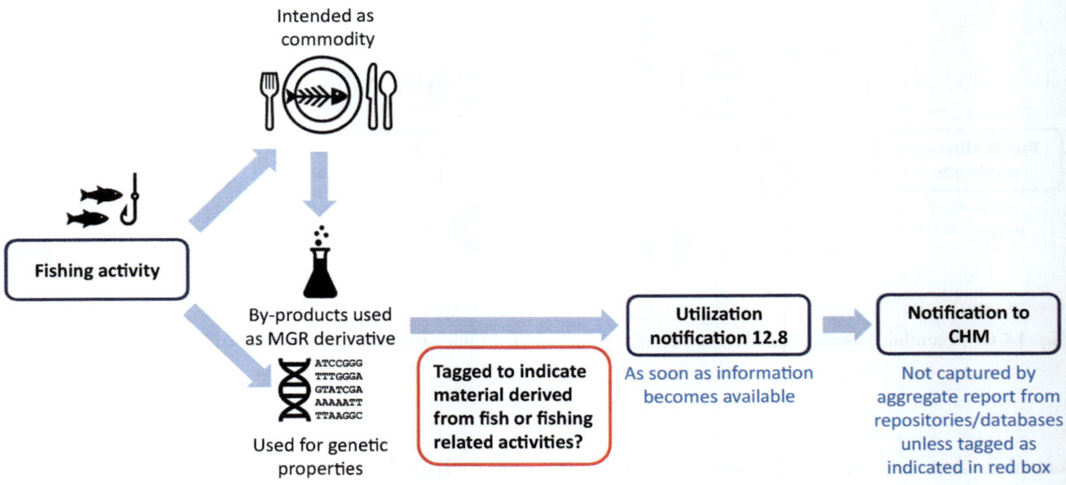

Fig. 14.7 Using MGR arising from fishing or fishing-related activities under the BBNJ agreement

commercial fishery from the high seas. It is sold at a fish market to a researcher who is interested in investigating its apparent unusual properties. There has been no pre-or post- collection notification because it was extracted from ABNJ under the fisheries exemption. Consequently, the harvest (collection) does not have an associated BBNJ Identifier (because there was no pre-collection notification) and the fish enters the R&D pipeline at the stage of "utilization". "Utilization" obligations are triggered if there is R&D 'on the genetic and/or biochemical composition' of MGR (art 1(14)). Genetic manipulation of the fish would clearly trigger the utilization notification, as would other investigations into its genetic or biochemical composition (see Sect. 2.3 above). This utilization notification to the CHM would apply in the same way as MGR that have been collected on a research cruise for example. The utilization notification is not dependent on having a BBNJ Identifier (it is only required "if available"). The aggregate report to the ABS Committee would not include access to MGR or DSI, which is dependent on the relevant MGR or DSI being linked to a BBNJ Identifier (art 12(7), see Sect. 2.4 above). This means that the only means of including the information about MGR in aggregate reports would be if the CoP designed another mechanism to "tag" this material as originating from ABNJ under fish or fishing-related activities, similar to potential tagging of "legacy" or pre-BBNJ MGR and DSI (scenarios 2 and 3) to link repository data to the CHM. The outcomes of research from its use, however, may be notified under the utilization notification to the CHM.

A less straightforward example would be harvesting marine living resources known to have been collected by fishing-related activities from ABNJ to increase biomass for pharmaceutical leads. For example, deep-sea sponges might be harvested from ABNJ or as bycatch in another ABNJ fishery such as halibut or cod fisheries (Munoz et al., 2020). To bring a commercial product to market may require large quantities of the original sponge (if it is not cost effective to synthesize the product) over a long period of time for different stages of R&D and commercialization. This might either be achieved from repeated collections from ABNJ or through aquaculture, which is a key activity for building biomass and has been used for producing sponges and metabolites for pharmaceutical purposes in AWNJ (Duckworth, 2009). Would obtaining the sponges under harvest fisheries be exempt so that the collection is not subject to pre-collection notification when the intent of the harvest is ultimately for R&D purposes? If sponges are originally collected from ABNJ and then farmed in AWNJ as a bulk commodity (increasing biomass) would it be exempt under article 10(2)? The operation of the article 10 exemption seems to require an element of intent, whereas the collection notification does not require "intent" of collection for R&D purposes (unlike the "utilization" notification) because the trigger for the pre-collection notification is simply collection or sampling of MGR of ABNJ (art 1(4)) (Humphries et al., 2025b). Although the sponges would be collected for fishing-related purposes (harvest or aquaculture), the ultimate use is for "utilization" and it is likely that it would not fall within the exemption and be subject to both pre-collection and "utilization" notifications.

The CoP may need to clarify whether intent is a relevant for the fishing-related exemption to apply and when intent must be ascertained. An example illustrating the possible need to ascertain the ultimate use of the harvested fish or marine living resources at the time of collection concerns harvesting fish to produce fish oil as a product for nutraceutical or medical applications. If the fish are collected (harvested) for their oil (as a bulk commodity), then article 10(2) exemption would apply (scenario A). If the oil was subsequently refined though industrial means to increase its strength (scenario B), would the exemption status change? Arguably, the fishing activity is still harvesting a fish to produce a (more refined) bulk commodity within the meaning of article 10. If the fish that were harvested or the derivatives from the fish (i.e. the oil) are subsequently subject to R&D on the genetic or chemical composition (e.g. ascertaining genes associated with higher oil content) to create an oil with new properties (scenario C), would the activity then fall within the "utilization" notification which applies to MGR or their derivatives (art 1(3) definition of biotechnology)? It is likely this would be viewed by scientists as R&D and not simple processing. However, when it comes to "utilization" of a derivative (as opposed to the MGR), it arguably falls within the notification trigger if it meets the higher threshold of "making or modifying products or processes", rather than simply investigating the genetic or chemical composition (see Sect. 2.3, Humphries, 2025). This indicates that in scenario C, if the fish are harvested for their oil (derivative), which is simply investigated for genes associated with higher oil content, it might not meet the "utilization" threshold but if they are harvested for their oil to create an oil with new properties (e.g. manipulate molecules to increase potency or work with other chemicals to increase storage life or minimize side effects), it might. In reality, whether a fish or fishing activity is in scope of the exemption will depend on the way a Party implements its treaty obligations under national law, but guidance from the CoP would be crucial for a consistent approach.

While the treaty is silent about "intent" for the exemption or utilization notification triggers to apply in a specific case,[10] in practice the above examples demonstrate that "intent" may indicate whether the harvest is for the purposes of bulk commodities (exemption is likely to apply) or R&D (exemption is unlikely to apply). The activities under the exemption are undefined and negotiators removed the qualifying term "commodity" which would have made it clearer that the activities under the exemption are not for the purpose of investigating genetic or biochemical composition. As the above examples demonstrate, the reality is that determining whether a fish or living marine resource or fishing/fishing-related activity is exempt or not may depend on the R&D activity being undertaken and whether the research relates to the MGR (investigating genetic/biochemical composition) or its derivatives (making or modifying products or processes). Figure 14.7 outlines the simple case above for determining whether the exemption applies and if not, how the information about its use would reach the CHM. It demonstrates that only the utilization notification to the CHM would be triggered but unless a new tag or identifier is created for material derived from fish/fishing-related activities (red box), the aggregate report to the ABS Committee for access to MGR or DSI from repositories and

[10] Intent is not relevant.

databases (art 12(7)) would not be able to pick up this data.

14.3.6 Scenario 6—Traditional Knowledge Associated with MGR in ABNJ

The BBNJ Agreement takes a similar approach to the CBD and Nagoya Protocol in regulating access to TK associated with MGR in ABNJ (see Pena-Neira & Coelho, 2025). Instead of TK being managed under the BBNJ Agreement's multilateral notification, monitoring and benefit sharing mechanisms, it will be governed by each Party "where relevant and as appropriate" under a bilateral approach of authorizations and contracts (Mutually Agreed Terms—MAT) with the knowledge holder. Parties have wide discretion about whether they take legislative, administrative or policy measures "with the aim of ensuring that" TK associated with MGRs in ABNJ held by IPLCs is only accessed with Free, Prior and Informed Consent (FPIC) or approval and involvement of the IPLC that holds the knowledge; "Access to and use of such TK shall be on mutually agreed terms" (art 13). This obligation means that each Party that decides to regulate TK may have different procedures and requirements for determining what TK is covered by the obligation (the scope), identifying the correct knowledge holders, obtaining FPIC and establishing MAT.

While the BBNJ Agreement does not directly regulate benefit sharing relating to access and use of TK, access to such knowledge "may be facilitated" by the CHM (art 13). However, the CHM is primarily a centralized open access platform (art 51), and aside from implications for potential confidentiality requirements of some TK, it is unlikely to be responsible for the accuracy of the information. Rather, it will be the responsibility for Parties to ensure correct and current information is on the CHM platform. The extent of the CHM role for facilitating access to TK will not be clear until the CoP has met. It is likely to include linking the public to relevant websites with information on procedures for identifying and approaching relevant knowledge holders or at least, how to approach governments and/or communities to find their information and procedures. Both access to and use of the TK must be on MAT, but the CHM does not have an explicit function to facilitate the "use of" the knowledge, indicating that regulating use and benefits from the use will be in accordance with national law. MAT may be affected under authorization systems or contract law or under other mechanisms such as registration systems already established to implement Nagoya Protocol obligations for TK. The extent to which the CHM will have a direct role in the exchange of information about benefit sharing is unclear, but it may have a passive role for linking databases where this information may be located (art 51(3)(c)).

In many cases, procedures for complying with article 13 are likely to be the same or similar to accessing TK associated with MGR in AWNJ. This is because TK systems and cosmologies are not bounded by legal fictions of boundaries and jurisdiction under international law (Menime & Bowrey, 2022; Mulalap et al., 2020). These common elements under national law may include the meaning of TK, IPLCs, FPIC and MAT. Definitions of TK and IPLCs are likely to be determined under national laws or by the IPLCs involved, sometimes on a case-by-case basis (see Humphries, 2025). There is a large body of work on the meaning of FPIC in the context of the CBD and Nagoya Protocol but again, there may be specific meanings under national laws (CBD, 2016). Mutually agreed terms usually means contractual mechanisms where both parties (the knowledge holder and the proposed knowledge recipient) agree on the terms and conditions of access to and use of the knowledge. This may or may not include monetary or non-monetary benefit sharing as there is nothing in the BBNJ Agreement that requires fair and equitable benefit sharing for TK, unlike article 5 of the Nagoya Protocol for TK associated with genetic resources from AWNJ. It would be up to each Party to decide how to deal with benefit sharing, either through contract law or multilateral benefit sharing funds.

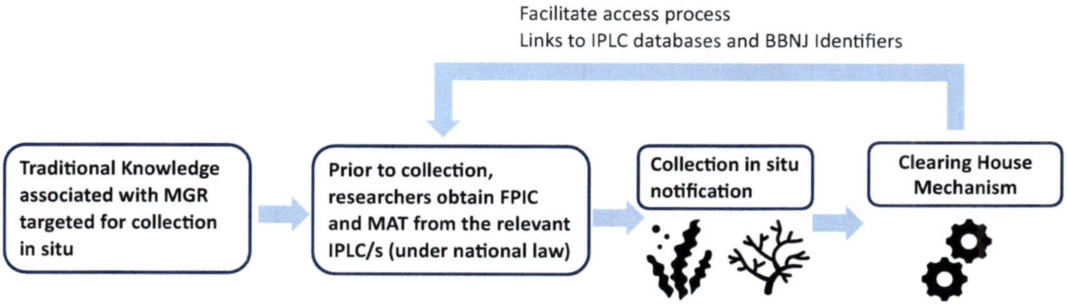

Fig. 14.8 Narrow interpretation—obligation is triggered when the traditional knowledge is used or proposed to be used to target MGR in ABNJ for their genetic material properties (prior to collection)

Although the TK obligation will be interpreted by Parties according to their interests and circumstances, there are interpretations that are unique to ABNJ that may require clarification by BBNJ Agreement bodies to avoid loopholes. These include clarifying the geographical, temporal and subject matter scope (nature of the knowledge) of the obligation. There is uncertainty about the types of TK that might fall within scope—whether it will be narrowly interpreted as only knowledge about the genetic attributes of MGR or more broadly includes knowledge about activities and observations associated with MGR (Mulalap et al., 2020). The obligation refers only to MGR *in* ABNJ, unlike other BBNJ provisions relating to MGR *of* ABNJ, suggesting that the geographical scope may be confined to specific MGR actually collected in situ in accordance with the BBNJ Agreement, rather than the known distribution of the MGR being ABNJ (Humphries, 2025). The activity of "access" to the TK is undefined and the term "use" is not the same as "utilization of MGR" defined under the BBNJ Agreement (art 1(14)). It is unclear whether the obligation extends to use of DSI on MGR in ABNJ associated with the knowledge and how it will relate to the retroactive application of the treaty to collection and utilization activities (scenarios 2 and 3).

Aside from interpretation, several gaps in procedures and processes unique to the context of ABNJ will also require guidance from treaty bodies. These include how to manage: identifying knowledge holders when there is no guidance under national measures; situations where there is more than one knowledge holder of the same TK; circumstances where new knowledge holders are identified after completing FPIC and MAT with another IPLC or knowledge holder; the link between the BBNJ Identifier of the MGR and the TK with which it is associated; procedures relating to secret and publicly available TK; whether the obligations extend to TK associated with DSI on MGR of ABNJ; and benefit sharing from TK of unknown origin (or where MAT is not possible).

Until there is further guidance from treaty bodies, it is unclear how the TK obligation will operate in practice and how it relates to MGR and DSI obligations. In the meantime, there is a narrow and a broad interpretation that may assist practitioners to align their practices with intent of the BBNJ Agreement. A narrow reading suggests that the obligation is only triggered by the collection activity of physical MGR within ABNJ (Fig. 14.8). This means that FPIC and MAT would only be required if the research project in the pre-collection notification intends to use TK associated with MGR in ABNJ. In other words, this applies in circumstances where the knowledge is used to target the MGR in ABNJ for their genetic material properties. As the BBNJ Identifier is automatically issued for the pre-collection activity, FPIC and MAT will pre-date the identifier, but the identifier could be subsequently linked to databases concerning TK associated with MGR of ABNJ from that particular collection. From a practical

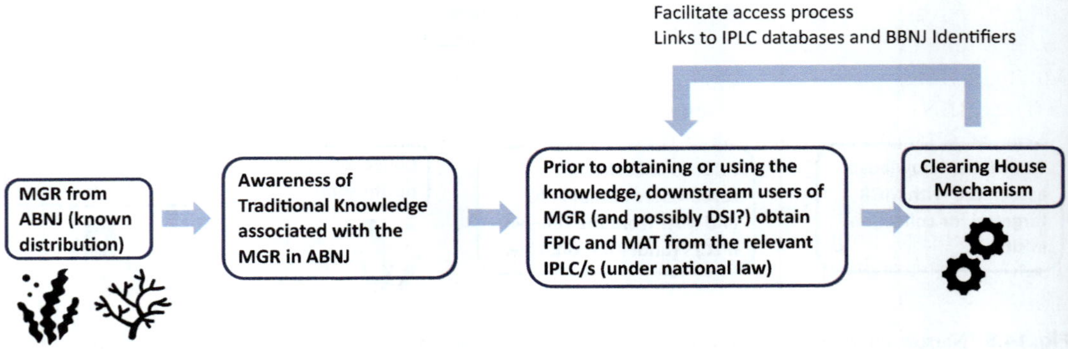

Fig. 14.9 Broad interpretation—obligation is triggered when someone seeks access to, or use of, traditional knowledge associated with MGR known to be located in ABNJ (irrespective of collection)

perspective, the timeframes for obtaining FPIC and MAT prior to collection may take months or years—long before the period of notification. This approach assumes that the researchers will know which MGRs they are targeting as part of the collection, which is often not the case for biodiscovery (scenario 1). It also assumes that the MGR will be where they expect them to be. Researchers and TK holders may go through years of negotiating FPIC and MAT without finding the targeted MGR. This narrow interpretation may promote misappropriation of TK (because it does not capture subsequent uses of collected MGR and associated knowledge) and it is also likely to delay or deter ABNJ research without any benefit to the TK holders. It puts the onus for engaging with TK holders for FPIC and MAT on the researchers who search for the samples, rather than downstream users that seek to utilize the MGR for economic or other benefits.

A broad interpretation suggests that the obligation is triggered by access to, or use of, TK associated with MGR known to be located in (rather than actually collected from) ABNJ (Fig. 14.9). This interpretation breaks the geographical and temporal link between the TK and the collection activity. The relevant time for seeking FPIC and MAT is when it becomes known that there is TK about the properties of MGR with a known distribution in ABNJ. In practice, the same knowledge may relate to the same MGR that travels within national jurisdiction. The regulatory mechanisms (FPIC and MAT) are the same under the Nagoya Protocol and BBNJ Agreement; therefore, one potential complication is that two separate FPIC and MAT processes for the same community may be required for the same research project, i.e. for TK associated with MGR both in ABNJ and AWNJ. There are many reputation and economic benefits for biotechnology companies to proactively seek FPIC and MAT and the broad interpretation ensures that the responsibility falls on these downstream users, rather than the researchers who are collecting the samples and who do not necessarily know what the "value" of the resource will be.

14.4 Conclusion

In general terms, all actors involved in acquiring, storing and utilizing MGR or associated DSI and TK, including academia, government and industry, need to understand BBNJ Agreement obligations and comply with the laws of the Parties that implement them. Given that there are practical aspects of the framework that are yet to be determined by the CoP, this will require proactive development of procedures and systems to compile, curate and provide necessary information to Parties, including DMPs and the BBNJ Identifier, when undertaking activities regulated under Part II. It is likely that this information will be provided at the first instance to the Party that has jurisdiction or control over the relevant activity, but there may be opportunities for directly sharing information with the CHM.

The BBNJ Agreement presents a linear vision of science (Lawson et al., 2025) which belies many inherent complexities. It is crucial that the R&D process for MGR is not imagined as a linear progression where such work would automatically result in commercialization. Most R&D pathways are non-linear with many side branches that may be abandoned or pursued, iterative loops and long breaks in the process. Often several research threads are pursued in parallel, and the intended application is completely changed between the start and end of the process. Although many existing research practices are consistent with the notification and information sharing requirements, many challenges arise for non-linear scenarios, including utilizing MGR and DSI from collections prior to the BBNJ Agreement, complex uses of multiple DSI, automation in collection and use, change of use from harvest fisheries to R&D and access and use of TK associated with MGR of ABNJ.

The negotiators of the BBNJ Agreement aimed for a balanced approach so that the MGR requirements could be "future proof" but also avoid unintended non-compliance from disproportionate or impractical requirements. To ensure that the negotiators concluded the work by the resumed fifth session, there was a delicate dance of determining the level of detail that would need to be included in the BBNJ Agreement and other matters that would be decided after entry into force. Collaboration and consultations between scientists, commercial end users and other stakeholders will be important to provide information to treaty bodies about the practical effects on the R&D process and innovation during the development of further procedures and guidance for implementation. These include considering current international scientific good practice, and building a timely, efficient and fit-for-purpose CHM that can evolve over time as technologies develop. Continued and robust engagement of a wide range of stakeholders on the importance of engaging with notification and benefit sharing may ensure more effective compliance.

Such consultations would benefit from timely information on emerging scientific advances and ensuring that "good scientific practices" represent a wide range of scientific disciplines. Aiming to support harmonization of scientific good practice could lead to institutions providing relevant data in a FAIR format, including information on the planned cruise, the eventual cruise report and information on location of MGR and DSI. The BBNJ Agreement demonstrates the importance of provenance of MGR good data management, diligent use of the BBNJ Identifier system, and FAIR data, including harmonization and standardization of approaches, and interoperability between datasets and repositories. The agreement provides an opportunity therefore to support data harmonization efforts across various repositories and databases. While many of the treaty requirements for MGR and DSI that demand robust data and sample management reflect existing good practice, there are likely to be cost implications. To that end, the financial mechanism could consider the need for increased funding for MGR repositories for accessioning and maintaining MGR collections long term. Consistency with other multilateral environmental agreements, such as the CBD, the Nagoya Protocol and GBF concerning DSI, will be essential for R&D, which uses data, sequences and samples/materials from multiple jurisdictions. It is critical that the scientific community strengthen the ongoing consultation and dialogue with various stakeholders, policymakers and entities that could be impacted by the future guidance on DSI.

Finally, there is a significant need for capacity building and the transfer of marine technology to foster scientific and technical advances. The BBNJ Agreement's fair and equitable sharing of benefits of MGR exemplifies the ways in which a wide range of stakeholders can engage in collaboration with the scientific community (Muraki Gottlieb & Girguis, 2022). In addition to the information that will be open to the public, there are additional opportunities that the scientific community can contribute (e.g. transfer of marine technology, technical and scientific cooperation, knowledge exchange) that go beyond providing information and samples. Institutions, philanthropies, private sector, NGOs, and academia have a crucial role in determining the need and allocating sustained and adequate resources. To ensure such strong partnerships, awareness raising is paramount and the scientific community can harness such opportunities to significantly contribute to the fair and equitable sharing of benefits.

References

Alcock, A. (2014). From seabed to World Wide Web: An overview of marine zoological sampling, data processing and potential production of digital marine Faunas Chapter 17. In M. F. Watson, C. H. C. Lyal, & C. A. Pendry (Eds.), *Descriptive taxonomy: The foundation of biodiversity research* (pp. 214–225). Cambridge University Press. https://doi.org/10.1017/CBO9781139028004.022

Broggiato, A., Dunshirn, P., Jaspars, M., & Pena-Neira, S. (2025). Monetary and non-monetary benefit sharing under the BBNJ Agreement. In F. Humphries (Ed.), *Decoding governance marine genetic resource governance under the BBNJ Agreement.* Springer.

Brown, A. E. L. (2025). The place of intellectual property under the BBNJ Agreement. In F. Humphries (Ed.), *Decoding marine genetic resource governance under the BBNJ Agreement.*. Springer.

CBD. (2016). *Report of the thirteenth meeting of the conference of the parties to the convention on biological diversity* CBD/COP/13/25 (2016) [210], Decision XIII/18) arts 10–11, 19, 28–29, 32 ('Mo'otz Kuxtal Voluntary Guidelines').

Chai, F., Johnson, K. S., Claustre, H., Xing, X., Wang, Y., Boss, E., Riser, S., Fennel, K., Schofield, O., & Sutton, A. (2020). Monitoring ocean biogeochemistry with autonomous platforms. *Nature Reviews Earth and Environment, 1*(6), 315–326. See Rudnick, D. L. (2016). Ocean research enabled by underwater gliders. *Annual Review of Marine Science 8*, 519–541.

Clark, M. R., Consalvey, M., & Rowden, A. A. (2016). *Biological sampling in the deep sea.* Wiley.

Collins, J. E., Rabone, M., Vanagt, T., Amon, D. J., Gobin, J., & Huys, I. (2021). Strengthening the global network for sharing of marine biological collections: Recommendations for a new agreement for biodiversity beyond national jurisdiction. *ICES Journal of Marine Science, 78*, 305–314.

COP. (2022). *Decision adopted by the conference of the parties to the convention on biological diversity, digital sequence information on genetic resources.* Conference of the Parties to the Convention on Biological Diversity. CBD/COP/15/9.

de la Concepción, R. T. (2024). Negotiating fair and equitable sharing of benefits in the BBNJ agreement: Role of the Group of 77 and China. *Marine Policy, 163*, 106085.

Díaz, S., et al. (2019). *Summary for policymakers of the global assessment report on biodiversity and ecosystem services of the intergovernmental science-policy platform on biodiversity and ecosystem services.* IPBES. https://zenodo.org/records/3553579

Droege, G., Barker, K., Seberg, O., Coddington, J., Benson, E., Berendsohn, W. G., Bunk, B., Butler, C., Cawsey, E. M., Deck, J., & Döring, M. (2016). The global genome biodiversity network (GGBN) data standard specification. *Database, 2016*, baw125.

Duckworth, A. (2009). Farming sponges to supply bioactive metabolites and bath sponges: A review. *Marine Biotechnology, 669*, 669–670.

Engel, M. S., Ceríaco, L. M., Daniel, G. M., Dellapé, P. M., Löbl, I., Marinov, M., Reis, R. E., Young, M. T., Dubois, A., Agarwal, I., & Lehmann, A. P. (2021). The taxonomic impediment: A shortage of taxonomists, not the lack of technical approaches. *Zoological Journal of the Linnean Society, 193*(2), 381–387.

Guralnick, R. P., Cellinese, N., Deck, J., Pyle, R. L., Kunze, J., Penev, L., Walls, R., Hagedorn, G., Agosti, D., Wieczorek, J., & Catapano, T. (2015). Community next steps for making globally unique identifiers work for biocollections data. *ZooKeys, 494*, 133–154.

Harden-Davies, H., Amon, D. J., Vierros, M., Bax, N. J., Hanich, Q., Hills, J. M., Guilhon, M., McQuaid, K. A., Mohammed, E., Pouponneau, A., & Seto, K. L. (2022). Capacity development in the Ocean Decade and beyond: Key questions about meanings, motivations, pathways, and measurements. *Earth System Governance, 12*, 100138.

Hardisty, A. R., Ellwood, E. R., Nelson, G., Zimkus, B., Buschbom, J., Addink, W., Rabeler, R. K., Bates, J., Bentley, A., Fortes, J. A., & Hansen, S. (2022). Digital extended specimens: Enabling an extensible network of biodiversity data records as integrated digital objects on the internet. *BioScience, 72*(10), 978–987.

Harrison, J. B., Sunday, J. M., & Rogers, S. M. (2019). Predicting the fate of eDNA in the environment and implications for studying biodiversity. *Proceedings of the Royal Society B., 286*, 20191409.

Humphries, F. (2025). Marine genetic resources beyond national jurisdiction: the expansive scope of the BBNJ Agreement. In F. Humphries (Ed.), *Decoding marine genetic resource governance under the BBNJ Agreement.* Springer.

Humphries, F., Berry, T., & Muraki Gottlieb, H. (2025a). Bridging divides: The evolution of marine genetic resource governance beyond national jurisdiction. In F. Humphries (Ed.), *Decoding marine genetic resource governance under the BBNJ Agreement.* Springer.

Humphries, F., Jaspars, M., Lavelle, J., & Kachelriess, D. (2025b). Accessing marine genetic resources:

the novel notification system under the BBNJ Agreement. In F. Humphries (Ed.), *Decoding governance marine genetic resource governance under the BBNJ Agreement.* Springer.

Humphries, F., Rabone, M., & Jaspars, M. (2021). Traceability approaches for marine genetic resources under the proposed ocean (BBNJ) treaty. *Frontiers in Marine Science, 8*, 661313.

International Maritime Organisation IMO. (2021). Outcome of the regulatory scoping exercise for the use of maritime autonomous surface ships (MASS). MSC.1/Corc/1638, 3 June 2021.

Islam, S., Beach, J., Ellwood, E. R., Fortes, J., Lannom, L., Nelson, G., & Plale, B. (2023). Assessing the FAIR digital object framework for global biodiversity research. *Research Ideas and Outcomes, 9*, e108808.

Juty, N., Wimalaratne, S. M., Soiland-Reyes, S., Kunze, J., Goble, C. A., & Clark, T. (2020). Unique, persistent, resolvable: Identifiers as the foundation of FAIR. *Data Intelligence, 2*(1–2), 30–39.

Kachelriess, D., Dunshirn, P., Langlet, A., Brown, A. E. L., & Scholz, A.H. (2025). Marine genetic resources and digital sequence information under the BBNJ Agreement: Interlinkages with other access and benefit sharing frameworks. In F. Humphries (Ed.), *Decoding marine genetic resource governance under the BBNJ Agreement.* Springer.

Koepper, S., Nuryati, S., Palm, H. W., Wild, C., Yulianto, I., & Kleinertz, S. (2022). Metazoan endoparasite fauna and feeding ecology of commercial fishes from Java, Indonesia. *Parasitology Research, 121*(2), 551–562.

Langlet, A., Dunshirn, P., Jaspars, M., Humphries, F., & Kachelriess, D. (2025) Monitoring and transparency aspects of MGR-utilization under the BBNJ Agreement. In F. Humphries (Ed.), *Decoding marine genetic resource governance under the BBNJ Agreement.* Springer.

Lavelle, J., & Wynberg, R. (2025) Benefit sharing under the BBNJ Agreement in practice. In F. Humphries (Ed.), *Decoding marine genetic resource governance under the BBNJ Agreement.* Springer.

Lawson, C., Humphries, F., Jaspars, M., & Rabone, M. (2025). Data management and the 'BBNJ Standardized Batch Identifier' under the BBNJ Agreement. In F. Humphries (Ed.), *Decoding marine genetic resource governance under the BBNJ Agreement.* Springer.

Menime, S. S. H. Y. F., & Bowrey, K. (2022). ABS or access before service: A Samoan perspective. In C. Lawson, M. Rourke, & F. Humphries (Eds.), *Access and benefit sharing of genetic resources, information and traditional knowledge* (pp. 209–220). Routledge.

Mulalap, C. Y., Frere, T., Huffer, E., Hviding, E., Paul, K., Smith, A., & Vierros, M. K. (2020). Traditional knowledge and the BBNJ instrument. *Marine Policy, 122*, 104103.

Munoz, P. D., Sacau, M., García-Alegre, A., & Román, E. (2020). Cold-water corals and deep-sea sponges by-catch mitigation: Dealing with groundfish survey data in the management of the northwest Atlantic Ocean high seas fisheries. *Marine Policy, 116*, 103712.

Muraki Gottlieb, H., & Girguis, P. (2022) Opportunities to foster conservation and sustainable use of biodiversity beyond national jurisdiction: a role for scientists. In *Limnology and oceanography bulletin*. Association for the Sciences of Limnology and Oceanography.

Muraki Gottlieb, H., Ardron, J., & Brown, A. E. L. (2025a). BBNJ agreement: A new infrastructure to foster benefit sharing of marine genetic resources. In F. Humphries (Ed.), *Decoding marine genetic resource governance under the BBNJ Agreement.* Springer.

Muraki Gottlieb, H., Kachelriess, D., & Slobodian, L. (2025b). Understanding the preamble, principles and objectives of the BBNJ Agreement: A focus on the fair and equitable sharing of benefits of marine genetic resources. In F. Humphries (Ed.), *Decoding marine genetic resource governance under the BBNJ Agreement.* Springer.

Oldham P., Thambisetty, S., (2023) *ONEST: The middle way for monetary benefit sharing in BBNJ negotiations*.

Page, R. (2023). Ten years and a million links: Building a global taxonomic library connecting persistent identifiers for names, publications and people. *Biodiversity Data Journal, 11*. https://doi.org/10.3897/BDJ.11.e107914.

Pena-Neira, S., & Coelho, L.F. (2025). Traditional knowledge associated with marine genetic resources in areas beyond national jurisdiction. In F. Humphries (Ed.), *Decoding marine genetic resource governance under the BBNJ Agreement.* Springer.

Rabone, M., Harden-Davies, H., Collins, J. E., Zajderman, S., Appeltans, W., Droege, G., Brandt, A., Pardo-Lopez, L., Dahlgren, T. G., Glover, A. G., & Horton, T. (2019). Access to marine genetic resources (MGR): Raising awareness of best-practice through a new agreement for biodiversity beyond national jurisdiction (BBNJ). *Frontiers in Marine Science, 6*, 520.

Rabone, M., Wiethase, J. H., Simon-Lled6, E., Emery, A. M., Jones, D. O., Dahlgren, T. G., Bribiesca-Contreras, G., Wiklund, H., Horton, T., & Glover, A. G. (2023a). How many metazoan species live in the world's largest mineral exploration region? *Current Biology, 33*(12), 2383–2396.

Rabone, M., Horton, T., Jones, D. O. B., Simon-Lled6, E., & Glover, A. G. (2023b) A review of the international seabed authority database DeepData from a biological perspective: challenges and opportunities in the UN Ocean Decade. *Database*, 2023, baad013

Regulation (EU) No 511/2014 of the European Parliament and of the Council of 16 April 2014 on compliance measures for users from the Nagoya Protocol on Access to Genetic Resources and the Fair and Equitable Sharing of Benefits Arising from their Utilization in the Union Text with EEA relevance.

Rohden, F., Huang, S., Dröge, G., & Hartman-Sholz, A. (2020) Combined study on digital sequence information (DSI) in public and private databases and traceability. Ad Hoc Technical Expert Group on Digital Sequence Information on Genetic Resources to the CBD (Report No. CBD/DSI/AHTEG/2020/1/4.

Schiller, E., Wiltschke-Schrotta, K., Häffner, E., Buschbom, J., Leliaert, F., Zimkus, B., Dickie, J., Gomes, S., Lyal, C., Mulcahy, D., & Paton, A. (2024). Permits, contracts and their terms for biodiversity specimens. *Research Ideas and Outcomes, 10*, e114366.

Scholz, A. H., Freitag, J., Lyal, C. H., Sara, R., Cepeda, M. L., Cancio, I., Sett, S., Hufton, A. L., Abebaw, Y., Bansal, K., & Benbouza, H. (2022). Multilateral benefit-sharing from digital sequence information will support both science and biodiversity conservation. *Nature Communications, 13*(1), 1086.

TRIPS: Agreement on Trade-Related Aspects of Intellectual Property Rights, Article 27. Final Act Embodying the Results of the Uruguay Round of Multilateral Trade Negotiations, Apr. 15, 1994, 1867 U.N.T.S. 14, 33 I.L.M. 1143 (1994)

UNEP. (2020). *Digital sequence information on genetic resources: concept, scope and current use, Ad hoc technical expert group on digital sequence information on genetic resources*, CBD/DSI/AHTEG/2020/1/3, 29 January 2020, annex.

UNEP. (2022). *Decision Adopted by the Conference of the Parties to the Convention on Biological Diversity, 15/9 Digital Sequence Information on Genetic Resources* Conference of the Parties to the Convention on Biological Diversity, (CBD/COP/DEC/15/9, 19 December 2022

UNGA. (2023). *Agreement under the United Nations Convention on the Law of the Sea on the conservation and sustainable use of marine biological diversity of areas beyond national jurisdiction,* Intergovernmental conference on an international legally binding instrument under the United Nations Convention on the Law of the Sea on the conservation and sustainable use of marine biological diversity of areas beyond national jurisdiction. Further resumed fifth session. A/CONF.232/2023/4. 19 June 2023

Wilkinson, M. D., Dumontier, M., Aalbersberg, I. J., Appleton, G., Axton, M., Baak, A., Blomberg, N., Boiten, J. W., da Silva Santos, L. B., Bourne, P. E., & Bouwman, J. (2016). The FAIR guiding principles for scientific data management and stewardship. *Scientific Data, 3*(1), 1–9.

WIPO. (2024). Diplomatic conference to conclude an international legal instrument relating to intellectual property, genetic resources and traditional knowledge associated with genetic resources. *WIPO Treaty on Intellectual Property, Genetic Resources and Associated Traditional Knowledge.* GRATK/DC/7.

WoRMS Editorial Board. (2023). *World register of marine species*. https://www.marinespecies.org. Accessed 12 Nov 2023

Muriel Rabone is a researcher based on the deep-sea ecology and systematics group of the Natural History Museum, London. She also studies the neglected tropical diseases: schistosomiasis and paragonimiasis.

Dr. Tammy Horton is an expert in deep-sea taxonomy, ecology and biodiversity, with 20 years' research experience, including participation in research cruises. She has extensive experience in the use and care of marine invertebrate specimen collections, having begun her academic career working at the Natural History Museum, London, and as manager of the Discovery Collections at NOC since 2009. Dr Horton has substantial experience of working with industry, particularly in providing advice to the oil and gas industry and the International Seabed Authority on the importance of robust taxonomy in faunal surveys.

Fran Humphries has specialized in marine and biodiversity law and policy for over two decades in government, academia and consultancies. She is an associate professor at Griffith Law School, Griffith University, Australia, with a background in fisheries management and marine law. She has led large international research consultancies on access and benefit sharing of biological resources for governments, UN organizations and other institutions. She was on the International Council of Environmental Law delegation for the Biodiversity Beyond National Jurisdiction (BBNJ) Agreement negotiations.

Christopher H. C. Lyal is a taxonomist with a strong level of policy engagement with the Convention on Biological Diversity, particularly regarding the Nagoya Protocol and Digital Sequence Information. He has undertaken international consultancies and been a delegate in CBD COPs representing the UK, and participated a several CBD expert groups. He has also advised on the implementation of the BBNJ agreement.

Hiroko Muraki Gottlieb is a licenced lawyer in the USA with a diversified career in global businesses, intergovernmental organizations, non-governmental organizations and academic institutions. She brings a unique blend of expertise in climate change and biodiversity conservation strategy, environmental law (domestic and international), policymaking, regulatory compliance and stakeholder engagement. She is the representative for the Ocean and led the International Council of Environmental Law to the BBNJ Agreement negotiations as the Head of Delegation and holds appointments with the Department of Organismic and Evolutionary Biology at Harvard University and Elisabeth Haub School of Law at Pace University.

Dr. Amber H. Scholz is a microbiologist and head of the Science Policy and Internationalization Department at the Leibniz Institute DSMZ in Braunschweig, Germany. She leads projects on international science policy especially on access and benefit sharing and digital sequence information and founded the DSI Scientific Network and Germany Nagoya Protocol HuB.

Thomas Vanagt is a marine biologist with a specific interest in marine genetic resources. He works towards closing the gap between policymakers, academic users and industrial actors.

Marcel Jaspars is a professor of Organic Chemistry at the University of Aberdeen where he leads the Marine Biodiscovery Centre which focusses on marine resources for novel pharmaceuticals, and to investigate fundamental questions in marine chemical ecology and biosynthesis. Marcel has been active at national and international levels to develop the science, its applications/industrial uptake and associated policy involved in marine biodiscovery and biotechnology. He provides scientific advice to the UK, EU and UN for global policy processes on ocean conservation and digital sequence information via reports, papers and taking part in discussion meetings.

Conclusions: Equity, Sustainability, and Transformation Under the BBNJ Agreement

15

Elisa Morgera

Abstract

This chapter reflects on the overarching contributions of the *Agreement Under the United Nations Convention on the Law of the Sea on the Conservation and Sustainable Use of Marine Biological Diversity of Areas Beyond National Jurisdiction* to inter- and intra-generational equity, environmental sustainability, and transformation. The chapter explains the areas of the Agreement that will require further development during its implementation in terms of incomplete theorization. The chapter will reflect on how the BBNJ Agreement can contribute to enhanced international cooperation to ultimately support the protection of everyone's human right to a healthy environment.

Keywords

BBNJ agreement · Inter-generational equity · Intra-generational equity · Environmental sustainability · Biodiversity conservation · Transformation · Human rights

The *Agreement Under the United Nations Convention on the Law of the Sea on the Conservation and Sustainable Use of Marine Biological Diversity of Areas Beyond National Jurisdiction* (BBNJ Agreement) created a new international regime on fair and equitable benefit-sharing from marine genetic resources of areas beyond national jurisdiction. It fills significant gaps in the international law of the sea with regard to critical research and innovation at the marine genetic level that can support more effective conservation and sustainable use of the ocean across marine areas within and outside national sovereignty. The BBNJ Agreement also addresses long-standing equity issues in international scientific, technological, and environmental management cooperation and could ultimately support the protection of everyone's human right to a healthy environment (Bennet et al., 2024). This chapter will reflect on the overarching contributions of the BBNJ Agreement to equity, sustainability, and transformation, while understanding the areas of the Agreement that will require further development during its implementation in terms of incomplete theorization.

Equity

The role of equity was unclear and controversial throughout the BBNJ negotiations: the mandate of the negotiations was even silent on

E. Morgera (✉)
University of Strathclyde, Glasgow, UK
e-mail: elisa.morgera@strath.ac.uk

F. Humphries (ed.), *Decoding Marine Genetic Resource Governance Under the BBNJ Agreement*, Sustainable Development Goals Series, https://doi.org/10.1007/978-3-031-72100-7_15

whether benefit-sharing was linked to equity and fairness (A/RES/72/249 (2017); Salpin, 2016). Fairness and equity, however, are otherwise a clear feature of benefit-sharing in international law (Morgera, 2016a, b). It was very clear, at least, from the outset that there were long-standing, deep equity issues in relation to marine science and bioprospecting in areas beyond national jurisdiction (Humphries et al., 2025a). They related to the fact that only a handful of countries, and very few companies within them (Blasiak et al., 2018),[1] have been able to file patents related to marine genetic resources (Arnaud-Haond et al., 2011).[2] On the other hand, the vast majority of developing countries are not part of these bioprospecting efforts and are greatly underrepresented in marine taxonomic research more generally (Broggiato et al., 2018; Hendriks & Duarte, 2008).

The BBNJ Agreement has certainly marked a step forward in addressing these intra-generational equity asymmetries. It has included equity among its general principles (Muraki Gottlieb et al., 2025b). It has also clarified that among the aims of its provisions on access to marine genetic resources and benefit-sharing, there is also the objective to build the capacity of, particularly, developing countries to carry out activities on these resources, transfer technology, and generate knowledge, scientific understanding, and technological innovation, as a 'fundamental contribution to the implementation of the Agreement' (art 7(c); Muraki Gottlieb et al., 2025b; Lavelle & Wynberg, 2025; Rabone et al., 2025). This ambition can be supported by the new system of notifications of access to marine genetic resources (MGRs; Humphries et al., 2025b; Langlet et al., 2025), which entails an obligation to include 'opportunities for scientists of all States, in particular developing states, to be involved or associated with the project' and the 'extent to which it is considered that States that may need and request technical assistance, in particular developing States, should be able to participate or be represented in the project' (art 12(2)(h)–(i)). Further, a plan for open and responsible data governance should be shared, and should later provide indications of the repository of marine genetic resources (art 12(2)(j); Humphries et al., 2025b; Lawson et al., 2025). As part of this notification process, fair and equitable sharing of non-monetary benefits can be considered implicit in terms of 'benefits to all humankind', notably 'advancing scientific knowledge' and promoting conservation and sustainable use, 'taking in particular consideration the interests and needs of developing States' (art. 11(6); Broggiato et al., 2025; Langlet et al., 2025; Lawson et al., 2025; Lavelle & Wynberg, 2025; Pena-Neira & Coelho, 2025; Rabone et al., 2025).

Equity issues can also be expected to be addressed through the BBNJ Clearinghouse (art 51; Muraki Gottlieb et al., 2025a) and the Access and Benefit-Sharing (ABS) Committee (art 12(7)). For instance, the BBNJ Agreement provision on fair and equitable benefit-sharing of non-monetary benefits refers to access to samples and sample collections, and open access to findable, accessible, interoperable and reusable data 'in accordance with current international practice', as well as technology transfer, capacity-building, and increased scientific and technical cooperation (art 14(2)(c)). But these are effectively left to bilateral deals on the basis of national legislation. The ABS Committee, therefore, can play a significant role in developing guidelines for benefit-sharing and ensuring transparency (art 15(1)). Further, based on Parties' self-reporting obligations and information on the clearing house, the ABS Committee will prepare periodic aggregate reports that will feed into the role of the COP to make recommendations on other forms of benefits (art 15(6)), and generally the implementation of the ABS provisions, taking into account 'the national capabilities and circumstances of Parties' (Art. 16(3); Humphries et al., 2025b; Muraki Gottlieb et al., 2025a, b).

[1] A 'single corporation registered 47% of all marine sequences including in gene patents, exceeding the combined share of 220 other companies (37%).

[2] Only 10 countries account for 90% of patents related to marine genetic resources (the US, Japan, certain EU countries, Switzerland and Norway).

In many ways, therefore, the BBNJ Agreement has been designed as a response to the ascertained limitations of pre-existing multilateral benefit-sharing instruments. As such, its design has specifically provided for an iterative, learning approach to fair and equitable benefit-sharing, while a multilateral approach that seeks to move away from a mere logic of exchange (Morgera, 2024a, b). This predominant frame that emphasizes transactional approaches seems to emerge as a source of continued ineffectiveness of international benefit-sharing regimes, as it inherently de-emphasizes the global benefits underlying specific benefit-sharing relationships, as well as the challenges for beneficiaries to exercise their agency in the context of power asymmetries (Morgera, 2024a, b). In the face of such continued challenges, all the pre-existing multilateral benefit-sharing approaches have increasingly devised ways to facilitate and broker, and also oversee and identify gaps in, an otherwise ad hoc flow of non-monetary benefits, such as information-sharing, scientific cooperation, and capacity-building activities (Broggiato et al., 2018; Morgera, 2016a, b). These adjustments to pre-existing multilateral benefit-sharing approaches have served to find common understanding in identifying and apportioning benefits to lay the foundation for a partnership among different actors in the context of power asymmetries. This is necessary to realize the underlying international objectives of the treaties/instruments in which benefits-sharing mechanisms are enshrined, as well as in coming to terms with a key lesson learnt across international instruments—namely, that monetary benefits are very difficult to be accrued in practice (Morgera, 2024a, b). Reliance on non-monetary benefits and voluntary financial contributions has rather become the norm across pre-existing benefit-sharing mechanisms, because of the challenges in linking monetary benefits to intellectual property rights, and because of the resulting paradox of restricting the use of genetic materials that provides other benefits to humanity (Tsioumani, 2022).

On the whole, iterative learning through some form of multilateral oversight or reflection on actual impacts on fairness and equity, and responsive re-design of multilateral benefit-sharing, has emerged as an essential approach to better understand how to generate and share global and local benefits in the achievement of international objectives of environmental protection, global food security, and global health security. Building on these lessons learnt across the international landscape, the BBNJ Agreement has embedded explicitly elements of iterative adaptation and oversight in its treaty design (Morgera, 2024a, b).

This contextual understanding is also relevant to reflect upon the fact that the BBNJ Agreement provides the first example of a treaty where ABS from 'digital sequence information' (DSI) has been specifically provided for. One of its objectives is sharing benefits fairly and equitably from DSI and enhance capacities to carry out activities on DSI, including from DSI generated before its entry into force, unless a Party makes an exception (arts 9(a) and 10(1); Humphries, 2025). Parties are then required to share the database in which DSI is being deposited, clarifying that they relate to areas beyond national jurisdiction, as well as sharing data management plans and their updates (Lawson et al., 2025). Benefit-sharing obligations then extend to access to DSI 'in accordance with current international practice' (art 14(3); Broggiato et al., 2025; Humphries et al., 2025b; Langlet et al., 2025; Lawson et al., 2025; Lavelle & Wynberg, 2025; Rabone et al., 2025). On the other hand, the BBNJ Agreement also allows for reasonable conditions to which access to DSI could be subject to, such as reasonable costs associated with maintaining database; as well as opportunities for access on fair, most favourable terms, including on concessional and preferential terms with researchers from developing countries (art 14(4)). Current scientific research and collaboration practices, however, entail their own justice dimensions, and it remains unclear how these will be tackled under the BBNJ Agreement (Morgera, 2024a, b). Once again, the role of the ABS Committee could be also that of assessing positive and negative impacts on different dimensions of equity of the DSI-relevant provisions of the BBNJ Agreement, and co-develop responses to them.

It remains to be seen to what extent the BBNJ Agreement has also addressed inter-generational equity. The Agreement does include as its overall objective the conservation and sustainable use of marine biodiversity 'for the present and in the long term' (art. 2), and as one of its general principles the common heritage of humankind. The Agreement also refers to future generations in its preamble in relation to States' desire to 'act as stewards of the ocean' in 'maintaining the integrity of the ocean ecosystems'. Even without these textual references, it is a scientific fact that the health of ocean ecosystems is essential for human survival and flourishing for generations to come and is severely threatened by the triple planetary crisis. So while the BBNJ Agreement, in and of itself, is relatively reticent on the question of inter-generational equity, it can and should be interpreted in the light of the UN Convention on the Rights of the Child (Morgera & Lennan, 2022, Morgera et al., 2023), as children are among the population groups most at risk from environmental harm and climate change, while contributing the least to environmental degradation (CRC/C/88/D/104/2019) and they represent new generations 'that are constantly arriving on this planet' (Knox, 2018). The 2023 UN General Comment 26 on children's human rights and a healthy environment, with a special focus on climate change, already helps clarify the obligations of States under law of the sea to put children's human rights at the heart of decision-making on the protection of the marine environment, notably to take immediate action to prevent marine pollution, transform industrial fisheries and conserve, protect, and restore biodiversity (CRC/C/GC/26 (2023)). In addition, the implementation of the BBNJ Agreement could consider the guidance provided by the Maastricht Principles on the Human Rights of Future Generations,[3] to ensure that burdens are not shifted to future generations. Children and young people have already expressed demands to address negative impacts in and around international processes to protect the marine environment (Morgera et al., 2022). The institutions of the BBNJ Agreement should then support inter-generational dialogue, mutual learning and partnership between international ocean decision-makers and children, by respecting and protecting children's right to be heard and drawing on existing children-centred methodologies to create platforms where children speak and their views and voices are actually listened to, rather than merely voiced (Shields et al., 2023). Finally, inter-generational equity will depend on the actual contributions of the implementation of the BBNJ Agreement to environmental sustainability, to which I turn to next.

Sustainability

Marine areas beyond national jurisdiction[4] (the high seas and the Area) represent '4 billion years of evolution' (Rogers et al., 2021) and 'contain 90% of the total biomass of the global ocean', encompassing a 'wide range of ecological processes and dynamics, from large-scale migrations by hundreds of species to low-productivity, highly stable deep-sea benthic ecosystems rich in biodiversity' (Crespo et al., 2020). But we have incomplete understanding of these dynamics, and of ecological impacts of human activities on them, which in itself undermines current conservation and sustainable use efforts. In other words, advancing knowledge of marine biodiversity at the genetic level is essential for environmental sustainability. And accordingly, the BBNJ provisions on fair and equitable sharing of benefits arising from activities with respect to MGRs and DSI are recognized as essential 'for the conservation and sustainable use of marine biological diversity (art 9(a)).

Knowledge of MGRs is what allows for enhanced understanding of the need for, and effectiveness of, conservation and

[3] https://www.ohchr.org/sites/default/files/documents/new-york/events/hr75-future-generations/Maastricht-Principles-on-The-Human-Rights-of-Future-Generations.pdf.

[4] This section draws on Morgera (2022).

sustainable use approaches not only in areas beyond national jurisdiction but also in areas within national jurisdiction due to the ecological connectivity of the ocean that relies on currents and the movement of migratory species (Popova et al., 2019; Yadav & Gjerde, 2020). This implies understanding MGRs in ABNJ as part of the 'ocean genome'—the whole of the genetic material present in all marine biodiversity, including both the genes and the information they encode—that is essential for 'determining the abundance and resilience of biological resources', 'increas[ing] awareness of the pressures facing marine biodiversity', and 'informing the designation of [marine protected areas] as well as innovative approaches to conservation' (Blasiak et al., 2018). In that connection, it has also been underscored that '[a]cknowledging the potential commercial value of biodiversity may lead to better funding for biodiversity surveys that access a broad range of marine life and assess these for bioactivity, which may lead to improved biodiversity conservation measures' (Blasiak et al., 2018).

Advancing knowledge of MGRs has, therefore, implications for the implementation of other parts of the BBNJ Agreement that are more widely associated with environmental sustainability. Knowledge of the ocean genome is essential to take effective and holistic decisions on environmental impact assessments (EIAs), marine protected areas, and other area-based management tools (ABMTs) in ABNJ, as well as enhance the capacities of States to manage sustainably marine spaces within national jurisdiction if they are among those with strongest connectivity to areas beyond national jurisdiction and the shortest timeframes of connectivity (Popova). In turn, it is down to decisions on the creation of area-based and other management tools in ABNJ to 'safeguard genetic diversity at the ecosystem level' (Blasiak et al., 2018).

It is therefore to be welcomed that under the BBNJ Agreement, MGRs have been linked to the provisions on capacity-building and technology transfer (art 42(f)). Advancing basic knowledge about MGRs of ABNJ more equitably, by genuinely partnering with scientist from the Global South and traditional knowledge holders, can support the co-production of knowledge (Lavelle & Wynberg, 2025) on more effective conservation and sustainable use measures and approaches, which take into account the interconnectivity of the ocean and its relevance for life and well-being on Earth. This, in turn, is essential for addressing the triple planetary crisis, including at the still-not-fully understood ocean-climate nexus (Morgera et al., 2023), and for to the realization of multiple economic and social benefits across all the elements of the BBNJ Agreement (Santos et al., 2022). On climate change, it cannot be overstated that the BBNJ Agreement provides for the creation of marine protected areas, environmental impact assessments and strategic environmental assessments, as well as capacity-building, technological, and scientific cooperation with developing countries, while considering the importance of conserving the carbon cycling services of ocean ecosystems (arts 7(h), 17, 27(c) and (f), 28 and 39–40).

With regard to the contribution of environmental sustainability in ABNJ to the broader Sustainable Development Agenda, resting on progress in conservation and sustainable use of marine biodiversity, it is worth pointing to the BBNJ Agreement provisions on ABMTs with the objective of supporting food security and other socioeconomic objectives, including the protection of cultural values (art 17(d)) and on EIAs in relation to economic, social, cultural, and human health impacts (arts 31(1)(b) and 35). The importance of these provisions also for the contributions of the law of the sea to clarify State obligations on climate change, and related human rights concerns, by the International Tribunal for the Law of the Sea in its 2014 Advisory Opinion (ITLOS Advisory Opinion No 31; A/HRC/56/45 (2024)).

Finally, ABS is expected to generate *economic* benefits to support biodiversity conservation. It has been argued that ABS is 'assumed' to create strong incentives for biodiversity conservation 'quasi-automatically' (Oberthür & Rosendal, 2013). In parallel, ABS seeks to enhance access for researchers and companies to quality samples of genetic resources, based on predictable access decisions at reasonably low transaction costs. This is expected to create new opportunities for nature-based research and development and the creation of innovative goods and services that help to meet societal challenges, including environmental challenges (e.g. renewable energy). However, there is very little evidence of a positive interaction between benefit-sharing and biodiversity conservation across pre-existing international ABS mechanisms (Oberthür & Rosendal, 2013) and of significant financial contributions and incentives to conservation (Laird et al., 2020).

Transformation

Transformative change was called for in the 2030 Agenda for Sustainable Development (A/RES/70/1 (2015)), and by the Intergovernmental Panel on Climate Change (IPCC, 2018), the Intergovernmental Science-Policy Panel on Biodiversity and Ecosystem Services (IPBES, 2019), and the 2022 UN Ocean Conference in relation to the need to 'halt and reverse the decline in the health of the ocean's ecosystems and biodiversity and to protecting and restoring its resilience and ecological integrity' (UN, 2022). The UN Decade of Ocean Science also called for 'transformative ocean science' (UN, 2021), to support a science-based engagement with equity in scientific cooperation, including with Indigenous knowledge holders and local knowledge holders (Broggiato et al., 2025; Humphries et al., 2025b; Langlet et al., 2025; Lavelle & Wynberg, 2025; Pena-Neira & Coelho, 2025; Rabone et al., 2025).

The key to transformative change is addressing inequalities, which also undermine the fair and equitable sharing of benefits arising from the use of biodiversity and its conservation. Transformation is understood, in biodiversity governance studies, as a shift from 'the technocratic and regulatory fix of environmental problems to more fundamental and transformative changes in social-political processes and economic relations', by preventing a shifting of the burden of response onto the vulnerable, paying attention to social differentiation, through the lens of non-discrimination, and addressing issues of power and legitimacy (Visseren-Hamakers & Kok, 2022).

In that connection, I have argued elsewhere that human rights can contribute to transformative change (Erinosho et al., 2022), through participatory processes that 'focus … on the rights-holder as [the] central concern in response to asymmetries and power imbalances' (Belinkx et al., 2022). In the particular context of the BBNJ Agreement, human rights implications should be connected to vulnerable communities in countries with strongest connectivity to areas beyond national jurisdiction (Popova et al., 2019). As our scientific understanding of other inter-connections between ecosystem services in areas beyond national jurisdiction and human well-being on Earth progresses, we can also expect to be able to point to human rights implications for other societies or vulnerable groups in different States, including land-locked States (Morgera, 2022).

Against that background, the implementation of the BBNJ Agreement can be transformative by supporting the co-identification with developing countries of the benefits of an integrated implementation of the ABS part of the BBNJ Agreement together with the capacity-building, technology transfer, scientific cooperation, and information-sharing obligations, even if these obligations are all dependent on resources in donor countries, who for that reason tend to 'call the shots' (Morgera, 2024a, b). Under the BBNJ Agreement, there is thus a possibility to support a transformative dialogue across different views of equity, benefits, and contributions to environmental sustainability that have already been voiced in the negotiations. For instance, the US and other developed States affirmed that research and development on MGR of ABNJ is a highly costly and time-consuming endeavour with uncertain results, that, when successful, would benefit humanity in the form of scientific advancements contributing to global public health, food security and environmental protection. On the other hand, developing countries have argued for fairly sharing opportunities to participate in scientific expeditions, follow-up research, relevant technology, and research results, could contribute in predictable ways to increasing developing countries' capacities to conduct marine scientific research and contribute to the protection of the marine environment and its sustainable use (Morgera, 2014).

In many respects, the BBNJ Agreement does support a concerted, institutionalized multilateral dialogue to ensure responsiveness to the needs of developing countries and provide oversight of the distribution of benefits across different regions and scales, which then support active participation in transformative conservation and sustainable use (Morgera, 2024a, b). Such an integrated approach to fair and equitable benefit-sharing from MGRs could foster a deeper form of cooperation (Morgera, 2018–2019), through the role and practices of BBNJ institutions (Muraki Gottlieb et al., 2025a, b). For instance, the Scientific and Technical Body, which is established to act 'in the best interest of the Agreement' and represent 'multidisciplinary expertise' (art 49(2)), could advance a shared understanding of changing scientific practices and their contributions to BBNJ conservation and sustainable use, fostering learning across the two distinct committees, one on ABS and one on capacity-building and technology transfer (Morgera, 2024a, b). The BBNJ Conference of the Parties (COP), in keeping under review and evaluation the implementation of the Agreement, could establish appropriate processes to promote coherence in efforts to conserve and sustainably use BBNJ (art 47(6)(c)), promote an enabling global environment for the inclusive advancement of ocean science and the enjoyment of the benefits of its applications, in the face of the current deep international disparities among countries in marine bio-based innovation (Morgera, 2022). In particular, the COP has an unprecedented power that could be significant in transforming current ABS practices and directly contribute to conservation and sustainable use: the COP can mandate a strategic environmental assessment (SEA) for an area or region to collate and synthesize the best available information, assess current and potential future impacts, and identify data gaps and research priorities (art 39(2)). Arguably, the planning of COP-mandated SEAs (and possibly that of State-led SEAs, with the due guarantees) could provide an ideal context for co-developing good practices in equitable MGR research and development. The resulting COP-mandated deep-sea research cruises could support capacity and technology co-development, taking into account the needs and priorities of developing States (Morgera et al., 2023) and of Indigenous knowledge holders and other knowledge holders (Erwin et al., 2022; Strand et al., 2022).

Indeed, the provisions across all parts of the BBNJ Agreement on Indigenous knowledge and local knowledge (arts 7(j), 13, 19(3), 21(1)(c), 24(3), 26(5), 31(1), 35, 37(4)), 41(2), 4491)(b), 48, 51(3), 52(6)) are all potentially suitable to support a process of iterative and participatory learning from the understanding of multiple dimensions of justice, as well as of the agency and evolving needs of those producing, sharing, and drawing on ocean knowledge, and increased cooperation on fair and equitable benefit-sharing from marine bio-based innovation, that could support transformation (Morgera, 2024a, b). Questions related to epistemic justice and recognition with regard to the notion of 'best scientific knowledge' used under the BBNJ Agreement should be addressed to assess the degree of effective participation of traditional knowledge holders with regard to EIAs and ABMTs, while respecting free prior informed consent (art 44(1)(b)). The key to transformative change here is ensuring the recognition of different knowledge systems equally as part of 'best scientific knowledge', to empower those whose interests are currently not being met and represent transformative sustainability values (Visseren-Hamakers & Kok, 2022). This, in turn, supports the development of transformative approaches to knowledge co-production and the co-development of sustainability solutions across cultures and worldviews to benefit of everyone's human right to a healthy environment (Morgera, 2024a, b). In addition, assumptions about benefits need to be tested: for instance, the assumption that the publication of genetic sequences in open access databases already represents a significant form of non-monetary benefit-sharing as it allows anyone, including researchers in provider countries and the Global South, to use the information. This argument does not take into account the limited capacity of different countries and different users to access and make use of the information contained in databases (CBS/SBSTTA/22/Inf/2 (2018)).

Similarly, power imbalances may also impact on the way actors can control their data once it has been publicly shared, or on the way research collaborations are structured and conducted. It is therefore essential for the BBNJ institutions to support dialogue among the range of actors involved in MGRs and DSI (marine biodiversity researchers, database managers, innovators, conservation experts, plural knowledge holders) and those negatively affected by the use of DSI, with a view to co-producing a more nuanced understanding of different dimensions of justice at stake and co-developing effective and fair solutions (Morgera, 2024a, b). Participatory governance under the BBNJ Agreement wound thus entail reaching and including different communities of practices involved in the use of DSI, including the meaningful and respectful integration of Indigenous and local knowledge in decision-making procedures, as well as scientists from different geographies and capacities, database managers, experts from different sectors. Crucially, what is learnt and co-developed from this participatory process should then be fed into the BBNJ international financial mechanism (art 52), as well as the other BBNJ institutions (Morgera, 2024a, b).

Further, the BBNJ Agreement provisions on transparency and review (arts 16, 26, 37, 45, 48; Muraki Gottlieb et al., 2025a, b), and on regime interaction (arts15(5), 22(3), 24(2), 26(2), 29(2)–(3), 47(6)(c), 50(4)(d), 51(4), 55(4); Kachelriess et al., 2025) can contribute to transformation. They can provide a way to institutionalize, at different scales, a new approach to international cooperation as a space to learn from a plurality of knowledge systems on an ongoing basis when and how benefit-sharing approaches can work (and when and why they do not) for scientists in the Global South and the Global North, as well as the implications for biodiversity conservation and sustainable use, global health, and global food security. This is essential to balance competing rights and interests, avoid discrimination, and respond to the needs of the vulnerable (Morgera, 2022).

Incomplete Theorization

There have been interesting academic reflections on international treaties that are 'incompletely theorized' where international agreement has been reached to a certain extent on a certain issue, but States have not necessarily agreed on what new international regulation entails to a full extent. Such an approach can be considered helpful because it allows for stability and flexibility over time, while concealing disagreement about particular specific issues (Sunstein, 2005; Switzer, 2018). Seeking fuller theorization in intergovernmental negotiations, on the other hand, would have exacerbated problems, bias, confusion, or inconsistency, potentially leading to unnecessary antagonism (Sunstein, 2005). Ultimately, incomplete theorization allows for the continuation of mutual respect and cooperation while allowing for 'moral evolution over time' and 'openness to new facts and perspectives' (Sunstein, 2005) that can help address the areas of persistent disagreement.

This perspective allows to understand that international benefit-sharing provisions are under-developed, in the BBNJ Agreement but also in other, pre-existing international treaties, because of lack of consensus due to fundamentally divergent agendas, as well as power and information asymmetries, among States, which will in turn affect ensuing implementation practices. It is also helpful to understand objective difficulties in fully theorizing benefit-sharing as a legal approach towards a more intense and cosmopolitan form of cooperation, which benefits from the creation of spaces for gradual development through learning by different actors at different levels, and across different regimes (Morgera, 2024a, b). Several chapters in this book have underscored the areas in which the BBNJ Agreement only provides partial approaches, such as how to share monetary benefits, how to relate to intellectual property rights (Brown, 2025), how to manage benefit-sharing from the use of the undefined 'digital sequence information' (DSI; Humphries, 2025; Kachelriess et al., 2025), and how to ensure the genuine and respectful engagement with Indigenous knowledge holders and local knowledge holders (Humphries, 2025; Muraki Gottlieb et al., 2025a, b; Pena-Neira & Coelho, 2025; Rabone et al., 2025). This is also the case of monetary benefit-sharing (Broggiato et al., 2025; Lavelle & Wynberg, 2025; Rabone et al., 2025): the BBNJ Agreement therefore, has sought to address the question of financial viability of a multilateral benefit-sharing approach, with annual contribution at a rate of 50% of a Party's assessed contributions to the budget adopted by the COP, while leaving the accruing of monetary benefits from the use of marine genetic resources to a future stage of negotiations (art 14(6)). Incomplete theorization has also been detected with regard to the connection between the notification system and the sharing of benefits (Humphries, 2025; Langlet et al., 2025).

One of the biggest areas of incomplete theorization is certainly that related to Indigenous knowledge and other traditional knowledge: from specific questions related to the governance of traditional knowledge associated with MGRs currently housed in laboratories or gene banks even if collected in ABNJ, or MGRs that may have been transported from ABNJ to national waters through natural processes, to broader questions about the burden of proof to demonstrate the validity and relevant of traditional knowledge for the purposes of the BBNJ Agreement (Pena-Neira & Coelho, 2025), to the underpinning suitability of international institutions to provide a genuine space for mutual learning across different knowledge systems (Morgera, 2024a, b). Fundamentally, it should be borne in mind that 'a seat at the table' is not necessarily a genuine opportunity for Indigenous knowledge to influence the process. Two underlying challenges must be considered. First, it is necessary to overcome 'prejudicial stereotypes' and 'misconceptions' about Indigenous peoples' authority and credibility to contribute to this area of international decision-making (Lupin & Towsend, 2023). Second, it is necessary to identify and address any 'unfair distribution of conceptual resources needed for speakers to have a say' (Lupin & Townsend, 2023). The current meeting formats and dominant understanding of science are inherently stacked against an open and meaningful exchange with different world views and knowledge systems. BBNJ experts may be unaware of Indigenous peoples' preferred communicative practices and 'fail to give appropriate uptake to their attempts to communicate' (Lupin & Townsend, 2023). To address these concerns, it is recommended that appropriate resources and expertise are put in place to support meaningful dialogue across knowledge systems, also taking into account the history of marginalization of Indigenous peoples and the current power imbalances in BBNJ. It would be essential to include experts in the social sciences and arts (or rely on their insights and approaches) to restructure the process and help build the capacity of existing BBNJ experts and decision-makers to meaningfully listen and respectfully engage with Indigenous peoples (Mthombeni et al., 2023). In addition, advice from the UN Special Rapporteur on Indigenous Peoples' Issues and the UN Office of the High Commissioner for Human Rights, as observers or advisors to the BBNJ bodies, would be beneficial to build a more human rights-cognizant institutional culture under the BBNJ Agreement. Human rights experts should also be considered suitable to sit in the BBNJ Scientific and Technical Body members, who are to serve 'in their expert capacity and in the best interest of the Agreement' and 'suitable qualifications, taking into account the need for multidisciplinary expertise, including ... expertise in relevant traditional knowledge of Indigenous Peoples and local communities' (art 49(2)).

Concluding Words of Hope

To a significant extent, the BBNJ Agreement includes innovative rules and an institutionalized approach to support iterative learning on fair and equitable benefit-sharing, with specific attention to equity and sustainability dimensions, which could have a transformative effect on the governance of the ocean and its essential benefits to humankind. As a whole, the BBNJ Agreement includes obligations that support the co-production of ocean science and solutions based on fair scientific collaboration and increased capacities in the Global South and among Indigenous knowledge holders and traditional knowledge holders across scales. The Agreement has certainly made progress in identifying and creating avenues to address injustices and power imbalances in the production and use of ocean knowledge across different knowledge systems, which has prevented more effective efforts to conserve and sustainably use BBNJ (Morgera, 2024a, b). There is thus a real chance for the BBNJ COP to identify collectively and prioritize areas of progress in ocean science and management to support to the greatest extent possible the protection of economic, social, and cultural rights of present and future generations, taking into account ecological connectivity between areas within and beyond national jurisdiction, as well as our evolving understanding of the benefits provided by BBNJ, including in terms of global climate regulation (Morgera et al., 2023).

Holistic, transformative action is needed to address the different dimensions of justice across the biodiversity, broader environmental (including climate change), agriculture, and health sectors, in the light of the alarming findings on the unprecedented rate of biodiversity loss and its wide-ranging implications for human well-being. The 2019 Global Assessment of Biodiversity and Ecosystems Services indicated that current negative trends in biodiversity and ecosystems are undermining progress towards 80% of targets assessed within the Sustainable Development Goals (IPBES, 2019). The implementation of the BBNJ Agreement rules on fair and equitable benefit-sharing, in their interactions with the other parts of the Agreement, can and must play a part in reversing this global trend.

References

Arnaud-Haond, S., Arrieta, J., & Duarte, C. (2011). Marine Biodiversity and Gene Patents. *Science, 331*, 1521.

Bellinkx, V., et al. (2022). Addressing climate change through international human rights law: From (extra) territoriality to common concern of Humankind. *Transnational Environmental Law, 11*, 69.

Bennet, N., Morgera, E., & Boyd, D. (2024). The human right to a clean, healthy and sustainable ocean. *NPJ Ocean Sustainability, 3*, 19.

Blasiak, R., et al. (2018). Corporate control and global governance of marine genetic resources. *Science Advances.*

Broggiato, A., Dunshirn, P., Jaspars, M., & Pena-Neira, S. (2025). Monetary and non-monetary benefit sharing under the BBNJ Agreement. In F. Humphries (Ed.), *Decoding governance marine genetic resource governance under the BBNJ Agreement.* Springer.

Broggiato, A., et al. (2018). *Mare Geneticum*: Balancing governance of marine genetic resources in international waters. *International Journal of Marine and Coastal Law, 33*, 3.

Brown, A. E. L. (2025). The place of intellectual property under the BBNJ Agreement. In F. Humphries (Ed.), *Decoding marine genetic resource governance under the BBNJ Agreement.* Springer.

CBD. (2018). Synthesis of views and information on the potential implications of the use of digital sequence information on genetic resources for the three objectives of the Convention and the objective of the Nagoya Protocol UN Doc CBS/SBSTTA/22/Inf/2.

Committee on the Rights of the Child, 'Decision adopted by the Committee on the Rights of the Child under the Optional Protocol to the Convention on the Rights of the Child on a communications procedure in respect of Communication No. 104/2019,' 8 October 2021, UN Doc CRC/C/88/D/104/2019.

Crespo, G., et al. (2020). Beyond static spatial management: Scientific and legal considerations for dynamic management in the high seas. *Marine Policy, 122*(104102), 1–2.

Erinosho, B., et al. (2022). Transformative governance for ocean biodiversity. In I. Visseren-Hamakers, and M. Kok (Eds.), *Transforming biodiversity governance.* Cambridge University Press.

Erwin, K., et al. (2022). Lalela uLwandle: An experiment in plural governance discussions. In R. Bosewell, D, O'Kane, & J. Hills (Eds.), *The Palgrave handbook of blue heritage.* Springer.

Hendriks, I. E., & Duarte, C. M. (2008). Allocation of effort and imbalances in biodiversity research. *Journal of Experimental Marine Biology and Ecology, 360*, 15.

Humphries, F. (2025). Marine genetic resources beyond national jurisdiction: The expansive scope of the BBNJ agreement. In F. Humphries (Ed.), *Decoding marine genetic resource governance under the BBNJ Agreement.* Springer.

Humphries, F., Berry, T., & Muraki Gottlieb, H. (2025a). Bridging divides: The evolution of marine genetic resource governance beyond national jurisdiction. In F. Humphries (Ed.), *Decoding marine genetic resource governance under the BBNJ Agreement.* Springer.

Humphries, F., Jaspars, M., Lavelle, J., & Kachelriess, D. (2025b). The novel notification information system for marine genetic resources under the BBNJ Agreement. In F. Humphries (Ed.), *Decoding governance marine genetic resource governance under the BBNJ Agreement.* Springer.

IPBES. (2019). Global assessment report of the intergovernmental science-policy platform on biodiversity and ecosystem services.

IPCC. (2018). Special report on the impacts of global warming of 1.5 °C above pre-industrial levels and related global greenhouse gas emission pathways, in the context of strengthening the global response to the threat of climate change, sustainable development, and efforts to eradicate poverty.

ITLOS Advisory Opinion No 31 (2024).

Kachelriess, D., Dunshirn, P., Langlet, A., Brown, A. E. L., & Scholz, A.H. (2025). Marine genetic resources and digital sequence information under the BBNJ Agreement: Interlinkages with other access and benefit sharing frameworks. In F. Humphries (Ed.), *Decoding marine genetic resource governance under the BBNJ Agreement.* Springer.

Knox, J. (2017). Report of the UN special rapporteur on human rights and the environment: Biodiversity and human rights, UN Doc A/HRC/34/49.

Knox, J. (2018). 'Report of the special rapporteur on the issue of human rights obligations relating to the enjoyment of a safe, clean, healthy and sustainable environment' UN Doc A/HRC/37/58.

Laird, S., et al. (2020). Rethink the expansion of access and benefit sharing. *Science, 367*(6483), 1200–1202.

Langlet, A., Dunshirn, P., Jaspars, M., Humphries, F., & Kachelriess, D. (2025). Monitoring and transparency aspects of MGR-utilization under the BBNJ Agreement. In F. Humphries (Ed.), *Decoding marine genetic resource governance under the BBNJ Agreement.* Springer.

Lavelle, J., & Wynberg, R. (2025). Benefit sharing under the BBNJ Agreement in practice. In F. Humphries (Ed.), *Decoding marine genetic resource governance under the BBNJ Agreement.* Springer.

Lawson, C., Humphries, F., Jaspars, M., & Rabone, M. (2025). Data management and the 'BBNJ Standardized Batch Identifier' under the BBNJ Agreement. In F. Humphries (Ed.), *Decoding marine genetic resource governance under the BBNJ Agreement.* Springer.

Lupin, D., & Townsend, L. (2023). The right to consultation is a right to be heard. In D. Lupin (Ed.), *A research agenda for human rights and the environment.* Edward Elgar.

Morgera, E. (2014). Benefit-sharing in marine areas beyond national jurisdiction: Where are we at? (Part I). BENELEX Blog. https://benelexblog.wordpress.com/2014/05/23/benefit-sharing-in-marine-areas-beyond-national-jurisdiction-where-are-we-at-part-i/

Morgera, E. (2016a). The need for an international legal concept of fair and equitable benefit-sharing. *European Journal of International Law, 27*, 353–383.

Morgera, E. (2016b). Study on experiences gained with the development and implementation of the Nagoya protocol and other multilateral mechanisms and the potential relevance of ongoing work undertaken by other processes, including case studies. UN Doc UNEP/CBD/ABS/A10/EM/2016/1/2.

Morgera, E. (2018–2019). Fair and equitable benefit-sharing in a new international instrument on marine biodiversity: A principled approach towards partnership building? *Maritime Safety and Security Law Journal, 5*, 5698.

Morgera, E. (2022). The relevance of the human right to science for the conservation and sustainable use of marine biodiversity of areas beyond national jurisdiction: A new legally binding instrument to support co-production of ocean knowledge across scales. In V. De Lucia, L. Nguyen, & A. Oude Elferink (Eds.), *International law and marine areas beyond national jurisdiction: Reflections on justice, space, knowledge and power.* Brill.

Morgera, E., Sweeney, M., & Shields, S. (2022). SDG14 and children's human rights. One Ocean Hub. https://pure.strath.ac.uk/ws/portalfiles/portal/142898190/Morgera_etal_OOH_2022_SDG14_and_childrens_human_rights.pdf

Morgera, E., et al. (2023). Addressing the ocean-climate nexus in the BBNJ agreement: Strategic environmental assessments, human rights and equity in ocean science. *The International Journal of Marine and Coastal Law, 38*, 447–479.

Morgera, E. (2024a). *Fair and equitable benefit-sharing in international law.* Oxford University Press.

Morgera, E. (2024b). Scene-setting report of the UN Special Rapporteur on Climate Change and Human Rights UN Doc A/HRC/56/45.

Morgera, E., & Lennan, M. (2022). Strengthening intergenerational equity at the ocean-climate nexus: Reflections on the UNCRC general comment No. 26. *Environmental Policy and Law., 52*, 445–459.

Morgera, E., & Lily, H. (2022). Public participation at the international seabed authority: An international human rights analysis. *Review of European, Comparative and International Environmental Law, 31*(3), 374–388.

Mthombeni, M., et al. (2023). *Deep sea decisions can consider Indigenous knowledge*, 360. https://360info.org/deep-sea-decisions-can-consider-indigenous-knowledge

Muraki Gottlieb, H., Ardron, J., & Brown, A. E. L. (2025a). BBNJ Agreement: A new infrastructure to foster benefit sharing of marine genetic resources. In F. Humphries (Ed.), *Decoding marine genetic resource governance under the BBNJ Agreement.* Springer.

Muraki Gottlieb, H., Kachelriess, D., & Slobodian, L. (2025b). Understanding the preamble, principles and objectives of the BBNJ Agreement: A focus on the fair and equitable sharing of benefits of marine genetic resources. In F. Humphries (Ed.), *Decoding marine genetic resource governance under the BBNJ Agreement.* Springer.

Oberthür, S., & Rosendal, K. (2013). Conclusions. In S. Oberthür, & K. Rosendal (Eds.), *Global governance of genetic resources: Access and benefit sharing after the nagoya protocol.* Routledge.

OHCHR. (2017). Right to access to justice under article 13 of the convention on the rights of persons with disabilities UN Doc A/HRC/RES/37/25.

OHCHR. (2018). Promotion and protection of human rights and the implementation of the 2030 Agenda for Sustainable Development. UN Doc A/HRC/RES/37/24.

Pena-Neira, S., & Coelho, L.F. (2025). Traditional knowledge associated with marine genetic resources in areas beyond national jurisdiction. In F. Humphries (Ed.), *Decoding marine genetic resource governance under the BBNJ Agreement.* Springer.

Popova, E., et al. (2019). So far, yet so close: ecological connectivity between ABNJ and territorial waters. IIED Policy Brief. https://pubs.iied.org/17500iied

Rabone, M., Horton, T., Humphries, F., Lyal, C., Muraki Gottlieb, H., Scholz, A. H., Vanagt, T., & Jaspars, M. (2025). BBNJ Agreement: Considerations for scientists and commercial end users of MGR at research, development and commercialisation stages. In F. Humphries (Ed.), *Decoding marine genetic resource governance under the BBNJ Agreement.* Springer.

Rogers, A., et al. (2021). Marine genetic resources in areas beyond national jurisdiction: Promoting marine scientific research and enabling equitable benefit-sharing. *Frontiers in Marine Science, 8*, 667274.

Salpin, C. (2016). Marine genetic resources of areas beyond national jurisdiction: Soul searching and the art of balance. In E. Morgera, & K. Kulovesi (Eds.), *Research handbook on international law and natural resources*. Edward Elgar.

Santos, B. S., Devereaux, S. G., Gjerde, K., Chand, K., Martinez, J., & Crowder, L. B. (2022). The diverse benefits of biodiversity conservation in global ocean areas beyond national jurisdiction. *Frontier Marine Science, 9*, 1001240. https://doi.org/10.3389/fmars.2022.1001240

Shields, S., et al. (2023). Children's human right to be heard at the ocean-climate nexus. *The International Journal of Marine and Coastal Law, 38*(3), 545–580.

Strand, M., et al. (2022). Reimagining ocean stewardship: Arts-based methods to "hear" and "see" indigenous and local knowledge in ocean management. *Frontiers in Marine Science, 9*, 1.

Sunstein, C. R. (2005). Incompletely theorized agreements. *Harvard Law Review, 108*, 1733.

Switzer, S., Morgera, E., Lavalle, J., & Wynberg, R. (2023). What does the 2022 UN biodiversity summit outcome on digital sequence information mean for the ocean and ocean research? (Part 2). One Ocean Hub blog post. https://oneoceanhub.org/what-does-the-2022-un-biodiversity-summit-outcome-on-digital-sequence-information-mean-for-the-ocean-and-ocean-research-part-2/

Switzer, S. (2018). Liminal spaces: Special and differential treatment as an incompletely theorized agreement. *Manchester Journal of International Economic Law, 15*, 62.

Tsioumani, E. (2022). *Fair and equitable benefit-sharing in agriculture: Reinventing agrarian justice.* Routledge.

UN General Assembly. (2017). Resolution on International legally binding instrument under the United Nations Convention on the Law of the Sea on the conservation and sustainable use of marine biological diversity of areas beyond national jurisdiction, UN Doc A/RES/72/249.

UN Ocean Conference, *Our Ocean, Our Future, Our Responsibility: Political Declaration for 2022 UNOC* (United Nations). Accessed 25 May 2022.

UN Committee on the Rights of the Child. (2023). General comment on children's human rights and a healthy environment, with a special focus on climate change, UN Doc CRC/C/GC/26.

UN. (2021). Summary of the implementation plan of the UN decade for ocean science for sustainable development 2021–2030.

UNGA. (2015). Transforming our world: The 2030 agenda for sustainable development, Resolution 70/1.

Visseren-Hamakers, I. J., & Kok, M. (2022). Introduction. In I. Visseren-Hamakers, & M. Kok, (Eds), *Transforming biodiversity governance.* Cambridge University Press.

Yadav, S., & Gjerde, K. (2020). The ocean, climate change and resilience: Making ocean areas beyond national jurisdiction more resilient to climate change and other anthropogenic activities. *Marine Policy, 104184*, 4–5.

Elisa Morgera is a professor of Global Environmental Law at the University of Strathclyde, Glasgow (UK), and an adjunct professor in International and European Union Environmental Law at the University of Eastern Finland. In May 2024, she was appointed as the UN Special Rapporteur on Climate Change and Human Rights by the Human Rights Council.

Elisa Morgera is a professor [illegible] [illegible] Law at the University of [illegible] [illegible] and an adjunct professor [illegible] School, and [illegible] Environmental Law [illegible] University [illegible]

GPSR Compliance

The European Union's (EU) General Product Safety Regulation (GPSR) is a set of rules that requires consumer products to be safe and our obligations to ensure this.

If you have any concerns about our products, you can contact us on ProductSafety@springernature.com

In case Publisher is established outside the EU, the EU authorized representative is:

Springer Nature Customer Service Center GmbH
Europaplatz 3
69115 Heidelberg, Germany

Batch number: 10731395

Printed by Printforce, the Netherlands